T0192415

Surveys in Geometry I

Athanase Papadopoulos
Editor

Surveys in Geometry I

 Springer

Editor
Athanase Papadopoulos
Institut de Recherche Mathématique
Avancée
Université de Strasbourg et CNRS
Strasbourg, France

ISBN 978-3-030-86694-5   ISBN 978-3-030-86695-2   (eBook)
https://doi.org/10.1007/978-3-030-86695-2

Mathematics Subject Classification: 30F10, 30F60, 32G15, 53C70, 51K05, 53A35, 57M60, 52B60, 30F10, 30F60, 32G15, 53C70, 51K05, 53A35, 57M60, 52B60

This Springer imprint is published by the registered company Springer Nature Switzerland AG.
The registered company address is: Gewerbestrasse 11, 6330 Cham, Switzerland

Preface

This is the first of a two-volume set of surveys on geometry in the broad sense (including group actions and topology). The surveys vary in their scope and degree of difficulty, but they all represent current research trends.

In planning this book, I naturally wanted to promote topics that I personally like. Some chapters involve only classical mathematics (e.g., geometry of finite-dimensional vector spaces and spherical geometry), while others are concerned with more recent topics (e.g., Gromov-hyperbolic spaces and Teichmüller spaces), but my personal feeling is that in the end there is little difference between classical and modern, or between elementary and advanced mathematics; all the ideas, old and new, are interrelated, and they form one single subject, geometry. Some of the surveys in this volume are based on lectures that were given by their authors to students who are at a middle-advanced level. In particular, three surveys consist of polished notes of lectures given at a CIMPA thematic school that I co-organized with Bankteshwar Tiwari in 2019, at the Banaras Hindu University in Varanasi. Other notes associated with lectures delivered at the same school will appear in the second volume.

When I asked the authors to write a survey for this collection, I knew what to expect in terms of content, but I emphasized the fact that the goal is to give the reader a real introduction to the subject, in an attractive way. Most of the authors succeeded in this, and I take this opportunity to thank them all for their contribution. My thanks also go to Elena Griniari, from Springer, for her support.

Strasbourg, France   Athanase Papadopoulos
September, 2021

Contents

1   **Introduction** 1
    Athanase Papadopoulos

2   **Spherical Geometry—A Survey on Width and Thickness of**
    Convex Bodies 7
    Marek Lassak

3   **Minkowski Geometry—Some Concepts and Recent**
    Developments 49
    Vitor Balestro and Horst Martini

4   **Orthogonality Types in Normed Linear Spaces** 97
    Javier Alonso, Horst Martini, and Senlin Wu

5   **Convex Bodies: Mixed Volumes and Inequalities** 171
    Ivan Izmestiev

6   **Compactness and Finiteness Results for Gromov-Hyperbolic**
    Spaces 205
    Gérard Besson and Gilles Courtois

7   **All 4-Dimensional Smooth Schoenflies Balls Are**
    Geometrically Simply-Connected 269
    Valentin Poénaru

8   **Classical Differential Topology and Non-commutative Geometry** 309
    Valentin Poénaru

9   **A Short Introduction to Translation Surfaces, Veech**
    Surfaces, and Teichmüller Dynamics 343
    Daniel Massart

**10   Teichmüller Spaces and the Rigidity of Mapping Class
      Group Actions**.. 389
      Ken'ichi Ohshika

**11   Holomorphic G-Structures and Foliated Cartan Geometries
      on Compact Complex Manifolds**.. 417
      Indranil Biswas and Sorin Dumitrescu

Index... 463

Contributors

Javier Alonso Instituto de Matemáticas de la UEx (IMUEx), Edificio Carlos Benítez, Badajoz, Spain

Vitor Balestro Instituto de Matemática e Estatística, Universidade Federal Fluminense, Niterói, Brazil

Gérard Besson Institut Fourier, Université de Grenoble, Saint Martin d'Hères, France

Indranil Biswas School of Mathematics, Tata Institute of Fundamental Research, Mumbai, India

Gilles Courtois Institut de Mathématiques de Jussieu – Paris Rive Gauche (IMJ-PRG), Paris, France

Sorin Dumitrescu Université Côte d'Azur, CNRS, LJAD, Nice, France

Ivan Izmestiev TU Wien, Institute of Discrete Mathematics and Geometry, Vienna, Austria

Marek Lassak Institute of Mathematics and Physics, University of Science and Technology, Bydgoszcz, Poland

Daniel Massart Institut Montpelliérain Alexander Grothendieck, CNRS, Université de Montpellier, Montpellier, France

Horst Martini Fakultät für Mathematik, Technische Universität Chemnitz, Chemnitz, Germany

Ken'ichi Ohshika Department of Mathematics, Faculty of Science, Gakushuin University, Toshima-ku, Tokyo, Japan

Athanase Papadopoulos Université de Strasbourg and CNRS, Strasbourg, France

Valentin Poénaru Université de Paris-Sud, Mathématiques, Orsay, France

Senlin Wu North University of China, Taiyuan, China

About the Editor

Athanase Papadopoulos (born 1957) is Directeur de Recherche at the French Centre National de la Recherche Scientifique. His main fields of interest are geometry and topology, history and philosophy of mathematics, and mathematics and music. He has held visiting positions at the Institute for Advanced Study, Princeton (1984–1985 and 1993–1994), USC (1998–1999), CUNY (Ada Peluso Professor, 2014), Brown University (Distinguished Visiting Professor, 2017), Tsinghua University, Beijing (2018), and Lamé Chair of the State University of Saint Petersburg (2019) and has made several month visits to the Max Planck Institute for Mathematics (Bonn), Erwin Schrödinger Institute (Vienna), Graduate Center of CUNY (New York), Tata Institute (Bombay), Galatasaray University (Istanbul), University of Florence (Italy), Fudan University (Shanghai), Gakushuin University (Tokyo), and Presidency University (Calcutta). He is the author of more than 200 published articles and 35 monographs and edited books.

Chapter 1
Introduction

Athanase Papadopoulos

Abstract This introductory chapter contains a general description of the subject matter and a detailed outline of the content of the book.

Keywords Spherical geometry · Minkowski geometry · Finiteness theorems in Gromov-hyperbolic space · Convex geometry · 4-Manifolds · Differential topology · Smooth Schoenflies ball · Geometrically simple connectivity · Translation surface · Veech surface · Teichmüller space · Mapping class group · Holomorphic structure · Foliated Cartan geometry

The present volume of surveys covers a large spectrum of current research topics in geometry in a broad sense, including spherical geometry, infinitesimal geometry (Riemannian and Finsler), metric spaces à la Gromov, Busemann spaces, convexity, singular flat structures on surfaces and their dynamics, Teichmüller spaces, Cartan geometries and generalizations, and the topology of 4-manifolds.

The first survey is concerned with the analogue in spherical geometry of a theory that was previously developed in the Euclidean setting. Working on such a subject follows the tradition of a series of efforts made by a number of mathematicians to adapt to a non-Euclidean setting notions and results that were known previously in Euclidean geometry. Among these authors, let me mention Leonhard Euler, who published a series of memoirs in which he presented theorems in spherical geometry that are analogues of results (some of which are classical and others due to him) that hold in the Euclidean setting. The next two surveys are concerned with the geometry of Minkowski spaces (finite-dimensional normed vector spaces). Again, the results are inspired by analogous results that hold in Euclidean vector spaces. Then comes a survey on convexity theory, and more precisely on the theory of mixed volumes for convex bodies in Euclidean space. The goal of the next survey is to give a set of comparison results and finiteness theorems, in the setting of

A. Papadopoulos (✉)
Université de Strasbourg and CNRS, Strasbourg, France
e-mail: papadop@math.unistra.fr

metric geometry (more precisely, Gromov hyperbolic spaces), that are analogues of classical results that hold for Riemannian manifolds. The next two surveys are of a topological character. A series of techniques are surveyed there, aimed to tackle classical problems in topology that are related to the 4-dimensional Schoenflies problem. Translation surfaces with their dynamics, a theory which generalizes in a very nontrivial manner the classical dynamics of linear flows on the Euclidean torus, constitute the subject of the next survey. This is followed by a survey which is concerned with rigidity theorems in Teichmüller spaces, equipped with their two most important Finsler metrics: the Teichmüller metric and the Thurston metric. The last survey in this volume is concerned with higher-dimensional complex geometry, and more especially, with generalizations of classical G-structures and Cartan geometries.

Each chapter is an illustration of how current research problems and novel theories are strongly rooted in classical mathematics.

In the rest of this introduction, I review in more detail the content of each of these chapters.

Chapter 2, by Marek Lassak, is titled *Spherical Geometry—A Survey on Width and Thickness of Convex Bodies*. It is a study of the geometry of convex bodies in d-dimensional spheres. The author develops the spherical analogue of the theory of width and thickness of convex bodies in Euclidean space. Supporting hemispheres and lunes play the same role as support hyperplanes and half-spaces in the Euclidean setting. Among the topics considered in the spherical case, we mention the notions of diameter, perimeter, circumradius, area, extreme point, reduced body, body of constant diameter, complete body and body of constant width. The chapter contains several examples and open questions.

Chapter 3, by Vitor Balestro and Horst Martini, titled *Minkowski Geometry—Some Concepts and Recent Developments*, is an introduction to Minkowski geometry, that is, the geometry of finite-dimensional normed spaces. The authors study triangles and the notions of orthocenter, circumcenter, circumradius, Euler line, Monge point and several others, in a general Minkowski space. They also survey area and volume, regular polygons, the geometry of circles and that of systems of circles, Feuerbach circles, equilateral sets and the equilateral dimension of a normed space, and they review Minkowskian analogues of results on intersections of circles associated with triangles and circle patterns in the Euclidean plane. They give several characterizations of Euclidean planes among normed planes. Orthogonality in Minskowski spaces, also considered in this chapter, is studied in more detail in the next one. In dimension 2, the notion of anti-norm leads to that of Radon plane (this is the case where the anti-norm is a multiple of the norm, or, equivalently, when Birkhoff orthogonality is symmetric).

In the same chapter, the authors study the differential geometry of curves in Minkowski planes, in particular the notions of Minkowski curvature, radius of curvature, circular curvature, normal curvature, arclength curvature and curves of constant width. They survey the differential geometry of surfaces, extending to the Minkowski setting classical notions such as the Gauss map, principal curvature, Gaussian curvature, mean curvature, normal curvature, umbilic, Dupin indicatrix,

Dupin metric, minimal surface and girth. They also review Busemann's work on isoperimetric problems. Finally, they study billiards in arbitrary convex bodies, a topic which they call *Minkowski billiards*. The chapter contains a number of open problems.

In Chapter 4, authored by Javier Alonso, Horst Martini and Senlin Wu and titled *Orthogonality Types in Normed Linear Spaces*, the authors review a large number of notions of orthogonality in normed vector spaces that generalize the classical notion of orthogonality in the Euclidean plane or in inner product spaces. Among the notions of orthogonality that are discussed, we mention Roberts orthogonality, Birkhoff orthogonality, James or isosceles orthogonality, Pythagorean orthogonality, Carlsson orthogonality, and there are several others. Symmetry, additivity and other properties of the various notions of orthogonality are discussed. Naturally, several characterizations of inner product spaces are obtained. Like in the previous chapter, a certain number of open problems are discussed.

Chapter 5, by Ivan Izmestiev, is titled *Convex Bodies: Mixed Volumes and Inequalities*. The author is motivated by the natural question: How does the volume of a convex set belonging to a certain class depend on the variables defining it? It turns out that for what concerns volume of convex bodies, an interesting object to study is the family of r-neighborhoods of such a body for variable r. A formula attributed to Jakob Steiner says that the volume of the r-neighborhood of a compact convex body in Euclidean d-space is a degree-d polynomial in r. In fact, two original versions of this formula were obtained by Steiner: they concern polytopes and smooth convex bodies in $\mathbb{R}^3$ respectively. The first one involves parameters such as edge lengths and exterior angles, and the second one involves area and the principal curvatures of the boundary. Izmestiev, in Chap. 5, reviews these formulae and their n-dimensional generalizations, leading to the notion of general *Steiner polynomial* which establishes a relation between volume and the average volumes of projections of the convex body to lower-dimensional subspaces. He surveys several geometrical concepts such as average width and total mean curvature and he presents n-dimensional generalizations of classical formulae attributed to Cauchy and Crofton. At the same time, the chapter contains an introduction to some basic notions in convexity theory, starting with elementary concepts such as support hyperplane, support function, Minkowski functional, Minkowski sum, the Blaschke selection theorem, the Hausdorff metric on the space of convex bodies and the volume function on this space, and continuing with more involved topics such as mixed volumes and their properties (mixed volume inequalities) and in particular the Alexandrov–Fenchel and Minkowski inequalities that establish relations between the volume of the Minkowski sum of two compact convex subsets of $\mathbb{R}^n$ and the volumes of the original two subsets. The author also reviews properties of the Steiner symmetrization of convex bodies, the Blaschke–Santalo inequality and a related conjecture by Mahler involving the volume of the polar dual of a convex body. He mentions the relation with isoperimetric inequalities and the spectrum of the Laplacian.

Chapter 6, by Gérard Besson and Gilles Courtois, is titled *Compactness and Finiteness Results for Gromov-Hyperbolic Spaces*. It is a survey of recent results obtained by the authors together with Sylvestre Gallot and Andrea Sambusetti in which they give analogues, in the setting of Gromov-hyperbolic metric spaces, of classical results due to Bishop and Gromov on the growth of balls in Riemannian manifolds with bounds on their Ricci curvature or on their entropy. The conclusion in the results of Bishop and Gromov is formulated as a comparison property: under the curvature bound condition, the volume of a ball of a certain radius at an arbitrary point is compared with the volume of a ball of the same radius in a simply-connected Riemannian manifold of constant sectional curvature. Important consequences of the main results in the Gromov-hyperbolic setting are obtained in the form of compactness and finiteness theorems.

In this chapter, the reader is led progressively from elementary notions to deep results in geometry and topology, illustrating important relations between these two fields and in particular the restrictions that geometry imposes on topology. The background material presented includes an introduction to Gromov-hyperbolic spaces equipped with measures and with isometric group actions, entropy in metric spaces, CAT(0)-spaces, families of metric spaces endowed with Gromov–Hausdorff distances, Busemann metric spaces, marked groups, growth of groups, the Margulis Lemma concerning thin-thick decompositions of manifolds, and systoles.

The next two surveys (Chaps. 7 and 8) are by Valentin Poénaru. Before describing their content, I would like to say a few words on the notion of geometric simple connectivity which plays a central role there.

A manifold (of arbitrary dimension, compact or not, possibly with or without boundary) is said to be geometrically simply connected if it admits a proper Morse function with no critical points of Morse index 1. Such a notion can also be defined in the combinatorial category, although it is more complicated to state there; it uses a handlebody decomposition, and it is a property of the relative positions of 1- and 2-handles. Roughly speaking, it says that "2-handles cancel 1-handles," a property that appears in Smale's proof of the high-dimensional Poincaré conjecture. Geometric simple connectivity implies simple connectivity in the usual sense (triviality of the fundamental group). The converse statement involves delicate questions, and in all generality it is false; for instance, it is known that it is false in dimension 4, and it is also false for noncompact manifolds with nonempty boundary. But this converse is true for instance in the case of compact manifolds of dimension ≥ 5, by a result of Smale, a fact which was a crucial step in the latter's proof of the high-dimensional Poincaré conjecture. It is also true in dimension 3, by Perelman's proof of the Poincaré conjecture in this dimension. Poénaru has developed during several decades an approach to the 3-dimensional Poincaré conjecture based on 4-dimensional topological constructions which involve in a crucial way the notion of geometric simple connectivity for smooth 4-manifolds.

Chapter 7 is titled *All 4-Dimensional Smooth Schoenflies Balls Are Geometrically Simply-Connected—A Fast Survey of the Proof*. We recall that a *Schoenflies ball* is any one of the two compact bounded smooth manifolds into which an arbitrary smooth embedding of the sphere S^{n-1} into S^n divides the S^n. Poénaru gives in this chapter a survey of the main steps of the proof of the result stated in the title. The 4-dimensional smooth Schoenflies problem is in the background. This problem asks whether any smooth 3-dimensional submanifold of the 4-sphere which is diffeomorphic to the 3-sphere separates it into two 4-manifolds that are both diffeomorphic to the 4-ball. The question is motivated by an analogue in dimension two, where one form of the Schoenflies theorem says that for any simple closed curve in a 2-sphere, any complementary component of this curve union the curve itself is homeomorphic to a closed disc. Without further restrictions, the n-dimensional analogue of this theorem is false.

The same author, in Chap. 8, titled *Classical Differential Topology and Non-commutative Geometry*, starts by reviewing some standard constructions in the theory of smooth 4-manifolds, in particular, spaces that are not geometrically simply connected. He then surveys some connections between various notions in the geometry and topology of 4-manifolds. This work is part of the author's project of understanding the structure at infinity of noncompact smooth 4-manifolds with non-empty boundary. He promotes the idea that these spaces lead to non-commutative spaces in the sense of Connes and that this theory leads to interesting questions in geometric group theory.

Chapter 9, by Daniel Massart, is titled *A Short Introduction to Translation Surfaces, Veech Surfaces and Teichmüller Dynamics*. The author starts with an exposition of some basic background material, namely, the various ways in which translation surfaces are defined: planar polygons with sides pairwise identified by translations, atlases with appropriate transition maps, holomorphic (or Abelian) differentials, etc. Half-translation surfaces are associated with quadratic differentials. Each definition of a translation surface emphasizes a particular point of view on this theory (combinatorial, geometric or complex-analytic). A translation surface has an associated flow, defined outside the singular locus. The point of view of Abelian differential induces another flow parametrized by the circle, namely, turning the differential by an angle $\theta \in [0, 2\pi]$. The existence of these two dynamical systems leads to the familiar questions of counting the periodic orbits, studying their distribution, describing associated invariant measures, etc. After reviewing these questions, Massart surveys properties that are more specific to translation surfaces, such as the Veech dichotomy: each direction is either completely periodic (all orbits are periodic or saddle connections) or uniquely ergodic (all infinite trajectories are uniformly distributed). An important question is to find classes of surfaces satisfying this dichotomy. The author then introduces the notions of moduli space of translation surfaces, of local coordinates given by the relative periods of the Abelian differentials that define them, and of the stratification of this moduli space by the type of zeroes of the differentials. He discusses relations with Teichmüller spaces and with the geodesic flow on moduli spaces. He surveys the basic properties of Teichmüller discs equipped with their $GL_2^+(\mathbb{R})$ and Veech group actions and he reports on McMullen's classification of $GL_2^+(\mathbb{R})$-orbit closures in genus two. The chapter ends with some notes on what is known in higher genus.

Chapter 10 by Ken'ichi Ohshika is titled *Teichmüller Spaces and the Rigidity of Mapping Class Action*. The author reviews several rigidity theorems, first in the setting of the classical Teichmüller metric, and then in the setting of the more recently introduced Thurston asymmetric metric. He starts with Royden's theorem stating that every isometry of Teichmüller space equipped with its Teichmüller metric is induced by the action of a unique element of the extended mapping class group (with few exceptional cases). He reviews at the same time a strongly related result of Royden, which he calls an infinitesimal rigidity result, stating that every complex-linear isometry between cotangent spaces of Teichmüller space at any two points, is a scalar multiple of an action induced by a mapping class. He presents then an analogue of this infinitesimal rigidity theorem in the setting of Thurston's asymmetric metric (recent work by Huang–Ohshika–Papadopoulos). The work involves a detailed analysis of the combinatorial structure of the unit sphere in the cotangent space at each point of Teichmüller space. At this occasion, several features of the Finsler metric geometry of Teichmüller space equipped with the Teichmüller and the Thurston metrics are highlighted.

Chapter 11, by Indranil Biswas and Sorin Dumitrescu, is titled *Holomorphic G-Structures and Foliated Cartan Geometries on Compact Complex Manifolds*. The subject is holomorphic geometric structures on compact complex manifolds, and more especially, holomorphic G-structures and holomorphic Cartan geometries and their generalizations. We recall that a Cartan geometry on a complex manifold is an infinitesimal structure modeled on a homogeneous space G/H where G is a complex Lie group and H a closed complex Lie subgroup of G. The precise definition is given in terms of holomorphic principal bundles. A Cartan geometry is equipped with a curvature tensor which vanishes when the manifold is modeled (not only infinitesimally) on G/H. Biswas and Dumitrescu develop the notion of branched holomorphic Cartan geometry on a complex manifold of any dimension, and a notion of generalized Cartan geometry, inspired by that of branched projective structure on Riemann surfaces introduced by R. Mandelbaum. Cartan and generalized Cartan geometries behave well with respect to holomorphic ramified maps. The authors then develop foliated versions of branched and generalized Cartan geometries, providing many examples and presenting classification results for these geometries. A special emphasis is placed on the GL(2, $\mathbb{C}$) and SL(2, $\mathbb{C}$) cases. In particular, with any holomorphic SL(2, $\mathbb{C}$) structure on a compact complex manifold of odd dimension is associated a holomorphic Riemannian metric. In dimension three the two notions are equivalent. The classification of compact complex manifolds with holomorphic Riemannian metrics is an open question. In the last section of this chapter, a certain number of commented open problems and conjectures on holomorphic G-structures and holomorphic (foliated) Cartan geometries on compact complex manifolds are presented.

Let me close this introduction by a personal note. Reading the original works of authors from the past was one of the most important parts of my mathematical education (I am talking especially of Euler, Lobachevsky and Riemann, but also of ancient authors like Apollonius, Menelaus and Pappus). But I also think that reading and writing surveys is a very rewarding activity, and that the mathematical community is always in need of good surveys of old and recent works.

Chapter 2
Spherical Geometry—A Survey on Width and Thickness of Convex Bodies

Marek Lassak

Abstract This chapter concerns the geometry of convex bodies on the d-dimensional sphere S^d. We concentrate on the results based on the notion of width of a convex body $C \subset S^d$ determined by a supporting hemisphere of C. Important tools are the lunes containing C. The supporting hemispheres take over the role of the supporting half-spaces of a convex body in Euclidean space, and lunes the role of strips. Also essential is the notion of thickness of C, i.e., its minimum width. In particular, we describe properties of reduced spherical convex bodies and spherical bodies of constant width. The last notion coincides with the notions of complete bodies and bodies of constant diameter on S^d. The results reminded and commented on here concern mostly the width, thickness, diameter, perimeter, area and extreme points of spherical convex bodies, reduced bodies and bodies of constant width.

Keywords Spherical geometry · Hemisphere · Lune · Convex body · Convex polygon · Width · Thickness · Reduced body · Diameter · Perimeter · Area · Extreme point · Constant width · Complete body · Constant diameter

Mathematical Subject Classification (2010) 52A55, 97G60

2.1 Introduction

Spherical geometry has its sources in the research on spherical trigonometry which in turn was motivated mostly by navigation needs and astronomic observations. This research started in ancient times mostly by Greeks and continued in the Islamic World, and also by ancient Chinese, Egyptians and Indians. The next important stage began in the XVIII-th century by mostly Euler, and by his collaborators and followers. Descriptions of these achievements is presented by Papadopoulos [73].

M. Lassak (✉)
Institute of Mathematics and Physics, University of Science and Technology, Bydgoszcz, Poland
e-mail: lassak@utp.edu.pl

© The Author(s), under exclusive license to Springer Nature Switzerland AG 2022
A. Papadopoulos (ed.), *Surveys in Geometry I*,
https://doi.org/10.1007/978-3-030-86695-2_2

A historical look to spherical geometry is given by Van Brummelen in his half-and-half historical and popular science book [85], by Rosenfeld [77] and Whittlesey [86].

There are many textbooks on spherical trigonometry, as for instance those from the XIX-th century by Todhunter [83] and Murray [69]. The oldest important research works on spherical convex bodies that the author of this chapter found are those of Kubota [44] and Blaschke [10].

The aim of this chapter is to summarize and comment on properties of spherical convex bodies, mostly these properties which are consequences of the notion of the width of a spherical convex body C determined by a supporting hemisphere of C. This notion of width is introduced in [49]. We omit the proofs besides those of four proofs of theorems which have many consequences.

In Sect. 2.2 we recall a number of elementary definitions of some special subsets of the sphere S^d. In particular, the notions of hemisphere, lune, k-dimensional great subsphere and spherical ball. Basic properties of convex bodies on S^d are presented in Sect. 2.3. We find there the notion of supporting hemisphere of a spherical convex body, as well as the definitions of convex polygons and Reuleaux odd-gons on S^2 and rotational convex bodies on S^d. In Sect. 2.4 we recall some simple claims on lunes and their corners. In Sect. 2.5 we review the notions of width and thickness of a spherical convex body $C \subset S^d$ and a number of properties of them. There is a useful theorem which says how to find the width of C determined by a supporting hemisphere in terms of the radii of the concentric balls touching C from inside or outside. In Sect. 2.6 we find the definition, examples and properties of reduced spherical convex bodies on S^d. First of all a helpful theorem says that through every extreme point e of a reduced body $R \subset S^d$ a lune $L \supset R$ of thickness $\Delta(R)$ passes with e as the center of one of the two $(d-1)$-dimensional hemispheres bounding L. There is also a theorem stating that every reduced spherical body of thickness greater than $\frac{\pi}{2}$ is smooth, and a few corollaries. Section 2.7 describes the shape of the boundaries of reduced spherical convex bodies on S^2 and shows a few consequences. A theorem and a problem concern the circumradius. Section 2.8 is on reduced polygons on S^2. In particular, it appears that every reduced spherical polygon is an odd-gon of thickness at most $\frac{\pi}{2}$. A theorem permits to recognize whether a spherical convex polygon is reduced. Some estimates of the diameter, perimeter and area of reduced polygons are also considered. Section 2.9 concerns the diameter of convex bodies, and in particular reduced bodies. Especially, we show the relationship between the diameter and the maximum width of a convex body on S^d and we find an estimate of the diameter of reduced bodies on S^2 in terms if their thickness. The notion of body of constant diameter is recalled. Section 2.10 is devoted to spherical bodies of constant width. We recall that if a reduced body $R \subset S^d$ has thickness at least $\frac{\pi}{2}$, then R is a body of constant width, and that every body of constant width smaller than $\frac{\pi}{2}$ is strictly convex. Section 2.11 concerns complete spherical convex bodies. It appears that complete bodies, constant width bodies and constant diameter bodies on S^d coincide. Section 2.12 compares the notions of spherically convex bodies and their width, as well as the notion of constant width discussed in this chapter, with some earlier analogous notions considered by other authors.

2.2 Elementary Notions

Let S^d be the unit sphere in the $(d + 1)$-dimensional Euclidean space E^{d+1}, where $d \geq 2$. The intersection of S^d with an $(m+1)$-dimensional subspace of E^{d+1}, where $0 \leq m < d$, is called an m-**dimensional great subsphere of** S^d. In particular, if $m = 0$, we get the 0-dimensional subsphere being **a pair of antipodal points** or **antipodes** in short, and if $m = 1$ we obtain the so-called **great circle**. Important for us is the case when $m = d - 1$. In particular, for S^2 the $(d - 1)$-dimensional great spheres are great circles.

Observe that if two different points are not antipodes, there is exactly one great circle containing them. If different points a, $b \in S^d$ are not antipodes, by the **arc** ab connecting them we mean the shorter part of the great circle containing a and b.

By the **spherical distance** $|ab|$, or, in short, **distance**, of these points we understand the length of the arc connecting them. Moreover, we put π, if the points are antipodes and 0 if the points coincide. Clearly $|ab| = \angle aob$, where o is the center of E^{d+1}. The **distance from a point** $p \in S^d$ **to a set** $A \subset S^d$ is understood as the infimum of distances from p to points of A.

Consider a non-empty set $A \subset S^d$. Its interior with respect to the smallest great subsphere of S^d which contains A is called the **relative interior**.

The **diameter** diam(A) of a set $A \subset S^d$ is the supremum of the spherical distances between pairs of points of A. Moreover, we agree that the empty set and any one-point set have diameter 0. Clearly, if A is closed and has a positive diameter, then the diameter of A is realized for at least one pair of points of C.

By a **spherical ball of radius** $\rho \in (0, \frac{\pi}{2}]$, or shorter, **a ball**, we understand the set of points of S^d having distances at most ρ from a fixed point, called the **center** of this ball. An **open spherical ball**, or shortly **open ball** is the set of points of S^d having distance smaller than ρ from a point. Balls on S^2 are called **disks**.

Spherical balls of radius $\frac{\pi}{2}$ are called **hemispheres**. In other words, by a **hemisphere** of S^d we mean the common part of S^d with any closed half-space of E^{d+1}. We denote by $H(p)$ the hemisphere whose center is p. Two hemispheres whose centers are antipodes are called **opposite hemispheres**. By an **open hemisphere** we mean the set of points having distance less than $\frac{\pi}{2}$ from a fixed point. Hemispheres of S^d play the role of half-spaces of E^d.

By a **spherical** $(d - 1)$-**dimensional ball of radius** $\rho \in (0, \frac{\pi}{2}]$ we mean the set of points of a $(d - 1)$-dimensional great sphere of S^d which are at distance at most ρ from a fixed point, called the **center** of this ball. The $(d - 1)$-dimensional balls of radius $\frac{\pi}{2}$ are called $(d - 1)$-**dimensional hemispheres**. If $d = 2$, we call them **semi-circles**.

If an arc ab is a subset of a hemisphere H with a in the boundary bd(H) of H and ab orthogonal to bd(H), we say that ab and bd(H) are orthogonal at a, or in short, that they are **orthogonal**.

We say that a set $C \subset S^d$ is **convex** if it does not contain any pair of antipodes and if together with every two points of C the whole arc connecting them is a subset of C.

Clearly, for any non-empty convex subset A of S^d there exists a smallest, in the sense of inclusion, non-empty subsphere containing A. It is unique.

Of course, the intersection of every family of convex sets is also convex. Thus for every set $A \subset S^d$ contained in an open hemisphere of S^d there exists a unique smallest convex set containing A. It is called **the convex hull of** A and it is denoted by $\operatorname{conv}(A)$. The following fact from [49] results by applying an analogous theorem for compact sets in E^{d+1}.

Claim 2.1 *If $A \subset S^d$ is a closed subset of an open hemisphere, then $\operatorname{conv}(A)$ is also closed.*

By the well-known fact that a set $C \subset S^d$ is convex if and only if the cone generated by it in E^{d+1} is convex, and by the separation theorem for convex cones in E^{d+1} we obtain, as observed in [49], the following analogous fact for S^d. By the way, the statement is later derived also by Zălinescu [89] from his Theorem 2.

Claim 2.2 *Every two convex sets on S^d with disjoint interiors are subsets of two opposite hemispheres.*

2.3 Convex Bodies

By a **convex body** on S^d we mean a closed convex set with non-empty interior.

If a $(d-1)$-dimensional great sphere G of S^d has a common point t with a convex body $C \subset S^d$ and if its intersection with the interior of C is empty, we say that G is a **supporting** $(d-1)$**-dimensional great sphere of** C **passing through** t. We also say that G **supports** C at t. If H is the hemisphere bounded by G and containing C, we say that H is a **supporting hemisphere of** C and that H **supports** C at t.

If at a boundary point p of a convex body $C \subset S^d$ exactly one hemisphere supports C, we say that p is a **smooth point** of the boundary $\operatorname{bd}(C)$ of C. If all boundary points of C are smooth, then C is called a **smooth body**.

Let $C \subset S^d$ be a convex body and let $Q \subset S^d$ be a convex body or a hemisphere. We say that C **touches** Q **from outside** if $C \cap Q \neq \emptyset$ and $\operatorname{int}(C) \cap \operatorname{int}(Q) = \emptyset$. We say that C **touches** Q **from inside** if $C \subset Q$ and $\operatorname{bd}(C) \cap \operatorname{bd}(Q) \neq \emptyset$. In both cases, points of $\operatorname{bd}(C) \cap \operatorname{bd}(Q)$ are called **points of touching**.

The convex hull V of $k \geq 3$ points on S^2 such that each of them does not belong to the convex hull of the remaining points is named a **spherical convex** k**-gon**. The points mentioned here are called the **vertices** of V. We write $V = v_1 v_2 \ldots v_k$ provided $v_1, v_2, \ldots, v_k$ are successive vertices of V when we go around V on the boundary of V. In particular, when we take $k \geq 3$ successive points in a spherical circle of radius less than $\frac{\pi}{2}$ on S^2 with equal distances of every two successive points, we obtain a **regular spherical** k**-gon**.

Take a regular spherical k-gon $v_1 v_2 \ldots v_k \subset S^2$, where $k \geq 3$ is odd. Clearly, all the distances $|v_i v_{i+\frac{k-1}{2}}|$ and $|v_i v_{i+\frac{k+1}{2}}|$ for $i = 1, \ldots, k$ are equal (the indices are taken modulo k). Denote them by δ. Assume that $\delta \leq \frac{\pi}{2}$. Let B_i, where $i = 1, \ldots, k$, be the disk with center v_i and radius δ. The set $B_1 \cap \cdots \cap B_k$ is called a **spherical Reuleaux k-gon**. Clearly, it is a convex body.

We say that e is an **extreme point** of a convex body $C \subset S^d$ provided the set $C \setminus \{e\}$ is convex. From the analogue of the Krein-Milman theorem for convex cones (e.g., see [25]) its analogue for spherical convex bodies formulated on p. 565 of [49] follows.

Claim 2.3 *Every convex body $C \subset S^d$ is the convex hull of its extreme points.*

This and the fact that the intersection of any closed convex body $C \subset S^d$ with any of its supporting $(d-1)$-dimensional great spheres is a closed convex set imply the following fact from [49].

Claim 2.4 *The boundary of every supporting hemisphere of a convex body $C \subset S^d$ passes through an extreme point of C.*

The next property is shown in [58].

Claim 2.5 *Let $C \subset S^d$ be a convex body. Every point of C belongs to the convex hull of at most $d+1$ extreme points of C.*

Let us add that Shao and Guo [82] proved analytically that every closed set $C \subset S^d$ is the convex hull of its extreme points.

We say that two sets on S^2 are **symmetric with respect to a great circle** if they are symmetric with respect to the plane of E^3 containing this circle. More generally, two sets on S^d are called **symmetric with respect to a $(d-1)$-dimensional subsphere** S^{d-1} if they are symmetric with respect to the hyperplane of E^{d+1} containing S^{d-1}. If a set coincides with its symmetric, we say that the set is **symmetric with respect to a great circle**.

Let us construct a rotational body on S^3. Take a two-dimensional subsphere S^2 of S^3 and a convex body C symmetric with respect to a great circle of S^2. Denote by A the common part of this great circle and C, and call it an **arc of symmetry of** C. Take all subspheres S^2_λ of S^3 containing A, and on each S^2_λ take the copy C_λ of C at the same place with respect to A. The union $\mathrm{rot}_A(C)$ of all these C_λ (including C) is called a **rotational body** on S^3, or the **body of rotation of C around** A. An analogous construction can be provided on S^{d+1} in place of S^3; this time we "rotate" a convex body $C \subset S^d$ having a $(d-1)$-dimensional subsphere S^{d-1} of symmetry. We rotate it around the common part A of S^{d-1} and C. Again we denote the obtained body by $\mathrm{rot}_A(C)$.

Since S^d with the distance of points defined in the preceding section is a metric space, we may consider the Hausdorff distance on S^d. Applying the classical Blaschke selection theorem (see e.g. p. 64 of the monograph [21] by Eggleston and p. 50 of the monograph by Schneider [81]) in E^{d+1} we easily obtain the following spherical version of Blaschke spherical theorem mentioned in [46] before Lemma 4. See also the paper of Hai and An [34], where Corollary 4.11 says that *if X is a proper uniquely geodesic space, then every uniformly bounded sequence of nonempty geodesically convex subsets of X contains a subsequence which converges to a nonempty compact geodesically convex subset of X.* For the context see the book

[72] by Papadopoulos. Moreover look to Theorem 3.2 of the paper [13] by Böroczky and Sagemeister.

Theorem 2.1 *Every sequence of convex bodies on S^d contains a subsequence convergent to a convex body on S^d.*

Corollary 2.1 *Every sequence of convex bodies being subsets of a fixed convex body $C \subset S^d$ contains a subsequence convergent to a convex body being a subset of C.*

Corollary 2.2 *For every convex body $C \subset S^d$ there exist a ball of minimum radius containing C and a ball of maximum radius contained in C.*

Proofs of these corollaries are similar to proofs of analogous facts in E^d. For instance, see Section 18 of the book [4] by Benson and Section 16 of the book [60] by Lay.

2.4 Lunes

Recall the classical notion of lune. For any distinct non-opposite hemispheres G and H of S^d the set $L = G \cap H$ is called **a lune** of S^d. Lunes of S^d play the role of strips in E^d. The $(d-1)$-dimensional hemispheres bounding the lune L which are contained in G and H, respectively, are denoted by G/H and H/G. Here is a claim from [49].

Claim 2.6 *Every pair of different points $a, b \in S^d$ which are not antipodes determines exactly one lune L such that a, b are the centers of the $(d-1)$-dimensional hemispheres bounding L.*

Since every lune L determines exactly one pair of centers of the $(d-1)$-dimensional hemispheres bounding L, from Claim 2.6 we see that there is a one-to-one correspondence between lunes and pairs of different points, which are not antipodes, of S^d.

Observe that $(G/H) \cup (H/G)$ is the boundary of the lune $G \cap H$ (in particular, every lune of S^2 is bounded by two different semi-circles). Denote by $c_{G/H}$ and $c_{H/G}$ the centers of G/H and H/G, respectively. By **corners** of the lune $G \cap H$ we mean points of the set $(G/H) \cap (H/G)$. Of course, $r \in (G/H) \cup (H/G)$ is a corner of $G \cap H$ if and only if r is equidistant from $c_{G/H}$ and $c_{H/G}$. Observe that the set of corners of a lune is a $(d-2)$-dimensional subsphere of S^d. In particular, every lune of S^2 has two corners. They are antipodes.

By the **thickness** $\Delta(L)$ **of a lune** $L = G \cap H \subset S^d$ we mean the spherical distance of the centers of the $(d-1)$-dimensional hemispheres G/H and H/G bounding L. Observe that it is equal to each of the non-oriented angles $\angle c_{G/H} r c_{H/G}$, where r is any corner of L.

Here is a fact shown in [49].

Claim 2.7 *Let H and G be different and not opposite hemispheres. Consider the lune $L = H \cap G$. Let $x \neq c_{G/H}$ belong to G/H. If $\Delta(L) < \frac{\pi}{2}$, we have $|xc_{H/G}| > |c_{G/H}c_{H/G}|$. If $\Delta(L) = \frac{\pi}{2}$, we have $|xc_{H/G}| = |c_{G/H}c_{H/G}|$. If $\Delta(L) > \frac{\pi}{2}$, we have $|xc_{H/G}| < |c_{G/H}c_{H/G}|$.*

Similarly to Claim 2.1 we obtain the following Claim formulated in [49] and needed for the proof of Theorem 2.4.

Claim 2.8 *Every sequence of lunes of a fixed thickness t of S^d contains a subsequence of lunes convergent to a lune of thickness t.*

The following two claims are from [51].

Claim 2.9 *Consider a hemisphere $H(c)$ of S^d. Then any $(d-1)$-dimensional subsphere of S^d containing c dissects $H(c)$ into two lunes of thickness $\frac{\pi}{2}$.*

Claim 2.10 *Let $L \subset S^d$ be a lune and let $C \subset L$ be a convex body such that the set $F = C \cap \mathrm{corn}(L)$ is non-empty. Then at least one extreme point of C is a corner point of L.*

The next claim and remark are taken from [49].

Claim 2.11 *Let $\mathrm{diam}(C) \leq \frac{\pi}{2}$ for a convex body $C \subset S^d$ and assume that $\mathrm{diam}(C) = |ab|$ for some points $a, b \in C$. Denote by L the lune such that a and b are the centers of the $(d-1)$-dimensional hemispheres bounding L. We have $C \subset L$.*

In general, this claim does not hold without the assumption that $\mathrm{diam}(C) \leq \frac{\pi}{2}$. A simple counterexample is the triangle $T = abc$ with $|ab| = \frac{2}{3}\pi \approx 2.0944$, $|bc| = \frac{\pi}{6} \approx 0.5236$ and $\angle abc = 95°$. From the law of cosines for sides, called also Al'Battani formulas in [69], p. 45, we get $|ac| \approx 2.0609$. Consequently, $|ab| = \frac{2}{3}\pi$ is the diameter of T. Since $\angle abc = 95°$, the lune with centers a and b of the semi-circles bounding it does not contain c. Still its thickness is $\frac{2}{3}\pi$. Thus this lune does not contain T.

The subsequent claim is shown in [52].

Claim 2.12 *Let K be a hemisphere of S^d and let $p \in \mathrm{bd}(K)$. Moreover, let $pq \subset K$ be an arc orthogonal to $\mathrm{bd}(K)$ with q in the interior of K and $|pq| < \frac{\pi}{2}$. Then amongst all the lunes of the form $K \cap M$, with q in the boundary of the hemisphere M, only the lune $K \cap K_\dashv$ such that pq is orthogonal to $\mathrm{bd}(K_\dashv)$ at q has the smallest thickness.*

2.5 Width and Thickness of a Convex Body

We say that **a lune passes through a boundary point** p **of a convex body** $C \subset S^d$ if the lune contains C and if the boundary of the lune contains p. If the centers of both the $(d-1)$-dimensional hemispheres bounding a lune belong to C, then we call such a lune an **orthogonally supporting lune of** C.

For any convex body $C \subset S^d$ and any hemisphere K supporting C we define **the width of C determined by K** as the minimum thickness of a lune $K \cap K^*$ over all hemispheres $K^* \neq K$ supporting C. We denote it by $\text{width}_K(C)$. By compactness arguments we immediately see that at least one such hemisphere K^* exists, and thus at least one corresponding lune $K \cap K^*$ exists.

This notion of width of $C \subset S^d$ is an analogue of the notion of width of a convex body $C \subset E^d$ between a pair of parallel supporting hyperplanes H_1, H_2 supporting C. Our $K \subset S^d$ takes over the role of one of the half-spaces bounding H_1 or H_2 and containing C and our K^* takes over the role of the other half-space, which in E^d is unique.

The following lemma from [49] is needed for the proof of the forthcoming Theorem 2.2.

Lemma 2.1 *Let G and H be different and not opposite hemispheres, and let g denote the center of G. If $g \notin \text{bd}(H)$, then by B denote the ball with center g which touches H (from inside or outside) and by t the point of touching. If $g \in \text{bd}(H)$, we put $t = g$. We claim that t is always at the center of the $(d-1)$-dimensional hemisphere H/G.*

Figure 2.1 shows the two situations when B touches H from inside and outside as in the text of the lemma. Here and in later figures we adopt an orthogonal look to the sphere from outside.

The theorem from [49] given below and its proof present a procedure for establishing the width $\text{width}_K(C)$ of a convex body $C \subset S^d$ in terms of the radii of balls concentric with K and touching C.

Theorem 2.2 *Let K be a hemisphere supporting a convex body $C \subset S^d$. Denote by k the center of K.*

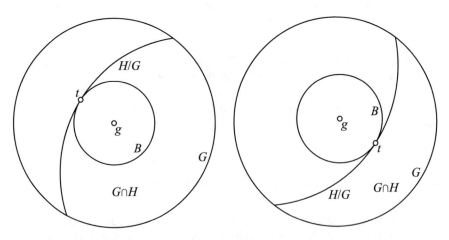

Fig. 2.1 Illustration to Lemma 2.1

I. *If $k \notin C$, then there exists a unique hemisphere K^* supporting C such that the lune $L = K \cap K^*$ contains C and has thickness $\text{width}_K(C)$. This hemisphere supports C at the point t at which the largest ball B with center k touches C from outside. We have $\Delta(K \cap K^*) = \frac{\pi}{2} - \rho_B$, where ρ_B denotes the radius of B.*

II. *If $k \in \text{bd}(C)$, then there exists at least one hemisphere K^* supporting C such that $L = K \cap K^*$ is a lune containing C which has thickness $\text{width}_K(C)$. This hemisphere supports C at $t = k$. We have $\Delta(K \cap K^*) = \frac{\pi}{2}$.*

III. *If $k \in \text{int}(C)$, then there exists at least one hemisphere K^* supporting C such that $L = K \cap K^*$ is a lune containing C which has thickness $\text{width}_K(C)$. Every such K^* supports C at exactly one point $t \in \text{bd}(C) \cap B$, where B denotes the largest ball with center k contained in C, and for every such t this hemisphere K^*, denoted by K_t^*, is unique. For every t we have $\Delta(K \cap K_t^*) = \frac{\pi}{2} + \rho_B$, where ρ_B denotes the radius of B.*

Proof In Fig. 2.2 we see two illustrations to this theorem and its proof. They, respectively, concern Parts I and III.

Part I. Since C is a convex body and B is a ball, we see that B touches C from the outside and the point of touching is unique. Denote it by t (see the first illustration of Fig. 2.2). By Lemma 2.2, the bodies C and B are in some two opposite hemispheres. What is more, since B is a ball touching C from the outside, this pair of hemispheres is unique. Denote by K_t^* the one of them which contains C. We intend to show that K_t^* is nothing else but the promised K^*.

Denote by k^* the center of K_t^*. Since k is also the center of B and since B and K_t^* touch from the outside at t, we have $t \in kk^*$. From Lemma 2.1 we see that t is the center of the $(d-1)$-dimensional hemisphere K_t^*/K. Analogously, from this lemma we conclude that the common point u of kk^* and the boundary of

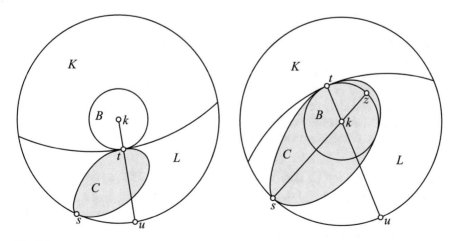

Fig. 2.2 Illustration to Theorem 2.1 and its proof

K is the center of K/K_t^*. Since t and u are centers of the $(d-1)$-dimensional hemispheres bounding the lune $K \cap K_t^*$, we have $|tu| = \Delta(K \cap K_t^*)$. This and $|kt| + |tu| = |ku| = \frac{\pi}{2}$ imply $\Delta(K \cap K_t^*) = \frac{\pi}{2} - \rho_B$.

If we assume that there exists a hemisphere $M \supset C$ with $\Delta(K \cap M) < \frac{\pi}{2} - \rho_B$, then the lune $K \cap M$ must be disjoint from B, and hence it does not contain C. A contradiction. Thus $K \cap K_t^*$ is a narrowest lune of the form $K \cap N$ containing C. It is the unique lune of this form in virtue of the uniqueness of t and K_t^* explained at the beginning of the proof of Part I

Part II.   Clearly, there is at least one hemisphere K^* supporting C at k. Of course, $\Delta(K \cap K^*) = \frac{\pi}{2}$. By Lemma 2.1 we see that k is the center of K^*/K.

Part III.   Take the largest ball $B \subset C$ with center k. Clearly, there is at least one boundary point t of C which is also a boundary point of B (see the second illustration of Fig. 2.2). We find a hemisphere K_t^* which supports C at t. Of course, it also supports B and thus, for given t, it is unique.

For every t there is a unique point $u \in K/K_t^*$ such that $k \in tu$. Thus, $|ku| = \frac{\pi}{2}$ and $|kt| = \rho_B$ imply $|tu| = \frac{\pi}{2} + \rho_B$. Hence the facts, resulting from Lemma 2.1, that t is the center of K_t^*/K and that u is the center of K/K_t^* give $\Delta(K \cap K_t^*) = \frac{\pi}{2} + \rho_B$.

If we assume that there exists a hemisphere $M \supset C$ such that the lune $K \cap M$ is narrower than $\frac{\pi}{2} + \rho_B$, then this lune does not contain B, and hence it also does not contain C, a contradiction. Thus the narrowest lunes of the form $K \cap N$ containing C are of the form $K \cap K_t^*$. $\square$

Let us point out that in Part I, since the center k of K does not belong to C, the lune $K \cap K^*$ is unique. In Part II this narrowest lune $K \cap K^*$ containing C sometimes is unique and sometimes not. This depends on the point $k = t$ of C which belongs to the boundary of B. In Part III for any given point t of touching C by B from the inside (we may have one, or finitely many, or infinitely many such points t), the lune $K \cap K_t^*$ is unique.

For instance, if $C \subset S^2$ is a regular spherical triangle of sides $\frac{\pi}{2}$ and the circle bounding a hemisphere K contains a side of this triangle, then $K \cap K^*$ is not unique. Namely, as K^* we may take any hemisphere containing C, whose boundary contains this vertex of C which does not belong to K. The thickness of every such lune $K \cap K^*$ is equal to $\frac{\pi}{2}$. If C is a regular spherical triangle of sides greater than $\frac{\pi}{2}$ and the boundary of K contains a side of this triangle, then $K \cap K^*$ is also not unique. This time the boundary of K^* contains a side of C different from the side which is contained in K. So we have exactly two positions of K^*.

Here are two corollaries from Theorem 2.2. For the second we additionally apply Lemma 2.1.

Corollary 2.3 *If $k \notin C$, then* $\mathrm{width}_K(C) = \frac{\pi}{2} - \rho_B$. *If $k \in \mathrm{bd}(C)$, we have* $\mathrm{width}_K(C) = \frac{\pi}{2}$. *If $k \in \mathrm{int}(C)$, then* $\mathrm{width}_K(C) = \frac{\pi}{2} + \rho_B$.

Corollary 2.4 *The point t of support in Theorem 2.2 is the center of the $(d-1)$-dimensional hemisphere K^*/K.*

The following theorem is also proved in [49].

Theorem 2.3 *As the position of the* $(d - 1)$*-dimensional supporting hemisphere of a convex body* $C \subset S^d$ *changes, the width of C determined by this hemisphere changes continuously.*

Proof We keep the notation of Theorem 2.2. Of course, the positions of k and thus that of B depend continuously on K. Hence $\frac{\pi}{2} - \rho_B$ and $\frac{\pi}{2} + \rho_B$ change continuously. This fact and Corollary 2.3 imply the thesis of our theorem. It does not matter here that for a fixed K sometimes the lunes $K \cap K^*$ are not unique; still they are all of equal thickness.   □

Compactness arguments lead to the conclusion that for every convex body $C \subset S^d$ the supremum of width$_H(C)$ over all hemispheres H supporting C is realized for a supporting hemisphere of C, that is, we may take here the maximum instead of supremum.

We define the **thickness** $\Delta(C)$ **of a convex body** $C \subset S^d$ as the minimum of the thickness of the lunes containing C (by compactness arguments, this minimum is realized), or even more intuitively as the thickness of each "narrowest" lune containing C. In other words,

$$\Delta(C) = \min\{\text{width}_K(C); \ K \text{ is a supporting hemisphere of } C\}.$$

Example 2.1 Applying Theorem 2.2 we easily find the thickness of any regular triangle $T_\alpha \subset S^2$ of angles α. Formulas of spherical trigonometry imply that $\Delta(T_\alpha) = \text{arc cos} \frac{\cos \alpha}{\sin \alpha / 2}$ for $\alpha < \frac{\pi}{2}$. If $\alpha \geq \frac{\pi}{2}$ (but, of course, $\alpha < \frac{2}{3}\pi$), then $\Delta(T_\alpha) = \alpha$. In both cases $\Delta(T_\alpha)$ is realized for width$_K(T_\alpha)$, where K is a hemisphere whose bounding semicircle contains a side S of T_α. In the first case T_α is symmetric with respect to the great circle containing the arc kk^* connecting the centers k and k^* of K/K^* and K^*/K, respectively, while in the second T_α is symmetric with respect to the great circle passing through the middle of kk^* and having endpoints at the corners of the lune $K \cap K^*$ (observe that this time we have exactly two positions of K^*, each with bd(K^*) containing a side of T_α different from S). For $\alpha = \frac{\pi}{2}$ there are infinitely many positions in K^*; just these hemispheres which support T_α at k^*.

The first thesis of the following fact was first proved in [49]. The last statement was given in [57] for S^2 and in [58] for S^d.

Claim 2.13 *Consider a convex body* $C \subset S^d$ *and any lune L of thickness* $\Delta(C)$ *containing C. Both centers of the* $(d - 1)$*-dimensional hemispheres bounding L belong to C. Moreover, if* $\Delta(C) > \frac{\pi}{2}$*, then both centers are smooth points of the boundary of C.*

2.6   Reduced Spherical Bodies

Reduced bodies in the Euclidean space E^d are defined by Heil [40] and next considered in very many papers (references of many of them can be found in the survey articles [55] and [56]). In analogy to this notion we define reduced convex bodies on S^d. Namely, after [49] we say that a convex body $R \subset S^d$ is **reduced** if $\Delta(Z) < \Delta(R)$ for every convex body $Z \subset R$ different from R.

It is easy to show that all regular odd-gons on S^2 of thickness at most $\frac{\pi}{2}$ are reduced bodies. The assumption that the thickness is at most $\frac{\pi}{2}$ matters here. For instance take the regular triangle T_α of angles $\alpha > \frac{\pi}{2}$ (see Example 2.1). Take the hemisphere K whose boundary contains a side of T_α and apply Part III of Theorem 2.2. The corresponding ball $B \subset T_\alpha$ touches T_α from the inside at exactly two points t_1, t_2. Cutting off a part of T_α by the shorter arc of the boundary of B between t_1 and t_2 we obtain a convex body $Z \subset T_\alpha$. We have $\Delta(Z) = \Delta(T_\alpha)$, which implies that T_α is not reduced.

Of course, the spherical bodies of constant width on S^d considered later in Sect. 2.10 are reduced bodies. In particular, every Reuleaux polygon is a reduced body on S^2.

Dissect a disk on S^2 by two orthogonal great circles through its center. The four parts obtained are called **quarters of a disk**. In particular, the triangle of sides and angles $\frac{\pi}{2}$ is a quarter of a disk. It is easy to see that every quarter of a disk is a reduced body and that its thickness is equal to the radius of the original disk. More generally, each of the 2^d parts of a spherical ball on S^d dissected by d pairwise orthogonal great $(d-1)$-dimensional spheres through the center of this ball is a reduced body on S^d. We call it a $\frac{1}{2^d}$-**th part of a ball**. Clearly, its thickness is equal to the radius of the above ball.

Example 2.2 On S^2 consider arcs a_1b_1 and a_2b_2 of equal lengths and with a common point c such that $ca_2 \perp a_2a_1$, $cb_1 \perp b_1b_2$ and $|ca_2| = |cb_1|$. As a consequence, $|ca_1| \geq |cb_1|$ and $|cb_1| \leq |cb_2|$. Provide the shorter piece P_1 of the circle with center c connecting a_1 and b_2. Also provide the shorter piece P_2 of the circle with center c connecting a_2 and b_1. Let $P = \mathrm{conv}(P_1 \cup P_2)$ (see Fig. 2.3). Observe that P is a reduced body of thickness $|a_1b_1| = |a_2b_2|$.

Consider a reduced body $R \subset S^{d-1}$ having a $(d-2)$-dimensional subsphere S^{d-2} of symmetry. Put $A = S^{d-2} \cap R$. Then $\mathrm{rot}_A(R)$ is a reduced body on S^d. For instance, when we take as $R \subset S^2$ a triangle with an axis of symmetry of it, we get a "spherical cone" on S^3, which is a spherical reduced body. We can also rotate the reduced body P from Example 2.2 taking as A its axis of symmetry, again obtaining a reduced body on S^3.

The following two theorems are proved in [49]. Having in mind the intuitive ideas of the first proof, we recall it here.

Theorem 2.4 *Through every extreme point e of a reduced body $R \subset S^d$ a lune $L \supset R$ of thickness $\Delta(R)$ passes with e as the center of one of the two $(d-1)$-dimensional hemispheres bounding L.*

Fig. 2.3 The reduced body
on S^2 from Example 2.2

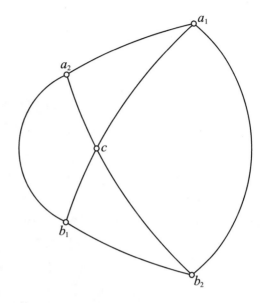

Proof Let B_i be the open ball of radius $\Delta(R)/i$ centered at e and let $R_i =$ conv$(R \setminus B_i)$ for $i = 2, 3, \ldots$ (see Fig. 2.4). By Claim 2.1 every R_i is a convex body. Moreover, since e is an extreme point of R, then R_i is a proper subset of R. So, since R is reduced, $\Delta(R_i) < \Delta(R)$. By the definition of thickness of a convex body, R_i is contained in a lune L_i of thickness $\Delta(R_i)$.

From Claim 2.8 we conclude that there exists a subsequence of the sequence $L_2, L_3, \ldots$ converging to a lune L. Since $R_i \subset L_i$ for $i = 2, 3, \ldots$, we obtain that $R \subset L$. Since $\Delta(L_i) = \Delta(R_i) < \Delta(R)$ for every i, we get $\Delta(L) \leq \Delta(R)$. This and $R \subset L$ imply $\Delta(L) = \Delta(R)$.

Let m_i, m_i' be the centers of the $(d-1)$-dimensional hemispheres H_i, H_i' bounding L_i. We maintain that at least one of these two centers, say m_i, belongs to the closure of $R \setminus R_i$. The reason is that in the opposite case, there would be a neighborhood N_i of m_i such that $N_i \cap R_i = N_i \cap R$, which would imply that H_i supports R at m_i. Moreover, H_i' supports R at m_i'. Hence $\Delta(R) = \Delta(L_i) = \Delta(R_i)$, in contradiction with $\Delta(R_i) < \Delta(R)$.

Since $m_i \in R \setminus R_i$ for $i = 2, 3, \ldots$, we conclude that the sequence of points $m_2, m_3, \ldots$ tends to e. Consequently, e is the center of a $(d-1)$-dimensional hemisphere bounding L. $\qquad\square$

As noted in Remark 2 of [49], besides the lune from Theorem 2.4, sometimes we have additional lunes $L' \supset R$ of thickness $\Delta(R)$ through e. Then e is not in the middle of a $(d-1)$-dimensional hemisphere bounding L'. This happens, for instance, when R is a spherical regular triangle T_α with $\alpha \leq \frac{\pi}{2}$.

By Theorem 2.4 we obtain the following spherical analog of a theorem from the paper [29] by Groemer (see also Corollary 1 in [45] and [55]). This theorem is presented in [51] and earlier for $d = 2$ in [57].

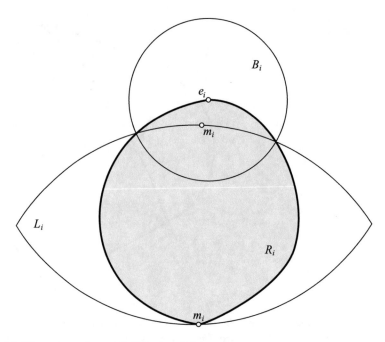

Fig. 2.4 Illustration to the proof of Theorem 2.4

Theorem 2.5 *Every reduced spherical body on S^d of thickness greater than $\frac{\pi}{2}$ is smooth.*

The thesis of Theorem 2.5 does not hold for thickness at most $\frac{\pi}{2}$. For instance, the regular spherical triangle of any thickness less than or equal to $\frac{\pi}{2}$ is reduced but not smooth.

2.7   More About Reduced Bodies on S^2

The notion of order of supporting hemispheres of a convex body $C \subset S^2$ presented in the next paragraph is needed for the later description of the boundaries of reduced bodies on S^2. For this definition the classical notion of the polar set $F^\circ = \{p : F \subset H(p)\}$ of a set $F \subset S^d$ is needed. It is known that if C is a spherical convex body, then C° is also a convex body. Observe that $\mathrm{bd}(C^\circ) = \{p : H(p)$ is a supporting hemisphere of $C\}$.

Let $C \subset S^2$ be a spherical convex body and $X = H(x), Y = H(y), Z = H(z)$ be different supporting hemispheres of C. If x, y, z are in this order on the boundary of $\mathrm{bd}(C^\circ)$ of C when viewed from the inside of this polar, then (after [57]) we write $\prec XYZ$ and say that X, Y, Z **support C in this order**. Clearly, $\prec XYZ$, $\prec YZX$ and $\prec ZXY$ are equivalent. The symbol $\preceq XYZ$ means that $\prec XYZ$ or $X = Y$ or

$Y = Z$ or $Z = X$. If $\prec XYZ$ (respectively, $\preceq XYZ$), then we say that Y **supports** C **strictly between** (respectively **between**) X **and** Z.

Let X and Z be hemispheres supporting C at p and let $\preceq XYZ$ for every hemisphere Y supporting C at p. Then X is said to be the **right supporting hemisphere at** p and Z is said to be the **left supporting hemisphere at** p. The left and right supporting hemispheres of C are called **extreme supporting hemispheres of** C.

Here are a lemma and a theorem from [57].

Lemma 2.2 *Let* $C \subset S^2$ *be a convex body with* $\Delta(C) < \frac{\pi}{2}$ *and let hemispheres* N_i *with* $\text{width}_{N_i} = \Delta(C)$ *for* $i = 1, 2, 3$ *support* C*. Then* $\prec N_1 N_2 N_3$ *if and only if* $\prec N_1^* N_2^* N_2^*$.

Theorem 2.6 *Let* $R \subset S^2$ *be a reduced spherical body with* $\Delta(R) < \frac{\pi}{2}$*. Let* M_1 *and* M_2 *be supporting hemispheres of* R *such that* $\text{width}_{M_1}(R) = \Delta(R) = \text{width}_{M_2}(R)$ *and* $\text{width}_M(R) > \Delta(R)$ *for every hemisphere* M *satisfying* $\prec M_1 M M_2$*. Consider lunes* $L_1 = M_1 \cap M_1^*$ *and* $L_2 = M_2 \cap M_2^*$*. Then the arcs* $a_1 a_2$ *and* $b_1 b_2$ *are in* $\text{bd}(R)$*, where* a_i *is the center of* M_i / M_i^* *and* b_i *is the center of* M_i^* / M_i *for* $i = 1, 2$*. Moreover,* $|a_1 a_2| = |b_1 b_2|$*.*

The union of triangles $a_1 a_2 c$ and $b_1 b_2 c$, where c denotes the intersection point of $a_1 a_2$ and $b_1 b_2$, is called a **butterfly** and the pair $a_1 a_2$, $b_1 b_2$ is called **a pair of arms** of this butterfly (for instance see Example 2.2 with Fig. 2.3). These names are analogous to the names from [47] for E^2, and for normed planes from [23] and [56].

Theorem 2.6 allows us to describe the structure of the boundary of every reduced body on S^2 with $\Delta(R) < \frac{\pi}{2}$. Namely, we conclude that the boundary consists of pairs of arms of butterflies (in particular, the union of some arms of two butterflies may form a longer arc) and from some "opposite" pieces of spherical curves of constant width (considered in Sect. 2.10). We obtain these "opposite" pieces always when $\text{width}_M(R) = \Delta(R)$ for all M fulfilling $\preceq M_1 M M_2$, where M_1 and M_2 are two fixed supporting semi-circles of R.

The following proposition and two theorems are also taken from [57].

Proposition 2.1 *Let* $R \subset S^2$ *be a reduced body with* $\Delta(R) < \frac{\pi}{2}$*. Assume that* $\text{width}_{M_1}(R) = \Delta(R) = \text{width}_{M_2}(R)$*, where* M_1 *and* M_2 *are two fixed supporting hemispheres of* R*. Denote by* a_i *and* b_i *the centers of respectively the semi-circles* M_i / M_i^* *and* M_i^* / M_i *for* $i = 1, 2$*. Assume that* $a_1 = a_2$ *and that* $\preceq M_1 M M_2$ *for every* M *supporting* R *at this point. Then the shorter piece of the spherical circle with the center* $a_1 = a_2$ *and radius* $\Delta(R)$ *connecting* b_1 *and* b_2 *is in* $\text{bd}(R)$*. Moreover,* $\text{width}_M(R) = \Delta(R)$ *for all such* M*.*

Theorem 2.7 *Let* R *be a reduced body with* $\Delta(R) < \frac{\pi}{2}$*. Assume that* M *is a supporting hemisphere of* R *such that the intersection of* $\text{bd}(M)$ *with* $\text{bd}(R)$ *is a non-degenerate arc* $x_1 x_2$*. Then* $\text{width}_M(R) = \Delta(R)$*, and the center of* M / M^* *belongs to* $x_1 x_2$*.*

Theorem 2.8 *If* M *is an extreme supporting hemisphere of a reduced spherical body* $R \subset S^2$*, then* $\text{width}_M(R) = \Delta(R)$*.*

After this theorem, the following problem is quite natural.

Problem 2.1 *Define the notion of extreme supporting hemispheres of a convex body $C \subset S^d$ and generalize Theorem 2.8 to reduced convex bodies on S^d.*

For any maximum piece $\overset{\frown}{gh}$ of the boundary of a reduced body $R \subset S^2$ with $\Delta(R) < \frac{\pi}{2}$ which does not contain any arc the following claim from [54] holds true.

Claim 2.14 *If a hemisphere K supports R at a point of $\overset{\frown}{gh}$, then* $\mathrm{width}_K(R) = \Delta(R)$.

By Lemma 2.2 and Claim 2.14 we see that all the points at which our hemispheres K^* touch R form a curve $\overset{\frown}{g'h'}$ in $\mathrm{bd}(R)$. We call it the curve *opposite to the curve* $\overset{\frown}{gh}$. Vice-versa, from this lemma we obtain that $\overset{\frown}{g'h'}$ determines $\overset{\frown}{gh}$. So we say that $\overset{\frown}{gh}$ and $\overset{\frown}{g'h'}$ is a **pair of opposite curves of constant width** $\Delta(R)$. As a recapitulation we obtain the following claim from [54]. There the next theorem (analogous to Theorem 3.3 from [48]) is also given.

Claim 2.15 *The boundary of a reduced body $R \subset S^2$ of thickness below $\frac{\pi}{2}$ consists of countably many pairs of arms of butterflies and of countably many pairs of opposite pairs of curves of constant width $\Delta(R)$.*

Theorem 2.9 *Let $R \subset S^2$ be a reduced body $R \subset S^2$ of thickness at most $\frac{\pi}{2}$. For arbitrary $\varepsilon > 0$ there exists a reduced body $R_\varepsilon \subset S^2$ with $\Delta(R_\varepsilon) = \Delta(R)$ whose boundary consists only of pairs of arms of butterflies and pieces of circles of radius $\Delta(R)$, such that the Hausdorff distance between R_ε and R is at most ε.*

Theorem 2.4 leads to the following questions asked in [49] for S^2, which is rewritten here more generally for S^d.

Problem 2.2 *Is it true that through every boundary point p of a reduced body $R \subset S^2$ passes a lune $L \supset R$ of thickness $\Delta(R)$?*

A consequence of a positive answer would be that every reduced body $R \subset S^2$ is an intersection of lunes of thickness $\Delta(R)$.

The example of the regular spherical triangle of thickness less than $\frac{\pi}{2}$ shows that the answer is negative when we additionally require that p is the center of one of the two $(d-1)$-dimensional hemispheres bounding L.

Theorem 2.26 shows that for bodies of constant width on S^d the answer to Problem 2.2 is positive.

After [46] recall that any reduced convex body $R \subset E^2$ is contained in a disk of radius $\frac{1}{2}\sqrt{\Delta(R)}$. Its spherical analog is given in the main result of the paper [70] by Musielak.

Theorem 2.10 *Every reduced spherical body R of thickness at most $\frac{\pi}{2}$ is contained in a disk of radius* $\mathrm{arc}\tan\left(\sqrt{2} \cdot \tan\frac{\Delta(R)}{2}\right)$.

The proof of this theorem uses Theorem III from the paper [68] by Molnár concerning a spherical variant of Helly's theorem. This permits to avoid using the following consequence of our Corollary 2.2.

Recall that every reduced polygon $R \subset E^2$ is contained in a disk of radius $\frac{2}{3}\Delta(R)$ as shown in Proposition from [46]. This generates the subsequent problem for the two dimensional sphere.

Problem 2.3 *What is the smallest radius of a disk which contains every reduced polygon of a given thickness on S^2?*

A general question is what properties of reduced bodies in E^d, and especially in E^2, (see [45] and [55]) can be reformulated and proved for reduced spherical convex bodies.

2.8   Spherical Reduced Polygons

In this section we recall a number of facts on reduced polygons. But the following theorem and propositions from [50] are true for the more general situation of all spherical convex polygons.

Theorem 2.11 *Let $V \subset S^2$ be a spherically convex polygon and let L be a lune of thickness $\Delta(V)$ containing V. Then at least one of the two semicircles bounding L contains a side of V. If $\Delta(V) < \frac{\pi}{2}$, then the center of this semicircle belongs to the relative interior of this side.*

Proof Our lune L has the form $G \cap H$, where G and H are different non-opposite hemispheres (see Fig. 2.5).

Let us prove the first statement. Assume the contrary, i.e., that each of the two semicircles bounding L contains exactly one vertex of V. Denote by g the vertex contained by G/H, and by h the vertex contained by H/G. Apply Claim 2.13. We get that g is the center of G/H and h is the center of H/G. Rotate G around g, and H around h, both according to the same orientation and by the same angle. Since

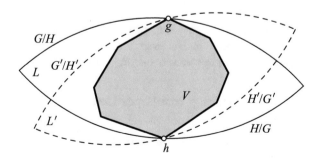

Fig. 2.5 Rotation of G around g and rotation of H around h

G/H and H/G strictly support V, after any sufficiently small rotation of both with the same orientation, we obtain a new lune $L' = G' \cap H'$, where G' and H' are the images of G and H under corresponding rotations (see Fig. 2.5). It still contains V and has in common with the boundary of V again only g and h. But since now g and h are not the centers of the semicircles G'/H', H'/G' which bound L' and since they are symmetric with respect to the center of L', we get $\Delta(L') < \Delta(L)$. This contradicts the assumption $\Delta(L) = \Delta(V)$. We see that the first part of our theorem holds true.

For the proof of the second part, assume the opposite, that is, that the center of the semicircle bounding L does not belong to the relative interior of the side S contained in this semicircle. This and Corollary 2.4 imply that it must be an end-point of S, and thus a vertex of V. Apply $\Delta(V) < \frac{\pi}{2}$. Let us rotate G around g, and H around h, both with the same orientation, such that after sufficiently small rotations we get a lune narrower than L still containing V. This contradiction shows that our assumption at the beginning of this paragraph is false.   □

If $\Delta(V) = \frac{\pi}{2}$, then the center of the semicircle in the formulation of Theorem 2.11 may be an end-point of a side of V. This holds, for instance, for the regular triangle of thickness $\frac{\pi}{2}$.

Below we see a proposition needed for the proof of the subsequent theorem, both from [50].

Proposition 2.2 *Let $V \subset S^2$ be a spherically convex polygon of thickness greater than $\frac{\pi}{2}$ and let L be a lune of thickness $\Delta(V)$ containing V. Then each of the two semicircles bounding L contains a side of V. Moreover, both centers of these semicircles belong to the relative interiors of the contained sides of V.*

Theorem 2.12 *Every reduced spherical polygon is of thickness at most $\frac{\pi}{2}$.*

For a convex odd-gon $V = v_1 v_2 \ldots v_n$ by the **opposite side to the vertex** v_i we mean the side $v_{i+(n-1)/2} v_{i+(n+1)/2}$. The indices are taken modulo n.

The following fact is proved for any reduced polygon V with $\Delta(V) < \frac{\pi}{2}$ in [50]. In Remark on p. 377 of [51], applying Theorem 2.9, this fact is transferred also for the case when $\Delta(V) = \frac{\pi}{2}$. Let us add that a little later this fact for $\Delta(V) = \frac{\pi}{2}$ is given again in Theorem 3.1 of [15]. In summary, below we formulate this fact for every convex odd-gon on S^2.

Theorem 2.13 *A spherically convex odd-gon V is reduced if and only if the projection of every vertex of V on the great circle containing the opposite side belongs to the relative interior of this side and the distance of this vertex from this side is $\Delta(V)$.*

Figure 2.6 shows a spherically convex pentagon. Theorem 2.13 permits to recognize that it is reduced.

Corollary 2.5 *Every spherical regular odd-gon of thickness at most $\frac{\pi}{2}$ is reduced.*

Fig. 2.6 A reduced spherical
pentagon

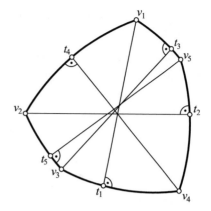

Corollary 2.6 *The only reduced spherical triangles are the regular triangles of
thickness at most $\frac{\pi}{2}$.*

Corollary 2.7 *If K is a supporting hemisphere of a reduced spherical polygon V
whose bounding circle contains a side of V, then* $\text{width}_K(V) = \Delta(V)$.

The above three corollaries from our Theorem 2.13 are given in [50]. The
following fact is also proved in [50] for $\Delta(V) < \frac{\pi}{2}$ and additionally for $\Delta(V) = \frac{\pi}{2}$
in the Remark of [51]. It is proved again in [15].

Corollary 2.8 *For every reduced odd-gon* $V = v_1 v_2 \ldots v_n$ *with* $\Delta(V) \leq \frac{\pi}{2}$
we have $|v_i t_{i+(n+1)/2}| = |t_i v_{i+(n+1)/2}|$, *for* $i = 1, \ldots, n$, *where* t_i *denotes the
projection of the vertex* v_i *on the opposite side of V.*

In addition, the paper [15] by Chang, Liu and Su gives the following converse
implication.

Proposition 2.3 *Every convex odd-gon* $V = v_1 v_2 \ldots v_n$ *on* S^2 *satisfying*
$|v_i v_{i+(n+1)/2}| = |v_i v_{i+(n+1)/2}| = \frac{\pi}{2}$ *for* $i = 1, \ldots, n$ *is reduced.*

Its proof is based on the fact given later in Proposition 2.9 that every reduced
body of thickness $\frac{\pi}{2}$ is a body of constant width. Thus this proof applying to bodies
of constant width cannot be repeated for reduced bodies of thickness below $\frac{\pi}{2}$. So
let us formulate the following problem.

Problem 2.4 *Consider an arbitrary convex odd-gon* $V = v_1 v_2 \ldots v_n$ *on* S^2 *of
thickness below* $\frac{\pi}{2}$ *which satisfies* $|v_i t_{i+(n+1)/2}| = |t_i v_{i+(n+1)/2}|$ *for every* $i \in
\{1, \ldots, n\}$. *Is V a reduced polygon?*

Theorem 2.12 and Corollary 2.8 together permit to construct reduced spherical
polygons of arbitrary thickness at most $\frac{\pi}{2}$. For instance see the reduced spherical
pentagon in Fig. 2.6.

In the proof of Corollary 2.8 it is shown that the triangles $v_i o_i t_{i+(n+1)/2}$
and $v_{i+(n+1)/2} o_i t_i$, where o_i is the common point of the arcs $v_i t_i$ and

$v_{i+(n+1)/2}t_{i+(n+1)/2}$, are symmetric with respect to a great circle. This implies the following corollary.

Corollary 2.9 *In every reduced odd-gon* $V = v_1 v_2 \ldots v_n$, *for every* $i \in \{1, \ldots, n\}$ *we have* $\angle t_{i+(n+1)/2} v_i t_i = \angle t_i v_{i+(n+1)/2} t_{i+(n+1)/2}$.

Also the following three corollaries are obtained in [50] for $\Delta(V) < \frac{\pi}{2}$ and in Remark [51] expanded also for $\Delta(V) = \frac{\pi}{2}$.

By Corollary 2.8 applied n times, the sum of the lengths of the boundary spherical arcs of V from v_i to t_i (with the positive orientation) is equal to the sum of the boundary spherical arcs of V from t_i to v_i (with the positive orientation). Let us formulate this statement as the following corollary.

Corollary 2.10 *Let V be a reduced spherical n-gon and let $i \in \{1, \ldots, n\}$. For every $i \in \{1, \ldots, n\}$ the spherical arc $v_i t_i$ divides in half the perimeter of V.*

Corollary 2.11 *For every reduced polygon and every i we have $\alpha_i > \frac{\pi}{6} + E$ and $\beta_i < \frac{\pi}{6} + E$, where E denotes the excess of the triangle $v_i t_i v_{i+(n+1)/2}$.*

In order to formulate the next corollary, for every $i \in \{1, \ldots, n\}$ let us put $\alpha_i = \angle v_{i+1} v_i t_i$ and $\beta_i = \angle t_i v_i v_{i+(n+1)/2}$ in a spherically convex odd-gon $V = v_1 v_2 \ldots v_n$.

The angle α_i is greater, but the one for β_i is smaller than these given in Theorem 8 of [45] for a reduced n-gon in E^2. For instance, in the regular spherical triangle of sides and thickness $\pi/2$ we have $\alpha_i = \beta_i = \pi/4$.

Corollary 2.12 *If $V = v_1 v_2 \ldots v_n$ is a reduced spherical polygon with $\Delta(V) < \frac{\pi}{2}$, then $\beta_i \leq \alpha_i$ for every $i \in \{1, \ldots, n\}$.*

Example 3.3 from [15] constructs a spherically convex pentagon of thickness $\frac{\pi}{2}$ answering the question from [50] about the existence of non-regular reduced polygons of thickness $\frac{\pi}{2}$.

The paper [50] presents two conjectures on the perimeter of a reduced spherical polygon. The first says that the perimeter of every reduced spherical polygon V is not larger than that of the spherical regular triangle T of the same thickness, so at most 6 arc $\cos \frac{\cos \Delta(T) + \sqrt{8 + \cos^2 \Delta(T)}}{4}$, and that it is attained only for this regular triangle. The second says that from amongst all reduced spherical polygons of fixed thickness and with at most n vertices only the regular spherical n-gon has minimal perimeter. For the special case when $\Delta(V) = \frac{\pi}{2}$ both conjectures are confirmed by Chang, Liu and Su [15] in the following two theorems.

Theorem 2.14 *The perimeter of every reduced spherical polygon of thickness $\frac{\pi}{2}$ is less than or equal to that of the regular spherical triangle of thickness $\frac{\pi}{2}$ and the maximum value is attained only for this regular spherical triangle.*

Theorem 2.15 *The regular spherical n-gon of thickness $\frac{\pi}{2}$ has minimum perimeter among all reduced spherical k-gons of thickness $\frac{\pi}{2}$.*

Since the second conjecture is not proved yet for thickness below $\frac{\pi}{2}$, we propose the next problem.

Problem 2.5 *Is it true that the perimeter of every reduced spherical polygon of thickness below $\frac{\pi}{2}$ is less than or equal to that of the regular spherical triangle of the same thickness? Is the maximum value attained only for this regular spherical triangle? Is it true that the regular spherical n-gon of any fixed thickness below $\frac{\pi}{2}$ has minimum perimeter among all reduced spherical n-gons of the same thickness?*

The following two facts are also presented by Chang, Liu and Su [15].

Claim 2.16 *The regular spherical n-gon of thickness $\frac{\pi}{2}$ has minimum perimeter among all regular spherical n-gons of thickness $\frac{\pi}{2}$.*

Corollary 2.13 *The regular spherical n-gon of thickness $\frac{\pi}{2}$ has minimum perimeter among all reduced spherical k-gons of thickness $\frac{\pi}{2}$, where k and n are odd natural numbers such that $3 \leq k \leq n$.*

Finally in this section, let us recall a few facts about the area of spherical convex polygons and especially if they are reduced.

A generalization of the classical formula of Girard for the area of a spherical triangle is presented by Todhunter in Part 99 of [83]. In particular, he says that the area of an arbitrary convex n-gon on S^2 is equal to the sum of its internal angles diminished by $(n-2)\pi$.

The following theorem is proved for polygons of thickness $\frac{\pi}{2}$ by Chang, Liu and Su in [15], while for the polygons of any thickness smaller than $\frac{\pi}{2}$ by Liu, Chang and Su in [63].

Theorem 2.16 *The regular spherical n-gon has maximum area amongst all regular spherical k-gons with odd numbers k, n and $3 \leq k \leq n$.*

The paper [63] by Liu, Chang and Su confirms two conjectures from [50] by proving the following two facts.

Theorem 2.17 *Every reduced spherical non-regular n-gon of any fixed thickness has area smaller than the regular spherical n-gon of this thickness.*

Corollary 2.14 *The area of arbitrary reduced spherical polygon V is smaller than $2(1 - \cos\frac{\Delta(V)}{2})\pi$.*

2.9   Diameter of Convex Bodies and Reduced Bodies

The following theorem and proposition from [49] are spherical analogs of the classical facts in the Euclidean space.

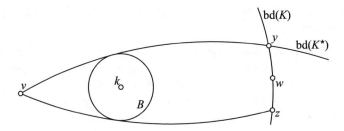

Fig. 2.7 Illustration to Example 2.3

Theorem 2.18 *Let* $\mathrm{diam}(C) \leq \frac{\pi}{2}$ *for a convex body* $C \subset S^d$. *We have*

$$\max\{\mathrm{width}_K(C);\ K \text{ is a supporting hemisphere of } C\} = \mathrm{diam}(C).$$

The following example from [49] shows that Theorem 2.18 does not hold without the assumption $\mathrm{diam}(C) \leq \frac{\pi}{2}$.

Example 2.3 Let T be an isosceles triangle with base of length λ close to 0 and the altitude perpendicular to it of length $\mu \in (\frac{\pi}{2}, \pi)$ (see Fig. 2.7). Denote by w the center of the base, by v the opposite vertex of T and by y, z the remaining vertices of T. Claim 2.7 implies that wv is the diametrical arc of T. Take the hemisphere K orthogonal to vw supporting T at w. Denote by k the center of K. Clearly, $k \in wv$, so k is in the interior of T. Let ρ be the radius of the largest disk B with center k contained in T. The radius ρ of B is arbitrarily close to 0, as λ is sufficiently small. Applying Part III of Theorem 2.2 we conclude that the width of T determined by K is $\frac{\pi}{2} + \rho$ which is realized for the hemisphere K^* whose boundary contains vy (that is, $\mathrm{width}_K = \angle vyz$). Hence it may be arbitrarily close to $\frac{\pi}{2}$, as λ is sufficiently small. On the other hand, the diameter $|wv|$ of T may be arbitrarily close to π, as μ is sufficiently close to π.

Proposition 2.4 *Let* $\mathrm{diam}(C) > \frac{\pi}{2}$ *for a convex body* $C \subset S^d$. *We have*

$$\max\{\mathrm{width}_K(C);\ K \text{ is a supporting hemisphere of } C\} \leq \mathrm{diam}(C).$$

The paper [51] presents the following corollary from this proposition.

Corollary 2.15 *For an arbitrary convex body* $C \subset S^d$ *we have* $\Delta(C) \leq \mathrm{diam}(C)$.

Let us apply Claim 2.18 for a convex body C of diameter larger than $\frac{\pi}{2}$. Having in mind that the center k of K is in pq and thus in C, by Part III of Theorem 2.2 we obtain $\Delta(K \cap K^*) > \frac{\pi}{2}$. This gives the following corollary from [58], which implies the second one.

Corollary 2.16 *Let $C \subset S^d$ be a convex body of diameter larger than $\frac{\pi}{2}$ and let* diam(C) *be realized for points* $p, q \in C$. *Take the hemisphere K orthogonal to pq at p which supports C. Then* width$_K(C) > \frac{\pi}{2}$.

Corollary 2.17 *Let $C \subset S^d$ be a convex body of diameter larger than $\frac{\pi}{2}$ and let K denote the family of all hemispheres supporting C. Then we have* max$_{K \in K}$ width$_K(C) > \frac{\pi}{2}$.

The set of extreme points of C is denoted by $E(C)$. Here is a property established in [59].

Claim 2.17 *For every convex body $C \subset S^2$ of diameter at most $\frac{\pi}{2}$ we have* diam$(E(C))$ = diam(C).

The assumption that diam$(C) \leq \frac{\pi}{2}$ is essential in Claim 2.17, as it follows from the example of a regular triangle of any diameter greater than $\frac{\pi}{2}$. The weaker assumption that $\Delta(C) \leq \frac{\pi}{2}$ is not sufficient, which we see taking in the part of C any isosceles triangle T with $\Delta(T) \leq \frac{\pi}{2}$ and legs longer than $\frac{\pi}{2}$ (so with base shorter than $\frac{\pi}{2}$). The diameter of T is equal to the distance between the midpoint of the base and the opposite vertex of T. So diam(T) is greater than the length of each of the sides.

The above claim was established as a lemma needed in Theorem 2.19 for S^2 only, and thus it concerns only bodies on S^2. We expect that the d-dimensional variant of Claim 2.17 holds as well.

From Theorem 2.5 and Corollary 2.17 we get obtain following three corollaries for reduced bodies on S^d (not only for bodies of constant width, as formulated in [58]).

Corollary 2.18 *If* diam$(R) > \frac{\pi}{2}$ *for a reduced body of $R \subset S^d$, then $\Delta(R) > \frac{\pi}{2}$.*

Corollary 2.19 *For every reduced body of $R \subset S^d$ of thickness at most $\frac{\pi}{2}$ we have* diam$(R) \leq \frac{\pi}{2}$.

Corollary 2.20 *Let p be a point of a reduced body $R \subset S^d$ of thickness at most $\frac{\pi}{2}$. Then $R \subset H(p)$.*

By the way, the last corollary was shown earlier for S^2 in [57].

Note the obvious fact that the diameter of a convex body $C \subset S^d$ is realized only for some pairs of points of bd(C). In the following four claims we recall some facts from [49, 51] and [58]. Two first two concern the realization of the diameter of any spherical convex body.

Claim 2.18 *Assume that the diameter of a convex body $C \subset S^d$ is realized for points p and q. The hemisphere K orthogonal to pq at p and containing $q \in K$ supports C.*

Claim 2.19 *Let $C \subset S^d$ be a convex body. If* diam$(C) < \frac{\pi}{2}$, *then every two points of C at the distance* diam(C) *are extreme. If* diam$(C) = \frac{\pi}{2}$, *then there are two points of C at distance $\frac{\pi}{2}$ such that at least one of them is an extreme point of C.*

Clearly, the first thesis in not true for diam$(C) = \frac{\pi}{2}$. After the proof of this claim in [51] the following amplification is also shown: *every convex body $C \subset S^2$ of diameter $\frac{\pi}{2}$ contains a pair of extreme points distant by $\frac{\pi}{2}$*. Now, we also observe that this stronger statement is also true for $C \subset S^d$, as formulated in next claim.

Claim 2.20 *Every convex body $C \subset S^d$ of diameter $\frac{\pi}{2}$ contains a pair of extreme points distant by $\frac{\pi}{2}$.*

Proof Consider a point $a \in C$ and an extreme point $b \in C$ guaranteed by the second part of Claim 2.19. If a is also extreme, the thesis is true. In the contrary case take the $(d-1)$-dimensional hemisphere K bounding the lune L from Claim 2.11 whose center is a. By Claim 2.5 there are at most d extreme points of $C \cap K$ whose convex hull contains a. Let a' be one of them. By the second part of Lemma 2.7, we have $|a'b| = \frac{\pi}{2}$ as required.   □

The following theorem from [59] is analogous to the first part of Theorem 9 from [45] and confirms the conjecture from the bottom of p. 214 of [50].

Theorem 2.19 *For every reduced spherical body $R \subset S^2$ with $\Delta(R) < \frac{\pi}{2}$ we have*

$$\text{diam}(R) \leq \arccos(\cos^2 \Delta(R)).$$

This value is attained if and only if R is the quarter of a disk of radius $\Delta(R)$. If $\Delta(R) \geq \frac{\pi}{2}$, then diam$(R) = \Delta(R)$.

From L'Hospital's rule we conclude that if R is just a quarter of disk, then the ratio $[\arccos(\cos^2 \Delta(R))]/\Delta(R)$ tends to $\sqrt{2}$ as $\Delta(R)$ tends to 0. Consequently, the limit factor $\sqrt{2}$ is like in the planar case in Theorem 9 of [45].

By the way, the weaker estimate diam$(R) \leq 2 \arctan\left(\sqrt{2} \tan \frac{\Delta(R)}{2}\right)$ than the one in Theorem 2.19 is a consequence of Theorem 2.10.

Theorem 2.19 implies the following proposition proved also in [59].

Proposition 2.5 *Let $R \subset S^2$ be a reduced body. Then diam$(R) < \frac{\pi}{2}$ if and only if $\Delta(R) < \frac{\pi}{2}$. Moreover, diam$(R) = \frac{\pi}{2}$ if and only if $\Delta(R) = \frac{\pi}{2}$.*

This proposition gives a more precise statement than the one diam$(R) \leq \frac{\pi}{2}$ following from Corollary 2.20.

An example from [46] shows that in E^d, where $d \geq 3$, there are reduced bodies of thickness 1 and arbitrary large diameter. Below we formulate an analogous question for reduced bodies on S^d.

Problem 2.6 *Is it true that for any number $q > 1$ there exists a reduced body on S^3 with the quotient of its diameter to its thickness at least q? Construct, if possible, such an example on S^d, with any $d \geq 3$.*

The next theorem is proved in [51].

Theorem 2.20 *For every reduced body* $R \subset S^d$ *such that* $\Delta(R) \leq \frac{\pi}{2}$ *we have* $\mathrm{diam}(R) \leq \frac{\pi}{2}$. *Moreover, if* $\Delta(R) < \frac{\pi}{2}$, *then* $\mathrm{diam}(R) < \frac{\pi}{2}$.

The special case of the first assertion of this theorem for $d = 2$ is stated in the observation just before Proposition 1 of [59].

By the way, for $d = 2$ the first assertion of our Proposition 2.5 is a special case of the first statement of Theorem 2.21 and the second assertion appears to be a consequence of just Theorem 2.20. Let us add that the approach of the proof of Theorem 2.20 is different from the argumentation in the proof of Proposition 2.5 which applies Theorem 2.19 proved only for $d = 2$.

From Theorems 2.27 and 2.23 the following proposition and corollary from [51] result.

Proposition 2.6 *For every reduced body* $R \subset S^d$ *satisfying* $\Delta(R) \geq \frac{\pi}{2}$ *we have* $\Delta(R) = \mathrm{diam}(R)$.

Observe that this proposition does not hold without the assumption that the body is reduced.

Corollary 2.21 *If* $\Delta(R) < \mathrm{diam}(R)$ *for a reduced body* $R \subset S^d$, *then both numbers are below* $\frac{\pi}{2}$. *Moreover, R is not a body of constant width.*

Here are five equivalences from [51].

Theorem 2.21 *Let* $R \subset S^d$ *be a reduced body. Then*

(a) $\Delta(R) = \frac{\pi}{2}$ *if and only if* $\mathrm{diam}(R) = \frac{\pi}{2}$,
(b) $\Delta(R) \geq \frac{\pi}{2}$ *if and only if* $\mathrm{diam}(R) \geq \frac{\pi}{2}$,
(c) $\Delta(R) > \frac{\pi}{2}$ *if and only if* $\mathrm{diam}(R) > \frac{\pi}{2}$,
(d) $\Delta(R) \leq \frac{\pi}{2}$ *if and only if* $\mathrm{diam}(R) \leq \frac{\pi}{2}$,
(e) $\Delta(R) < \frac{\pi}{2}$ *if and only if* $\mathrm{diam}(R) < \frac{\pi}{2}$.

Let us add that (a) and (e) were established earlier for $d = 2$ in [59].

As in Section 4 of [58], we say that a convex body $D \subset S^d$ of diameter δ is **of constant diameter** δ provided for arbitrary $p \in \mathrm{bd}(D)$ there exists $p' \in \mathrm{bd}(D)$ such that $|pp'| = \delta$ (more generally, this notion makes sense for a closed set D of diameter δ in a metric space M, such that for all $x, z \in D$ and $y \in M$ with $|xy| + |yz| = |xz|$ we have $y \in D$, so for instance when M is a Riemannian manifold). This is an analog of the notion of a body of constant diameter in Euclidean space considered by Reidemeister [76].

We get some bodies of constant diameter on S^2 as particular cases of the later Example 2.6 by taking there any non-negative $\kappa < \frac{\pi}{2}$ and $\sigma = \frac{\pi}{4} - \frac{\kappa}{2}$. The following example from [51] presents a wider class of bodies of constant diameter $\frac{\pi}{2}$.

Example 2.4 Take a triangle $v_1 v_2 v_3 \subset S^2$ of diameter at most $\frac{\pi}{2}$. Put $\kappa_{12} = |v_1 v_2|, \kappa_{23} = |v_2 v_3|, \kappa_{31} = |v_3 v_1|, \sigma_1 = \frac{\pi}{4} - \frac{\kappa_{12}}{2} + \frac{\kappa_{23}}{2} - \frac{\kappa_{31}}{2},$ $\sigma_2 = \frac{\pi}{4} - \frac{\kappa_{12}}{2} - \frac{\kappa_{23}}{2} + \frac{\kappa_{31}}{2}, \sigma_3 = \frac{\pi}{4} + \frac{\kappa_{12}}{2} - \frac{\kappa_{23}}{2} - \frac{\kappa_{31}}{2}.$ Here we consent only to triangles with the sum of lengths of two shortest sides at most the length of the longest side plus $\frac{\pi}{2}$ (equivalently: with $\sigma_1 \geq 0, \sigma_2 \geq 0$ and $\sigma_3 \geq 0$). Extend the

Fig. 2.8 A spherical body of constant diameter

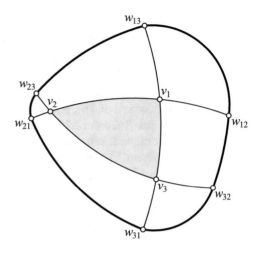

following $v_1 v_2$ up to $w_{12} w_{21}$ with $v_1 \in w_{12} v_2$, $v_2 v_3$ up to $w_{23} w_{32}$ with $v_2 \in w_{23} v_3$, $v_3 v_1$ up to sides: $w_{31} w_{13}$ with $v_3 \in w_{31} v_1$ such that $|v_1 w_{12}| = |v_1 w_{13}| = \sigma_1$, $|v_2 w_{21}| = |v_2 w_{23}| = \sigma_2$, $|v_3 w_{31}| = |v_3 w_{32}| = \sigma_3$ (see Fig. 2.8). Draw six pieces of circles: with center v_1 of radius σ_1 from w_{12} to w_{13} and of radius $\frac{\pi}{2} - \sigma_1$ from w_{21} to w_{31}, with center v_2 of radius σ_2 from w_{23} to w_{21} and of radius $\frac{\pi}{2} - \sigma_2$ from w_{32} to w_{12}, with center v_3 of radius σ_3 from w_{31} to v_{32} and of radius $\frac{\pi}{2} - \sigma_3$ from w_{13} to w_{23}. Clearly, the convex hull of these six pieces of circles is a body of constant diameter $\frac{\pi}{2}$.

Generalizing, take a convex odd-gon $v_1 \ldots v_n \subset S^2$ of diameter at most $\frac{\pi}{2}$. For $i = 1, \ldots, n$ put $\kappa_{i\ i+(n-1)/2} = |v_i v_{i+(n-1)/2}|$ (here and later we mean indices modulo n). Let $\sigma_i = \frac{\pi}{4} + \Sigma_{i=1}^{n} s_i \kappa_{i\ i+(n-1)/2}$, where $s_i = \frac{1}{2}$ if $|\frac{n+1}{2} - i| \le \frac{n-3}{4}$ for n of the form $3+4k$ and $|\frac{n}{2} - i| \le \frac{n-1}{4}$ for n of the form $5+4k$ (where $k = 0, 1, 2, \ldots$), and $s_i = -\frac{1}{2}$ in the opposite case. We agree only for odd-gons with $\sigma_i \ge 0$ for $i = 1, \ldots, n$. Prolong each diagonal $v_i v_{i+(n-1)/2}$ up to the arc $w_{i\ i+(n-1)/2} w_{i+(n-1)/2\ i}$ such that $v_i \in w_{i\ i+(n-1)/2} v_{i+(n-1)/2}$ and $|v_i w_{i\ i+(n-1)/2}| = |v_i w_{i\ i+(n+1)/2}| = \sigma_i$. For $i = 1, \ldots, n$ we draw the piece of the circle with center v_i of radius σ_i from $w_{i\ i+(n-1)/2}$ to $w_{i\ i+(n+1)/2}$ and the piece of circle of radius $\frac{\pi}{2} - \sigma_i$ from $w_{i+(n-1)/2\ i}$ to $w_{i+(n+1)/2\ i}$. The convex hull of the union of these $2n$ pieces is a body of constant diameter $\frac{\pi}{2}$.

The proposition below from [52] presents a property of bodies of constant diameter, and consequently also for bodies of constant width and complete considered in the two next sections. It does not hold for S^d with $d \ge 3$. A not very complicated example of such a body of constant diameter on S^3 is the body obtained by the rotation of a quarter of a disk on S^2 around its axis of symmetry.

Proposition 2.7 *Let $D \subset S^2$ be a body of constant diameter. Then every two diametral chords of D intersect.*

Note that Proposition 2.8 and Theorem 2.22 below give some information on the diameter of reduced polygons.

The paper [45] addresses the problem of whether there exist reduced polytopes in E^d for $d \geq 3$. It is positively answered by Gonzalez Merino, Jahn, Polyanskii and Wachsmuth in [28]. For S^d the analogous question has a specific positive answer. The spherical simplex being the $\frac{1}{2^d}$-part of the ball of radius $\frac{\pi}{2}$ is a reduced spherical polytope (by the way, in E^d with $d \geq 3$ all simplices are not reduced, as shown by Martini and Swanepoel [66]). This follows from the fact that it is a spherical body of constant width. If we disregard this example, we still have the following problem.

Problem 2.7 *Do there exist spherical d-dimensional polytopes (possibly some simplices?) on S^d, where $d \geq 3$, different from the $\frac{1}{2^d}$-part of S^d ?*

Let us recall a proposition and a theorem from [50].

Proposition 2.8 *The diameter of any reduced spherical n-gon is realized only for some pairs of vertices whose indices (modulo n) differ by $\frac{n-1}{2}$ or $\frac{n+1}{2}$.*

Here is a sharper estimate for reduced polygons in place of reduced bodies on S^2 than the estimate from Theorem 2.19.

Theorem 2.22 *For every reduced spherical polygon $V \subset S^2$ we have*

$$\operatorname{diam}(V) \leq \arccos\left(\sqrt{1 - \frac{\sqrt{2}}{2} \sin \Delta(V) \cdot \cos \Delta(V)}\right),$$

with equality for the regular spherical triangle in the part of V.

By the way, from Theorem 2.22 we get a slightly better estimate of the radius of a disk covering a reduced spherical polygon than the one in Theorem 2.10. The resulting formula is very complicated, so we omit it here. Let us give only an example; for $\Delta(R) = 60°$ it gives the estimate about $(38, 38)°$ for the radius, while the estimate from Theorem 2.10 gives about $(39, 23)°$.

Problem 2.8 *We conjecture that the equality in Theorem 2.22 holds only for regular triangles.*

By the way, recently Horváth [42] showed that every body of constant diameter in the hyperbolic plane is a body of constant width.

2.10 Bodies of Constant Width

We say that a convex body $W \subset S^d$ is of **constant width** w provided that for every supporting hemisphere K of W we have $\operatorname{width}_K(W) = w$.

This notion is analogous to the classical notion of bodies of constant width in E^n. The oldest paper on this subject we found is the famous paper of Barbier [3] which proves that the area of all planar bodies of the same constant width is the same. The book [65] by Martini, Montejano and Oliveros gives a wide survey of results on bodies of constant width in various structures. Also Section 11 of the book [60] by Lay and Part 2.5 of the book [84] by Toth, besides classical facts on bodies of constant width recall a number of less known properties and exercises of them. Many properties of bodies of constant width in the plane are recalled in the popular book [88] by Yaglom and Boltyanskij.

In particular, spherical balls of radius smaller than $\frac{\pi}{2}$ are spherical bodies of constant width. Also every spherical Reuleaux odd-gon is a convex body of constant width. Each of the $\frac{1}{2^d}$-th part of any ball on S^d is a spherical body of constant width $\frac{\pi}{2}$, which easily follows from the definition of a body of constant width.

After [58] we remind the reader an example of a spherical body of constant width on the sphere S^3.

Example 2.5 Take a circle $X \subset S^3$ (i.e., a set congruent to a circle in E^2) of a positive diameter $\kappa < \frac{\pi}{2}$, and a point $y \in S^3$ at the distance κ from every point $x \in X$. Prolong every spherical arc yx by a distance $\sigma \leq \frac{\pi}{2} - \kappa$ up to points a and b so that a, y, x, b are on one great circle in this order. All these points a form a circle A, and all points b form a circle B. On the sphere on S^3 of radius σ whose center is y take the "smaller" part A^+ bounded by the circle A. On the sphere on S^3 of radius $\kappa + \sigma$ with center y take the "smaller" part B^+ bounded by B. For every $x \in X$ denote by x' the point on X such that $|xx'| = \kappa$. Prolong every xx' up to points d, d' so that d, x, x', d' are in this order and $|dx| = \sigma = |x'd'|$. For every x provide the shorter piece C_x of the circle with center x and radius σ connecting the b and d determined by x and also the shorter piece D_x of the circle with center x of radius $\kappa + \sigma$ connecting the a and d' determined by x. Denote by W the convex hull of the union of A^+, B^+ and all pieces C_x and D_x. This is a body of constant width $\kappa + 2\sigma$ (hint: for every hemisphere H supporting W and every H^* the centers of H/H^* and H^*/H belong to bd(W) and the arc connecting them passes through one of our points x, or through the point y).

Recall the following related question from p. 563 of [49].

Problem 2.9 *Is a convex body $W \subset S^d$ of constant width provided every supporting hemisphere G of W determines at least one hemisphere H supporting W such that $G \cap H$ is a lune with the centers of G/H and H/G in* bd(W)?

Let us formulate a comment from [49], p. 562–563 as the following claim.

Claim 2.21 *If $W \subset S^d$ is a body of constant width, then every supporting hemisphere G of W determines a supporting hemisphere H of C for which $G \cap H$ is a lune such that the centers of G/H and H/G belong to the boundary of C.*

Problem 2.10 *Is the contrary true? More precisely, is a convex body $C \subset S^d$ of constant width provided every supporting hemisphere G of C determines at least*

*one hemisphere H supporting C such that G ∩ H is a lune with the centers of G/H
and H/G in the boundary of C?*

The above claim implies the next one from [49].

Claim 2.22 *Every spherical body W of constant width is an intersection of lunes
of thickness $\Delta(W)$ such that the centers of the $(d-1)$-dimensional hemispheres
bounding these lunes belong to* bd(W).

The following Theorem from [51] generalizes a theorem from [57] for $d = 2$.
The proof presented below is analogous, but this time we apply Theorem 2.5 in the
proof.

Theorem 2.23 *If a reduced convex body $R \subset S^d$ satisfies $\Delta(R) \geq \frac{\pi}{2}$, then R is a
body of constant width $\Delta(R)$.*

Let us return for a while to Example 2.2. Having in mind the above theorem we
observe that if $|ab'| = \frac{\pi}{2}$, then P is a body of constant width $\frac{\pi}{2}$. In particular, we
may take $|ac| = \frac{\pi}{6}$ and $|cb'| = \frac{\pi}{3}$. Clearly, if $|ca| = |cb'|$, then our P is a disk. If
$a = c = a'$, then P is a quarter of a disk.

Recall that every spherical convex body of constant width is reduced. Theo-
rem 2.23 says that the opposite implication holds for bodies of thickness at least $\frac{\pi}{2}$,
i.e. that every reduced spherical convex body of thickness at least $\frac{\pi}{2}$ is of constant
width. This does not hold for reduced spherical convex bodies of thickness below
$\frac{\pi}{2}$ since every regular spherical triangle of thickness below $\frac{\pi}{2}$ is reduced but not of
constant width.

Here is a claim from [51].

Claim 2.23 *If for a convex body $C \subset S^d$ the numbers $\Delta(C)$ and* diam(C) *are equal
and at most $\frac{\pi}{2}$, then C is of constant width $\Delta(C) =$ diam(C).*

The following two theorems are from [58].

Theorem 2.24 *At every boundary point p of a body $W \subset S^d$ of constant width
$w > \pi/2$ we can inscribe a unique ball $B_{w-\pi/2}(p')$ touching W from inside at
p. Moreover, p' belongs to the arc connecting p with the center of the unique
hemisphere supporting W at p, and $|pp'| = w - \frac{\pi}{2}$.*

Theorem 2.25 *Every spherical convex body of constant width smaller than $\frac{\pi}{2}$ on
S^d is strictly convex.*

This theorem was earlier shown in [57] for $d = 2$ only. Let us add that the thesis
of this theorem does not hold for spherical convex bodies of constant width at least
$\frac{\pi}{2}$, as we conclude from the following example given in [57].

Example 2.6 Take a spherical regular triangle abc of sides of length $\kappa < \frac{\pi}{2}$ and
prolong them by the same distance $\sigma \leq \frac{\pi}{2} - \kappa$ in both "directions" up to points
d, e, f, g, h, i, so that i, a, b, f are on a great circle in this order, e, b, c, h are on a
great circle in this order, and g, c, a, d are on a great circle in this order. Consider
three pieces of circles of radius $\kappa + \sigma$: with center a from f to g, with center b from

h to i, with center c from d to e. Consider also three pieces of circles of radius σ: with center a from i to d, with center b from e to f, with center c from g to h. The convex hull U of these six pieces of circles is a spherical body of constant width $\kappa + 2\sigma$. In particular, when $\kappa + \sigma = \frac{\pi}{2}$, three boundary circles of U become arcs; namely de, fg and hi.

The following theorem from [58] gives a positive answer to Problem 2.2 in the case of spherical bodies of constant width. It is a generalization of the version for S^2 given as Theorem 5.3 in [57]. The idea of the proof of our theorem below for S^d substantially differs from the one mentioned for S^2.

Theorem 2.26 *For every body $W \subset S^d$ of constant width w and every $p \in \mathrm{bd}(W)$ there exists a lune $L \supset W$ satisfying $\Delta(L) = w$ with p as the center of one of the two $(d-1)$-dimensional hemispheres bounding this lune.*

If the body W from Theorem 2.26 is of constant width greater than $\frac{\pi}{2}$, then by Theorem 2.5 it is smooth. Thus at every $p \in \mathrm{bd}(W)$ there is a unique supporting hemisphere of W, and so the lune L from the formulation of this theorem is unique. If the constant width of W is at most $\frac{\pi}{2}$, there are non-smooth bodies of constant width (e.g., a Reuleaux triangle on S^2) and then for non-smooth $p \in \mathrm{bd}(W)$ we may have more such lunes.

Theorem 2.26 implies the first statement of the following corollary. The second statement follows from Theorem 2.5 and the last part of Lemma 2.7 from [49].

Corollary 2.22 *For every convex body $W \subset S^d$ of constant width w and for every $p \in \mathrm{bd}(W)$ there exists $q \in \mathrm{bd}(W)$ such that $|pq| = w$. If $w > \frac{\pi}{2}$, this q is unique.*

Theorem 2.27 *If $W \subset S^d$ is a body of constant width w, then $\mathrm{diam}(W) = w$.*

The above theorem is proved in [58] and the following one in [49].

Theorem 2.28 *Every smooth reduced body on S^d is of constant width.*

Having in mind that the definition of constant width by Araújo is equivalent to our definition of constant width (see the last section), we may present the case of his Theorem B from [1] for S^d as follows.

Theorem 2.29 *If a body $W \subset S^2$ of constant width w has perimeter p and area a, then $p = (2\pi - a) \tan \frac{w}{2}$.*

The next proposition proved in [51] is applied in the proof of the forthcoming Theorem 2.30.

Proposition 2.9 *The following conditions are equivalent:*

(1) $C \subset S^d$ *is a reduced body with $\Delta(C) = \frac{\pi}{2}$,*
(2) $C \subset S^d$ *is a reduced body with $\mathrm{diam}(C) = \frac{\pi}{2}$,*
(3) $C \subset S^d$ *is a body of constant width $\frac{\pi}{2}$,*
(4) $C \subset S^d$ *is of constant diameter $\frac{\pi}{2}$.*

By part (b) of Theorem 2.21, we conclude the following variant of Theorem 2.23 presented in [51].

Corollary 2.23 *If a reduced convex body $R \subset S^d$ satisfies* $\mathrm{diam}(R) \geq \frac{\pi}{2}$, *then R is a body of constant width w equal to* $\mathrm{diam}(R)$. *It is also a body of constant diameter w.*

Let us comment on an application of spherical bodies of constant width in the research on the Wulff shape. Recall that Wulff [87] defined a geometric model of a crystal equilibrium, later named **Wulff shape**. The literature concerning this and related subjects is very comprehensive. For a survey see the monograph [75] by Pimpinelli and Vilain, also the article [37] by Han and Nishimura. In [36] and [38] the authors consider the dual Wulff shape and the **self-dual Wulff shape** as a Wulff shape which is equal to its dual. They apply the classical notion of central projection from the open hemisphere centered at a point $n \in S^d$ into the hyperplane of E^{d+1} supporting S^d at n. This hyperplane may be treated as E^d with origin n. The image of a Wulff shape under the inverse projection is called the **spherical convex body induced by this Wulff shape**. The paper [36] presents a proof that a Wulff shape is self-dual if and only if the spherical convex body induced by it is a spherical body of constant width $\frac{\pi}{2}$. Hence from Proposition 2.9 we obtain the following theorem. In particular, the equivalence with the third condition gives a positive answer to a question by Han and Nishimura considered at the end of their paper [37].

Theorem 2.30 *Each of the following conditions is equivalent to the statement that the Wulff shape W_γ is self-dual:*

- *the spherical convex body induced by W_γ is of constant width $\frac{\pi}{2}$,*
- *the spherical convex body induced by W_γ is a reduced body of thickness $\frac{\pi}{2}$,*
- *the spherical convex body induced by W_γ is a reduced body of diameter $\frac{\pi}{2}$,*
- *the spherical convex body induced by W_γ is a body of constant diameter $\frac{\pi}{2}$.*

Some more applications of the papers [49, 52, 58] and [70] on the width of spherical convex bodies to Wulff shape are in the paper [35].

The last theorem in this section, being a spherical version of the theorem of Blaschke [11], is a special case of Theorem 2.9 for reduced bodies.

Theorem 2.31 *Let $W \subset S^2$ be a body of constant width $w \leq \frac{\pi}{2}$. For arbitrary $\varepsilon > 0$ there exists a body $W_\varepsilon \subset S^2$ of constant width w whose boundary consists only of circles of radius w such that the Hausdorff distance between W_ε and W is at most ε.*

2.11   Complete Spherical Convex Bodies

Similarly to the traditional notion of a complete set in the Euclidean space E^d (for instance, see [12, 14, 21] and [65]) we say that any subset of a hemisphere of S^d

which is a largest (in the sense of inclusion) set of a given diameter $\delta \in (0, \frac{\pi}{2})$ is a **complete set of diameter** δ, or for brevity, a **complete set**. By the way, the above definition adds the leaking assumption that our set "is a subset of a hemisphere" to the definition given in [53]. This correction is also added in the arXiv version mentioned in [53].

Theorem 2.32 *An arbitrary set of diameter $\delta \in (0, \pi)$ on the sphere S^d is a subset of a complete set of diameter δ on S^d.*

The proof of this theorem from [53] is similar to the proof by Lebesgue [61] for E^d (it is recalled in Part 64 of [12]). Let us add that earlier Pál [71] proved this for E^2 by a different method.

The following fact from [53] permits to use the term **complete convex body** for a complete set.

Claim 2.24 *Any complete set of diameter δ on S^d coincides with the intersection of all balls of radius δ centered at points of this set. As a consequence, every spherical complete set is a convex body.*

Below we see two lemmas from [53] needed for the proof of Theorem 2.33.

Lemma 2.3 *If $C \subset S^d$ is a complete body of diameter δ, then for every $p \in \mathrm{bd}(C)$ there exists $p' \in C$ such that $|pp'| = \delta$.*

For distinct points $a, b \in S^d$ at distance $\delta < \pi$ from a point $c \in S^d$ we define the following piece $P_c(a, b)$ of a circle as the set of points $v \in S^d$ such that cv has length δ and intersects ab. See Fig. 2.9 with $P_c(a, b)$ drawn by a broken line. Also the second lemma, presented below, is formulated for S^d despite the fact that we apply it in the proof of the forthcoming Theorem 2.33 only for the case of S^2.

Lemma 2.4 *Let $C \subset S^d$ be a complete convex body of diameter δ. Take $P_c(a, b)$ with $|ac|$ and $|bc|$ equal to δ such that $a, b \in C$ and $c \in S^d$. Then $P_c(a, b) \subset C$.*

The following theorem from [53] presents the spherical version of the classical theorem in E^d proved by Meissner [67] for $d = 2, 3$ and by Jessen [43] for arbitrary d.

Fig. 2.9 Illustration to Lemma 2.4

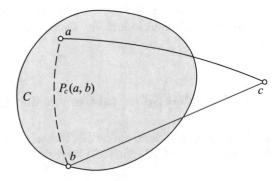

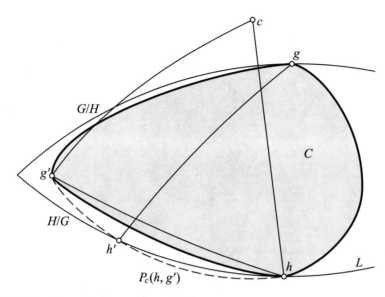

Fig. 2.10 First illustration to the proof of Theorem 2.33

Theorem 2.33 *A convex body of diameter δ on S^d is complete if and only if it is of constant width δ.*

Proof In the first part of the proof let us prove that if a body $C \subset S^d$ of diameter δ is complete, then it is of constant width δ.

Suppose the contrary, i.e., that width$_J(C) \neq \delta$ for a hemisphere J supporting C. Our aim is to get a contradiction.

By Theorem 2.18 and Proposition 2.4 we have width$_J(C) \leq \delta$, and so by the fact that width$_J(C) \neq \delta$ we obtain $\Delta(C) < \delta$. Recall that $\Delta(C)$ is the thickness of each narrowest lune containing C. Take such a lune $L = G \cap H$. Denote by g, h the centers of G/H and H/G, respectively (see Fig. 2.10). Of course, $|gh| < \delta$. By Claim 2.13 we have $g, h \in C$. By Lemma 2.3 there exists a point $g' \in C$ at distance δ from g. If $\Delta(L) \geq \frac{\pi}{2}$, then by $C \subset L$ we get $g' \in L$, which contradicts with the last part of Claim 2.7.

Now assume that $\Delta(L) < \frac{\pi}{2}$. Since the triangle ghg' is non-degenerate, there is a unique two-dimensional sphere $S^2 \subset S^d$ containing g, h, g'. Clearly, ghg' is a subset of $M = C \cap S^2$. Hence M is a convex body on S^2. Denote by F this hemisphere of S^2 such that $hg' \subset \mathrm{bd}(F)$ and $g \in F$. There is a unique $c \in F$ such that $|ch| = \delta = |cg'|$. By Lemma 2.4 for $d = 2$ we have $P_c(h, g') \subset M$ (see Fig. 2.10).

We intend to show that c is not on the great circle J of S^2 through g and h. In order to see this, suppose for a while the opposite, i.e. that $c \in J$. Then from $|g'g| = \delta, |g'c| = \delta$ and $|hc| = \delta$ we conclude that $\angle gg'c = \angle hcg'$. So the spherical triangle $g'gc$ is isosceles, which together with $|gg'| = \delta$ gives $|cg| = \delta$. Since

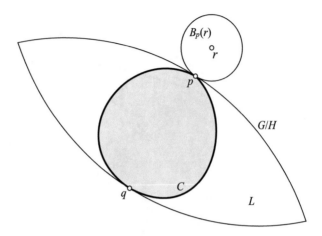

Fig. 2.11 Second illustration to the proof of Theorem 2.33

$|gh| = \Delta(L) = \Delta(C) > 0$ and g is a point of ch different from c, which is impossible. Hence $c \notin J$.

By the preceding paragraph $P_c(h, g')$ intersects bd(M) at a point h' different from h and g'. So the open piece of $P_c(h, g')$ between h and h' is out of L and thus also out of M (see again Fig. 2.10). This gives a contradiction with the result of the paragraph before the last. Consequently, C is a body of constant width δ.

In the second part of the proof let us prove that if C is a spherical body of constant width δ, then C is a complete body of diameter δ.

In order to prove this, it is sufficient to take any point $r \notin C$ and to show that diam$(C \cup \{r\}) > \delta$.

Take the largest ball $B_p(r)$ disjoint from the interior of C (see Fig. 2.11). Since C is convex, $B_p(r)$ has exactly one point p in common with C. By Theorem 2.26 there exists a lune $L = G \cap H$ containing C of thickness δ with p as the center of G/H. Denote by g the center of H/G. By Claim 2.13 we also have $q \in C$. From the fact that rp and pq are orthogonal to G/H at p, we see that $p \in rq$. Moreover, p is not an endpoint of rq and $|pq| = \delta$, Hence $|rq| > \delta$. Thus diam$(C \cup \{r\}) > \delta$. Since $r \notin C$ is arbitrary, C is complete. □

The above theorem is applied by Bezdek [8] for a generalization of the theorem of Leichtweiss [62] saying that amongst all bodies of constant width w on S^2 only the Reauleaux triangle has the smallest possible area, as conjecture by Blaschke [11]).

Here is next theorem from [53].

Theorem 2.34 *Bodies of constant diameter on S^d coincide with complete bodies.*

Proof Take a complete body $D \subset S^d$ of diameter δ. Let $g \in$ bd(D) and G be a hemisphere supporting D at g. By Theorem 2.33 the body D is of constant width δ. So width$_G(D) = \delta$ and there exists a hemisphere H such that the lune $G \cap H \supset D$

has thickness δ. By Claim 2.13 the centers h of H/G and g of G/H belong to D. So $|gh| = \delta$. Thus D is of constant diameter δ.

Consider a body $D \subset S^d$ of constant diameter δ. Let $r \notin D$. Take the largest $B_\rho(r)$ whose interior is disjoint from D. Denote by p the common point of $B_\rho(r)$ and D. A unique hemisphere J supports $B_\rho(r)$ at p. Observe that $D \subset J$ (if not, there is a point $v \in D$ outside J; clearly vp passes through $\mathrm{int}(B_\rho(r))$, a contradiction). Since D is of constant diameter δ, there is $p' \in D$ with $|pp'| = \delta$. Observe that $\angle rpp' \geq \frac{\pi}{2}$. If it is equal to $\frac{\pi}{2}$, then $|rp'| > \delta$. If it is larger than $\frac{\pi}{2}$, the triangle rpp' is obtuse and then by the law of cosines $|rp'| > |pp'|$ and hence $|rp'| > \delta$. Since $|rp'| > \delta$ in both cases we see that D is complete. $\square$

Theorem 2.33 permits to change "complete" into "constant width" in Theorem 2.34, so we get the following one proved in [53]. It is analogous to the result in E^d given by Reidemeister [76].

Theorem 2.35 *Bodies of constant diameter on S^d coincide with bodies of constant width.*

Independently, this fact is also proved by Han and Wu in their forthcoming paper [39] providing a quite different consideration using the polarity on S^d. Let us add that earlier same partial facts were established. Namely, in [58] it is shown that any body of constant width δ on S^d is of constant diameter δ and the inverse is argued for $\delta \geq \frac{\pi}{2}$, and in [52] for $\delta < \frac{\pi}{2}$ if $d = 2$.

Only here after presenting a number of facts, we are ready to discuss our latest Problem. It is a spherical analog of a conjecture from [46] for reduced bodies in E^2. There such a conjecture for E^d with $d \geq 3$ is not true by the example on p. 558 in [45] of a reduced body of thickness 1 and arbitrarily large fixed diameter.

Problem 2.11 *Is it true that every reduced body $R \subset S^d$, where $\Delta(R) < \frac{\pi}{2}$, is a subset of a ball of radius $\Delta(R)$ centered at a boundary point of R?*

The answer to this problem is positive for each body of constant width on S^d and, what is more, for every its boundary point. This results by Theorem 2.33 (it results by Claim 2.24 as well). So by Theorem 2.23 this is also true for every reduced $R \subset S^d$ with $\Delta(R) \geq \frac{\pi}{2}$. By the way, Michał Musielak (the author of [70] and coauthor of [57]-[59]) has got, an unpublished yet, positive answer for reduced polygons on S^2.

2.12   Final Remarks

Convex spherical sets are considered by many authors. Mostly they go with the same natural notion of convexity as in this chapter. Other definitions are considered very seldom in the literature. Three soft variations of this natural notion are recalled in Section 9.1 of the article [16] by Danzer, Grünbaum and Klee. The definitions recalled there of a spherically convex set are slightly weaker than the definition of convex sets on S^d used in the present chapter (and named strongly convex sets in [16]), i.e., the obtained families of convex sets in these three senses are "slightly" wider than the family of convex sets resulting from the above natural definition.

We know the following two earlier approaches to define the width of a spherically convex body.

Santaló [78] (see also [79] and [80]) defines a notion of breadth of a convex body $C \subset S^2$ corresponding to a boundary point of it. We surmise that he tacitly assumes that C is a smooth body because his definition is as follows. He takes a boundary point p of C, the supporting unique hemisphere K of C at p and the great circle P orthogonal to K at p. Next he finds the other hemisphere H, which is orthogonal to P and supports C. By the breadth of C corresponding to p, Santaló understands the distance from p to the great circle bounding H. Even if we mean this definition as the width generated by the unique hemisphere K supporting C, this notion differs from the notion $\text{width}_K(C)$ considered by us; just look to Fig. 2.2. Gallego, Reventós, Solanes, and Teufel [26] continue the research on the same notion calling it width in place of breadth.

Analogously, Hernandez Cifre and Martinez Fernández [41] define a notion of width of $C \subset S^2$ at a smooth point p of bd(C). Next they add that if p is an acute point of the boundary of C, so if there is a family of more than one supporting hemispheres of C at p, then the width of C at p is meant to be the minimum of the above widths for all hemispheres H supporting C at p. This improved notion still does not have the advantage to be continuous as p moves on the boundary of C. A simple example for this is any regular triangle on S^2 whose sides are of length less than $\frac{\pi}{2}$. Clearly, this notion, as well as the preceding one by Sántalo, can be defined also for S^d.

Recall that our definition of a body of constant width $C \subset S^d$ in Sect. 2.5 is based on the notion of the width $\text{width}_K(C)$ of C determined by a supporting hemisphere K of C. Surprisingly, some authors define bodies of constant width on the sphere without using any notion of width as the base. Namely, each of them takes into account a property of a convex body in E^d equivalent of the definition of a body of constant width. Such a property is used as the condition defining a convex body on S^d (sometimes only for $d = 2$). Let us recall some such definitions.

In 1915 Santálo [78], who defined the breadth of spherically convex body at a smooth boundary point (see three paragraphs before), using this notion defines a convex body of constant breadth. Namely, he says that a convex body $W \subset S^2$ is of constant breadth if for every boundary point its minimum breadth corresponding to this point and the maximum breadth corresponding to the same point are equal. Since he tacitly assumes that W is smooth, below we also make this assumption. Clearly, this definition is equivalent to the statement that at every boundary point of W the breadth is the same. If a smooth convex body W is of constant width in our sense, then by Claim 2.21 it is of constant breadth in the sense of Santálo. On the other hand, if W is of constant breadth w in the sense of Santálo, then for the unique (by the smoothness) supporting hemisphere G at any fixed boundary point there is a hemisphere H supporting W such that the lune $G \cap H$ has width w. We know that

H touches bd(W), but we do not know if the midpoints of G/H and H/G are in W. So we do not know the answer if every body of constant breadth in the sense of Santáló is of constant width in our sense. Gallego, Reventós, Solanes and Teufel [26] consider the same notion (which the call constant width) for any smooth convex body in a Riemannian space, so in particular for S^d.

In 1996 Araújo [1] presents his own proof of the planar case of the theorem of Meissner [67] that bodies of constant diameter coincide with bodies of constant width. Next, by analogy he defines bodies of constant width on S^2 just as bodies of constant diameter. By our Theorem 2.34 his notion is equivalent to our notion of constant width.

In 1983 Chakerian and Groemer [14] calls a convex body $W \subset S^2$ to be of constant width w if it can be rotated in a lune L of thickness w such that W always touches both the bounding semicircles of L at it midpoints (they need this notion in order to recall a result of Blaschke from [10] on a spherical analog of the Barbier's theorem, who rather intuitively tells on constant width on S^2). In order to avoid the term "rotate", let us rephrase this condition into "for every hemisphere G supporting W there exists a hemisphere H such that the lune $G \cap H$ is an orthogonally supporting lune of W". By Claim 2.22 we conclude that every body of constant width in our sense is also of constant width in the sense of Chakerian and Groemer. Is the converse true? Taking into account the above rephrased condition, we see that the answer is positive if the answer to the Problem 2.10 is positive.

In 1995 Dekster [17] (see also his paper [20] with Wegner) calls a convex body $C \subset S^d$ of constant width $w > 0$ if for any $p \in$ bd(W) and any normal m to C at p (by such m Dekster means the orthogonal direction to a particular supporting hemisphere of C at p) there exists an arc pq of length w having the direction $-m$ such that $pq \subset C$ but C does not contain any longer arc $pq' \supset pq$. If W is a body of constant width w in our sense, then by Claim 2.21 it is of constant width w in the sense of Dekster. Take a body of constant width W in the sense of Dekster. Assume that diam(W) $\leq \frac{\pi}{2}$. His Theorem 2 from [19] implies that at the ends of the above arc pq the body W is supported by two hemispheres orthogonal to pq, which by Claim 2.11 and Theorem 2.18 leads to the conclusion that W is of constant width in our sense. To sum up, the families of bodies of constant width in the sense of Dekster and in our sense are identical.

In 2005 Leichtweiss [62] considers a so called strip region T_G of width b on S^2 determined by a great circle G as the set consisting of all points distant by at most $\frac{b}{2}$ from G. If a convex body C is contained in a T_G and touched by the two circles bounding it, he says that T_G is a supporting strip region of C. He calls a convex body $W \subset S^2$ of constant width b if all supporting strip regions T_G of W have width b. If a body W is of constant width w in our sense, then from Claim 2.21 it follows that W is the intersection of the family $\mathcal{L}(W)$ of all lunes supporting W, and that the midpoints of the semicircles bounding every lune from $\mathcal{L}(W)$ belong to W. Of course all these lunes have thickness w. Take into account all great circles G of S^2 passing through the corners of every lune from $\mathcal{L}(W)$. Clearly all these strip regions T_G have the same width w. Consequently, W is of constant width in the sense of Leichtweiss. Conversely, assume that W is of constant width in the sense of Leichtweiss. By virtue of the convexity of W, any strip region T_G of W has exactly one point on each of the two bounding circles of T_G, and so their distance is w. Hence adding any extra point to W increases the diameter. Thus W is complete, and by Theorem 2.33 it is of constant diameter w. We conclude that the family of bodies of constant width in the sense of Leichtweiss coincides with the family of bodies of constant width in our sense. Observe that all this can be generalized to S^d. Namely, in place of G take a $(d-1)$-dimensional great subsphere of S^d and as the strip region $T(G)$ of width b, again take the set of all points distant by at most $\frac{b}{2}$ from G.

In 2012 Bezdek [7] calls a convex body $W \subset S^d$ of constant width if the smallest thickness of a lune containing W is equal to diam(W). If W is of constant width w in the sense of this chapter, then it is of constant width in the sense of Bezdek. Still the narrowest lune containing W has thickness w, and diam$(W) = w$ by Theorem 2.27. On the other hand, let us show that if a convex body $W \subset S^d$ is of constant width w in the sense of Bezdek, then W is of constant width in our sense. Assume the contrary, that is that W is not of constant width in our sense. Then not all widths of W are equal. Thus there are hemispheres H and H' supporting W such that width$_H(W) <$ width$_{H'}(W)$. By Theorem 2.18 and Proposition 2.4 we have width$_{H'}(W) \leq$ diam(W). So width$_H(W) <$ diam(W). This contradicts Bezdek's condition that the narrowest lune containing W has thickness diam(W). Both implications show that the families of constant width in Bezdek's sense and in our sense coincide.

The literature concerning the subject of this paper is very wide. Here we chronologically list a few valuable papers on this subject not quoted earlier in this chapter: [2, 5, 6, 9, 18, 22, 24, 27, 30–33, 64, 74, 90].

References

1. P.V. Araújo, Barbier's theorem for the sphere and the hyperbolic plane. L'Enseign. Math. **42**, 295–309 (1966)
2. M.J.C. Baker, A spherical Helly-type theorem. Pac. J. Math. **23**, 1–3 (1967)
3. E. Barbier, Note sur le problème de l'aiguille et le jeu du joint couvert. J. Math. Pures Appl. **5**, 273–286 (1860)
4. R.V. Benson, *Euclidean Geometry and Convexity* (McGraw-Hill Book Company, New York, 1966)
5. F. Besau, T. Hack, P. Pivovarov, F.E. Schuster, *Spherical Centroid Bodies* (2019). arXiv:1902.10614
6. F. Besau, S. Schuster, Binary operations in spherical convex geometry. Indiana Univ. Math. J. **65**(4), 1263–1288 (2016)
7. K. Bezdek, Illuminating spindle convex bodies and minimizing the volume of spherical sets of constant width. Discrete Comput. Geom. **47**(2), 275–287 (2012)
8. K. Bezdek, A new look at the Blaschke-Leichtweiss theorem. arXiv:101.00538v1
9. K. Bezdek, Z. Lángi, From spherical to Euclidean illumination. Monatsh. Math. **192**(3), 483–492 (2020)
10. W. Blaschke, Einige Bemerkungen über Kurven and Flaschen von konstanter Breite. Ber. Verh. Sächs. Akad. Leipzig **67**, 290–297 (1915)

11. W. Blaschke, Konvexe Bereiche gegebener konstanter Breite und kleinsten Inthalts. Math. Ann. **76**, 504–513 (1915)

12. T. Bonnesen, T.W. Fenchel, *Theorie der konvexen Körper*. Springer, Berlin (1934) (English translation: Theory of Convex Bodies, BCS Associates, Moscow, Idaho, 1987)

13. K.J. Böroczky, A. Sagemeister, The isodiametric problem on the sphere and the hyperbolic space. Acta Math. Hungar. **160**(1), 13–32 (2020)

14. G.D. Chakerian, H. Groemer, Convex bodies of constant width, in The *Collection of Surveys "Convexity and Its Applications*, ed. by P.M. Gruber, J.M. Wills (Birkhäuser, Basel 1983), pp. 49–96

15. C.Y. Chang, C. Liu, Z. Su, The perimeter and area of reduced spherical polygons of thickness $\pi/2$. Results Math. **75**(4), 135 (2020)

16. L. Danzer, B. Grünbaum, V. Klee, Helly's theorem and its relatives, in *Proc. of Symp. in Pure Math.*, vol. VII, ed. by V. Klee (Convexity, 1963), pp. 99–180

17. B.V. Dekster, Completness and constant width in spherical and hyperbolic spaces. Acta Math. Hungar. **67**, 289–300 (1995)

18. B.V. Dekster, The Jung theorem for spherical and hyperbolic space. Acta Math. Hungar. **67**(4), 315–331 (1995)

19. B.V. Dekster, Double normals characterize bodies of constant width in Riemannian manifolds, in *Geometric Analysis and Nonlinear Partial Differential Equations*. Lecture Notes in Pure an Applied Mathematics, vol. 144 (New York, 1993), pp. 187–201

20. B.V. Dekster, B. Wegner, Constant width and transformality in spheres. J. Geom. **56**(1–2), 25–33 (1996)

21. H.G. Eggleston, *Convexity* (Cambridge University Press, Cambridge, 1958)

22. L. Euler, it De curvis triangularibus (on triangular curves). Acta Akad. Sci. Imper. Petropolitinae **1778**(II), 3–30 (1781); Opera Omnia, Series, Vol. 28, pp. 298–321

23. E. Fabińska, M. Lassak, Reduced bodies in normed planes. Israel J. Math. **161**, 75–87 (2007)

24. O.P. Ferreira, A.N. Iusem, S.Z. Németh, Projections onto convex sets on the sphere. J. Global Optim. **57**, 663–676 (2013)

25. B. Fuchssteiner, W. Lucky, *Convex Cones*. North-Holland Mathematics Studies, vol. 56 (North-Holland, Amsterdam, 1981)

26. E. Gallego, A. Reventós, G. Solanes, E. Teufel, Width of convex bodies in spaces of convex curvature. Manuscr. Math. **126**(1), 115–134 (2008)

27. F. Gao, D. Hug., R. Schneider, Intrinsic volumes and polar sets in spherical space. Math. Notae **41**, 159–176 (2003)

28. B. Gonzalez Merino, T. Jahn, A. Polyanskii, G. Wachsmuth, Hunting for reduced polytopes. Discrete Comput. Geom. **60**(3), 801–808 (2018)

29. H. Groemer, Extremal convex sets. Monatsh. Math. **96**, 29–39 (1983)

30. Q. Guo, *Convexity Theory on Spherical Spaces* (Scientia Sinica Mathematica, 2020)

31. Q. Guo, Y. Peng, Spherically convex sets and spherically convex functions. J. Convex Anal. **28**, 103–122 (2021)

32. H. Hadwiger, Kleine Studie zur kombinatorischen Geometrie der Sphäre. Nagoya Math. J. **8**, 45–48 (1955)

33. H. Hadwiger, Ausgewählte Probleme der kombinatorischen Geometrie des Euklidischen und Sphärischen Raumes. L'Enseign. Math. **3**, 73–75 (1957)

34. N.N. Hai, P.T. An, *A Generalization of Blaschke's convergence theorem in metric spaces*. J. Conv. Anal. **4**, 1013–1024 (2013)

35. H. Han, Maximum and minimum of support functions. arXiv:1701.08956v2

36. H. Han, T. Nishimura, Self-dual shapes and spherical convex bodies of constant width $\pi/2$. J. Math. Soc. Jpn. **69**, 1475–1484 (2017)

37. H. Han, T. Nishimura, Spherical method for studying Wulff shapes and related topics, in *The volume "Singularities in Generic Geometry"* ed. by Math. Soc. Japan. Adv. Stud. in Pure Math., vol. 78 (2018), pp. 1–53

38. H. Han, T. Nishimura, Wulff shapes and their duals—RIMS, Kyoto University. http://www.kurims.kyoto-u.ac.jp/~kyodo/kokyuroku/contents/pdf/2049-04.pdf~

39. H. Han, D. Wu, Constant diameter and constant width of spherical convex bodies. Aequat. Math. **95**, 167–174 (2021)
40. E. Heil, *Kleinste konvexe Körper gegebener Dicke*, Preprint No. 453, Fachbereich Mathematik der TH Darmstadt (1978)
41. M.A. Hernandez Cifre, A.R. Martinez Fernández, The isodiametric problem and other inequalities in the constant curvature 2-spaces. Rev. R. Acad. Cienc. Exactas Fis. Nat. Ser. A Mat. RACSAM **109**(2), 315–325 (2015)
42. A.G. Horváth, Diameter, width and thickness in hyperbolic plane. arXiv:2011.14739v1
43. B. Jessen, Über konvexe Punktmengen konstanter Breite. Math. Z. **29**(1), 378–380 (1929)
44. T. Kubota, On the maximum area of the closed curve of a given perimeter. Tokyo Matk. Ges. **5**, 109–119 (1909)
45. M. Lassak, Reduced convex bodies in the plane. Israel J. Math. **70**(3), 365–379 (1990)
46. M. Lassak, On the smallest disk containing a planar reduced convex body. Arch. Math. **80**, 553–560 (2003)
47. M. Lassak, Area of reduced polygons. Publ. Math. **67**, 349–354 (2005)
48. M. Lassak, Approximation of bodies of constant width and reduced bodies in a normed plane. J. Convex Anal. **19**(3). 865–874 (2012)
49. M. Lassak, Width of spherical convex bodies. Aequat. Math. **89**, 555–567 (2015)
50. M. Lassak, Reduced spherical polygons. Colloq. Math. **138**, 205–216 (2015)
51. M. Lassak, Diameter, width and thickness of spherical reduced convex bodies with an application to Wulff shapes. Beitr Algebra Geom. **61**, 369–379 (2020) (earlier in arXiv:193.041148)
52. M. Lassak, When is a spherical convex body of constant diameter a body of constant width? Aequat. Math. **94**, 393–400 (2020)
53. M. Lassak, Complete spherical convex bodies. J. Geom. **111**(2), 35 (2020), 6p (a corrected version is in arXiv:2004.1011v.2)
54. M. Lassak, Approximation of reduced spherical convex bodies. J. Convex Anal. (to appear). arXiv:2106.00118v1
55. M. Lassak, H. Martini, Reduced convex bodies in Euclidean space—a survey. Expositiones Math. **29**, 204–219 (2011)
56. M. Lassak, H. Martini, Reduced convex bodies in finite-dimensional normed spaces—a survey. Results Math. **66**(3–4), 405–426 (2014)
57. M. Lassak M., M. Musielak, Reduced spherical convex bodies. Bull. Pol. Ac. Math. **66**, 87–97 (2018)
58. M. Lassak, M. Musielak, Spherical bodies of constant width. Aequat. Math. **92**, 627–640 (2018)
59. M. Lassak, M. Musielak, Diameter of reduced spherical convex bodies. Fasciculi Math. **61**, 83–88 (2018)
60. S.B. Lay, *Convex Sets and Its Applications*. A Willey-Interscience Publication (Wiley, New York, 1982)
61. H. Lebesgue, Sur quelques questions de minimum, relatives aux courbes orbiformes, et sur leurs rapports avec le calcul des variations. J. Math. Pures Appl. **8**(4), 67–96 (1921)
62. K. Leichtweiss, Curves of constant width in the non-Euclidean geometry. Abh. Math. Sem. Univ. Hamburg **75**, 257–284 (2005)
63. C. Liu, Y. Chang, Z. Su, The area of reduced spherical polygons. arXiv:2009.13268v1
64. H. Maehara, H. Martini, Geometric probability on the sphere. Jahresber. Dtsch. Math.-Ver. **119**(2), 93–132 (2017)
65. M. Martini, L. Montejano, D. Oliveros, *Bodies of Constant Width. An Introduction to Convex Geometry with Applications* (Springer Nature Switzerland AG, 2019)
66. H. Martini, K.J. Swanepoel, Non-planar simplices are not reduced. Publ. Math. Debrecen **64**, 101–106 (2004)
67. E. Meissner, Über Punktmengen konstanter Breite, Vjschr. Naturforsch. Ges. Zürich **56**, 42–50 (1911)
68. J. Molnár, Über eine Übertragung des Hellyschen Satzes in sphärischen Räume. Acta Mat. Hungar. **8**, 315–318 (1957)

69. D.A. Murray, *Spherical Trigonometry* (Longmans Green, London, Bombay and Calcutta, 1900)
70. M. Musielak, Covering a reduced spherical body by a disk. Ukr. Math. J. **72**(10), 1400–1409 (2020)
71. J. Pál, Über ein elementares Variationsproblem. Bull. de l'Acad. de Dan. **3**(2), 35 (1920)
72. A. Papadopoulos, *Metric Spaces, Convexity and Nonpositive Curvature*, 2nd edn. IRMA Lectures in Mathematics and Theoretical Physics. European Mathematical Society (Zürich, 2014)
73. A. Papadopoulos, On the works of Euler and his followers on spherical geometry. Ganita Bharati **38**(1), 53–108 (2014)
74. A. Papadopoulos, Three theorem of Menelaus. Am. Math. Monthly **126**(7), 610–619 (2019)
75. A. Pimpinelli, J. Vilain, *Physics of Crystal Growth*. Monographs and Texts in Statistical Physics (Cambridge University Press, Cambridge, New York, 1998)
76. K. Reidemeister, Über Körper konstanten Durchmessers. Math. Z. **10**, 214–216 (1921)
77. B.A. Rosenfeld, *A History of Non-Euclidean Geometry* (Springer, New York, 2012)
78. L.A. Santaló, Note on convex spherical curves. Bull. Am. Math. Soc. **50**, 528–534 (1944)
79. L.A. Santaló, Properties of convex figures on a sphere. Math. Notae **4**, 11–40 (1944)
80. L.A. Santaló, Convex regions on the n-dimensional spherical surface. Ann. Math. **47**, 448–459 (1946)
81. R. Schneider, *Convex Bodies: The Brunn-Minkowski Theory*, second expanded edition, Encyclopedia of Mathematics and its Applications, vol. 151 (Cambridge University Press, 2013, Cambridge, 2014)
82. Y. Shao, Q. Guo, An analytic approach to spherically convex sets in S^{n-1}. J. Math. (PRC) **38**(3), 473–480 (2018)
83. I. Todhunter, *Spherical Trigonometry*, 5th edn. (Macmillan, London, 1886)
84. G. Toth, *Measures of Symmetry for Convex Sets and Stability* (Springer, Cham, 2015)
85. G. Van Brummelen, *Heavenly Mathematics. The Forgotten Art of Spherical Trigonometry* (Princeton University Press, Princeton, 2013)
86. M.A. Whittlesey, *Spherical Geometry and Its Applications* (CRC Press, Taylor and Francis Group, 2020)
87. G. Wulff, Zur Frage der Geschwindigkeit des Wachstrums und der Auflösung der Krystallfläaschen. Z. Kryst. Mineral. **34**, 449–530 (1901)
88. I.M. Yaglom, V.G. Boltyanskij, *Convex Figures* (Moscow 1951). (English translation: Holt, Rinehart and Winston, New York 1961)
89. C. Zălinescu, On the separation theorems for convex sets on the unit sphere. arXiv:2103.04321v1
90. X. Zhou, Q. Guo, Compositions and valuations on spherically convex sets. Wuhan Univ. J. Nat. Sci. (2020)

Chapter 3
Minkowski Geometry—Some Concepts and Recent Developments

Vitor Balestro and Horst Martini

Abstract The geometry of finite-dimensional normed spaces (= Minkowski geometry) is a research topic which is related to many other fields, such as convex geometry, discrete and computational geometry, subfields of functional analysis, Finsler geometry, and optimization. For that reason, Minkowski geometry is connected with a lot of interesting and recent research directions. The main objective of this exposition is to discuss some recent progress in various ramifications of the subject area, as well as to present some open problems and research directions. We also present brief historical overviews and some basic concepts of the theory.

Keywords Angular measures · Birkhoff orthogonality · Curvature · Curvature types · Differential geometry · Distance geometry problem · Elementary geometry · Equilateral set · Fermat–Torricelli problem · Finsler geometry · Geometric constants · Hadwiger's covering conjecture · Mahler's conjecture · Minkowski billiards · Minkowski geometry · Normed spaces · Reduced bodies · Viterbo's conjecture

MSC 2010 52A21, 52A10, 52A15, 52A20, 52A38, 52A40, 46B20, 46B85, 53A15, 53A35, 53B40

3.1 Introduction

Our main concern is to give a (quite informal) perspective about the research area of *Minkowski geometry* (= geometry of normed real finite dimensional vector spaces, not to be confused with Minkowski's space-time), partly discussing its origins and

V. Balestro
Instituto de Matemática e Estatística, Universidade Federal Fluminense, Niterói, Brazil
e-mail: vitorbalestro@id.uff.br

H. Martini (✉)
Fakultät für Mathematik, Technische Universität Chemnitz, Chemnitz, Germany
e-mail: martini@mathematik.tu-chemnitz.de

early developments, and partly presenting the current state of research. As basic references close to the spirit of our paper here we mention the monograph [214] as well as the expositions [156, 159, 161], and [160].

With this text, we want to introduce any interested researcher to the basic definitions, tools and main references of important selected parts of the area, as well as to provide an overview where interesting current research might be started. This means that we shall walk through different parts (and viewpoints) of the theory that motivated also recent development, but we do not have the intention of being exhaustive. After all, Minkowski geometry is not a "centralized" research area, with classical theory and problems somehow "guiding" the research efforts, but rather a quite "sparse" topic, with interested mathematicians from many different areas and backgrounds working on it (as it may be clear by a quick look at the references list). This possibly happens because the natural objects of study in Minkowski geometry are (or can be) naturally related to various other areas, such as *convex geometry, functional analysis, discrete and computational geometry, Finsler geometry*, and *optimization*. For us, this is what makes the geometry of normed spaces so interesting and inspiring.

Due to our specified knowledge and natural limitations of interests, various topics are missing or will be only briefly mentioned (always with good references, we hope) and some others will be introduced in a more detailed manner. This does not imply, by any means, a hierarchic classification of the topics mentioned here. For example, the recent progress in the differential geometry of curves and surfaces in normed spaces will be carefully treated for the simple reason that this is one of our main current research interests, whereas the large field of orthogonality concepts is discussed only briefly.

Some historical background is due. The *axioms of Minkowski spaces* were introduced by Minkowski [178] when studying problems from number theory, even though the earliest reference to a non-Euclidean geometry in the sense of Minkowski geometry seems to be the mention of the l_4 norm by Riemann in his Habilitationsvortrag [197] (this norm in the plane is illustrated in Fig. 3.3). Later, Gołab [99] and Gołab and Härlen [100] studied Minkowski geometry from a geometrical point of view. Already the monograph [49] gives early references to the field (see the end of Sections 14 and 38 there). In his book [214], Thompson says that Gołab obtained the possibly oldest result in Minkowski geometry. It states that the perimeter of the unit circle of any normed plane must lie in the interval [6, 8], and the extremal cases characterize, respectively, a normed plane whose unit ball is an affine regular hexagon, and a normed plane whose unit ball is a parallelogram (also called *rectilinear*, see Example 3.2.1 below). We refer to [139], p. 236, where Laugwitz showed in an elementary way that each real value between 6 and 8 can be reached already by a hexagonal norm (the parallelogram then presenting a borderline case). To see this immediately, one can use our Fig. 3.5 below (right hand side): a continuous transform of the hexagonal norm circle there into the blue square shown in Fig. 3.2 (bringing the aslant segments step by step into perpendicular position) yields obviously any real value of the interval [6, 8].

The relation of Minkowski geometry to differential geometry, which led also to the development of Finsler geometry, started directly after Minkowski's fundamental work in terms of basic ideas of relative differential geometry; cf. the end of Section 38 in [49] and the introduction to [36]. In the middle of the twentieth century, Busemann intensively studied Minkowski geometry as, in his own words, "a necessary preliminary step for any progress in the theory of *Finsler spaces*" (see [60] and [61], where various examples for the local Minkowskian situation, a discussion of measures and isoperimetry in Minkowski spaces, and of the relations to the field of relative differential geometry are given). From the viewpoint of Minkowski geometry itself, a major contribution by him was the settlement of the isoperimetric problem for normed spaces (see [58] and [59]). A few years later, Busemann's student Petty studied curvature types for curves in normed planes, see [189] and also [191]. At the same time, Biberstein's doctoral thesis [45] came to be, as the possibly first reference treating the differential geometry of surfaces in normed spaces. The *differential geometry of curves and surfaces in Minkowski spaces* was later rediscovered by Guggenheimer in the papers [105] and [106]. Another line of extending Minkowski geometry started even earlier, namely that one leading to *functional analysis* and especially to *Banach space theory*. The history of this broad development is excellently presented in the monographs [81, 179], and [192]. It is also very important to mention the *local theory of Banach spaces*, whose milestone is Dvoretzky's theorem [86] (see also [176] for Milman's remarkable proof). We refer the reader to [177, 193] and [16]. And here we should also mention that some of our results or references even refer to *gauges* which are also described by notions like *convex distance functions, Minkowski functionals, asymmetric distances*, or *generalized Minkowski spaces*. Indeed, already Minkowski himself introduced this concept at the end of the nineteenth century as that of "einhellige Strahldistanzen" in [178], calling the symmetric subcase "wechselseitige Strahldistanzen". At the beginning of the twentieth century this subcase was renamed "norms" and became very popular, especially among analysts leading finally to notions like Finsler spaces or Banach spaces.

The text is organized as follows. In Sect. 3.2, we introduce the basic definitions, and also highlight some early differences between the geometry of Euclidean spaces and general Minkowski spaces. One of these differences is that there is no immediate orthogonality concept in a Minkowski space, since we do not have an inner product. However, we can define several orthogonality concepts, each of which generalizes the Euclidean orthogonality. As one may expect, these concepts are related to lots of other geometric ideas, some of which are still active research topics. We deal with orthogonality concepts in Sect. 3.3, and Sect. 3.4 is dedicated to a variety of theorems in the spirit of elementary geometry, also in higher dimensions. Section 3.5 aims to present some long standing problems regarding the distance induced by a norm. In Sect. 3.6 we study both classical and differential curve theories in normed planes, and Sect. 3.7 will focus on recent work on the geometry of surfaces immersed in three-dimensional normed spaces. Section 3.8 deals with the recent and exciting topic of Minkowski billiards, featuring a very nice relation between famous conjectures in the fields of symplectic geometry and convex geometry (namely, Viterbo's conjecture and Mahler's conjecture) discovered by Artstein-Avidan, Karasev and Ostrover [17]. Last but not least, Sect. 3.9 will deal with other research directions, not covered in the previous sections.

3.2 First Steps

What we call *Minkowski geometry* is the geometry given by *norms* in finite dimensional vector spaces. Formally, a *norm* in a vector space X is a map $|| \cdot || : X \to \mathbb{R}$ which satisfies, for any $x, y \in X$ and $\lambda \in \mathbb{R}$,

(a) $||x|| \geq 0$, with equality if and only if $x = 0$,
(b) $||\lambda x|| = |\lambda| \cdot ||x||$, and
(c) $||x + y|| \leq ||x|| + ||y||$.

Properties (b) and (c) will sometimes be referred to as *homogeneity* and *triangle inequality*. Of course, the most common examples of norms in vector spaces are the norms given by inner products (we call such norms *Euclidean*, like also the respective spaces). One of the interesting sides of Minkowski geometry is to explore the similitudes and differences between Euclidean and non-Euclidean thinking. A first immediate illustration is that the distance induced by a norm is always translation-invariant, but not $SO(n)$-invariant. A norm in a vector space defines naturally a metric between vectors as

$$d(x, y) := ||x - y||,$$

for $x, y \in X$. For finite dimensional spaces, this metric is equivalent to the one induced by any inner product (and hence both induce the same topology on X).

Example 3.2.1 The norm in $\mathbb{R}^2$ given by

$$||(x, y)||_1 := |x| + |y|$$

is known as the *taxicab norm* (or *driving in Manhattan*). The reason is that this norm models traveled distances in neighborhoods with "perfectly organized" blocks (see Fig. 3.1). This norm is also known as the l_1 norm.

The space of norms in a given vector space X is related to the space of centrally symmetric convex bodies in X via the *Minkowski functional*, which is defined as

$$\rho_K(x) := \inf\{\lambda \geq 0 : x \in \lambda K\},$$

for any $x \in X$, where K is a convex body centered at the origin. It is not hard to see that ρ_K is a norm for each such K. Conversely, given a norm $|| \cdot ||$ in X, and letting $B := \{x \in X : ||x|| \leq 1\}$ be its *unit ball*, it is clear that $|| \cdot || = \rho_B(\cdot)$. A norm is called a *smooth norm* if at each point of the *unit sphere* (in two dimensions

Fig. 3.1 In the taxicab norm, the distance between $(1, 1)$ and $(3, 3)$ is 4

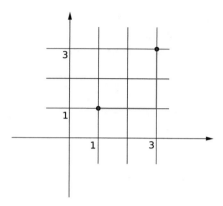

Fig. 3.2 The unit balls of the norms l_1 (in red) and l_∞ (in blue)

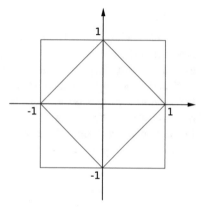

unit circle) $S := \{x \in X : ||x|| = 1\}$ there is a unique supporting hyperplane of B. Also, a norm is said to be a *strictly convex norm* if the unit circle does not contain non-degenerate line segments (or, equivalently, if the triangle inequality is strict for linearly independent vectors). Notice that the l_1 norm, defined in Example 3.2.1, is neither smooth nor strictly convex. Indeed, it is easy to see that its unit circle is the square with vertices $(\pm 1, 0)$ and $(0, \pm 1)$, see Fig. 3.2.

Here a further comment on the class of norms in a given finite dimensional vector space is due. The geometry induced by a given norm is indistinguishable up to a linear transformation. Formally, if B_1 and B_2 are the unit balls of given norms $|| \cdot ||_1$ and $|| \cdot ||_2$ in a vector space X, and if there exists a linear map $T : X \to X$ such that $T(B_1) = B_2$, then $|| \cdot ||_1$ and $|| \cdot ||_2$ yield the same geometry on X. Indeed, the image of the unit circle S_1 must be the unit circle S_2, and for any $x \in S_1$ and $\alpha \in \mathbb{R}$ we have

$$||T(\alpha x)||_1 = |\alpha| \cdot ||Tx||_1 = |\alpha| = ||\alpha x||_2.$$

Intuitively, a linear map only produces a difference in the geometry that can be "seen from outside". As an example, we have that all ellipsoids in a given vector space yield the same Euclidean geometry. Also, the l_∞ *norm* in $\mathbb{R}^2$ defined as

$$||(x, y)||_\infty := \max\{|x|, |y|\}$$

gives the same geometry as the l_1 norm. See Fig. 3.2, where the unit balls of both norms are illustrated.

Thus far we only gave examples of norms which are neither smooth nor strictly convex. Quite standard examples of smooth and strictly convex norms are the l_p *norms* for $1 < p < +\infty$, defined in $\mathbb{R}^2$ as

$$||(x, y)||_p := \left(|x|^p + |y|^p\right)^{1/p}.$$

Figure 3.3 illustrates the unit circle of the l_p norm where $p = 4$. The reader may notice that the l_p norms that we have defined for $1 \le p \le +\infty$ can be naturally extended to any finite dimensional vector space $\mathbb{R}^n$.

When replacing the Euclidean norm by an arbitrary Minkowski norm, even very elementary questions might have not so trivial answers. One example is the question whether there is always a circle through three given non-collinear points. In Minkowski geometry, the answer is positive if and only if the norm is smooth. However, we might not have uniqueness if the norm is not strictly convex. For a

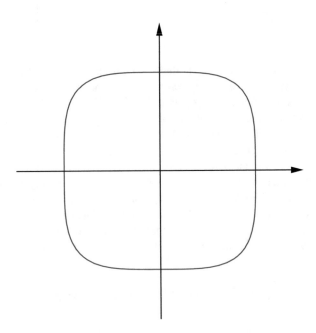

Fig. 3.3 The unit circle of l_4

Fig. 3.4 The red and blue circles of the l_∞ norm pass through the points p, q and r. The blue and green circles are the unique circles through p and x, and none of them passes through y

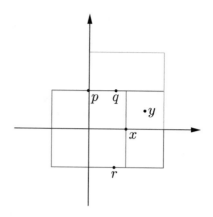

discussion on these results, we refer the reader to [161, § 7.1]. Figure 3.4 illustrates the situation.

We continue with discussing the concept of *area* in a Minkowski space. It is clear that a norm in a vector space endows it with a natural length structure, as explained before. However, an area element does not follow directly from this structure. When dealing with an n-dimensional Minkowski space $(X, || \cdot ||)$, the problem is to introduce a measure on each k-dimensional subspace of X which is, somehow, related to the geometry given by the norm $|| \cdot ||$. First of all, we demand all these measures to be Haar measures, and since two Haar measures in a finite dimensional vector space are equal up to constant multiplication, the problem becomes to find the "correct" normalization for each subspace. For that sake, it is usual to consider the sections of such subspaces with the unit ball of the norm. For a very good account of area definitions in Minkowski spaces, we refer the reader to Chapter 5 of Thompson's book [214]; see also the surveys [161] and [13].

For our purposes, we just need to clarify the simplest case where the dimension of the space equals 2. In this case, we just endow the plane with an *area element*, which is a nondegenerate skew-symmetric bilinear form $\omega : X \times X \to \mathbb{R}$. For example, if $X = \mathbb{R}^2$, then ω can be taken as the usual determinant. Since the 1-dimensional "area" is simply the length given by the norm, and there is only one 2-dimensional subspace in this case, the normalization is not important in general (only in the particular case of Radon planes, as we will briefly comment later). Also, it is easy to see that any two such bilinear forms are equal up to constant multiplication.

To deal with the general even-dimensional case, recall that the *dual space* X^* of a vector space X is the space of *linear functionals* defined on X:

$$X^* := \{f : X \to V : f \text{ is linear}\}.$$

Given a norm $|| \cdot ||$ in X, we may endow X^* with a natural norm, called *dual norm* and defined as

$$||f||^* := \sup\{|f(x)| : ||x|| = 1\}.$$

Any symplectic form ω fixed in X gives a natural isomorphism $T : X \to X^*$ by setting

$$Tx(\cdot) = \iota_x \omega(\cdot) = \omega(x, \cdot),$$

and hence, for each $x \in X$, we may calculate the dual norm of Tx as $||Tx|| = \sup\{|\omega(x, y)| : ||y|| = 1\}$. Therefore, the dual norm in X^* can be regarded as a norm in X by setting

$$||x||_a := \sup\{|\omega(x, y)| : ||y|| = 1\}, \quad x \in X.$$

This new norm $|| \cdot ||_a : X \to \mathbb{R}$ is called the *anti-norm* associated to $|| \cdot ||$. The balls and circles of the anti-norm will be called *anti-balls* and *anti-spheres* (or *anti-circles*, in two dimensions), respectively. Also, the anti-norm is dual to the norm in the sense that the anti-norm of the anti-norm is the norm (with respect to the same symplectic bilinear form).

Now we return to the two-dimensional case. In this case, if the anti-norm is a multiple of the norm, then the plane is called a *Radon plane* (this is the same as stating that the *Birkhoff orthogonality* given by the norm is symmetric, see the next section). Of course, the bilinear skew-symmetric form ω can be rescaled in a way that we have $|| \cdot || = || \cdot ||_a$ (indeed, any rescaling in ω produces the same rescaling in $|| \cdot ||_a$). This approach of the dual norm follows from [160], which is the main reference to anti-norms and Radon planes, but see also the respective sections in [161] and [31].

Example 3.2.2 Let $\mathbb{R}^2$ be endowed with the l_p norm $|| \cdot ||_p$, for some $p < 1 < \infty$. Also, let ω be the standard determinant in $\mathbb{R}^2$. It is not difficult to see that the anti-norm of $|| \cdot ||_p$ is the l_q norm where $1/p + 1/q = 1$. Moreover, for such numbers p, q we may define the $l_p - l_q$ *norm* in $\mathbb{R}^2$ as

$$||v||_{p,q} := \begin{cases} ||v||_p, & \text{if } v \text{ is in the first or third quadrant} \\ ||v||_q, & \text{if } v \text{ is in the second or fourth quadrant} \end{cases}.$$

These are examples of smooth and strictly convex Radon norms. Extending this, it is easy to see that the norms l_1 and l_∞ are dual to each other in the same way, and that the $l_1 - l_\infty$ norm is affinely equivalent to a norm given by a regular hexagon. Figure 3.5 illustrates this, and also the unit ball of the $l_p - l_q$ norm where $p = 3$ and $q = 3/2$.

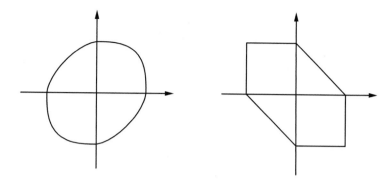

Fig. 3.5 The unit balls of the Radon planes $l_3 - l_{3/2}$ and $l_1 - l_\infty$, respectively

3.3 Orthogonality Concepts

As the reader may have noticed, without an inner product we have no immediate definition of orthogonality. However, the notion of "least distance" coming from usual orthogonality in Euclidean spaces can be directly translated to a general normed space $(X, || \cdot ||)$. We say that a vector $x \in X$ is *Birkhoff orthogonal* to a vector $y \in X$ whenever $||x + \lambda y|| \geq ||x||$ for any $\lambda \in \mathbb{R}$. We denote this relation by $x \dashv_B y$, see, e.g., the survey [7] by Alonso, Martini and Wu. The geometric interpretation of this orthogonality concept is described by the equivalence of the following three claims for non-zero vectors $x, y \in X$:

(i) $x \dashv_B y$,
(ii) the line $t \mapsto x/||x|| + ty$ supports the unit ball at $x/||x||$,
(iii) fixing any $p, q \in X$ such that p does not belong to the line $l : t \mapsto q + ty$, the distance from p to l (in the norm) is assumed in a point $z \in l$ such that the segment $seg[p, z]$ is in the direction of x.

The proof is straightforward, and Fig. 3.6 illustrates the concept. Notice that in a non-strictly convex normed plane the distance from a point to a line can be reached for more than one point in the line (this happens when the line is parallel to a segment of the unit circle).

It is clear that Birkhoff orthogonality is not necessarily symmetric. (We may ensure the existence of a pair of mutually orthogonal directions, though. Such a pair is called a *conjugate pair*.) Indeed, it is only symmetric in dimension 2 if the unit circle is a *Radon curve* (see [31] and [160]). For dimension greater than or equal to 3, the symmetry of the Birkhoff orthogonality characterizes Euclidean norms (see Day's paper [77] for a proof). Moreover, in a space whose dimension is greater than 2, Birkhoff orthogonality naturally extends to a relation between vectors and hyperplanes. We say that a vector $x \in X$ is *Birkhoff orthogonal* to a hyperplane H whenever $x \dashv_B H$ for any $v \in H$ (we denote simply $x \dashv_B H$). Geometrically, if $x \neq 0$, then we have $x \dashv_B H$ if and only if H supports the unit ball B at $x/||x||$.

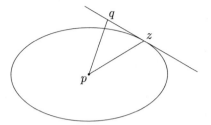

Fig. 3.6 A homothet of the unit ball centered at p, yielding $||q - p|| > ||z - p||$

There are other orthogonality concepts that generalize the usual orthogonality of Euclidean spaces, and we restrict ourselves to Birkhoff orthogonality in this text merely to avoid this introduction to be very lengthy. However, for the sake of completeness we briefly introduce two other important orthogonality types: two vectors $x, y \in X$ are said to be *isosceles orthogonal* (or *James orthogonal*) whenever $||x - y|| = ||x + y||$, and *Roberts orthogonal* if $||x + \lambda y|| = ||x - \lambda y||$ for any $\lambda \in \mathbb{R}$. A very important early reference on orthogonality in normed spaces is James' paper [126], followed by the basic book [14] of Amir (see Sections 3, 4, 7, 18, and 19 there) and the surveys [1, 2], and [44]. We also refer to the respective chapter [8] in the present book. A very good reference on Birkhoff orthogonality and isosceles orthogonality is the survey [7] by Alonso, Martini and Wu.

Orthogonality concepts in normed spaces (and related ideas) are an active topic of research, and we will comment on some recent progress. In [30], Roberts orthogonality was related to *bisectors of segments*, which is the set defined as

$$\text{bis}(x, y) := \{z \in X : ||x - z|| = ||y - z||\},$$

for a given segment seg$[x, y] \in X$. Bisectors of segments were studied by Jahn and Spirova [125] for the planar case, and by Horváth [117] for the general case, see also [120]. They prove, in particular, that for the strictly convex case, the bisectors of segments are homeomorphic to hyperplanes (which, in the planar case, are lines). In the mentioned paper, Jahn and Spirova also described completely the structure of the bisector of a segment which is parallel to a straight line segment of the unit circle. Another proof can be found in [30]. It is worth mentioning that understanding bisectors of segments in normed spaces is a necessary step to study *Voronoi diagrams* for convex functions, which is an active topic of research in the fields of computational and stochastic geometry (see, e.g., [20] and [43], and the monographs [185] and [21]). For a survey on bisectors in normed spaces, we refer the reader to [159, Section 4]. Later results describe the complicated structure that bisectors can have via their radial projections; see [170] and [120], where Birkhoff orthogonality and shadow boundaries of unit balls play a role. In [115] the authors show the following result holding for normed planes and even for gauges: if the set of points x in the boundary of the unit ball, that create the bisector $B(-x, x)$ as straight line, has non-empty interior with respect to this boundary, then the norm is

Euclidean. An extension to higher dimensions is also derived. And in [42] oscillation properties of bisectors (and also trisectors) in a Minkowski space are studied. It turns out that, in the sense of Baire category results, geometrically irregular behavior is typical.

Another research direction that has seen development during the past few years is that of approximate orthogonality concepts. As references to that we may cite Chmieliński's paper [67] on approximate Birkhoff orthogonality, and also the paper [226], by Zamani and Moslehian, dealing with two approximations of Roberts orthogonality. Also, in a very recent paper, Jahn [121] extended Birkhoff orthogonality and isosceles orthogonality to *generalized Minkowski spaces*. Furthermore, Birkhoff orthogonality in the normed spaces of linear operators between vector spaces has been extensively studied in the past few years. Among others, we may cite the works of Hait et al. [112], Paul et al. [187], Ghosh et al. [97], and Sain [198].

Remark 3.3.1 As opposed to the Euclidean plane geometry, in a normed plane whose unit ball is an affine regular hexagon it is possible to construct a triangle whose sides are all mutually (Birkhoff) orthogonal. It was proved in [34] that this is the unique type of norm with this property. See Fig. 3.7.

To finish this section, it is worth mentioning that the Birkhoff orthogonality concept can be used to develop the classical theory of convex functions replacing the inner product of the domain by a norm. This is done by introducing analogues of *subgradients* and *subdifferentials* via the *Legendre transform* associated to the norm. Interestingly, the geometric behavior of these extensions is completely analogous to that of the classical theory. In this direction, we refer the reader to [38] (and for the classical approach we mention [200]).

Fig. 3.7 If the unit ball is an affine regular hexagon, then a triangle whose sides are parallel to three consecutive segments of the unit circle is such that all of its sides are mutually orthogonal

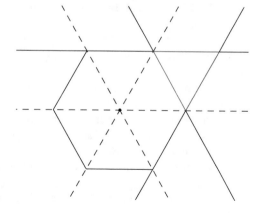

3.4   Some Elementary Geometry

It seems that there is no exact definition of the field *elementary geometry* (see, e.g., [111], the introduction of [225], and [224] for different and similar opinions in this direction). If one takes the intersection of opinions of these and further authors roughly, then elementary geometry consists of a *large variety of nontrivial theorems describing geometric properties of triangles and polygons, of circles and systems of circles.* In this sense, also monographs entitled "college geometry" (or containing similar material), with [129] as prominent example, and also books like [71] and large parts of the Japanese temple geometry (cf. [91]) belong to this field. Studying this framework, we do not agree with Dieudonné's related saying "Down with Euclid!" (meant also in the sense of "Death to triangles!") since we believe that also this field still yields many sufficiently deep and interesting problems and ideas, e.g. mainly referring to interesting higher dimensional and/or non-Euclidean extensions. And sometimes respective theorems bring forward insights in the initial character of Euclidean geometry when looking suitably back. Elementary geometry also motivates important results in neighbouring disciplines, like in discrete and computational geometry (e.g., Voronoi diagrams in normed spaces, cf. [20] and [21]) or in Banach space theory (for example, many characterizations of inner product spaces are based on elementary properties of certain polygons, see the monographs [14] and [10]). A modern and broad collection of theorems from Euclidean elementary geometry, many of which are not extended to higher dimensions and/or non-Euclidean geometries until now, is given in the two surveys [146] and [147]. On the other hand, in some parts of the surveys [107] (treating even gauges), [153, 156, 161], and [5] such extensions to Minkowski planes and spaces are studied. Related problem collections and papers with educational backgrounds are [203, 223], and [93]. In this section, we try to continue and to complete these collections.

We start with the *geometry of triangles and simplices.* In the survey [161] many basic and elementary results on Minkowski planes (and spaces) are presented, sometimes with new and streamlined proofs. Regarding *triangles in normed planes*, the following topics are discussed there: in- and circumcircles of triangles, importance of the triangle inequality (metric segments, strict convexity, diagonals of quadrilaterals, the monotonicity lemma), equilateral triangles and affine regular hexagons, equilateral sets, and area of triangles in inequalities. Further contributions to this little field follow now.

Results discussed now refer directly or indirectly to the notion of *orthocenters of triangles.* One way to introduce this concept for Minkowski geometry is the following, see [219] and [136]: In a strictly convex and smooth normed plane, the parallels h to a tangent of the unit circle are called left-orthogonal to the parallels g of the corresponding radius, and if a line h is left-orthogonal to another line g, then g is right-orthogonal to h. Correspondingly, left- and right-altitudes of a triangle can be defined which, in general, are not concurrent defining then an altitude-triangle. Also with methods from projective geometry, it is shown in [219] that the respective normed plane is Euclidean if and only if each triangle has concurrent left-altitudes (and thus a so-defined left-orthocenter). Further on, questions of the following type are discussed in [219]: In a given normed plane, find all triangles with concurrent altitudes! And if a triangle has a non-degenerate altitude-triangle, one may construct the altitudes of it again and study the iteration procedure thus started. It turns out that the generated sequence of triangles does not always have a limit left- (or right-) orthocenter. Continuing these investigations, also Kozma [136] deals with left- and right-orthocenters (as well as left- and right-circumcenters) of Minkowskian triangles; he obtains in all four cases analogous characterizations of the Euclidean case. Similarly defining height vectors of a triangle T in normed planes (e.g., based on Birkhoff orthogonality), *characterizations of Radon planes or the Euclidean plane* can be obtained. Busemann shows in § 50 of [62] that, defining the *"area"* of a triangle T as half product of one side-length and the corresponding height length, then the "area" of T is independent of the chosen side if and only if the considered plane is a Radon plane. Tamassy [212] confirms the same characterization of Radon planes assuming that the unit circle is an oriented closed curve of class C^1 which is star-shaped with respect to the origin, thus considering a strong generalization of the usual Minkowski metric. And the authors of [101] and [11] introduce the notion of *orthocenter* of T, namely for the case that all three altitudes of T pass through this point. They show that the existence of an orthocenter for a certain class of triangles is necessary and sufficient for the Euclidean case, and in [11] they also study the situation when these three lines intersect in three different points. Also for this general situation they obtain characterizations of the Euclidean plane, and the notion of Euler line is dealt with, too. And in [9], characterizations of inner product spaces are obtained via successive usage of the length of the height and the length of the bisectrix in terms of the side-lengths in a triangle. Here we also mention once more the book [10].

Classifying all shapes of the intersection of two norm circles, the authors of [3] describe all possible locations of *circumcenters* (and therefore also of all *circumcircles*) of non-degenerate triangles in arbitrary normed planes. With similar methods, in [4] all locations of *centers of minimal enclosing circles* of arbitrary triangles in normed planes are clarified. It turns out that this set is either a single point or a line segment. Its location depends on the type of the given triangle with respect to "angles": acute, obtuse, or right, all three suitably defined in terms of the norm. Therefore the results on minimal enclosing circles are obtained via an equally interesting and completely new norm-classification of triangles in such planes. These investigations were continued in [124, 145], and [102], also treating n-dimensional normed spaces. The authors refer to the sets of centers of minimal enclosing balls of convex bodies, called *Chebyshev sets* of these bodies. In the first paper, these sets are for instance combined with the notion of completions of convex bodies (a set is called complete if any proper superset of it has larger diameter; and a complete superset of a bounded set having the same diameter is called a completion of it). From the second paper one gets that the condition "the unit ball is not strictly convex" is equivalent to each of the following two properties: there exists a convex body whose Chebyshev set is of dimension larger than zero; every two-elements set has a Chebyshev set of dimension larger than zero. In [102] this is generalized to the following equivalence: For $k \in \{0, \ldots, n - 2\}$ there is a convex body whose Chebyshev set has dimension larger than k if and only if each set of cardinality at most n has a Chebyshev set of dimension larger than k. The two papers [215] and [216] of Väisälä should be also mentioned here. The author treats alternative approaches to results given in [3] and [4].

It is well known that a convex body is called *reduced body* if any convex body properly contained in it has strictly smaller minimal width (see also our Sect. 3.9 here). The paper [6] shows how interesting already the geometry of reduced triangles in normed planes is. While the class of *reduced triangles* in the Euclidean plane coincides with that of equilateral triangles, it is much richer in non-Euclidean normed planes. E.g., it is shown that every reduced triangle (normalized in the sense that its centroid is the origin and its incircle has radius one) is equilateral if and only if the norm is Euclidean. Bounds on the side lengths and vertex norms of such triangles are given, and various further (uniqueness and existence) results on reduced triangles in normed planes are obtained.

We continue with results from the *geometry of the Minkowskian simplex* in higher dimensions. First we note that our subsection on equilaterality in Sect. 3.5 below should also be mentioned here, in particular the contributions on equilateral simplices (like, e.g., the paper [53]). In [66], an extension of Viviani's property (referring to polytopes with the property that the sum of the Minkowskian distances to the affine hulls of their facets is the same for all their interior and boundary points) is used to derive several results on Fermat-Torricelli points in the Euclidean case, and also for smooth and strictly convex normed spaces. Besides results on this famous location problem also the geometry of simplices is emphasized, and characteristic properties of special types of simplices are derived (e.g., when the Minkowskian incenter coincides with the centroid, or the Minkowskian heights are all equal). Also the reflection principle of Heron is generalized in this direction.

Fixing a Lebesgue measure in a Minkowski space playing the role of volume there, the authors of [24] determine for the planar case the complete system of inequalities for *normalized area* and *side lengths of triangles*, considering equality cases. They also discuss possible extensions to higher dimensions and obtain a tight lower bound for the normalized volume of an arbitrary simplex.

In [22] theorems on special points (centers), in- and escribed hyperspheres and their radii, and Minkowskian heights of simplices in n-dimensional Minkowski spaces are obtained. These results enable the author to go on with characterization theorems for special norms as well as for special types of simplices. Extending the concept of trilinear coordinates, he successfully uses Minkowskian multilinear coordinates. In the papers [140] and [141], Leopold and Martini systematically extend results on Euclidean n-dimensional simplices to normed spaces, thus presenting Minkowskian analogues of notions like *circumcenter, Euler line, Monge point*, and *Feuerbach hypersphere*. They derive new related theorems which in several cases hold only for simplices with circumcenters (note that simplices in Minkowski spaces can have also infinitely many or no circumcenters), first with circumcenters different from the vertex centroids (e.g., for considering Euler lines and interesting points on them), and later supposing their coincidence (getting so-called quasiregular simplices). The coincidence of more special simplex centers in quasiregular simplices is assumed or derived. Such investigations show the difficulties and also the varieties of opportunities to introduce notions like *regular simplices* and *regular polytopes in Minkowski spaces*. And Feuerbach hyperspheres, occuring as $2(n + 1)$-point hyperspheres, are also studied.

We continue with results on *polygons (and polytopes)* and start again with the survey [161]. There, e.g., Dowker-type results on m-gons in- and circumscribed to convex curves, and also regular m-gons are studied. In the two papers [19] and [152], which will be comprehensively discussed below, one finds theorems of the following type: Let $p_1 p_2 \ldots p_m$, $m > 2$, be an m-gon inscribed to a Minkowskian circle; then the Feuerbach circles of all $(m - 1)$-gons obtainable from $p_1, p_2, \ldots, p_m$ have a common point which is the center of the Feuerbach circle of the m-gon $p_1 p_2 \ldots p_m$.

Similarly, also Section 5 of the paper [141] treats results on *Minkowskian cyclic polygons*, i.e., polygons having all vertices on their (thus existing) circumcircle. The theorems there refer to respective extensions of the notions of Monge point, vertex centroid, Euler line, Feuerbach circle and others.

In [113], only the *Manhattan norm* (i.e., taxicab geometry) is studied. The author defines *regular polygons* via equal side lengths and equal angles (the latter measured in the Euclidean sense) and shows that within this framework no regular triangles and no regular pentagons exist, but regular quadrilaterals can be constructed.

The paper [93] is an excellent example for pointing out that the tools of undergraduate mathematics and elementary geometry can be very helpful in modern research. Namely, the parallelogram identity implies that a rhombus inscribed in the unit circle of the Euclidean norm always has side-lengths $\sqrt{2}$. Borwein and Keener asked whether the Euclidean norm is the only norm with this property, and in [93] this is elegantly disproved; if the unit circle is invariant under rotation by $45°$, then every inscribed rhombus has the mentioned property (hence regular $8m$-gons yield this). Going a bit in another direction, but staying with rhombi, we mention [167] where it is shown that any convex Jordan curve in a normed plane admits an inscribed Minkowskian square, and that no two different Minkowskian rhombi with the same direction of one diagonal can be inscribed into the same strictly convex Jordan curve.

Now we discuss results on the *Minkowskian geometry of circles and systems of circles*. In [161], the following related themes are presented: in- and circumcircles of triangles, characterizations of circles, self-circumference and area of unit circles (e.g., in several inequalities).

It is well known that the "re-entrant property" of the Euclidean circle C (saying that there are six points on C successively being the centers of six circles congruent to C and always passing through the two neighbours of their own center taken from this six-point set) yields the famous pattern "flower of life"; in [133] the analogue for strictly convex normed planes is derived. For related results on equilateral triangles and sets see Petty [190], Section 4.1 in [214], Section 5 in the survey [161, 206], and our discussion in the next section here. From these considerations it follows that the Euclidean construction of the regular triangle (based on regular hexagons inscribed to the circle) works in any normed plane, and Ceder [64] shows that this property like also the three-circles theorem given below do not depend on symmetry (cf. also [65]). Spirova [204] gives further nice extensions for strictly convex norms and presents also Minkowskian analogues of other famous circle patterns like Yin and Yang, the Arbelos, Apollonius' problem on touching circles, and an interesting Sangaku-circle problem from the Japanese temple geometry, see [91]. This last problem is particularly interesting since its Euclidean subcase requires the use of circle inversion and Pythagoras' theorem (these tools have no immediate analogues in normed planes). Important relations to Möbius geometry are also discussed in [204].

Asplund and Grünbaum wrote the nice paper [19] mainly studying the extension of the so-called three-circles theorem (§ 104 of [129] and pp. 50–51 of [195]) and its consequences in smooth, strictly convex normed planes. This paper is continued by [152], where also the condition of smoothness is relaxed. The following results are taken from [19, 152], and [165], see also Section 4.1 and p. 130 in [214]. We start with the popular *three-circles theorem* itself which holds in the following form in any strictly convex normed plane and for any triangle with circumcircle (note that triangles in such planes have *at most* one circumcircle):

Theorem 3.4.1 *Let p_1, p_2, p_3 be pairwise distinct points forming a circle $C(x, \lambda)$ with center x and radius λ, and $C(x_i, \lambda)$, $i = 1, 2, 3$, be three circles different from $C(x, \lambda)$ each of which contains two of the three points p_i. Then the intersection of the latter three circles consists precisely of one point p.*

This point p is called the *C-orthocenter* of the triangle $p_1 p_2 p_3$, and one has $p = p_1 + p_2 + p_3 - 2x$. This name "*C*-orthocenter" makes really sense since in fact p plays the role of an orthocenter regarding $p_1 p_2 p_3$ if the notion of *James orthogonality* is applied to the quadruple p, p_1, p_2, p_3, and of course analogously to x, x_1, x_2, x_3! Both quadruples form *C*-orthocentric systems, meaning that (in both cases) any point can be interpreted as the *C*-orthocenter of the other three! The circumradii of the four triangles formed by the four points of a *C*-orthocentric system are equal, and the centroids of any three points of such a system form also a *C*-orthocentric system, similar to the first by having a third of its size. In [165] some characterizations of the Euclidean plane in terms of *C*-orthocentric systems are derived, e.g.: A strictly convex Minkowski plane is Euclidean if and only if for any *C*-orthocentric system p_1, p_2, p_3, p_4 satisfying $\|p_3 - p_1\| = \|p_3 - p_2\|$ the point p_4 lies in the line passing through p_3 and $1/2(p_1 + p_2)$. Various further consequences of the three-circles theorem are presented in [19] and [152], for example also *Miquel's theorem* for strictly convex normed planes:

Theorem 3.4.2 *If the points p_1, p_2, p_3, p_4 lie on the circle $C(O, \lambda)$ and four circles of the same size are drawn through p_1 and p_2, p_2 and p_3, p_3 and p_4, as well as p_4 and p_1, respectively, then the other four intersection points of these new circles lie on the circle $C(p_1 + p_2 + p_3 + p_4, \lambda)$.*

In the Euclidean case, the participating circles can even have different radii, in strictly convex normed planes in general not. This observation (that theorems from

Euclidean circle geometry allow the participating circles to have different radii, but their Minkowskian analogues demand equal radii) was made repeatedly, e.g. also in case of Clifford configurations, which are the subject of another generalization of the three-circles theorem, see below and [154]. Further notions and concepts from the geometry of polygons successfully extended in [19] and [152] to strictly convex normed planes are Feuerbach circles of triangles, of cyclic quadrangles and of cyclic m-gons (e.g., for triangles in all non-Euclidean normed planes they occur no longer as nine-point circles but as six-point circles), theorems on quadrangles whose diagonals are James orthogonal, and the already mentioned extensions of Clifford configurations derived in [154]. They relate $2m$ circles passing through a given point suitably to $2m$ points lying on a circle, all participating circles having equal radii.

In [155] the authors use Minkowskian bisectors, different orthogonality types, and further notions from Minkowski geometry to construct lattice coverings of normed planes, and they study related Voronoi regions. They discuss relations between *Voronoi diagrams*, the bundle theorem, and Miquel's six-circles theorem, the latter two presenting $(8_3, 6_4)$-configurations formed by 8 points and 6 circles, any of these points lying on 3 circles and any of the circles passing through 4 of these points. Using Möbius geometry and some type of extended circle inversion, also in [205] the norm-extension of Miquel's six-circles theorem for circles of equal radii is derived; additionally it is shown that this theorem holds for circles with not necessarily equal radii only in the Euclidean case.

In [220] it is shown that a normed plane is Euclidean if and only if for any point x exterior to its unit circle S the two tangent segments connecting x and S have equal Minkowskian lengths. The author derives also an n-dimensional analogue of this theorem. And in [168] the following properties are shown to characterize the Euclidean plane: the two tangent segments of the unit circle of a normed plane from each point of a Minkowskian disc centered at the origin and having a sufficiently large diameter have equal lengths; the lengths of the tangent segments from each point of a fixed circle centered at the origin are determined only by the radius of this circle. Three further and similar characterizations of the Euclidean plane follow, and in one case a new and natural orthogonality type (*arc-length orthogonality*) is introduced. In addition, a relation between arc-length orthogonality and Birkhoff orthogonality is also used to characterize the Euclidean plane.

The authors of [162] prove that a normed plane is Euclidean if and only if the midpoint of any arc on its unit circle has the same distance to the endpoints of that arc. It is even sufficient to check this for any arc having as length an irrational multiple of the circumference of the unit circle. The authors obtain these results with the help of characterizations of the Euclidean case which are also new. E.g., a normed plane is Euclidean if and only if the length of any (minor) arc of the unit circle can be presented as a function of the distance of its endpoints, this function being independent of the chosen arc. The authors point out that in higher dimensions these results yield characterizations of inner product spaces.

A chord of a Minkowskian ball is called a *minimum chord* with respect to a point p in the ball if it contains p and no other chord through p is shorter (in the norm, of course). In [164] a thorough study of properties of minimum chords in normed planes is given. Motivated by the Euclidean subcase, relationships to various types of orthogonality (such as Birkhoff and James orthogonality) are investigated, and also related characterizations of the Euclidean case within arbitrary, strictly convex or smooth norms, or of smooth norms within arbitrary norms, are presented. But also theorems like this one are proved: A strictly convex normed plane is Euclidean if and only if the common chord of any two circles intersecting in exactly two points is the minimum chord of these two circles with respect to the point of intersection of the chord and the line passing through their centers.

In [166], the following results are derived: Let x be a point from the unit ball B of a Minkowski space, and let $M(x)$ denote the set of midpoints of all chords of the unit sphere S containing x. In the Euclidean situation, $M(x)$ is clearly a homothet of S, but in the general case $M(x)$ might even be non-convex. It is proved that the norm is Euclidean if and only if S and $M(x)$ are homothets. Other properties are also shown to characterize the Euclidean case, e.g. the following one: for any two equally oriented parallel chords $[p, p']$ and $[q, q']$ of S the identity $\|p - p'\| = \|q - q'\|$ holds.

3.5 Metric Problems

As we saw, a norm in a vector space induces a translation-invariant distance. Hence a lot of problems coming from distance geometry can be investigated for normed spaces, and in this section we outline some of the research that has been made in this direction, as well as some problems.

As a first interesting question, we can mention the following problem posed by Oikhberg and Rosenthal in [184]: given a vector space equipped with a metric d, what conditions on d guarantee that d is induced by a norm? They proved the following theorem.

Theorem 3.5.1 *Let X be a vector space over $\mathbb{R}$, and assume that $d : X \times X \to \mathbb{R}$ is a metric on it. If*

(i) *d is translation invariant, that is, $d(x+z, y+z) = d(x, y)$ for any $x, y, z \in X$,*
(ii) *for all $x \in X$ the map $[0, 1] \ni t \mapsto tx \in X$ is continuous (with respect to the topology induced by d), and*
(iii) *each affine one-dimensional subspace of X is isometric to $\mathbb{R}$,*
        then d is induced by a norm.

Remark 3.5.1 The authors proved also a version for vector spaces over $\mathbb{C}$, but this is beyond the scope of the present exposition.

In [217], Šemrl proved that the condition (ii) can be dropped if $\dim X \geq 2$. In [218], the same author proves that conditions (ii) and (iii) can be replaced by

(ii') *for every $x \in X$, the set $\{tx : t \in [0, 1]\}$ is bounded in (X, d),*

(iii') the algebraic midpoint of any points $x, y \in X$ is their metric midpoint.

A set S of points in a Minkowski space $(X, || \cdot ||)$ is called an *equilateral set* (or a 1-*distance set*) if, for some $\lambda > 0$, we have that $||x - y|| = \lambda$ for any distinct $x, y \in S$. Equilateral sets in normed spaces are discussed in Section 4.1 of [214] and Section 5 of [161], and a very broad overview on related problems and results is the exposition [206] by Swanepoel. We denote by $e(X)$ the size of the largest equilateral set in X, and it is easy to see that $e(X) < \infty$ if X is finite dimensional. The number $e(X)$ is called the *equilateral dimension* of X. Petty [190] proved the following upper bound for $e(X)$.

Theorem 3.5.2 *Let $(X, || \cdot ||)$ be an n-dimensional normed space. Then $e(X) \le 2^n$, and equality holds if and only if the unit ball of the norm is a parallelotope. In this case, any equilateral set of size 2^n is the vertex set of some ball.*

To improve this bound is a difficult problem, even for extensively studied families of normed spaces, such as the l_p^n spaces (which are the natural n-dimensional extensions of the two-dimensional l_p spaces discussed in Sect. 3.1). We can mention two problems by Kusner, which are described in [206] (see also [110]).

Question 1 Is $e(l_1^n) = 2n$ for $n \ge 5$?

Question 2 Is $e(l_p^n) = n + 1$ for $p > 2$ $(p \ne 4)$?

In [92], Füredi, Lagarias and Morgan proved that there exist strictly convex normed spaces with equilateral sets whose sizes are at least exponential in the dimension. They also posed the following conjecture.

Conjecture 1 There exists some constant $\varepsilon > 0$ such that if $(X, || \cdot ||)$ is a strictly convex n-dimensional normed space, then $e(X) \le (2 - \varepsilon)^n$.

It is also an interesting problem to find equilateral sets in general Minkowski spaces, providing lower bounds for their equilateral dimensions. It is known that if $\dim(X) \ge 2$, then $e(X) \ge 3$, and the following conjecture was posed in, e.g., [104] and [180].

Conjecture 2 If $(X, || \cdot ||)$ is an n-dimensional normed space, then $e(X) \ge n + 1$.

This is known to be true if $n = 2$ or $n = 3$. For $n \ge 4$, the best known result is due to Brass [53] and Dekster [78], who proved that $e(X) \ge c(\log n)^{1/3}$ for some constant $c > 0$ and n sufficiently large.

Almost nothing is known for infinite dimensional spaces. According to Swanepoel [206], in this case the most important open question is the following.

Question 3 Does any separable infinite-dimensional normed space contain an infinite equilateral set?

Here one should also mention that *2-distance sets* in normed spaces (i.e., sets of points between which only two distances occur) are a promising topic of investigation. For the planar case we refer to the comprehensive work [85].

Somewhat generalizing the problem of finding equilateral sets, Averkov and Düvelmeyer [23] studied *metric capacities* of normed spaces, which are defined to be the largest integer numbers k such that any metric space with k elements can be isometrically embedded in the normed space.

As another problem that can be very naturally studied in the Minkowski context we can mention the classical *Fermat–Torricelli problem*. This can be stated in a normed space $(X, || \cdot ||)$ as follows: let $\{x_1, \ldots, x_m\}$ be a given set of points. What are the points of X minimizing $x \mapsto \sum_{j=1}^{m} ||x - x_j||$? This problem was studied for Minkowski spaces by Durier and Michelot in [83], and by Martini, Swanpoel and Weiss in [157], see also Part II of the monograph [48]. And more generally we should add the reference [122], in which the Fermat–Torricelli problem for normed spaces is extended in even two ways: the distance function is generalized to a gauge (i.e., as convex body the unit ball is no longer necessarily centered at the origin), and the m given objects can be arbitrary convex sets (instead of points). The authors derive numerous analogues to the normed subcase, and also a number of statements holding in the normed subcase, but not in general. On the other hand, one might modify the shape of the searched, point-shaped object (also called "searched facility") of the m given points: if this searched facility is a hyperplane minimizing the distance sum to m given points, we get the *minsum or median hyperplane problem* which, for instance, is important in numerical approximation, computational geometry and robust statistics. And if the searched object is an $(n-1)$-dimensional sphere, the *minsum or median hypersphere problem* is obtained. In the first case, the two halfspaces created by an optimal hyperplane have to satisfy some halving criterion regarding the m given points, and for smooth norms each optimal hyperplane has to be the affine hull of given points, cf. [150]. Clearly, this property makes the problem of searching minsum hyperplanes to a discrete problem. For general norms, there *exists* at least one optimal hyperplane as affine hull of given points. For these and many more (also algorithmical) results in this direction see [149, 151, 201], and [194]. In the second case, only weaker incidence properties of optimal hyperspheres can be proved; we refer to [181] and [135].

Finally, we mention that the fundamental problem in the field of distance geometry, known as the *distance geometry problem* (*DGP*), is usually stated in terms of normed spaces (see, e.g., [143]).

The Distance Geometry Problem Let $G = (V, E, d)$ be a connected simple edge-weighted graph, where $d(u, v)$ denotes the length of the edge $[u, v]$ connecting two vertices $u, v \in V$. For a given $n > 0$, we ask if there is a *realization* $x : V \to (X, || \cdot ||)$ satisfying

$$||x(u) - x(v)|| = d(u, v),$$

for any $[u, v] \in E$.

Distance geometry is a very important research area which has a lot of applications in many areas, such as biology, molecular modeling, material sciences and, more recently, big data. For more on this topic, we refer the reader to the

excellent exposition [142] by Liberti and Lavor, and references therein. See also
the survey [143] by Liberti, Lavor, Maculan and Mucherino. An interesting survey
on combinatorial geometry in normed spaces emphasizing problems which involve
distances is [207]. The author treats problems on unit-distance graphs, minimum-
distance graphs, diameter graphs, minimum spanning trees, and Steiner minimum
trees, leading also to location problems, translative kissing numbers, blocking
numbers, and partition problems (e.g., Borsuk). Historically important references
treating distance geometry are [174] and [46], and also the monograph [80] should
be mentioned here.

3.6 Classical and Differential Theory of Planar Curves

It is a little surprising that there are not too many results on curve theory in normed
planes (and spaces). In this section, we give a landscape of what has been done,
with special emphasis on recent research progress. The content of this section can
be, roughly speaking, divided into two parts: the classical theory and the differential
theory. For the first, the main reference is the expository paper [172], and for the
second we refer the reader to [29], which is also an expository paper.

By a *curve* in a normed plane $(X, || \cdot ||)$ we mean the image of a continuous map
$\gamma : J \to X$, where $J = [a, b] \subseteq \mathbb{R}$ is a bounded closed interval (the same definition
holds for space curves, but we are not interested in them here). Given such a curve,
the norm in the plane provides immediately a *length* of curves defined as

$$l(\gamma) := \sup \left\{ \sum_{j=1}^n ||\gamma(t_j) - \gamma(t_{j-1})|| : a = t_0 < t_1 < \ldots < t_n = b \text{ is a partition of } J \right\}.$$

The curve $\gamma(J)$ is said to be *rectifiable* whenever this is a (finite) number. The
map γ is called a *parametrization* of the curve (sometimes we will abuse of the
notation and call a parametrization a curve). The length of a rectifiable curve is
independent of the parametrization, and every rectifiable curve admits an *arc-length
parametrization*, which is a continuous map $\gamma : [0, c] \to X$ such that $l\left(\gamma|_{[0,t]}\right) = t$
for each $t \in [0, c]$. For a proof of these two claims, we refer the reader to [55]. We
start with some inequalities for curves in normed planes.

A curve $\gamma : [a, b] \to X$ is said to be *closed* when $\gamma(a) = \gamma(b)$, and *simple* if
it has no self-intersections other than that. A *halving pair* of γ is a pair of points
dividing its length into two equal parts. The *geometric dilation* of γ is the number

$$\delta_X(\gamma) := \sup \left\{ \frac{d_\gamma(p, q)}{||p - q||} : p, q \in \gamma, \ p \neq q \right\},$$

where $d_\gamma(p, q)$ is the minimum among the lengths of the two parts of γ determined by p and q. In [163], Martini and Wu gave a lower bound for the geometric dilation of a curve in a Minkowski plane, extending a known result of Euclidean geometry.

Theorem 3.6.1 *The geometric dilation* $\delta_X(\gamma)$ *of a closed, simple and rectifiable curve* γ *in a normed plane* $(X, \|\cdot\|)$ *satisfies the inequality*

$$\delta_X(\gamma) \geq l(S)/4,$$

where $l(S)$ *denotes the length of the unit circle of the norm.*

In the mentioned paper, they also prove that $\delta_X(\gamma)$ is attained for a halving pair. Moreover, they prove that if $(X, \|\cdot\|)$ is strictly convex, then the unique closed, simple, rectifiable curves whose geometric dilation attains the optimal value $l(S)/4$ are the Minkowskian circles. This is not the case if the strict convexity hypothesis is dropped, though. Related to this is the notion of Zindler curve. Namely, the perimeter-halving chords of a planar convex curve C are those chords that half the perimeter of C regarding length; C is called a *Zindler curve* if the perimeter-halving chords all have the same length. Recently, Zindler curves are also studied in computational geometry, e.g. in view of the geometric dilation. In [169], several results on Zindler curves in normed planes are obtained; for example, these curves are characterized in terms of Birkhoff orthogonality. Also the paper [114] should be cited here. The relations between the length of a closed curve and the length of its midpoint curve as well as the length of its image under the so-called halving pair transformation are studied there, yielding a sufficient condition for the geometric dilation of a closed convex curve to be larger than a quarter of the perimeter of the unit circle. In addition, various inequalities between typical quantities of rectifiable (convex) curves C, well known in convex geometry, are derived. E.g., these are: the length of C, its in- and circumradius, its minimal and maximal width (the latter being the diameter), the geometric dilation of C and the length of its midpoint curve. And finally we mention here, of course still missing, the popular notion of *curves of constant width*, see also our final part in this section here. The broad material about this notion is comprehensively collected in the monograph [148], and Chapter 10 is particularly dedicated to such curves in Minkowski planes and higher dimensional analogues.

Much of the early development of Minkowski geometry is related to Busemann's solution to the *isoperimetric problem in Minkowski spaces* (see [58] and [59]). In the planar situation this problem asks which convex bodies have maximal area for a given fixed perimeter. We endow the plane with an area element ω (see Sect. 3.2) and obtain the isomperimetric inequality

$$\frac{l(\partial K)^2}{\lambda(K)} \geq 4\lambda(B_a),$$

for any convex body $K \subseteq \mathbb{R}^2$, where λ denotes the area with respect to ω, and B_a is the unit anti-ball (with the anti-norm obtained via ω, of course). Moreover, equality holds if and only if $K = B_a$.

Still within the classical theory, another topic of research which has been active in the past few years is that of the special classes of curves in normed planes. Since many special curves in the Euclidean plane are (or can be) defined in terms of the Euclidean metric, their definitions can be immediately carried to general normed planes. For example, we can easily define an *ellipse* in a Minkowski plane X as the curve $\mathcal{E}(p, q, c) := \{x \in X : ||x - p|| + ||x - q|| = 2c\}$, for given fixed distinct points $p, q \in X$ (the *foci* of the ellipse), and a suitable number $c > 0$. This and two other definitions of ellipses (each of which extending the Euclidean subcase) were introduced and studied by Martini and Horváth in [119], and it is interesting to notice that they do not necessarily coincide in a normed plane (only two of them, actually). In the mentioned paper, the authors also introduce and study analogues of hyperbolas. In this direction, we refer the reader also to [89] and [213]. In the latter paper, two Euclidean approaches to ellipses (via the sum of distances to the foci, and via the ratio of distances to a focus and the respective directrix) are analogized for norms, and their coincidence is shown to characterize the Euclidean subcase. In [221] metrically defined ellipses in normed planes are investigated; they present strictly convex curves if and only if the norm is strictly convex (and a parallelogram can never occur as such a metric ellipse). Relations to the modulus of smoothness of the norm and characterizations of the Euclidean plane are given. In the paper [137], the Euclidean subcase is characterized by geometric properties of "quadrics" in normed spaces which can be of elliptic, parabolic and hyperbolic type. Treating the Fermat–Torricelli problem in normed spaces, Gross and Strempel [103] investigate the corresponding level curves and surfaces, which occur as *multifocal generalized conics and quadrics*. Connections to numerical analysis and physics are investigated. This setting is even more generalized in [123], e.g. to multifocal elliptic hypersurfaces with respect to gauges in n-dimensional spaces. In a similar way, in [171] Martini and Wu extended the concept of *Cassini curves* to normed planes and the multifocal generalization. And also in [123], Jahn, Martini and Richter studied these curves as bi- and multifocal ones for planes endowed with gauges. In addition, they extended this framework to analogous *bi-* and *multifocal Cassini hypersurfaces* in higher dimensions, and the same is done there for (also higher dimensional) gauge analogues of bi- and multifocal hyperbolas and Apollonius circles.

Now we go to the differential theory, introducing the concepts of *curvature of curves* in normed planes. From now on we assume that all the curves involved (including the unit circle of the plane) are C^2 curves which are *regular* (meaning that each curve admits a parametrization whose first derivative does not vanish). In this case, an arc-length parametrization is simply a parametrization where the tangent vector is unit (in the norm) at each point. To the best of the knowledge of the authors, the first curvature concept in Minkowski planes was introduced by Biberstein in [45]. An early reference to the topic, very complete, is the paper [189] by Petty. Later, these concepts were revisited by Guggenheimer, see [106] and [105].

As more recent developments, Ait-Haddou, Biard and Slawinski used this theory to study general offset curves, and in [95] and [96] Ghandehari extended classical problems on curvature to the Minkowski context. Also, Craizer studied involutes of constant width curves in [72] (in view of discrete differential geometry, the polygonal case was studied in [73]), and Craizer, Teixeira and Balestro investigated cycloids in normed planes in [74]. Finally, Artstein-Avidan, Florentin, Ostrover and Rosen used curvature concepts to study Minkowski billiards in [15].

An idea to "capture" curvature intuition is to measure "how fast" the tangent vector of a curve rotates. In a curve parametrized by arc-length, the tangent vector can be identified within the unit ball, and this rotation can be measured in terms of areas in the unit ball and lengths in the unit circle. This leads to several different definitions of curvature which agree in the Euclidean plane, but not necessarily in an arbitrary Minkowski plane. From now on, denote by $\varphi : \mathbb{R} \bmod 2\lambda(B) \to S$ a positively oriented (meaning $\omega(\varphi, \varphi') > 0$) parametrization of the unit circle by twice the area of sectors (recall that we denote by λ the area given by the fixed area element ω). For a given smooth curve $\gamma : [0, l(\gamma)] \to X$ parametrized by arc-length we clearly may choose a smooth function $u(s) : [0, l(\gamma)] \to \mathbb{R}$ such that

$$\gamma'(s) = \varphi(u(s)), \quad s \in [0, l(\gamma)],$$

and any two of such functions differ by a constant. We define the *Minkowski curvature* of γ at $\gamma(s)$ to be the number

$$k_m(s) := u'(s), \quad s \in [0, l(\gamma)].$$

The intuitive reason for this definition is that it measures "how quick" the unit tangent vector of γ sweeps the area of the unit ball, with respect to the variation of the arc-length (see Fig. 3.8).

Other curvature concepts are obtained similarly. For the so called *normal curvature*, one considers the *right normal field* n_γ of γ, which is the (unique) vector field such that $\gamma' \dashv_B n_\gamma$ and $\omega(\gamma', n_\gamma) = 1$, varying within the unit anti-ball. Letting $\psi : \mathbb{R} \bmod 2\lambda(B_a) \to S_a$ be a parametrization of the unit anti-circle by twice the area of the sectors, we choose a function $v : [0, l(\gamma)] \to \mathbb{R}$ such that $n_\gamma(s) = \psi(v(s))$, and define

$$k_n(s) := v'(s), \quad s \in [0, l(\gamma)]$$

Fig. 3.8 The unit tangent vector of γ sweeps an area within B

to be the *normal curvature* of γ at $\gamma(s)$. Obviously, these curvature concepts agree in the Euclidean plane. In a Minkowski plane, we have immediately the *Frenet formulas*:

$$\gamma''(s) = k_m(s)n_\gamma(s), \text{ and}$$

$$n'_\gamma(s) = -k_n(s)\gamma'(s).$$

We notice carefully from the last equality that $|k_n(s)| = ||n_\gamma(s)||$. None of these curvatures, however, is related to the *radius of curvature* of the curve, defined as the radius of an osculating circle attached to the curve at a given point. Related to that concept is the *circular curvature*, which we define in the following way. Let $\phi : \mathbb{R} \bmod l(S) \to S$ be an arc-length parametrization of the unit circle, and choose a smooth function $t : [0, l(\gamma)] \to \mathbb{R}$ such that

$$\gamma'(s) = \frac{d\phi}{dt}(t(s)).$$

The function t is, therefore, the length traveled along the unit circle when $\gamma'(s)$ varies as its tangent field. The *circular curvature* of γ at $\gamma(s)$ is the number

$$k_c(s) := t'(s),$$

and it is not difficult to see that, whenever it does not vanish, the absolute value of this number is the inverse of the radius of a second order contact Minkowski circle at $\gamma(s)$, called an *osculating circle* (see [29, Section 4]). The circular curvature is dual to the normal curvature, in the sense that the first is the latter in the anti-norm (and vice-versa). In [29], it was introduced a curvature concept which is dual to the Minkowski curvature in this same sense. This is the *arc-length curvature*, and it is defined as follows. We let $\varphi : \mathbb{R} \bmod l(S) \to S$ be an arc-length parametrization of the unit circle, and choose a smooth function $t : [0, l(\gamma)] \to \mathbb{R}$ such that $\gamma'(s) = \varphi(t(s))$. Then the number

$$k_l(s) := t'(s)$$

is the *arc-length curvature* of γ at the point $\gamma(s)$. As one may expect, Euclidean planes are obtained when different curvature types coincide. As an example, we can mention the following theorem (for a proof, we refer the reader to [29, Theorem 5.1]).

Theorem 3.6.2 *If there exists a curve γ in $(X, || \cdot ||)$ whose tangent field goes through every direction of the plane, and whose Minkowski curvature and normal curvature coincide in each point, then the norm $|| \cdot ||$ is Euclidean. The same conclusion holds if we replace these curvature concepts by normal curvature and arc-length curvature.*

The curves of constant curvature can also be characterized. As one may expect from the relation with osculating circles, the curves whose circular (normal) curvature is constant must be an arc of a Minkowski circle (anti-circle) in the plane. Also, Petty showed in [189] that a curve with constant Minkowski curvature must be a homothet of the *centroid curve*, which is the locus of the centroids of the regions determined in the unit ball by the lines through the origin. Of course, the curves with constant arc-length curvature are the translated homothets of the centroid curve of the unit anti-ball. In the mentioned paper, Petty noticed, without a proof, the following theorem (a proof is given in [29, Theorem 6.2]).

Theorem 3.6.3 *If a circle of a normed plane has constant Minkowski curvature, then the plane is Euclidean. By duality, the same holds if an anti-circle has constant arc-length curvature.*

Since in a non-Euclidean Minkowski plane we have that the group SO(2) is not an isometry group, one may expect that the curvature of a curve is not invariant under an orthogonal transformation, and this is indeed true. Consequently, in contrast to the Euclidean case, two curves (parametrized by arc-length) with the same curvature function are not necessarily equal up to translation and an orthogonal transformation mapping the initial tangent vector of one onto the initial tangent vector of the other. However, we still can prove fundamental theorems for the curvature types in the following sense.

Theorem 3.6.4 *Given a C^1 function $k : [0, c] \to X$, there exists a curve $\gamma : [0, c] \to X$ whose (Minkowski, normal, circular or arc-length) curvature in the arc-length parameter is k. Such curve is unique up to the choice of an initial point and an initial tangent vector. Moreover, any curvature type is (up to the sign) invariant under isometries of the plane.*

The sign of the curvature is preserved if and only if the isometry is orientation preserving with respect to the fixed bilinear form ω.

The classical *four-vertex theorem*, which states that the curvature function of a closed, simple, strictly convex curve of class C^2 has at least four extrema (two maxima and two minima), can be extended for every curvature type in a normed plane. Petty proved it for the Minkowski curvature in [189], but his method works also for the other three curvature types (see [29, Section 8]). This theorem was also studied by Tabachnikov in [209] from the viewpoint of contact geometry (for the circular curvature). To the authors best knowledge, the converse of the four vertex theorem, as proved by Gluck in [98], is not yet proved for the Minkowski context.

From the positional viewpoint, a compactness argument gives global maximum and minimun for the radius of curvature of a closed, simple and convex curve γ, and a natural question appears: is γ located in the region bounded by its smallest and greatest circles of curvature? The positive answer comes from the more general following result, which is proved in [29, Theorem 8.3].

Theorem 3.6.5 *Assume that $\gamma, \sigma : S^1 \to X$ are simple, closed, strictly convex curves of class C^2, which are parametrized by the angle of the tangent vector with a* *given fixed direction. If both curves have the same initial point and the same initial tangent directions, and if their circular curvature functions satisfy $k_{c,\gamma} \leq k_{c,\sigma}$, then σ is cointained in the region bounded by γ.*

Notice that the same holds for the largest and smallest anti-circles of normal curvature. The reader may notice that the importance of assuming that the curves have the same initial tangent direction comes due to the geometry of the plane being not invariant under the action of SO(2). This also makes it difficult to obtain some analogue to Schur's theorem (see [188, p. 406]), and as far as the authors know, there is no such analogue yet in the literature. With regard to the relative positions of the largest and smallest circles of curvature we refer the reader to [106], where Guggenheimer proves that a closed, simple and strictly convex curve of class C^2 contains at least two of its circles of curvature.

The very important topic of curves of constant width (see the monograph [148], containing a whole chapter on constant width and related notions in Minkowski geometry, and also the second part of the survey [159]) can also be tackled from the viewpoint of differential theory. A closed, simple and convex curve is said to have *constant width* if the (Minkowski) distance between its pairs of parallel tangent lines is constant. Analogously to the Euclidean case, one can prove that for a given C^2 curve γ of constant width d the sum of the curvature radii at any pair of points with a common tangent direction is equal to d. As a consequence, we obtain that the length $l(\gamma)$ of each such curve satisfies the equality $l(\gamma) = d\frac{l(S)}{2}$. This result is a version of the classical Barbier theorem for normed planes. Still in this direction, the following extension of the Rosenthal-Szasz inequality for Radon planes was proved in [28].

Theorem 3.6.6 *Let X be a Radon plane and $\gamma : S^1 \to X$ be a simple, closed and convex curve. Then we have the inequality*

$$l(\gamma) \leq \text{diam}\{\gamma\} \cdot \frac{l(S)}{2},$$

where $\{\gamma\}$ denotes the range of γ and $\text{diam}\{\gamma\} = \max\{\|p - q\| : p, q \in \{\gamma\}\}$ is its diameter.

An analogous result is derived for general norms, but there the perimeters are measured in the antinorm.

Among other topics in the differential geometry of curves in normed planes, we can mention that evolutes and involutes of curves have been studied, and that they have a lot of similarities with the Euclidean subcase. For more in this direction, we refer the reader to [29, Section 9], [72] and [74]. Also, discrete analogues of these concepts were studied in [73] and [75], and an extension of circular curvature for *Legendre curves* was investigated in [33].

3.7   Differential Geometry of Surfaces

The differential geometry of surfaces immersed in a normed space $(\mathbb{R}^3, || \cdot ||)$ is related to Finsler geometry in the same way as the classical differential geometry of surfaces is related to Riemannian geometry. Since the 1950's there have been some sparse contributions to this topic, which are due to Busemann [60] and Guggenheimer [105], for example. In the mentioned paper, Busemann states that

> The exploration of Minkowski (or finite dimensional Banach) spaces is a necessary preliminary step for any progress in the theory of Finsler spaces. They are the local spaces which belong to any finitely compact space with geodesics and a minimum of differentiability properties. Therefore they are natural geometric objects to study.

Recently, an attempt to construct a systematic theory of differential geometry in normed spaces has been made. This was done by Balestro, Martini and Teixeira in the papers [35–37] and [40] (regarding another approach to the differential geometry of surfaces in Minkowski spaces we refer to [118]). The novelty in this approach is the construction of an analogue of the Gauss map using the Birkhoff orthogonality concept. Assume that $|| \cdot ||$ is a norm in $\mathbb{R}^3$ whose unit circle ∂B is a smooth surface with strictly positive Gaussian curvature (in the Euclidean sense), and let M be an orientable surface immersed in $(\mathbb{R}^3, || \cdot ||)$. The *Birkhoff–Gauss map* of M is a smooth map $\eta : M \to (\mathbb{R}^3, || \cdot ||)$ such that $\eta(p)$ is a (choice of a) vector which is unit and Birkhoff orthogonal to $T_p M$ (observe that there are exactly two possible choices of η).

In the language of affine differential geometry (see [182]), this is an *equiaffine vector field*. The Gauss equation reads

$$D_X Y = \nabla_X Y + h(X, Y)\eta, \qquad (3.7.1)$$

and since η is equiaffine, the derivative $D_X \eta$ is always tangential. Because of that, the image of the differential $d\eta_p$ is a subspace of $T_{\eta(p)} \partial B$, and the natural identification $T_p M \simeq T_{\eta(p)} \partial B$ allows us to regard $d\eta_p$ as an endomorphism of $T_p M$. Moreover, each differential map $d\eta_p$ is self-adjoint, and hence has two real eigenvalues, which are called the *(Minkowski) principal curvatures* of M at p (the corresponding eigenvectors are called the *principal directions* of M at p). Also extending concepts of classical differential geometry, we define the *(Minkowski) Gaussian curvature* and *(Minkowski) mean curvature* of M at p as

$$K(p) := \det(d\eta_p) \text{ and}$$

$$H(p) := \frac{1}{2}\text{tr}(d\eta_p),$$

where "det" and "tr" denote the usual determinant and the trace, respectively. We also endow M with a coherent *area element*, which is the 2-form defined as

$$\omega(X, Y) = \det(X, Y, \eta).$$

We call this the *Minkowski* area element. There is an analogue of normal curvature which can be constructed in a geometric way. A little surprising, this concept relates to the principal curvatures as in the Euclidean case. Let $p \in M$, $X \in T_pM$ be a non-zero vector, and denote by $\mathrm{pl}(\eta, X)$ the plane spanned by η and X. The translation of $\mathrm{pl}(\eta, X)$ passing through p intersects M in a curve which we will call γ. The *(Minkowski) normal curvature* of M at p and in the direction X is the circular curvature of γ at p measured in the planar geometry whose unit ball is $B \cap \mathrm{pl}(\eta, X)$. We denote this number as $k_{M,p}(X)$. The mentioned relation between Minkowski normal curvatures and Minkowski principal curvatures is the following result, which is proved in [36].

Proposition 3.7.1 *Assume that $\lambda_1 \leq \lambda_2$ are the principal curvatures of M at p. Then $\lambda_1 = \min_{X \in T_pM \setminus \{0\}} k_{M,p}(X)$ and $\lambda_2 = \max_{X \in T_pM \setminus \{0\}} k_{M,p}(X)$.*

A point $p \in M$ is said to be *umbilic* if the principal curvatures of M at p are equal or, equivalently, if the normal curvature of M is constant at p. As a similarity between the Euclidean case and the general Minkowski case we have the following result.

Proposition 3.7.2 *If M is a connected surface all whose points are umbilic, then M is contained in a plane or in a Minkowski sphere.*

One can also prove that a compact, connected immersed surface without boundary of constant Minkowski Gaussian curvature is a Minkowski sphere, but only under an additional hypothesis (see [36, Theorem 6.1]). Moreover, the following (completely analogous) version of Alexandrov's theorem for the Minkowski context was proved in [37].

Theorem 3.7.1 (Minkowskian Version of Alexandrov's Theorem) *An embedded compact surface without boundary whose Minkowski mean curvature is constant is a Minkowski sphere.*

One of the tools of the proof of this theorem is a Minkowskian version of the classical divergence theorem which uses projections on the outward-pointing Birkhoff normal vector field and the Minkowski area element, instead of the Euclidean counterparts. In the mentioned paper, several other global theorems are stated and proved. As an example, a lower bound for Willmore's energy functional is obtained in terms of the area of the unit sphere of the norm.

Theorem 3.7.2 *If M is a compact surface without boundary immersed in $(\mathbb{R}^3, ||\cdot||)$, then*

$$\int_M H^2 \omega \geq \lambda(\partial B),$$

where $\lambda(\partial B)$ is the Minkowski area of the unit sphere. Equality holds if and only if M is a Minkowski sphere.

Since we are assuming that ∂B is a smooth surface with strictly positive Gaussian curvature, we have that at each point $p \in \partial B$ the Dupin indicatrix of $T_p \partial B$ is an ellipse. This ellipse can be regarded as the unit circle of a norm derived from an inner product, and hence the identification between the tangent spaces of M with tangent spaces of ∂B induces a Riemannian metric on M. We call it the *Dupin metric* of M. It turns out that this metric "organizes" the directions in $T_p M$ in a sense made explicit below.

Theorem 3.7.3 *Let $p \in M$ be a non-umbilic point, and let $X, Y \in T_p M$ be non-zero vectors. Then*

$$k_{M,p}(X) + k_{M,p}(Y) = 2H$$

if and only if X and Y are orthogonal or complementary in the Dupin metric. Also, the Minkowski mean curvature can be calculated as the mean

$$H(p) = \frac{1}{2\pi} \int_{\mathcal{E}} k_{M,p},$$

where $\mathcal{E}$ is the Dupin indicatrix in $T_{\eta(p)} \partial B \simeq T_p M$.

Denote by ξ the Euclidean Gauss map of M. For some purposes, it is better to consider the so called *weighted Dupin metric*, which is obtained by dividing, at each $p \in M$, the Dupin metric by the factor $\langle \eta(p), \xi(p) \rangle$, where $\langle \cdot, \cdot \rangle$ is the usual inner product in $\mathbb{R}^3$. This metric is usually denoted by $b(\cdot, \cdot)$. The b-Hessian of the weighted Dupin metric is defined as

$$\text{hess}_b f(X, Y) = X(Yf) - (\hat{\nabla}_X Y)f,$$

for any smooth function $f : M \to \mathbb{R}$ and smooth sections X, Y of TM (notice that it actually depends only on the vectors, since the expression is tensorial). This concept can be used to characterize the Minkowski normal curvature as follows.

Theorem 3.7.4 *Given a fixed point $p \in M$, let $g : M \to \mathbb{R}$ be the map which associates to each $q \in M$ the distance in the norm from q to $T_p M$. If $X \in T_p M$ is unit in the metric b, then*

$$k_{M,p}(X) = \text{hess}_b g(X, X),$$

where the b-Hessian is calculated at p. Moreover, if $a \in \mathbb{R}^3 \setminus M$ and $D_a : M \to \mathbb{R}$ is the function defined as $D_a(q) = ||q - a||$, then we have the equivalence

$$\text{hess}_b D_a(X, X) = 0 \Leftrightarrow k_{M,p}(X) = \frac{1}{D_a(p)},$$

for each critical point p of D_a and $X \in T_p M$.

We refer the reader to [35] for the proof of this theorem, and also for related discussions. We also mention a further theorem from this paper. The *affine distance function* $\rho : M \to \mathbb{R}$ with respect to a given point $a \in \mathbb{R}^3 \setminus M$ is defined by the decomposition $p - a = \rho(p)\eta(p) + V(p)$, where V is a section of TM (see [182]).

Theorem 3.7.5 *Assume that M is nondegenerate, in the sense that the bilinear form h defined in (3.7.1) has rank 2. If $\rho : M \to \mathbb{R}$ is an affine distance function, then*

$$\Delta\rho = 2(H\rho - 1),$$

where Δ is the Laplacian operator defined in M as

$$\Delta f = \mathrm{div}_\nabla(\mathrm{grad}_h f) = \mathrm{tr}\{Y \mapsto \nabla_Y(\mathrm{grad}_h f)\},$$

with $\mathrm{grad}_h f$ being the vector field defined by $h(\mathrm{grad}_h f, \cdot) = df(\cdot)$.

A surface is said to be *minimal* if its Minkowski mean curvature vanishes everywhere. Thus the theorem above characterizes Minkowski minimal surfaces in terms of the elliptic differential equation $\Delta\rho = -2$, where ρ is the affine distance function from a given fixed point. There are other characterizations of Minkowski minimal surfaces which are similar to the Euclidean case. Maybe the most interesting of all are the following ones.

Theorem 3.7.6 *A surface is minimal if and only if it is (locally) a critical point of any (Minkowski) area variation.*

Theorem 3.7.7 *An immersed surface $f : M \to \mathbb{R}^3$ is minimal if and only if the coordinate functions of the immersion are harmonic with respect to the Laplace–Beltrami operator of the weighted Dupin metric.*

For a proof of these theorems, and also for other properties and characterizations of Minkowski minimal surfaces, we refer the reader to [40].

Of course, one can pose a lot of questions about the differential geometry of surfaces in normed spaces. At large, for example, an investigation of the case for hypersurfaces immersed in normed spaces with arbitrary dimension is yet to be done. Also, there is not much known about the Finsler metric induced by the ambient norm, and its connections with curvature. For advancing in this direction, it might be necessary to adapt the tools from Finsler geometry to this Minkowski context. As an example, doing that allows us to guarantee the existence of a smooth geodesic flow.

Two problems posed by Schäffer [199], which somehow might be related to this theory, are worth mentioning. The *girth* of a normed space is the infimum of the lengths of all closed, rectifiable, centrally symmetric curves on its unit sphere. The *perimeter* of a normed space is twice the supremum of the lengths of the curves connecting antipodal points on its unit sphere. In the remarkable paper [12], Álvarez-Paiva settled one of Schäffer's conjectures using methods from symplectic and Finsler geometries. He proved the following theorem.

Theorem 3.7.8 *The girth of a normed space is equal to the girth of its dual space.*

The other conjecture posed by Schäffer is, to the best knowledge of the authors, still open and states that the perimeter of any normed space whose dimension is greater than 2 is less than or equal to 2π. In this direction, the following estimate is proved in [40].

Theorem 3.7.9 *Let* $\| \cdot \|$ *be a norm in* $\mathbb{R}^3$ *whose unit sphere is a smooth surface with strictly positive usual Gaussian curvature* $K_{\partial B}$, *and denote the perimeter of* $(\mathbb{R}^3, \| \cdot \|)$ *by* $\mathrm{per}(\partial B)$. *Then we have the estimate*

$$\mathrm{per}(\partial B) \leq \frac{2\pi}{\sqrt{m}},$$

where $m = \inf_{p \in \partial B} K_{\partial B}(p)$.

We finish this section by mentioning that, recently, Burago and Ivanov made some very interesting progress concerning geodesics of surfaces in Minkowski spaces. In [57], they prove (among other results) that geodesics on *saddle surfaces* (which can be defined as surfaces having negative Minkowski Gaussian curvature) have no conjugate points. In [56] the same authors prove that every two-dimensional disc in a normed space is an area-minimizing surface among all immersed discs with the same boundary (with respect to Holmes–Thompson surface area, see [214]). Also the survey [63] is related to Holmes–Thompson measure in this framework. Its authors discuss different concepts of mean curvature in smooth, strictly convex Minkowski spaces, exploring various natural definitions of mean curvature as first variation of area. They show advances and disadvances of such definitions based on Hausdorff and Holmes–Thompson notions of volume. Last, we highlight that the second variation of the Minkowskian surface area of a immersed surface was studied by Silva in his PhD thesis [202]. Some applications to stability results were also derived.

3.8 Minkowski Billiards

One of the most interesting research directions related to Minkowski geometry is the study of *Minkowski billiards* whose geometry is ruled by an arbitrary convex body. This was first studied by Gutkin and Tabachnikov in [109], but we shall discuss here the more recent work by Artstein-Avidan, Ostrover and other researchers, starting by the definitions given in [18]. A nice fundamental reference treating billiards is [210].

Let $K \subseteq \mathbb{R}^n_q$ and $T \subseteq \mathbb{R}^n_p$ (where the subscripts display how we denote the coordinates of each space) be two convex bodies with smooth boundaries, and assume that T contains the origin in its interior. Denote by $g_T : \mathbb{R}^n_p \to \mathbb{R}$ and by $h_T : \mathbb{R}^n \to \mathbb{R}$ the gauge funcional and the support function of T, respectively (and recall that, since $0 \in \mathrm{int}(T)$, we have $h_T = g_{T^\circ}$). We consider the unit cotangent bundle

$$U_T^* K = K \times T = \{(q, p) : q \in K, \text{ and } g_T(p) \le 1\} \subseteq T^* \mathbb{R}_q^n = \mathbb{R}_q^n \times \mathbb{R}_p^n,$$

which is a smooth manifold with corners. A *closed* (K, T)-*billiard trajectory* is the image of a piecewise smooth curve $\gamma : S^1 \to \partial(K \times T)$ whose velocity vector is described as follows. If $t \notin \mathcal{B}_\gamma := \{t \in S^1 : \gamma(t) \in \partial K \times \partial T\}$, then

$$\dot{\gamma}(t) = c v(\gamma(t)),$$

for some positive constant c, where v is the vector field given by

$$v(q, p) = \begin{cases} (-\nabla g_T(p), 0), & \text{if } (q, p) \in \mathrm{int}(K) \times \partial T \\ (0, \nabla g_K(q)), & \text{if } (q, p) \in \partial K \times \mathrm{int}(T) \end{cases},$$

and if $t \in \mathcal{B}_\gamma$, the left and right derivatives of $\gamma(t)$ exist, and

$$\dot{\gamma}^\pm(t) \in \{\alpha(-\nabla g_T(p), 0) + \beta(0, \nabla g_K(q)) : \alpha, \beta \ge 0, \text{ and } (\alpha, \beta) \ne (0, 0)\}.$$

The points of $\mathcal{B}_\gamma$ are called the *bouncing points*. The geometric interpretation, illustrated in [18, Figure 1], is that the trajectory travels within $\mathrm{int}(K)$ in a direction given by some outer normal of T until it reaches the boundary of K (at a point q_1, say). Then, the trajectory travels within $\mathrm{int}(T)$ in the direction of the outer normal of K at q_1 until it reaches the boundary ∂T (at the point p_1, say). After this, the trajectory returns to travel in $\mathrm{int}(K)$, but now following the direction of the outer normal of T at p_1 (in the opposite orientation, because of the definition of v above). This process is repeated until the trajectory "closes". Notice that by this definition the roles of K and T are clearly dual. However, we shall interpret it as a billiard trajectory where K is the billiard table, and the geometry of the billiard is governed by T.

A closed trajectory is said to be a *proper trajectory* if the number of bouncing points is finite. That is, once the trajectory reaches a point of $\partial K \times \partial T$, it immediately exits it. Also, a trajectory is said to be a *gliding trajectory* if $\mathcal{B}_\gamma = S^1$, i.e., γ travels only along $\partial K \times \partial T$. In [18], the authors prove that every closed (K, T)-billiard trajectory is either proper or gliding.

Recall that $K \times T$ is a subset of $\mathbb{R}^{2n}$ whose boundary is a convex, non-smooth hypersurface. Let ω be the standard symplectic structure in $\mathbb{R}^{2n}$, defined as $\omega = \sum dq \wedge dp$. If $\Sigma \subseteq \mathbb{R}^{2n}$ is a subset whose boundary $\partial \Sigma$ is a closed connected smooth hypersurface, then $\ker(\omega|_{\partial \Sigma})$ is a 1-dimensional bundle over $\partial \Sigma$, and the trajectories of this bundle (i.e., the curves which are tangent to a direction of the bundle at each point) are called the *characteristics* of $\partial \Sigma$. A *closed characteristic* of $\partial \Sigma$ is a characteristic which is an embedded circle, and the *action* $A(\gamma)$ of a closed curve $\gamma : S^1 \to \partial \Sigma$ is defined as $A(\gamma) = \int_\gamma \lambda$, where $\lambda = pdq$ is the *Liouville form* (notice that $\omega = -d\lambda$). The *action spectrum* of Σ is the set

$$\mathcal{L}(\Sigma) = \{|A(\gamma)| : \gamma \text{ is a closed characteristic on } \partial \Sigma\},$$

and the infimum of $\mathcal{L}(\Sigma)$ is closely related to certain *symplectic capacities* of Σ. A *symplectic capacity* on $(\mathbb{R}^{2n}, \omega)$ is a functor c which associates to each subset $U \subseteq \mathbb{R}^{2n}$ a number $c(U) \in [0, \infty]$ with the following properties:

(i) $c(U) \le c(V)$ if $U \subseteq V$,
(ii) $c(\varphi(U)) = |\alpha| c(U)$ if $\varphi \in \mathrm{Diff}(\mathbb{R}^{2n})$ is such that $\varphi^* \omega = \alpha \omega$ for some $\alpha \in \mathbb{R}$ and
(iii) $c(B^{2n}(r)) = c(B^2(r) \times \mathbb{C}^{n-1}) = \pi r^2$,

where $B^2(r) \times \mathbb{C}^{n-1}$ is the symplectic cylinder $B^2(r) \times \mathbb{C}^{n-1} := \{(q, p) \in \mathbb{R}^{2n} : q_1^2 + p_1^2 \le r^2\}$. For a survey on the theory of symplectic capacities, see [69]. We are concerned with two particular symplectic capacities: the Ekeland–Hofer capacity c_{EH}, and the Hofer–Zehnder capacity c_{HZ}. For them, we have the following

Theorem 3.8.1 (Theorem 2.2 in [18]) *If $\Sigma \subseteq \mathbb{R}^{2n}$ is a convex body whose boundary $\partial \Sigma$ is a smooth hypersurface, then there exists a closed characteristic γ^* on $\partial \Sigma$ such that*

$$c_{EH}(\Sigma) = c_{HZ}(\Sigma) = A(\gamma^*) = \mathcal{L}(\Sigma).$$

In particular, the Ekeland–Hofer and the Hofer–Zehnder capacities coincide in smooth convex bodies, and for that reason and convenience, we shall speak from now on about the Ekeland–Hofer–Zehnder capacity. In [18], the definition of closed characteristics is properly extended to the class of arbitrary (not necessarily smooth) convex bodies, as well as the Ekeland–Hofer–Zehnder capacity (this extension is denoted by $\tilde{c}_{EHZ}$ in the mentioned paper). It turns out that the proper closed (K, T)-billiard trajectories are the (extended) closed characteristics of $\partial(K \times T)$. A way to interpret that, at least in the "smooth" part, is that in a point $(q, p) \in \mathrm{int}(K) \times \partial T$, then we clearly have that $\ker(\omega|_{\partial(K \times T)})$ at (q, p) is the line in the direction of $(-\nabla g_T(p), 0)$, and something similar occurs in the other smooth portion of $\partial(K \times T)$. We get

$$\tilde{c}_{EHZ}(K \times T) = \min\{A(\gamma) : \gamma \text{ is a } (K, T)-\text{billiard trajectory}\}.$$

Moreover, the action of the projection in K of a (K, T)-billiard trajectory is precisely its length in terms of the support function h_T. Indeed, if the trajectory travels from $q_0 \in K$ to $q_1 \in K$ in the direction of the outer normal of T at $p_0 \in \partial T$, we have

$$A(\gamma|_{q_0 \to q_1}) = \int_{t_0}^{t_1} p(t)\dot{q}(t)\, dt = p_0(q_1 - q_0) = h_T(q_1 - q_0).$$

Based on that, in [18] the authors prove the following Brunn–Minkowski inequality for the length $\xi_T(K)$ of the shortest closed (K, T)-billiard trajectory:

$$\xi_T(K_1 + K_2) \geq \xi_T(K_1) + \xi_T(K_2),$$

for any smooth convex bodies $K_1, K_2, T \in \mathbb{R}^n$. In the Euclidean case, obtained when T is the Euclidean unit ball B, they obtained the bounds

$$\xi_B(K) \leq C\sqrt{n}\text{Vol}(K)^{1/n},$$

for some constant $C > 0$, and

$$\xi_B(K) \leq 2(n + 1)r_{in}(K),$$

where $r_{in}(K)$ denotes the *inradius* of K, that is, the radius of the largest ball inscribed in K.

This approach was used by Artstein-Avidan, Karasev and Ostrover in the beautiful paper [17] to prove that a particular case of the Viterbo conjecture (also known as symplectic isoperimetric conjecture) implies the long-standing Mahler conjecture. We describe both conjectures below.

Conjecture 3 (Viterbo Conjecture) For any symplectic capacity c and any convex body $K \subseteq \mathbb{R}^{2n}$,

$$\frac{c(\Sigma)}{c(B)} \leq \left(\frac{\text{Vol}(\Sigma)}{\text{Vol}(B)}\right)^{1/n},$$

where B is the unit ball of the Euclidean metric in $\mathbb{R}^{2n}$.

Conjecture 4 (Mahler Conjecture) Let K be a centrally symmetric convex body in $\mathbb{R}^n$, and let K° denote its dual body. Then

$$\text{Vol}(K \times K^\circ) \geq \frac{4^n}{n!}.$$

Notice that since the volume is continuous and the smooth convex bodies are dense in the Hausdorff metric, we get that it is sufficient to prove Mahler's conjecture for smooth convex bodies.

What the authors remarkably proved in [17] is the following theorem.

Theorem 3.8.2 *If the Viterbo conjecture is true for the Hofer–Zehnder capacity when $\Sigma = K \times K^\circ$ for some smooth convex body $K \subseteq \mathbb{R}^{2n}$, then the Mahler conjecture is true.*

Idea of the Proof The characterization of the Hofer–Zehnder capacity in terms of closed billiard trajectories is used to prove that for any smooth centrally symmetric convex body $K \subseteq \mathbb{R}^n$ the following equality holds:

$$c_{HZ}(K \times K^{\circ}) = 4.$$

Notice that, in this particular case, the length of (the projection of) a billiard trajectory is calculated with respect to the Minkowski norm induced by K as unit ball. Indeed, the Hofer–Zehnder capacity of $K \times T$ equals the length of the minimum closed billiard trajectory, where the length is calculated with respect to the support function h_T. In the case where K is centrally symmetric and $T = K^{\circ}$, we get immediately $h_T = \|\cdot\|_K$.

Now, assuming that Viterbo's conjecture is true for the described particular case, we get

$$4^n = c_{HZ}(K \times K^{\circ})^n \leq c_{HZ}(B^{2n}(1))^n \frac{\text{Vol}(K \times K^{\circ})}{\text{Vol}(B^{2n}(1))}$$

$$= \pi^n \frac{\text{Vol}(K \times K^{\circ})}{\text{Vol}(B^{2n}(1))} = n!\text{Vol}(K \times K^{\circ}),$$

since $\text{Vol}(B^{2n}(1)) = \pi^n/n!$. $\qquad\qquad\qquad\qquad\qquad\qquad\qquad\qquad\qquad\quad$ □

Still regarding Minkowski billiards, it is worth mentioning that in [15] Artstein-Avidan, Florentin, Ostrover and Rosen studied *caustics* for Minkowski billiards, which are convex bodies with the property that, once a trajectory is tangent to it, it remains tangent after every bouncing in the boundary.

It is also worth mentioning that in [39] the closed characteristics of a smooth convex hypersurface of $\mathbb{R}^{2n}$ (endowed with a *gauge*, i.e., an asymmetric norm) were characterized as being the optimal cases of a certain isoperimetric inequality involving the geometry given by the gauge.

3.9   Other Topics

As already said, various contents of this exposition were selected since they are close to the interests of the authors. In this section, we discuss some interesting research directions that have not been addressed in the previous sections.

The first topic we will shed some light on is that of *geometric constants in normed spaces*. Such constants can be used to measure (or even to "capture") geometric properties of norms, such as how different orthogonality types really are, or how "far from being Euclidean" a certain norm is. Maybe the most studied geometric constants in normed spaces are the *von Neumann–Jordan constant* defined as

$$c_{NJ}(X, ||\cdot||) = \sup\left\{ \frac{||x+y||^2 + ||x-y||^2}{2(||x||^2 + ||y||^2)} : x, y \in X \setminus \{0\} \right\},$$

and the *James constant*, defined as

$$c_J(X, ||\cdot||) = \sup\{\min(||x+y||, ||x-y||) : x, y \in S_X\},$$

where S_X denotes the unit sphere of $(X, ||\cdot||)$. The first one was introduced by Clarkson in [70] (inspired by [130]), and the second one was introduced by Gao and Lau in [94]. These constants (and also some generalizations) were intensively studied in papers such as [76, 128, 131, 132, 134] and [211]. Also, geometric constants named after Ptolemy and Zbăganu were studied in [144]. It is worth mentioning that some of these constants are linked to fixed point properties.

In [32] and [34] geometric constants were defined to measure "how non-Euclidean" or "how non-Radon" a normed plane can be. Also, geometric constants that quantify the difference between orthogonality types were studied by Ji and Wu in the paper [127], by Papini and Wu in [186], and also in the paper [30]. Moreover, Chmieliński and Wójcik [68] introduced a geometric constant based on "how symmetric" Birkhoff orthogonality is in a normed space.

Another topic that has seen recent activity is the one of *angles and trigonometric functions in normed spaces* (see the survey [27]). It is clear that there is no immediate natural definition of angular measures in general normed spaces, and a reason is that the geometry induced by a general norm is anisotropic (in the sense that it is not invariant through rotations). We can mention two ways of introducing angle measures in normed spaces. The first of them, which seems to have been introduced for normed planes by Brass [52] when studying distance problems in normed spaces, considers a Borel measure μ in the unit circle S satisfying:

(i) $\mu(S) = 2\pi$,
(ii) for any Borel set $A \subseteq S$, $\mu(-A) = \mu(A)$, and
(iii) for each $p \in S$, $\mu(\{p\}) = 0$.
    Studying bisectors of angles, Düvelmeyer [84] added an additional non-degeneracy hypothesis:
(iv) A non-degenerate arc of the unit circle has positive measure.

As easy examples of measures of this kind, we have the length μ_l (in the norm) of arcs and the corresponding sector areas μ_a. It is known that μ_a is proportional to μ_l in the anti-norm (see [160] for a proof). Consequently, in a Radon plane the length of an arc is always proportional to the corresponding sector area. However, this is true not only for Radon norms, but for all norms whose unit circle is an *equiframed curve*, as proved in [84] (see also [158]). In [88] and [87], Fankhänel studied angle measures which have "good behavior" with respect to orthogonality types. Also, a similar approach was adopted by Böröczky and Soltan [50] for studying "snake packings" of convex domains.

The "somewhat standard" second way to define angle measures in normed spaces is to give immediate metric definitions. In this direction, we can mention the papers [175] by Miličić, [90] and [82] by Diminnie, Andalafte and Freese, [108] by Gunawan, Lindiarni and Neswan, [79] by Dekster, and, more recently, [183] by Obst. Regarding all these approaches for angles in normed planes and spaces we refer the reader to the recent survey [27]. Related to that, extensions of trigonometric functions for Minkowski spaces were studied by Szostok in [208], and also in the papers [32] and [41]. Further on, in [25] rotations were generalized to normed planes, and analogues of the Euler-Savary equations for roulettes were obtained, see also [26].

Another fruitful topic in Minkowski geometry consists of special classes of convex bodies in normed spaces, e.g., the class of *bodies of constant width*. Since the definition of these bodies in normed spaces is based on the constancy of the respective width function, Minkowski addition can be used to define them alternatively: A convex body K in a Minkowski plane is said to be of constant width if the vector sum $K + (-K)$ is a homothetical copy of the unit ball of the space. The knowledge on such bodies and related classes in Minkowski spaces can be taken from Section 2 of the survey [159] and from Chapter 10 of the monograph [148], where also various open problems are collected. We will mention three of them referring to the related class of reduced convex bodies in normed spaces (for a related survey we refer to [138]). A convex body in an n-dimensional normed space is said to be a *reduced convex body* if it does not properly contain a convex body of the same minimal width. In some dual sense, a convex body in such a space is called a *complete set* (or diametrically complete) if any proper superset of it has a larger diameter. (We note that a good survey on complete sets in normed spaces is missing; but see at least the pages 231–235 in [148]!) It is well known that in the Euclidean subcase completeness and constant width are even equivalent notions. In Minkowski spaces this is true only in the planar situation; in higher dimensions constant width still implies completeness, but the norms where the mentioned equivalence holds (called "perfect norms") are still not characterized! It is also known that any normed space contains reduced bodies that are not complete (see [196]), and that non-planar complete bodies need not be reduced (cf. [173]). Regarding the intersection of these set families one might ask: Is a convex body being complete and reduced with respect to some norm (or even gauge) necessarily of constant width? In [51] it is shown that this implication holds for simplices and also for all convex bodies having a smooth extreme point! The second problem is somehow dual to the famous isodiametric problem (which treats bodies of given diameter in a normed space which have maximal volume, where the ball of respective diameter is the unique solution): Which convex bodies of fixed minimal width have, for any Minkowski space, minimal volume? (There are certainly normed spaces for which the answer is easy, but the authors are not aware of a thorough study of this problem. In particular, the problem is open even in the non-planar Euclidean situation.) And the third problem is also open in the non-planar Euclidean situation (see Chapters 7 and 10 of [148]): Is a strictly convex reduced body necessarily of constant width? In the planar case this is confirmed, and replacing strict convexity by smoothness, the answer is "yes" in all dimensions. It is also interesting that in non-planar Minkowski spaces there exist reduced bodies of arbitrarily large ratio between diameter and minimal width (a property that complete sets cannot share, see again [196]).

The last topic we mention is that of the long standing covering conjecture of Hadwiger: given a convex body $K \subseteq \mathbb{R}^n$, denote by $c(K)$ the least number of translates of int(K) needed to cover K.

Conjecture 5 (Hadwiger's Covering Conjecture) For each convex body $K \subseteq \mathbb{R}^n$ we have $c(K) \leq 2^n$, and equality holds if and only if K is a parallelotope.

For more about this conjecture we refer the reader to the books [47] by Boltyanski, Martini and Soltan and [54] by Brass, Moser and Pach, as well as to the paper [227] by Zong. Interpreting K as unit ball of a gauge or, in the centrally symmetric case, of a norm, the relation of this conjecture to Minkowski geometry is obvious, see also [116] and [222].

References

1. J. Alonso, C. Benítez, Orthogonality in normed linear spaces: a survey. I. Main properties. Extracta Math. **3**, 1–15 (1988)
2. J. Alonso, C. Benítez, Orthogonality in normed linear spaces: a survey. II. Relations between main orthogonalities. Extracta Math. **4**, 121–131 (1989)
3. J. Alonso, H. Martini, M. Spirova, Minimal enclosing discs, circumcircles, and circumcenters in normed planes (Part I). Comput. Geom. **45**(5–6), 258–274 (2012)
4. J. Alonso, H. Martini, M. Spirova, Minimal enclosing discs, circumcircles, and circumcenters in normed planes (Part II). Comput. Geom. **45**(7), 350–369 (2012)
5. J. Alonso, H. Martini, M. Spirova, Discrete geometry in Minkowski spaces, in *Discrete Geometry and Optimization*. Fields Inst. Commun., vol. 69 (Springer, New York, 2013), pp. 1–15
6. J. Alonso, H. Martini, M. Spirova, On reduced triangles in normed planes. Results Math. **64**(3–4), 269–288 (2013)
7. J. Alonso, H. Martini, S. Wu, On Birkhoff orthogonality and isosceles orthogonality in normed linear spaces. Aequat. Math. **83**(1–2), 153–189 (2012)
8. J. Alonso, H. Martini, S. Wu, Orthogonality types in normed linear spaces, in *Surveys in Geometry*, ed. by A. Papadopoulos (2021)
9. C. Alsina, P. Guijarro, M.S. Tomás, Some remarkable lines of triangles in real normed spaces and characterizations of inner product structures. Aequat. Math. **54**(3), 234–241 (1997)
10. C. Alsina, J. Sikorska, M. S. Tomás, *Norm Derivatives and Characterizations of Inner Product Spaces* (World Scientific Publishing, Hackensack, 2010)
11. C. Alsina, M. S. Tomás, Orthocenters in real normed spaces and the Euler line. Aequat. Math. **67**(1–2), 180–187 (2004)
12. J.C. Álvarez Paiva, Dual spheres have the same girth. Am. J. Math. **128**(2), 361–371 (2006)
13. J.C. Álvarez Paiva, A.C. Thompson, Volumes on normed and Finsler spaces, in *A Sampler of Riemann–Finsler Geometry*. Math. Sci. Res. Inst. Publ., vol. 50 (Cambridge Univ. Press, Cambridge, 2004)
14. D. Amir, *Characterizations of Inner Product Spaces*. Operator Theory: Advances and Applications, vol. 20 (Birkhäuser, Basel, 1986)
15. S. Artstein-Avidan, D.I. Florentin, Y. Ostrover, D. Rosen, Duality of caustics in Minkowski billiards. Nonlinearity **31**(4), 1197–1226 (2018)

16. S. Artstein-Avidan, A. Giannopoulos, V. Milman, *Asymptotic Geometric Analysis. Part I. Mathematical Surveys and Monographs*, vol. 202 (American Mathematical Society, Providence, 2015)
17. S. Artstein-Avidan, R. Karasev, Y. Ostrover, From symplectic measurements to the Mahler conjecture. Duke Math. J. **163**, 2003–2022 (2014)
18. S. Artstein-Avidan, Y. Ostrover, Bounds for Minkowski billiard trajectories in convex bodies. Int. Math. Res. Not.(IMRN) **2014**(1), 165–193 (2014)
19. E. Asplund, B. Grünbaum, On the geometry of Minkowski planes. Enseign. Math. **2**, 299–306 (1961)
20. F. Aurenhammer, Voronoi diagrams—a survey of a fundamental geometric data structure. ACM Comput. Surv. **23**(3), 345–405 (1991)
21. F. Aurenhammer, R. Klein, D.-T. Lee, *Voronoi Diagrams and Delaunay Triangulations* (World Scientific Publishing, Hackensack, 2013)
22. G. Averkov, On the geometry of simplices in Minkowski spaces. Stud. Univ. Žilina Math. Ser. **16**(1), 1–14 (2003)
23. G. Averkov, N. Düvelmeyer, Embedding metric spaces into normed spaces and estimates of metric capacity. Monatsh. Math. **152**(3), 197–206 (2007)
24. G. Averkov, H. Martini, On area and side length of triangles in normed planes. Colloq. Math. **115**(1), 101–112 (2009)
25. V. Balestro, A.G. Horváth, H. Martini, Angle measures, general rotations, and roulettes in normed planes. Anal. Math. Phys. **7**(4), 656–671 (2017)
26. V. Balestro, A.G. Horváth, H. Martini, The inverse problem on roulettes in normed planes. Anal. Math. Phys. **9**(4), 2413–2434 (2019)
27. V. Balestro, A.G. Horváth, H. Martini, R. Teixeira, Angles in normed spaces. Aequat. Math. **91**(2), 201–236 (2017)
28. V. Balestro, H. Martini, The Rosenthal-Szasz inequality for Radon planes. Bull. Aust. Math. Soc. **99**, 130–136 (2018)
29. V. Balestro, H. Martini, E. Shonoda, Concepts of curvatures in normed planes. Expo. Math. **37**(4), 347–381 (2019)
30. V. Balestro, H. Martini, R. Teixeira, Geometric constants for quantifying the difference between orthogonality types. Ann. Funct. Anal. **7**(4), 656–671 (2016)
31. V. Balestro, H. Martini, R. Teixeira, A new construction of Radon curves and related topics. Aequat. Math. **90**(5), 1013–1024 (2016)
32. V. Balestro, H. Martini, R. Teixeira, Geometric properties of a sine function extendable to arbitrary normed planes. Monatsh. Math. **182**(4), 781–800 (2017)
33. V. Balestro, H. Martini, R. Teixeira, On Legendre curves in normed planes. Pac. J. Math. **297**, 1–27 (2018)
34. V. Balestro, H. Martini, R. Teixeira, Optimal constants in normed planes. J. Convex Anal. **26**(1), 89–104 (2019)
35. V. Balestro, H. Martini, R. Teixeira, Surface immersions in normed spaces from the affine point of view. Geom. Dedicata **201**(1), 21–31 (2019)
36. V. Balestro, H. Martini, R. Teixeira, Differential geometry of immersed surfaces in three-dimensional normed spaces. Abh. Math. Semin. Univ. Hambg. **90**, 111–134 (2020)
37. V. Balestro, H. Martini, R. Teixeira, On curvature of surfaces immersed in normed spaces. Monatsh. Math. **192**, 291–309 (2020)
38. V. Balestro, H. Martini, R. Teixeira, Convex analysis in normed spaces and metric projections onto convex bodies. J. Convex Anal. (2021), to appear
39. V. Balestro, H. Martini, R. Teixeira, Duality of gauges and symplectic forms in vector spaces. Collect. Math. **72**, 501–525 (2021)
40. V. Balestro, H. Martini, R. Teixeira, Some topics in differential geometry of normed spaces. Adv. Geom. **21**(1), 109–118 (2021)
41. V. Balestro, E. Shonoda, On a cosine function defined for smooth normed spaces. J. Convex Anal. **25**(1), 21–39 (2018)

42. I. Bárány, R. Schneider, Universal points of convex bodies and bisectors in Minkowski spaces. Adv. Geom. **14**(3), 427–445 (2014)
43. G. Barequet, M.T. Dickerson, M.T. Goodrich, Voronoi diagrams for convex polygon-offset distance functions. Discrete Comput. Geom. **25**, 271–291 (2001)
44. C. Benítez, Orthogonality in normed linear spaces: a classification of the different concepts and some open problems. Congress on functional analysis (Madrid, 1988). Rev. Mat. Univ. Complut. Madrid **2**(suppl.), 53–57 (1989)
45. O.A. Biberstein, *Elements de géométrie différentielle minkowskienne*. PhD Thesis, Université de Montreal (1957)
46. L.M. Blumenthal, *Theory and Applications of Distance Geometry* (Oxford University Press, Oxford, 1953)
47. V. Boltyanski, H. Martini, P.S. Soltan, *Excursions into Combinatorial Geometry*. Universitext (Springer, Berlin, 1997)
48. V. Boltyanski, H. Martini, V. Soltan, *Geometric Methods and Optimization Problems* (Kluwer Academic Publishers, Dordrecht, 1999)
49. T. Bonnesen, W. Fenchel, *Theory of Convex Bodies*. Translated from the German and edited by L. Boron, C. Christenson and B. Smith (BCS Associates, Moscow, 1987) (first edition in German: Springer, Berlin, 1934)
50. K. Böröczky, V. Soltan, Smallest maximal snakes of translates of convex domains. Geom. Dedicata **54**(1), 31–44 (1995)
51. R. Brandenberg, B. González Merino, T. Jahn, H. Martini, Is a complete, reduced set necessarily of constant width? Adv. Geom. **19**(1), 31–40 (2019)
52. P. Brass, Erdős distance problems in normed spaces. Comp. Geom. **6**, 195–214 (1996)
53. P. Brass, On equilateral simplices in normed spaces. Breiträge Algebra Geom. **40**(2), 303–307 (1999)
54. P. Brass, W. Moser, J. Pach, *Research Problems in Discrete Geometry* (Springer, New York, 2005)
55. D. Burago, Y. Burago, S. Ivanov, *A Course in Metric Geometry* (American Mathematical Society, Providence, 2001)
56. D. Burago, S. Ivanov, On asymptotic volume of Finsler tori, minimal surfaces in normed spaces, and symplectic filling volume. Ann. Math. **156**, 891–914 (2002)
57. D. Burago, S. Ivanov, On intrinsic geometry of surfaces in normed spaces. Geom. Topol. **15**, 2275–2298 (2011)
58. H. Busemann, The isoperimetric problem in the Minkowski plane. Am. J. Math. **69**(4), 863–871 (1947)
59. H. Busemann, The isoperimetric problem for Minkowski area. Am. J. Math. **71**, 743–762 (1949)
60. H. Busemann, The foundations of Minkowskian geometry. Comment. Math. Helvet. **24**(1), 156–187 (1950)
61. H. Busemann, The geometry of Finsler spaces. Bull. Am. Math. Soc. **56**, 5–16 (1950)
62. H. Busemann, *The Geometry of Geodesics* (Academic Press, New York, 1955)
63. E. Cabezas-Rivas, V. Miquel, Mean curvature In Minkowski spaces, in *Proceedings of the Conference on Contemporary Geometry and Related Topics, Belgrade (2005)* (University of Belgrade, Faculty of Mathematics, Belgrade, 2006), pp. 81–97
64. J. Ceder, A property of planar convex bodies. Israel J. Math. **1**, 248–253 (1963)
65. J. Ceder, B. Grünbaum, On inscribing and circumscribing hexagons. Colloq. Math. **17**, 99–101 (1967)
66. G.D. Chakerian, M.A. Ghandehari, The Fermat problem in Minkowski spaces. Geom. Dedicata **17**(3), 227–238 (1985)
67. J. Chmieliński, On an ε-Birkhoff orthogonality. J. Inequal. Pure Appl. Math. **6**(3), 7 pp. (2005)
68. J. Chmieliński, P. Wójcik, Approximate symmetry of Birkhoff orthogonality. J. Math. Anal. Appl. **461**(1), 625–640 (2018)
69. K. Cieliebak, H. Hofer, J. Latschev, F. Schlenk, Quantitative symplectic geometry, in *Dynamics, Ergodic Theory, and Geometry*. Math. Sci. Res. Inst. Publ., vol. 54 (Cambridge Univ. Press, Cambridge, 2007), pp. 1–44

70. J.A. Clarkson, The von Neumann-Jordan constant for the Lebesgue space. Ann. Math. **38**, 114–115 (1937)
71. H.S.M. Coxeter, S.L. Greitzer, *Geometry Revisited*. New Mathematical Library, vol. 19 (Random House, New York, 1967)
72. M. Craizer, Iteration of involutes of constant width curves in the Minkowski plane. Beitr. Algebra Geom. **55**(2), 479–496 (2014)
73. M. Craizer, H. Martini, Involutes of constant width polygons in the Minkowski plane. Ars Math. Contemp. **11**(1), 107–125 (2016)
74. M. Craizer, R. Teixeira, V. Balestro, Closed cycloids in a normed plane. Monatsh. Math. **185**(1), 43–60 (2018)
75. M. Craizer, R. Teixeira, V. Balestro, Discrete cycloids from convex symmetric polygons. Discrete Comput. Geom. **60**(4), 859–884 (2018)
76. Y. Cui, W. Huang, H. Hudzik, R. Kaczmarek, Generalized von Neumann-Jordan constant and its relationship to the fixed point property. Fixed Point Theory Appl. **2015**, 40 (2015)
77. M.M. Day, Some characterizations of inner-product spaces. Trans. Am. Math. Soc. **62**, 320–337 (1947)
78. B.V. Dekster, Simplexes with prescribed edge lengths in Minkowski and Banach spaces. Acta Math. Hungar. **86**(4), 343–358 (2000)
79. B.V. Dekster, An angle in Minkowski space. J. Geom. **80**(1–2), 31–47 (2004)
80. M. Deza, M. Laurent, *Geometry of Cuts and Metrics* (Springer, Berlin, 1997)
81. J. Dieudonné, *History of Functional Analysis*. North-Holland Mathematics Studies, vol. 49. Notas de Matemática, 77 (North-Holland Publishing, Amsterdam–New York, 1981)
82. C. Diminnie, E.Z. Andalafte, W.F. Raymond, Generalized angles and a characterization of inner product spaces. Hous. J. Math. **14**(4), 475–480 (1988)
83. R. Durier, C. Michelot, Geometrical properties of the Fermat-Weber problem. Eur. J. Oper. Res. **20**, 332–343 (1985)
84. N. Düvelmeyer, Angle measures and bisectors in Minkowski planes. Can. Math. Bull. **48**(4), 523–534 (2005)
85. N. Düvelmeyer, General embedding problems and two-distance sets in Minkowski planes. Beitr. Algebra Geom. **49**(2), 549–598 (2008)
86. A. Dvoretzky, Some results on convex bodies and Banach spaces, in *Proc. Internat. Sympos. Linear Spaces* (Jerusalem Academic Press, Jerusalem, 1960), pp. 123–160
87. A. Fankhänel, I-measures in Minkowski planes. Beitr. Algebra Geom. **50**, 295–299 (2009)
88. A. Fankhänel, On angular measures in Minkowski planes. Beitr. Algebra Geom. **52**(2), 335–342 (2011)
89. A. Fankhänel, On conics in Minkowski planes. Extracta Math. **27**, 13–29 (2012)
90. R.W. Freese, C.R. Diminnie, E.Z. Andalafte, Angle bisectors in normed linear spaces. Math. Nachr. **131**(1), 167–173 (1987)
91. H. Fukagawa, T. Rothman, *Sacred Mathematics. Japanese Temple Geometry.* With a Preface by Freeman Dyson (Princeton University Press, Princeton, 2008)
92. Z. Füredi, J.C. Lagarias, F. Morgan, *Singularities of Minimal Surfaces and Networks and Related Extremal Problems in Minkowski Space. Discrete and Computational Geometry (New Brunswick, NJ, 1989/1990).* DIMACS Ser. Discrete Math. Theoret. Comput. Sci., vol. 6 (Amer. Math. Soc., Providence, 1991), pp. 95–109
93. J. Gao, An application of elementary geometry in functional analysis. Coll. Math. J. **28**(1), 39–43 (1997)
94. J. Gao, K.S. Lau, On the geometry of spheres in normed linear spaces. J. Aust. Math. Soc. Ser. A **48**, 101–112 (1990)
95. M. Ghandehari, Controlling curvature in Minkowski planes. J. Math. Anal. Appl. **252**, 951–958 (2000)
96. M. Ghandehari, Total curvature in Minkowski planes. Libertas Math. **20**, 107–112 (2000)
97. P. Ghosh, D. Sain, K. Paul, On symmetry of Birkhoff-James orthogonality of linear operators. Adv. Oper. Theory **2**(4), 428–434 (2017)
98. H. Gluck, The converse to the Four Vertex Theorem. Enseign. Math. **17**, 295–309 (1971)

99. S. Gołab, Some metric problems in the geometry of Minkowski (Polish, French Summary). Prace Akademii Górniczej w Krakowie **6**, 1–79 (1932)
100. S. Gołab, H. Härlen, Minkowskische Geometrie I u. II. Monatsh. Math. Phys. **38**, 387–398 (1931)
101. S. Gołab, L. Tamássy, Eine Kennzeichnung der euklidischen Ebene unter den Minkowskischen Ebenen. Publ. Math. Debrecen **7**, 187–193 (1960)
102. B. González Merino, T. Jahn, C. Richter, Uniqueness of circumcenters in generalized Minkowski spaces. J. Approx. Theory **237**, 153–159 (2019)
103. C. Gross, T.-K. Strempel, On generalizations of conics and on a generalization of the Fermat-Torricelli problem. Am. Math. Monthly **105**(8), 732–743 (1998)
104. B. Grünbaum, On a conjecture of H. Hadwiger. Pac. J. Math. **11**, 215–219 (1961)
105. H. Guggenheimer, Pseudo-Minkowski differential geometry. Ann. Mat. Pura Appl. (4) **70**(1), 305–370 (1965)
106. H. Guggenheimer, On plane Minkowski geometry. Geom. Dedicata **12**(4), 371–381 (1982)
107. H.W. Guggenheimer, Elementary geometry of the unsymmetric Minkowski plane. Rev. Un. Mat. Argentina **29**, 270–281 (1984)
108. H. Gunawan, J. Lindiarni, O. Neswan, P-, I-, g-, and D-angles in normed spaces. J. Math. Fund. Sci. **40**(1), 24–32 (2008)
109. E. Gutkin, S. Tabachnikov, Billiards in Finsler and Minkowski geometries. J. Geom. Phys. **40**(3–4), 277–301 (2002)
110. R.K. Guy, An olla-podrida of open problems, often oddly posed. Am. Math. Monthly **90**, 196–199 (1983)
111. J. Hadamard, *Lessons in Geometry. I. Plane Geometry. Transl. from the 13th French edition from 1947 (Original Book from 1898)* (American Mathematical Society (AMS), Providence; Education Development Center, Newton, 2008)
112. S. Hait, K. Paul, D. Sain, Operator norm attainment and Birkhoff-James orthogonality. Linear Algebra Appl. **476**, 85–97 (2015)
113. J.R. Hanson, Regular polygons in taxicab geometry. Int. J. Math. Ed. Sci. Tech. **45**(7), 1084–1095 (2014)
114. C. He, H. Martini, S. Wu, Halving closed curves in normed planes and related inequalities. Math. Inequal. Appl. **12**(4), 719–731 (2009)
115. C. He, H. Martini, S. Wu, On bisectors for convex distance functions. Extracta Math. **28**(1), 57–76 (2013)
116. C. He, H. Martini, S. Wu, On covering functionals of convex bodies. J. Math. Anal. Appl. **437**(2), 1236–1256 (2016)
117. Á. G. Horváth, On bisectors in Minkowski normed spaces. Acta Math. Hungar. **89**(3), 233–246 (2000)
118. A.G. Horváth, Semi-indefinite inner product and generalized Minkowski spaces. J. Geom. Phys. **60**(9), 1190–1208 (2010)
119. Á.G. Horváth, H. Martini, Conics in normed planes. Extracta Math. **26**(1), 29–43 (2011)
120. A.G. Horváth, H. Martini, Bounded representation and radial projections of bisectors in normed spaces. Rocky Mount. J. Math. **43**(1), 179–191 (2013)
121. T. Jahn, Orthogonality in generalized Minkowski spaces. J. Convex Anal. **26**(1), 49–76 (2019)
122. T. Jahn, Y.S. Kupitz, H. Martini, C. Richter, Minsum location extended to gauges and to convex sets. J. Optim. Theory Appl. **166**(3), 711–746 (2015)
123. T. Jahn, H. Martini, C. Richter, Bi- and multifocal curves and surfaces for gauges. J. Convex Anal. **23**(3), 733–774 (2016)
124. T. Jahn, H. Martini, C. Richter, Ball convex bodies in Minkowski spaces. Pac. J. Math. **289**(2), 287–316 (2017)
125. T. Jahn, M. Spirova, On bisectors in normed spaces. Contrib. Discrete Math. **10**(2), 1–9 (2014)
126. R.C. James, Orthogonality in normed linear spaces. Duke Math. J. **12**(2), 291–302 (1945)
127. D. Ji, S. Wu, Quantitative characterization of the difference between Birkhoff orthogonality and isosceles orthogonality. J. Math. Anal. Appl. **323**(1), 1–7 (2006)

128. A. Jiménez-Melado, E. Llorens-Fuster, S. Saejung, The von Neumann-Jordan constant, weak orthogonality and normal structure in Banach spaces. Proc. Am. Math. Soc. **134**(2), 355–364 (2005)
129. R.A. Johnson, *Advanced Euclidean Geometry* (Dover, New York, 1960) (First edition: 1929)
130. P. Jordan, J. von Neumann, On inner products in linear metric spaces. Ann. Math. **36**, 719–723 (1935)
131. M. Kato, L. Maligandra, On James and Jordan-von Neumann constants of Lorentz sequence spaces. J. Math. Anal. Appl. **258**(2), 457–465 (2001)
132. M. Kato, L. Maligandra, Y. Takahashi, On James and Jordan-von Neumann constants and the normal structure coefficient of Banach spaces. Stud. Math. **144**(3), 275–295 (2001)
133. P.J. Kelly, A property of Minkowskian circles. Am. Math. Monthly **57**, 677–678 (1950)
134. N. Komuro, K.-S. Saito, R. Tanaka, On the class of Banach spaces with James constant $\sqrt{2}$. Math. Nachr. **289**(8–9), 1005–1020 (2016)
135. M.-C. Körner, H. Martini, A. Schöbel, Minsum hyperspheres in normed spaces. Discrete Appl. Math. **160**(15), 2221–2233 (2012)
136. J. Kozma, Characterization of Euclidean geometry by existence of circumcenter or orthocenter. Acta Sci. Math. (Szeged) **81**(3–4), 685–698 (2015)
137. A. Kurusa, Conics in Minkowski geometries. Aequat. Math. **92**(5), 949–961 (2018)
138. M. Lassak, H. Martini, Reduced convex bodies in finite dimensional normed spaces: a survey. Results Math. **66**(3–4), 405–426 (2014)
139. D. Laugwitz, Konvexe Mittelpunktsbereiche und normierte Räume. Math. Z. **61**, 235–244 (1954)
140. U. Leopold, H. Martini, Geometry of simplices in Minkowski spaces. Results Math. **73**(2), 17 pp. (2018)
141. U. Leopold, H. Martini, Monge points, Euler lines, and Feuerbach spheres in Minkowski spaces, in *Discrete Geometry and Symmetry*. Springer Proc. Math. Stat., vol. 234 (Springer, Cham, 2018), pp. 235–255
142. L. Liberti, C. Lavor, Open research areas in distance geometry, in *Open Problems in Mathematics, Optimization and Data Science*. Springer Optim. Appl., vol. 141 (Springer, Cham, 2018), pp. 183–223
143. L. Liberti, C. Lavor, N. Maculan, A. Mucherino, Euclidean distance geometry and applications. SIAM Rev. **56**, 3–69 (2014)
144. E. Llorens-Fuster, E.M. Mazcuñán, S. Reich, The Ptolemy and Zbăganu constants of normed spaces. Nonlinear Anal. **72**, 3984–3993 (2010)
145. P. Martín, H. Martini, M. Spirova, Chebyshev sets and ball operators. J. Convex Anal. **21**(3), 601–618 (2014)
146. H. Martini, Neuere Ergebnisse der Elementargeometrie, in *Geometrie und ihre Anwendungen*, ed. by O. Giering, J. Hoschek (Hanser, Munich, 1994), pp. 9–42
147. H. Martini, Recent results in elementary geometry. II, in *Proceedings of the 2nd Gauss Symposium. Conference A: Mathematics and Theoretical Physics* (Munich, 1993). Sympos. Gaussiana (de Gruyter, Berlin, 1995), pp. 419–443
148. H. Martini, L. Montejano, D. Oliveros, *Bodies of Constant Width. An Introduction to Convex Geometry* (Birkhäuser, Cham, 2019)
149. H. Martini, A. Schöbel, Median hyperplanes in normed spaces—a survey. Discrete Appl. Math. **89**(1–3), 181–195 (1998)
150. H. Martini, A. Schöbel, Two characterizations of smooth norms. Geom. Dedicata **77**(2), 173–183 (1999)
151. H. Martini, A. Schöbel, Median and center hyperplanes in Minkowski spaces—a unified approach. Selected papers in honor of Helge Tverberg. Discrete Math. **241**(1–3), 407–426 (2001)
152. H. Martini, M. Spirova, The Feuerbach circle and orthocentricity in normed planes. Enseign. Math. **53**(3–4), 237–258 (2007)
153. H. Martini, M. Spirova, Recent results in Minkowski geometry. East–West J. Math. Special 59–101 (2007)

154. H. Martini, M. Spirova, Clifford's chain of theorems in strictly convex Minkowski planes. Publ. Math. Debrecen **72**(3–4), 371–383 (2008)
155. H. Martini, M. Spirova, On regular 4-coverings and their application for lattice coverings in normed planes. Discrete Math. **309**, 5158–5168 (2009)
156. H. Martini, M. Spirova, K.J. Swanepoel, Geometry where direction matters—or does it? Math. Intelligencer **33**(3), 115–125 (2011)
157. H. Martini, K. Swanepoel, G. Weiss, The Fermat-Torricelli problem in normed planes and spaces. J. Optim. Theory Appl. **115**, 283–314 (2002)
158. H. Martini, K.J. Swanepoel, Equiframed curves—a generalization of Radon curves. Monatsh. Math. **141**(4), 301–314 (2004)
159. H. Martini, K.J. Swanepoel, The geometry of Minkowski spaces—a survey. Part II. Expo. Math. **22**(2), 93–144 (2004)
160. H. Martini, K.J. Swanepoel, Antinorms and Radon curves. Aequat. Math. **72**(1–2), 110–138 (2006)
161. H. Martini, K.J. Swanepoel, G. Weiß, The geometry of Minkowski spaces—a survey. Part I. Expo. Math. **19**(2), 97–142 (2001)
162. H. Martini, S. Wu, Halving circular arcs in normed planes. Period. Math. Hungar. **57**(2), 207–215 (2008)
163. H. Martini, S. Wu, Geometric dilation of closed curves in normed planes. Comput. Geom. **42**, 315–321 (2009)
164. H. Martini, S. Wu, Minimum chords in Minkowski planes. Results Math. **54**(1–2), 371–383 (2009)
165. H. Martini, S. Wu, On orthocentric systems in strictly convex normed planes. Extracta Math. **24**(1), 31–45 (2009)
166. H. Martini, S. Wu, Concurrent and parallel chords of spheres in normed linear spaces. Stud. Sci. Math. Hungar. **47**(4), 505–512 (2010)
167. H. Martini, S. Wu, Minkowksian rhombi and squares inscribed in convex Jordan curves. Colloq. Math. **120**(2), 249–261 (2010)
168. H. Martini, S. Wu, Tangent segments and orthogonality types in normed planes. J. Geom. **99**(1–2), 89–100 (2010)
169. H. Martini, S. Wu, On Zindler curves in normed planes. Can. Math. Bull. **55**(4), 767–773 (2012)
170. H. Martini, S. Wu, Radial projections of bisectors in Minkowski spaces. Extracta Math. **23**(1), 7–28 (2012)
171. H. Martini, S. Wu, Cassini curves in normed planes. Results Math. **63**(3–4), 1159–1175 (2013)
172. H. Martini, S. Wu, Classical curve theory in normed planes. Comput. Aided Geom. Des. **31**(7–8), 373–397 (2014)
173. H. Martini, S. Wu, Complete sets need not be reduced in Minkowski spaces. Beitr. Algebra Geom. **56**, 533–539 (2015)
174. K. Menger, Untersuchungen über allgemeine Metrik. Math. Ann. **100**, 75–163 (1928)
175. P.M. Miličič, On the B-angle and g-angle in normed spaces. J. Inequal. Pure Appl. Math **8**(3), 1–9 (2007)
176. V. Milman, A new proof of A. Dvoretzky's theorem on cross-sections of convex bodies (in Russian). Funkcional. Anal. i Prilozhen **5**(4), 28–37 (1971)
177. V.D. Milman, G. Schechtman, *Asymptotic Theory of Finite Dimensional Normed Spaces* (Springer, Berlin, 1986)
178. H. Minkowski, Sur les propriétés des nombres entiers qui sont dérivées de l'intuition de l'espace. Nouv. Ann. Math. **15**(3), 393–403 (1896)
179. A.F. Monna, *Functional Analysis in Historical Perspective* (Wiley, New York–Toronto, 1973)
180. F. Morgan, Minimal surfaces, crystals, shortest networks, and undergraduate research. Math. Intelligencer **14**(3), 37–44 (1992)
181. Y. Nievergelt, Median spheres: theory, algorithms, applications. Numer. Math. **114**(4), 573–606 (2010)

182. K. Nomizu, T. Sasaki, *Affine Differential Geometry* (Cambridge University Press, Cambridge, 1994)
183. M. Obst, A perimeter-based angle measure in Minkowski planes. Aequat. Math. **92**(1), 135–163 (2018)
184. T. Oikhberg, H. Rosenthal, A metric characterization of normed linear spaces. Rocky Mount. J. Math. **37**, 597–608 (2007)
185. A. Okabe, B. Boots, K. Sugihara, S.N. Chiu, *Spatial Tessellations: Concepts and Applications of Voronoi Diagrams*. With a foreword by D. G. Kendall. Second edition. Wiley Series in Probability and Statistics (Wiley, Chichester, 2000) (First edition: 1992)
186. P.L. Papini, S. Wu, Measurements of differences between orthogonality types. J. Math. Anal. Appl. **397**(1), 285–291 (2013)
187. K. Paul, D. Sain, A. Mal, Approximate Birkhoff-James orthogonality in the space of bounded linear operators. Linear Algebra Appl. **537**, 348–357 (2018)
188. M. Perdigão do Carmo, *Differential Geometry of Curves and Surfaces* (Prentice-Hall, New Jersey, 1976)
189. C.M. Petty, On the geometry of the Minkowski plane. Riv. Mat. Univ. Parma **6**, 269–292 (1955)
190. C.M. Petty, Equilateral sets in Minkowski spaces. Proc. Am. Math. Soc. **29**, 369–374 (1971)
191. C.M. Petty, J.E. Barry, A geometrical approach to the second-order linear differential equation. Can. J. Math. **14**(2), 349 (1962)
192. A. Pietsch, *History of Banach Spaces and Linear Operators* (Birkhäuser, Boston, 2007)
193. G. Pisier, *The Volume of Convex Bodies and Banach Space Geometry* (Cambridge University Press, Cambridge, 1989)
194. F. Plastria, E. Carrizosa, Gauge distances and median hyperplanes. J. Optim. Theory Appl. **110**(1), 173–182 (2001)
195. G. Polya, *Mathematical Discovery* (Wiley, New York, 1981)
196. C. Richter, The ratios of diameter and width of reduced and of complete convex bodies in Minkowski spaces. Beitr. Algebra Geom. **59**(2), 211–220 (2018)
197. B. Riemann, Über die Hypothesen, welche der Geometrie zu Grunde liegen. *Abh. Königlichen Gesellschaft Wiss. Göttingen*, 13, 1868. Expanded English translation of the German original: On the hypotheses which lie at the bases of geometry. Edited and with commentary by Jürgen Jost. Classic Texts in the Sciences (Birkhäuser/Springer, Cham, 2016)
198. D. Sain, Birkhoff-James orthogonality of linear operators on finite dimensional Banach spaces. J. Math. Anal. Appl. **447**(2), 860–866 (2017)
199. J.J. Schäffer, *Geometry of Spheres in Normed Spaces* (Marcel Dekker, New York and Basel, 1976)
200. R. Schneider, *Convex Bodies: The Brunn-Minkowski Theory* (Cambridge University Press, Cambridge, 2014)
201. A. Schöbel, *Locating Lines and Hyperplanes. Theory and Algorithms*. Applied Optimization, vol. 25 (Kluwer Academic Publishers, Dordrecht, 1999)
202. D.O. Silva, *Differential Geometry and Stability of Hypersurfaces in Minkowski Spaces*. PhD Thesis, Universidade Federal de Minas Gerais (UFMG), Belo Horizonte, Brazil, 2021
203. S.S. So, Recent developments in taxicab geometry. Cubo Mat. Educ. **4**(2), 79–96 (2002)
204. M. Spirova, Circle configurations in strictly convex normed planes. Adv. Geom. **10**(4), 631–646 (2010)
205. M. Spirova, On Miquel's theorem and inversions in normed planes. Monatsh. Math. **161**(3), 335–345 (2010)
206. K. Swanepoel, Equilateral sets in finite-dimensional normed spaces, in *Seminar of Mathematical Analysis*, ed. by D.G. Álvarez, G.L. Acedo, R.V. Caro. Secretariado de Publicaciones, Universidad de Sevilla, Seville (2004), pp. 195–237

207. K.J. Swanepoel, Combinatorial distance geometry in normed spaces, in *New Trends in Intuitive Geometry*. Bolyai Soc. Math. Stud., vol. 27 (János Bolyai Math. Soc., Budapest, 2018), pp. 407–458
208. T. Szostok, On a generalization of the sine function. Glas. Mat. **38**(1), 29–44 (2003)
209. S. Tabachnikov, Parameterized plane curves, Minkowski caustics, Minkowski vertices and conservative line fields. Enseign. Math. **43**, 3–26 (1997)
210. S. Tabachnikov, *Geometry and Billiards*. Student Mathematical Library, vol. 30 (American Mathematical Society, Providence; Mathematics Advanced Study Semesters, University Park, 2005)
211. Y. Takahashi, M. Kato, Von Neumann-Jordan constant and uniformly non-square Banach spaces. Nihonkai Math. J. **9**, 155–169 (1998)
212. L. Tamássy, Ein Problem der zweidimensionalen Minkowskischen Geometrie. Ann. Polon. Math. **9**, 39–48 (1960/1961)
213. L. Tamássy, K. Bélteky, On the coincidence of two kinds of ellipses in Minkowskian spaces and in Finsler planes. Publ. Math. Debrecen **31**(3–4), 157–161 (1984)
214. A.C. Thompson, *Minkowski Geometry* (Cambridge University Press, Cambridge, 1996)
215. J. Väisälä, Observations on circumcenters in normed planes. Beitr. Algebra Geom. **58**(3), 607–615 (2017)
216. J. Väisälä, Triangles in convex distance planes. Beitr. Algebra Geom. **59**(4), 797–804 (2018)
217. P. Šemrl, A characterization of normed spaces. J. Math. Anal. Appl. **343**, 1047–1051 (2008)
218. P. Šemrl, A characterization of normed spaces among metric spaces. Rocky Mount. J. Math. **41**(1), 293–298 (2011)
219. G. Weiss, The concept of triangle orthocenters in Minkowski planes. J. Geom. **74**(1–2), 145–156 (2002)
220. S. Wu, Tangent segments in Minkowski planes. Beitr. Algebra Geom. **49**(1), 147–151 (2008)
221. S. Wu, J. Donghai, J. Alonso, Metric ellipses in Minkowski planes. Extracta Math. **20**(3), 273–280 (2005)
222. S. Wu, Z. Ma, K. Xu, Covering unit spheres and balls of normed spaces by smaller balls. Math. Inequal. Appl. **21**(1), 139–154 (2018)
223. I.M. Yaglom, Plane Minkowski geometry: problems and results (in Russian), in *Studies in the Theory of Functions of Several Real Variables (in Russian)* (Yaroslav. Gos. Univ., Yaroslavl, 1976), pp. 90–103
224. I.M. Yaglom, Elementary geometry, then and now, in *The Geometric Vein—The Coxeter Festschrift* (Springer, New York–Heidelber–Berlin, 1981), pp. 253–269
225. M. Zacharias, *Elementargeometrie der Ebene und des Raumes* (Walter de Gruyter, Berlin, 1930)
226. A. Zamani, M.S. Moslehian, Approximate Roberts orthogonality. Aequat. Math. **89**(3), 529–541 (2015)
227. C. Zong, A quantitative program for Hadwiger's covering conjecture. Sci. China Math. **53**(9), 2551–2560 (2010)

Chapter 4
Orthogonality Types in Normed Linear Spaces

Javier Alonso, Horst Martini, and Senlin Wu

Abstract This chapter gives a comparing exposition of the large family of orthogonality concepts in normed linear spaces. With the help of fundamental properties that such concepts can have, or cannot have, we show their differences, similarities and direct connections. Based on this framework, we try to structurize this little subfield of the theory of real Banach spaces. For example, many characterizations of inner product spaces or of other special norm classes are presented. We also try to emphasize the geometric side of the given theoretical setting. Various open problems are posed, and a final survey can be taken as an update of the large amount of existing related literature. Hence our exposition should be useful for beginners in this research direction.

Keywords Characterizations of ellipsoids · Characterizations of inner product spaces · Finsler geometry · Inner product spaces · Joly's construction · Minkowski space · Normed linear space · Orthogonality in normed linear spaces · Radon curves · Zindler curves

MSC (2000) 46B20, 46Cxx, 46C15, 51F20, 52A20, 52A21, 53B40

4.1 Introduction

In a real inner product space, $(E, \langle \cdot, \cdot \rangle)$, the fact that two vectors are orthogonal can be expressed in many different ways in terms of the norm induced by the inner

J. Alonso
Instituto de Matemáticas de la Universidad de Extremadura (IMUEX), Badajoz, Spain
e-mail: jalonsoro@gmail.com

H. Martini (✉)
Technische Universität Chemnitz, Fakultät für Mathematik, Chemnitz, Germany
e-mail: martini@mathematik.tu-chemnitz.de

S. Wu
North University of China, Taiyuan, China
e-mail: wusenlin@nuc.edu.cn

product. For example, for $x, y \in E$, the following properties are equivalent: (a) $x \perp y$, i.e., $\langle x, y \rangle = 0$; (b) $\|x + y\| = \|x - y\|$; (c) $\|x + \lambda y\| \geq \|x\|$ for every $\lambda \in \mathbb{R}$; etc. Here the norm considered is $\|x\| = \langle x, x \rangle^{1/2}$, but it makes sense to study properties like (b) and (c) also in an arbitrary normed space. This allows to bring the definition of orthogonality, in principle described for inner product spaces, to the more general setting of normed linear spaces, leading to several definitions of generalized orthogonality.

Continuing the surveys [6, 8, 38], and [14] on orthogonalities in real normed linear spaces (presented also in Alonso's dissertation [1]), this chapter is an expository representation of the large variety of such kinds of orthogonalities (Sect. 4.2). In Sect. 4.3 we collect the general properties (symmetry, homogeneity, additivity, ...) that the orthogonality relation in an inner product space has, and later we will see that they are, in general, not inherited by generalized orthogonalities, unless the space possesses certain properties.

In particular (but also extending the viewpoints of the above mentioned surveys) we concentrate on notions like Roberts orthogonality (Sect. 4.4), Birkhoff orthogonality (Sect. 4.5), James or isosceles orthogonality (Sect. 4.6) and Pythagorean orthogonality (Sect. 4.7).

Although all the definitions of generalized orthogonalities coincide in an inner product space, they are, in general, different in a normed space. In fact, the coincidence of any two of them is, in most cases, only possible in inner product spaces. In Sect. 4.8 this question is treated in detail for the cases of Birkhoff orthogonality, isosceles orthogonality and Pythagorean orthogonality.

In Sect. 4.9 further types of orthogonality are discussed, mainly in survey-like style. The proper survey part (Sect. 4.10), included at the end of this chapter, can be taken as a comprehensive update of existing literature related to this field.

We hope, and this was what prompted us to write it, that this work can help those mathematicians who are beginning in this fascinating field of research.

In this chapter we denote by $(X, \| \cdot \|)$ a *real normed linear space of dimension* $\dim X \geq 2$, with *unit sphere* $S_X = \{x \in X : \|x\| = 1\}$ and X^* as *dual space of* X.

4.2 Definitions of Orthogonalities

The first definition of orthogonality for a normed space was given by Roberts [127] in 1934, thus called *Roberts orthogonality*:

$$x \perp_R y :\Leftrightarrow \quad \|x + \lambda y\| = \|x - \lambda y\| \text{ for every } \lambda \in \mathbb{R}.$$

In 1935, Birkhoff [41] gave what has revealed to be the most important definition of generalized orthogonality:

$$x \perp_B y :\Leftrightarrow \quad \|x + \lambda y\| \geq \|x\| \text{ for every } \lambda \in \mathbb{R}.$$

Birkhoff orthogonality mimics the perpendicularity of any radius of the circumference with the corresponding tangent line: a vector $x \in X \setminus \{0\}$ is *Birkhoff orthogonal*

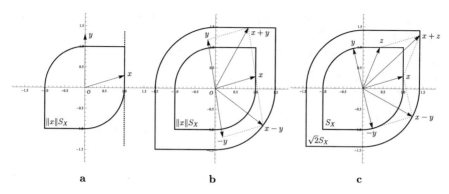

Fig. 4.1 (a) $x \perp_B y$. (b) $x \perp_I y$. (c) $x \perp_{P_-} y$, $x \perp_{P_+} z$

to $y \in X \setminus \{0\}$ if the line with the direction of y through x supports the sphere of radius $\|x\|$ at x (see Fig. 4.1a).

In 1945, James [86] observed that Roberts orthogonality has not the minimum of desirable properties (see Sect. 4.4) and introduced *isosceles orthogonality* (Fig. 4.1b) and *Pythagorean orthogonality* (Fig. 4.1c):

$$x \perp_I y \;:\Leftrightarrow\quad \|x - y\| = \|x + y\|,$$
$$x \perp_{P_-} y \;:\Leftrightarrow\quad \|x - y\|^2 = \|x\|^2 + \|y\|^2.$$

Sometimes, Pythagorean orthogonality is referred to as

$$x \perp_{P_+} y \;:\Leftrightarrow\quad \|x + y\|^2 = \|x\|^2 + \|y\|^2.$$

Isosceles and Pythagorean orthogonalities became particular cases of the family of orthogonalities introduced by Carlsson [50] in 1962, thus called *Carlsson orthogonalities*:

$$x \perp_C y \;:\Leftrightarrow\quad \sum_{i=1}^{n} a_i \|b_i x + c_i y\|^2 = 0,$$

where $a_i, b_i, c_i \in \mathbb{R}$, $i = 1, \dots, n$, are such that $\sum_{i=1}^{n} a_i b_i^2 = \sum_{i=1}^{n} a_i c_i^2 = 0$ and $\sum_{i=1}^{n} a_i b_i c_i = 1$. Note that if the norm is induced by an inner product then $\langle x, y \rangle = \frac{1}{2} \sum_{i=1}^{n} a_i \|b_i x + c_i y\|^2$.

Further particular cases of Carlsson orthogonality are the following ones:

$$x \perp_a y \;:\Leftrightarrow\quad (1 + a^2)\|x - y\|^2 = \|x - ay\|^2 + \|y - ax\|^2,$$

for fixed $a \neq 1$, introduced by Diminnie et al. [66] in 1983. It embraces isosceles and Pythagorean orthogonalities, and it was extended by the same authors [19] in 1985 to:

$$x \perp_{ab} y \;:\Leftrightarrow\quad \|x - y\|^2 + \|ax - by\|^2 = \|x - by\|^2 + \|y - ax\|^2,$$

for fixed $a, b \in \mathbb{R} \setminus \{1\}$. This last orthogonality was implicitly considered by Kapoor and Prasad [93] in 1978. Alonso and Benítez [7] considered in 1988 two Carlsson orthogonalities that also embrace isosceles and Pythagorean orthogonalities:

$$x \perp_{\lambda I} y :\Leftrightarrow \quad \|x + \lambda y\| = \|x - \lambda y\|,$$
$$x \perp_{\lambda P} y :\Leftrightarrow \quad \|x + \lambda y\|^2 = \|x\|^2 + \lambda^2 \|y\|^2,$$

both for fixed $\lambda \neq 0$.

In 1995 Boussouis [47, 48] extended the concept of Carlsson orthogonality, thus introducing the so-called *Boussouis orthogonality*:

$$x \perp_M y :\Leftrightarrow \quad \int_\Omega a(\omega) \|b(\omega)x + c(\omega)y\|^2 \, d\mu(\omega) = 0,$$

where (Ω, μ) is a positive measure space and a, b, c are μ-measurable functions from Ω to $\mathbb{R}$ such that $a \neq 0$, μ-a.e., ab^2 and ac^2 are μ-integrable, and $\int_\Omega ab^2 \, d\mu = \int_\Omega ac^2 \, d\mu = 0$ and $\int_\Omega abc \, d\mu = 1$. If the norm is induced by an inner product, then $\langle x, y \rangle = \frac{1}{2} \int_\Omega a(\omega) \|b(\omega)x + c(\omega)y\|^2 \, d\mu(\omega)$.

In 1957, Singer [135] restricted the definition of isosceles orthogonality to the unit sphere:

$$x \perp_S y :\Leftrightarrow \quad \text{either } \|x\|\|y\| = 0 \text{ or } \left\| \frac{x}{\|x\|} - \frac{y}{\|y\|} \right\| = \left\| \frac{x}{\|x\|} + \frac{y}{\|y\|} \right\|.$$

Similar to this *Singer orthogonality*, but based on Pythagorean orthogonality, is the one defined by Diminnie et al. [67] in 1986:

$$x \perp_{DP} y :\Leftrightarrow \quad \text{either } \|x\|\|y\| = 0 \text{ or } \left\| \frac{x}{\|x\|} - \frac{y}{\|y\|} \right\| = \sqrt{2}.$$

In [8] and [47] the two definitions above were generalized to:

$$x \perp_{UC} y :\Leftrightarrow \quad \text{either } \|x\|\|y\| = 0 \text{ or } \frac{x}{\|x\|} \perp_C \frac{y}{\|y\|}.$$

$$x \perp_{UM} y :\Leftrightarrow \quad \text{either } \|x\|\|y\| = 0 \text{ or } \frac{x}{\|x\|} \perp_M \frac{y}{\|y\|}.$$

In 1983 Diminnie [65] gave a definition of orthogonality based on the area of the parallelogram with vertices at $0, x, y$ and $x + y$, in the subspace (identified with $\mathbb{R}^2$) spanned by x and y:

$$x \perp_D y :\Leftrightarrow \quad \|x, y\| := \sup\{f(x)g(y) - f(y)g(x) : f, g \in S_{X^*}\} = \|x\|\|y\|.$$

Observe that if $u, v \in X$ are such that $x = x_1 u + x_2 v$, $y = y_1 u + y_2 v$, with $x_i, y_i \in \mathbb{R}$, $i = 1, 2$, then $\|x, y\| = |x_1 y_2 - x_2 y_1| \|u, v\|$. Also note that in an inner product space the identity $\|x, y\|^2 = \|x\|^2 \|y\|^2 - \langle x, y \rangle$ holds [78, p. 29], which proves that this *Diminnie orthogonality* (or *D-orthogonality*) coincides with usual orthogonality in such spaces.

A geometric interpretation of D-orthogonality was given in [9]. Let P be a minimum area parallelogram bounding S_X. Two vectors $x, y \in S_X$ are D-orthogonal if and only if they are midpoints of two concurrent sides of a parallelogram with the same area as P.

Another area based orthogonality is the *area orthogonality* introduced by Alonso [1] in 1984:

$x \perp_A y :\Leftrightarrow$ either $\|x\|\|y\| = 0$ or x and y are linearly independent and $\pm\frac{x}{\|x\|}$ and $\pm\frac{y}{\|y\|}$ determine, in the unit disk of the plane generated by them (identified with $\mathbb{R}^2$), four quadrants of equal area.

Recall that $\mathbb{R}^2$ endowed with a norm is Euclidean if and only if its unit sphere is an ellipse, and that any two conjugate diameters of an ellipse have the property that defines A-orthogonality.

Alsina et al. [16] introduced in 1995 *height orthogonality*, which is based on the fact that the height onto the hypotenuse in a right triangle divides it into two similar triangles:

$x \perp_H y :\Leftrightarrow$ either $\|x\|\|y\| = 0$ or $\|x - y\| = \left\|\|y\|\frac{x}{\|x\|} + \|x\|\frac{y}{\|y\|}\right\|$.

Note that in the Euclidean plane the height at 0 of the triangle $x0y$ is given by $z = \frac{(\|y\|^2 - \langle x,y \rangle)x + (\|x\|^2 - \langle x,y \rangle)y}{\|x\|^2 + \|y\|^2}$. Moreover, the triangle is right at 0 if and only if z divides the segment $[x, y]$ into two pieces such that the triangles $xz0$ and $yz0$ are similar, which is equivalent to $x \perp_H y$.

4.3   General Properties of Orthogonality

In a real inner product space $(E, \langle \cdot, \cdot \rangle)$ the orthogonality relation has the following well known properties:

Non-degeneracy: For $x \in E$ and $\lambda, \mu \in \mathbb{R}$, $\lambda x \perp \mu x$ if and only if $\lambda \mu x = 0$.
Simplification: If $x \perp y$, then $\lambda x \perp \lambda y$ for any $\lambda \in \mathbb{R}$.
Continuity: If $x_n \to x$, $y_n \to y$ and $x_n \perp y_n$ for all $n \in \mathbb{N}$, then $x \perp y$.
Symmetry: If $x \perp y$, then $y \perp x$.
Homogeneity: If $x \perp y$, then $x \perp \lambda y$ for any $\lambda \in \mathbb{R}$.
Additivity: If $x \perp y$ and $x \perp z$, then $x \perp y + z$.
Existence: For any $x, y \in E$, linearly independent, and any $\rho > 0$, there exists $z = \alpha x + \beta y$, with $\alpha, \beta \in \mathbb{R}$, $\beta > 0$, such that $\|z\| = \rho$ and $x \perp z$.
Uniqueness: The vector z in the existence property is unique.
α-Existence: For any $x, y \in E$ there exists $\alpha \in \mathbb{R}$ such that $x \perp \alpha x + y$.
α-Uniqueness: For any $x, y \in E \setminus \{0\}$ the above α is unique.
Orthogonal diagonals: For any $x, y \in E \setminus \{0\}$ there exists $\rho > 0$ such that $x + \rho y \perp x - \rho y$. Moreover, such ρ is unique.

When we move to a real normed space $(X, \|\cdot\|)$, the orthogonalities defined in Sect. 4.2 do not necessarily fulfill the above properties. Moreover, if an orthogonality is not symmetric (like Birkhoff orthogonality), then we must distinguish between additivity on the right (defined as above) and additivity on the left ($x \perp y$, $z \perp y$

imply $x + z \perp y$). Similarly with existence, uniqueness, etc. Note also that if an orthogonality is homogeneous, then existence and α-existence properties are equivalent; and that if it is, in addition, non-degenerate, then uniqueness and α-uniqueness are also equivalent. The next theorem makes this result more precise.

Theorem 4.3.1 *Assume that the orthogonality $\perp$ is positively homogeneous (i.e., $x \perp y \Rightarrow x \perp \lambda y$, $\forall \lambda \geq 0$). The following properties are equivalent:*

(i) *$\perp$ has the existence property in the sphere (i.e., when $\rho = \|x\|$ in the existence property).*
(ii) *$\perp$ has the α-existence property.*

If, in addition, $\perp$ has the non-degeneracy property, then uniqueness in the sphere and α-uniqueness are equivalent properties.

Proof (i)$\Rightarrow$(ii) Let $x, y \in E$. By (i) there exist $\bar{\alpha} \in \mathbb{R}$ and $\bar{\beta} > 0$ such that $x \perp \bar{\alpha}x + \bar{\beta}y$ and $\|\bar{\alpha}x + \bar{\beta}y\| = \|x\|$. Then $x \perp \frac{\bar{\alpha}}{\bar{\beta}}x + y$.

(ii)$\Rightarrow$(i) Let $x, y \in E$ be linearly independent. By (ii) there exists $\bar{\alpha} \in \mathbb{R}$ such that $x \perp \bar{\alpha}x + y$. Let $\alpha = \frac{\|x\|\bar{\alpha}}{\|\bar{\alpha}x+y\|}$, $\beta = \frac{\|x\|}{\|\bar{\alpha}x+y\|}$. Then $x \perp \alpha x + \beta y$, $\beta > 0$ and $\|\alpha x + \beta y\| = \|x\|$.

The proofs of the uniqueness properties are completely similar. $\qquad\square$

All the orthogonalities considered in Sect. 4.2 have the properties of non-degeneracy, simplification and continuity. Next, we will embark on the study of the other properties of each of these orthogonalities. We will do this in detail with respect to Roberts, Birkhoff, isosceles and Pythagorean orthogonality, the more widely studied, and we will summarize what is known about the other orthogonalities.

4.4 Roberts Orthogonality

Definition 4.4.1 Two vectors x and y are said to be *Roberts orthogonal* ($x \perp_R y$) if

$$\|x + \lambda y\| = \|x - \lambda y\|, \ \forall \lambda \in \mathbb{R}.$$

The definition of Roberts orthogonality was the first one given for a generalized orthogonality (see [127]). It is symmetric and homogeneous, and it is easy to see (e.g., by considering a norm in $\mathbb{R}^3$ with unit sphere "not sufficiently symmetric") that it is in general not additive. But this orthogonality has serious deficiencies with respect to existence properties. James [86] gave an example of a normed space in which two vectors are Roberts orthogonal if and only if one of them is zero (which implies that Roberts orthogonality is additive in such space). This was James's motivation to introduce isosceles and Pythagorean orthogonalities.

Moreover, Ficken [74] proved that a normed space $(X, \| \cdot \|)$ is an inner product space if and only if the following property holds (see Theorem 4.6.12):

$$x, y \in X, \; \|x\| = \|y\| \quad \Rightarrow \quad \|\alpha x + y\| = \|x + \alpha y\| \; \forall \alpha \in \mathbb{R}. \tag{4.1}$$

From this result, James [86] got that isosceles orthogonality is homogeneous only in inner product spaces, and then he concluded that Roberts orthogonality has the α-existence property (and then the existence property, because it is homogeneous) only in such spaces. For his part, Del Rio [62] proved that the same occurs with the existence of orthogonal diagonals. Let us show all this in detail.

First, note that in any normed space Roberts orthogonality implies Birkhoff orthogonality: let $x, y \in X$, $x \perp_R y$. Since the function $\lambda \in \mathbb{R} \to f(\lambda) = \|x + \lambda y\|$ is convex, we have that for any $\lambda \in \mathbb{R}$, $\|x\| = f(0) = f\left(\frac{1}{2}\lambda + \frac{1}{2}(-\lambda)\right) \leq \frac{1}{2}f(\lambda) + \frac{1}{2}f(-\lambda) = \frac{1}{2}(\|x + \lambda y\| + \|x - \lambda y\|) = \|x + \lambda y\|$, i.e., $x \perp_B y$.

Theorem 4.4.2 *A normed space X is an inner product space if and only if Roberts orthogonality has the existence (or α-existence) property.*

Proof The necessity is evident. To prove the sufficiency we can assume that $\dim X = 2$. Assume that Roberts orthogonality has the existence property. We will see first that then X is strictly convex. Assume, on the contrary, that there is a segment $[u, v] \subset S_X$, $u \neq v$, that we assume to be maximal. Let $x = (u + v)/2$, and let $y \in S_X$ be such that $x \perp_R y$. Then $\|x + \lambda y\| = \|x - \lambda y\|$ for every $\lambda \in \mathbb{R}$, which means that S_X is symmetric with respect to the line $\langle -x, x \rangle$ in the direction of y. This implies that y has the direction of $u - v$. Let $\bar{x} = \frac{1}{4}u + \frac{3}{4}v$, and let $\bar{y} \in S_X$ be such that $\bar{x} \perp_R \bar{y}$. Since $x \perp_B y$ and $\bar{x} \perp_B \bar{y}$, then either $\bar{y} = y$ or $\bar{y} = -y$. Moreover, S_X is symmetric with respect to the line $\langle -\bar{x}, \bar{x} \rangle$ in the direction of y, which implies that $\bar{v} := -\frac{1}{2}u + \frac{3}{2}v \in S_X$. Then $[u, \bar{v}] \subset S_X$, contradicting the maximality of $[u, v]$.

We will see in Theorem 4.6.6 and Theorem 4.6.11 that since X is strictly convex, isosceles orthogonality has the uniqueness and α-uniqueness properties.

Now we will apply Ficken's characterization (4.1). So let $x, y \in X$, $\|x\| = \|y\|$, and $\alpha \in \mathbb{R}$. Without loss of generality we can assume $\alpha \neq -1$. Let $u = x + y$ and $v = x - y$. Then $u \perp_I v$. Let $\bar{\alpha} \in \mathbb{R}$ be such that $u \perp_R \bar{\alpha}u + v$. Then $u \perp_I \bar{\alpha}u + v$. The uniqueness of isosceles orthogonality implies that $\bar{\alpha} = 0$, and then $u \perp_R v$, i.e., $\|u + \lambda v\| = \|u - \lambda v\|$ for every $\lambda \in \mathbb{R}$. Taking $\lambda = \frac{\alpha-1}{\alpha+1}$, we get $\|\alpha x + y\| = \|x + \alpha y\|$. □

When Roberts orthogonal vectors exist, they have the uniqueness property.

Theorem 4.4.3 *If $x, y \in X \setminus \{0\}$, $\alpha, \beta \in \mathbb{R}$ are such that $x \perp_R \alpha x + y$ and $x \perp_R \beta x + y$, then $\alpha = \beta$.*

Proof Assume that $\alpha < \beta$, and let $u = \alpha x + y$, $v = \beta x + y$. Then $\|x + \lambda u\| = \|x - \lambda u\|$ and $\|x + \lambda v\| = \|x - \lambda v\|$ for every $\lambda \in \mathbb{R}$. We will get an absurdity. The idea of the proof is to use the symmetry of S_X with respect to the line $\langle -x, x \rangle$, which we refer to in the proof of Theorem 4.4.2, to get that S_X is not bounded. For

$n \in \mathbb{N}$, let $\lambda_n = 1/(1 + n(\beta - \alpha))$. Then, for $n = 0, 1, \ldots$ the following identities hold:

$$x + \lambda_n v = \frac{\lambda_n}{\lambda_{n+1}}(x + \lambda_{n+1}u), \qquad x - \lambda_n u = \frac{\lambda_n}{\lambda_{n+1}}(x - \lambda_{n+1}v).$$

Applying them recursively from $n = 0$, we get that

$$\|x + v\| = \frac{1}{\lambda_1}\|x + \lambda_1 u\| = \frac{1}{\lambda_1}\|x - \lambda_1 u\| = \frac{1}{\lambda_1}\frac{\lambda_1}{\lambda_2}\|x - \lambda_2 v\| = \frac{1}{\lambda_2}\|x + \lambda_2 v\|$$

$$= \frac{1}{\lambda_2}\frac{\lambda_2}{\lambda_3}\|x + \lambda_3 u\| = \frac{1}{\lambda_3}\|x - \lambda_3 u\| = \frac{1}{\lambda_3}\frac{\lambda_3}{\lambda_4}\|x - \lambda_4 v\|$$

$$= \frac{1}{\lambda_4}\|x + \lambda_4 v\| = \cdots,$$

which leads to the identity $\|x + v\| = \frac{1}{\lambda_{2n}}\|x + \lambda_{2n} v\|$, for $n = 0, 1, \ldots$, but this is absurd because the right term goes to infinity as n increases. □

Theorem 4.4.4 *A normed space X is an inner product space if and only if Roberts orthogonality has the existence of diagonals property.*

Proof Assume that X has the existence of diagonals property. Let $x, y \in X$, $\|x\| = \|y\|$, and let $\rho > 0$ be such that $x + \rho y \perp_R x - \rho y$. Then $\|(1+\lambda)x + \rho(1-\lambda)y\| = \|(1 - \lambda)x + \rho(1 + \lambda)y\|$ for every $\lambda \in \mathbb{R}$. Taking $\lambda = 1$, we get that $\rho = 1$. For $\alpha \in \mathbb{R}$, $\alpha \neq -1$, taking $\lambda = (\alpha - 1)/(\alpha + 1)$ we obtain $\|\alpha x + y\| = \|x + \alpha y\|$. Ficken's characterization (4.1) concludes the proof. □

When Roberts orthogonal diagonals exist, they are unique. In fact, if $x + \rho y \perp_R x - \rho y$, then $\rho = \pm\|x\|/\|y\|$.

4.5 Birkhoff Orthogonality

Definition 4.5.1 A vector x is said to be *Birkhoff orthogonal to* another vector y $(x \perp_B y)$ if

$$\|x + \lambda y\| \geq \|x\|, \ \forall \lambda \in \mathbb{R}.$$

Without any doubt, the wealth of properties of Birkhoff orthogonality, introduced in [41], makes it the possibly most interesting definition of orthogonality. First, note that it is *homogeneous*, which allows to consider B-orthogonal directions. Prior to walking to the other properties, let us recall the functional version of the Hahn–Banach Theorem, which is closely related to this orthogonality type:

Let $L \subset X$ be a linear subspace. For any $f \in L^*$ there exists $\bar{f} \in X^*$ such that $\bar{f}|_L = f$ and $\|f\|_{L^*} = \|\bar{f}\|_{X^*}$.

As a corollary, we get the functional characterization of the Birkhoff orthogonality of two vectors.

Theorem 4.5.2 *For $x, y \in X$, $x \perp_B y$ holds if and only if there exists a continuous linear functional $f \in X^* \setminus \{0\}$ such that $f(x) = \|f\| \|x\|$ and $f(y) = 0$.*

Proof Let $f \in X^* \setminus \{0\}$ satisfy the hypothesis. Then, for any $\lambda \in \mathbb{R}$, $\|f\| \|x + \lambda y\| \geq |f(x + \lambda y)| = |f(x)| = \|f\| \|x\|$, which implies $x \perp_B y$. Conversely, assume that $x \perp_B y$. Let L be the subspace spanned by x and y (that we can assume to be linearly independent), and consider the functional $f : \alpha x + \beta y \in L \mapsto \alpha \in \mathbb{R}$. Since $x \perp_B y$, it follows that $\|f\|_{L^*} = \sup_{\alpha, \beta \in \mathbb{R}} |\alpha| / \|\alpha x + \beta y\| \leq 1/\|x\| \leq \|f\|_{L^*}$. Thus, $1 = f(x) = \|f\|_{L^*} \|x\|$, and $f(y) = 0$. By the Hahn–Banach Theorem, f extends to the whole space X. $\qquad\square$

4.5.1 Existence and Uniqueness

Also from the Hahn–Banach Theorem it follows that for any $x \in X$ there is at least a continuous linear functional that attains its norm on it, which, a fortiori, gives the *existence* (equivalently, the α-*existence*) *on the right* property for Birkhoff orthogonality [88].

Theorem 4.5.3 *For any $x \in X$ there exists $f \in X^* \setminus \{0\}$ such that $f(x) = \|f\| \|x\|$.*

Proof Let L be the subspace spanned by x, and consider the functional $f : \alpha x \in L \mapsto \alpha \|x\| \in \mathbb{R}$. Then $f(x) = \|x\|$ and $\|f\|_{L^*} = 1$. Extend f to X. $\qquad\square$

Recall that X is said to be *smooth* if for any $x \in X \setminus \{0\}$ the functional f in Theorem 4.5.3 is unique, supposed $\|f\| = 1$.

Theorem 4.5.4 *For any $x, y \in X$ there exists $\alpha \in \mathbb{R}$ such that $x \perp_B \alpha x + y$. Moreover, if $x \neq 0$, $|\alpha| \leq \|y\| / \|x\|$.*

Proof We can assume that x and y are linearly independent. Take (by Theorem 4.5.3) $f \in X^* \setminus \{0\}$ such that $f(x) = \|f\| \|x\|$. Consider the closed hyperplane $H = \{z \in X : f(z) = 0\}$ and take $z \in H \cap \langle \{x, y\} \rangle$, $z \neq 0$. Then $z = \alpha_1 x + \alpha_2 y$, with $\alpha_2 \neq 0$. Take $\alpha = \alpha_1 / \alpha_2$. From Theorem 4.5.2 it follows that $x \perp_B \alpha x + y$ and $|\alpha| \leq \|y\| / \|x\|$. $\qquad\square$

The scalar α in Theorem 4.5.4 may be not unique. A direct proof of the next theorem can be found in [88]. The proof given here uses Theorem 4.5.2 and provides additional information.

Theorem 4.5.5 *If $x \perp_B \alpha x + y$ and $x \perp_B \beta x + y$, with $\alpha \leq \beta$, then $x \perp_B \gamma x + y$ for $\alpha \leq \gamma \leq \beta$.*

Proof Let $f, g \in S_{X^*}$ be such that $f(x) = \|x\| = g(x)$ and $f(\alpha x + y) = 0 = g(\beta x + y)$. Let $\alpha \leq \gamma \leq \beta$, and let $\mu \in [0, 1]$ be such that $\gamma = \mu\alpha + (1 - \mu)\beta$. Take $h = \mu f + (1 - \mu)g$. Then $h(x) = \|x\|$, $\|h\| = 1$ and $h(\gamma x + y) = \mu h(\alpha x + y) + (1 - \mu)h(\beta x + y) = \mu(1 - \mu)(g(\alpha x + y) + f(\beta x + y)) = 0$. $\qquad\square$

Theorem 4.5.5 is really a corollary of a deeper result (Theorem 4.5.6), proved also by James [88], that leads to the characterization of the uniqueness on the right of Birkhoff orthogonality. Recall that for any $x, y \in X$ the limits

$$N_{\pm}(x; y) := \lim_{\lambda \to 0^{\pm}} \frac{\|x + \lambda y\| - \|x\|}{\lambda}$$

always exist, and $N_-(x, y) \le N_+(x, y)$. Moreover, X is smooth if and only if for any $x, y \in X$, $x \ne 0$, $N_-(x; y) = N_+(x; y) := N(x, y)$, i.e., the norm is *Gâteaux differentiable* at any $x \ne 0$.

Theorem 4.5.6 $x \perp_B \alpha x + y$ *if and only if* $N_-(x; y) \le -\alpha\|x\| \le N_+(x; y)$.

Thus, for $x, y \in X \setminus \{0\}$, $-N_+(x; y)/\|x\|$ and $-N_-(x; y)/\|x\|$ are, respectively, the least and largest scalars α such that $x \perp_B \alpha x + y$.

If X is smooth and $x \in X \setminus \{0\}$, then the functional $y \in X \mapsto N(x, y)$ is the only functional $f \in X^\star$ such that $f(x) = \|x\|$ and $\|f\| = 1$.

Theorem 4.5.7 *In a normed space X, Birkhoff orthogonality is unique on the right if and only if X is smooth.*

The way to see that Birkhoff orthogonality has also the *existence* (equivalently, α-*existence*) *on the left* property is much more simple. It follows directly from the fact that for any $x, y \in X \setminus \{0\}$, the function $f : \lambda \in \mathbb{R} \mapsto \|\lambda x + y\|$ is convex and such that $\lim_{\lambda \to \pm\infty} f(\lambda) = +\infty$, and thus, it attains the absolute minimum at some point (or interval). This function is also the tool to characterize the *uniqueness on the left*.

Theorem 4.5.8 *For any $x, y \in X$ there exists $\alpha \in \mathbb{R}$ such that $\alpha x + y \perp_B x$. Moreover, if $x \ne 0$, $|\alpha| \le 2\|y\|/\|x\|$. If $\alpha x + y \perp_B x$ and $\beta x + y \perp_B x$, with $\alpha \le \beta$, then $\gamma x + y \perp_B x$ for $\alpha \le \gamma \le \beta$.*

Proof Just consider the convexity of the function $\lambda \in \mathbb{R} \mapsto \|\lambda x + y\|$. The bound for α follows from Theorem 4.5.2. Let $f \in X^* \setminus \{0\}$ such that $f(\alpha x + y) = \|f\|\|\alpha x + y\|$ and $f(x) = 0$. Then $\|f\|\|y\| \ge f(y) = \|f\|\|\alpha x + y\| \ge \|f\|(|\alpha|\|x\| - \|y\|)$, from which it follows that $|\alpha| \le 2\|y\|/\|x\|$. $\qquad\square$

Now recall that a normed space X is said to be *strictly convex* if $z, z' \in S_X$, $\|z + z'\| = 2$ imply $z = z'$. The next lemma gives one of the most frequently used characterizations of strict convexity.

Lemma 4.5.9 *A normed space X is strictly convex if and only if, for any $x, y \in X \setminus \{0\}$, the function $f(\lambda) = \|x + \lambda y\|$, $\lambda \in \mathbb{R}$, is not constant in any non-trivial interval.*

Proof First, note that $f(\lambda)$ is a convex function. Assume that X is not strictly convex. Take $u, v \in S_X$, $u \ne v$, such that $\|u + v\| = 2$. Then, the function $f(\lambda) = \|u + \lambda(v - u)\|$ satisfies $f(0) = f(1/2) = f(1)$, and therefore it must be constant in $[0, 1]$. Conversely, assume that there exist $x, y \in X \setminus \{0\}$ such that

$f(\lambda) = \|x + \lambda y\|$ is constant in $[a, b]$, $a < b$. Take $u = (x + ay)/\|x + ay\|$, $v = (x + by)/\|x + by\|$. Then $u, v \in S_X$, $u \neq v$, and $\|u + v\| = 2$. □

Theorem 4.5.10 *In a normed space X, Birkhoff orthogonality is unique on the left if and only if X is strictly convex.*

Proof Assume that there exist $x, y \in X \setminus \{0\}$ and $\alpha, \beta \in \mathbb{R}$, $\alpha < \beta$, such that $\alpha x + y \perp_B x$ and $\beta x + y \perp_B x$. Then, by Theorem 4.5.8, $\gamma x + y \perp_B x$, for $\alpha \leq \gamma \leq \beta$. Therefore, $\|\alpha x + y\| = \|\gamma x + y + (\alpha - \gamma)x\| \geq \|\gamma x + y\| = \|\alpha x + y + (\gamma - \alpha)x\| \geq \|\alpha x + y\|$, which implies that $f(\lambda) = \|y + \lambda x\|$ is constant in $[\alpha, \beta]$. By Lemma 4.5.9, X is not strictly convex.

Conversely, assume that X is not strictly convex, and let $x, y \in X \setminus \{0\}$, $\alpha < \beta$, be such that $f(\lambda) = \|y + \lambda x\|$ is constant in $[\alpha, \beta]$. Since $f(\lambda)$ is convex, it attains the minimum in such an interval, from which it follows that $y + \gamma x \perp_B x$ for any $\alpha \leq \gamma \leq \beta$. □

4.5.2 Symmetry

Birkhoff orthogonality is, in general, not symmetric. A long process, performed by Birkhoff [41], James [87, 88] and Day [59], gave rise to the following strong theorem.

Theorem 4.5.11 *A real normed space of dimension ≥ 3 is an inner product space if and only if Birkhoff orthogonality is symmetric.*

There are many important characterizations of inner product spaces that orbit around this theorem. As an example of geometric characterization we have the Theorem of Blaschke [42]:

Blaschke's characterization. *A centrally symmetric convex body $S \subset \mathbb{R}^3$ is an ellipsoid if and only if the set of contact points between S and any circumscribed cylinder is flat.*

And as functional characterization, we have the Theorem of Kakutani [92]:

Kakutani's characterization. *A real normed space of dimension ≥ 3 is an inner product space if and only if there is a linear projection of norm one on every closed subspace.*

In two-dimensional spaces, the symmetry of Birkhoff orthogonality has other consequences. It can be symmetric in non-inner product spaces, e.g., in $\mathbb{R}^2$ endowed with the norm (see Fig. 4.2a)

$$\|(x_1, x_2)\|_{\infty,1} = \begin{cases} \max\{|x_1|, |x_2|\} & \text{if } x_1 x_2 \geq 0, \\ |x_1| + |x_2| & \text{if } x_1 x_2 \leq 0. \end{cases}$$

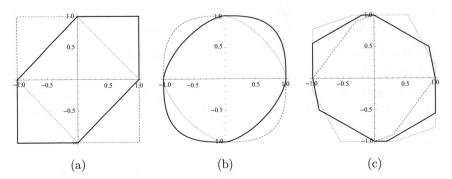

Fig. 4.2 Three examples of norms (continuous line) for which Birkhoff orthogonality is symmetric. Dashed line = original norm; dotted line = dual norm

James [87] observed that there is symmetry with any norm

$$\|(x_1, x_2)\|_{p,q} = \begin{cases} (|x_1|^p + |x_2|^p)^{1/p} & \text{if } x_1 x_2 \ge 0, \\ (|x_1|^q + |x_2|^q)^{1/q} & \text{if } x_1 x_2 \le 0, \end{cases}$$

with $\frac{1}{p} + \frac{1}{q} = 1$ (see Fig. 4.2b). Furthermore, Day [59] proved that the shape of such norms is the one of any norm for which Birkhoff orthogonality is symmetric. Let us clarify what this means.

First note that in any two-dimensional normed space X (that we can identify with $\mathbb{R}^2$) it is possible to find two unit vectors, $\bar{x}, \bar{y} \in S_X$, such that $\bar{x} \perp_B \bar{y}$ and $\bar{y} \perp_B \bar{x}$. Just consider the map $(x, y) \in S_X \times S_X \to A(x, y) := x_1 y_2 - x_2 y_1 \in \mathbb{R}$; take $\bar{x}, \bar{y} \in S_X$ such that $A(\bar{x}, \bar{y}) = \sup\{A(x, y) : x, y \in S_X\}$. Then, for any $\lambda \in \mathbb{R}$, $A(\bar{x}, \bar{y}) \ge A\left(\frac{\bar{x}+\lambda\bar{y}}{\|\bar{x}+\lambda\bar{y}\|}, \bar{y}\right) = \frac{1}{\|\bar{x}+\lambda\bar{y}\|} A(\bar{x}, \bar{y})$, which implies that $\bar{x} \perp_B \bar{y}$. Similarly, we get also $\bar{y} \perp_B \bar{x}$.

Now consider the dual space X^* (also identified to $\mathbb{R}^2$) as the set of linear functionals $(y_1, y_2) : x_1\bar{x} + x_2\bar{y} \in X \mapsto y_1 x_1 + y_2 x_2 \in \mathbb{R}$, with the norm $\|(y_1, y_2)\|^* = \sup_{(x_1,x_2)\ne(0,0)} \frac{|y_1 x_1 + y_2 x_2|}{\|(x_1, x_2)\|}$.

Finally, define in X the norm

$$\|(x_1, x_2)\| = \begin{cases} \|(x_1, x_2)\| & \text{if } x_1 x_2 \ge 0, \\ \|(x_2, -x_1)\|^* & \text{if } x_1 x_2 \le 0. \end{cases}$$

In $(X, \|\cdot\|)$ Birkhoff orthogonality is symmetric. But what is surprising is that the norm in any two-dimensional normed space with Birkhoff orthogonality being symmetric can be constructed in the above way. Such norms are usually known as *Radon norms*, and their unit circles are called *Radon curves*; see the exposition [109].

4.5.3 Additivity

The non-symmetry of Birkhoff orthogonality also forces to distinguish between right and left additivity. In fact, both properties have a completely different behavior.

Theorem 4.5.12 *In a normed space X, Birkhoff orthogonality is additive on the right if and only if X is smooth.*

Proof Assume that Birkhoff orthogonality is additive on the right. To see that X is smooth we will apply Theorem 4.5.7. Let $x, y \in X \setminus \{0\}$ and $\alpha, \beta \in \mathbb{R}$ be such that $x \perp_B \alpha x + y$ and $x \perp_B \beta x + y$. Then, from the homogeneity and right additivity, it follows that $x \perp_B (\alpha - \beta)x$, and non-degeneracy implies $\alpha = \beta$. Conversely, assume that X is smooth, and let $x, y, z \in X$ be such that $x \perp_B y, x \perp_B z$. Then, by Theorem 4.5.2, there exist $f, g \in S_{X^*}$ such that $f(x) = \|x\| = g(x)$, $f(y) = 0 = g(z)$. Since X is smooth, $f = g$, and then $f(y + z) = 0$. From Theorem 4.5.2, we have $x \perp_B y + z$. □

Theorem 4.5.13 *Let X be a normed space. Then we have*

1. *If Birkhoff orthogonality is additive on the left in X, then X is strictly convex.*
2. *If $\dim X = 2$ and X is strictly convex, then Birkhoff orthogonality is additive on the left in X.*

Proof (1) Completely similar to the first part of the proof of Theorem 4.5.12, we get this by applying Theorem 4.5.10. (2) Assume that $x, y \in X$ are linearly independent, and that $x \perp_B y$ and $z = \alpha x + \beta y \perp_B y$, with $\alpha, \beta \in \mathbb{R}, \alpha \neq 0$. By Theorem 4.5.10, $\beta = 0$, and then $x + z = (1 + \alpha)x \perp_B y$. □

Theorem 4.5.14 *A normed space X, with $\dim X \geq 3$, is an inner product space if and only if Birkhoff orthogonality is additive on the left in X.*

Proof If X is an inner product space, then it is clear that Birkhoff orthogonality is additive in X. Conversely, assume that $\dim X \geq 3$ and that Birkhoff orthogonality is additive on the left in X. To prove that X is an inner product space we can assume, without loss of generality, that $\dim X = 3$, and consider $X \equiv \mathbb{R}^3$. We will see that the unit sphere S_X is an ellipsoid by applying Blaschke's characterization. Let C be a cylinder circumscribed about S_X with axis following the direction of $y \in S_X$. Let $x, z \in C \cap S_X$ be linearly independent. Then $x \perp_B y, z \perp_B y$, and $\{x, y, z\}$ is a basis in X. Let $w = \alpha x + \beta y + \gamma z \in C \cap S_X$, with $\alpha, \beta, \gamma \in \mathbb{R}$. Then, $w \perp_B y$. Since Birkhoff orthogonality is homogeneous, we have $-\alpha x \perp_B y$ and $-\gamma z \perp_B y$, and since we are assuming that it is additive on the left, we obtain $\beta y \perp_B y$, which implies $\beta = 0$. Therefore, $C \cap S_X$ is flat. □

The requirements on Theorem 4.5.14 can be weakened [4, 105]. One can replace additivity on the left by *additivity on the left for biorthogonal pairs of vectors*: $x \perp_B y, y \perp_B x, x \perp_B z, y \perp_B z \Rightarrow x + y \perp_B z$.

4.5.4 Orthogonal Diagonals

The original proof of the next theorem and other related results can be found in [37].

Theorem 4.5.15 *Let X be a normed space. For any $x, y \in X \setminus \{0\}$ there exists a unique $\rho > 0$ such that $x + \rho y \perp_B x - \rho y$ (Fig. 4.3).*

Proof If $x = \mu y$ for some $\mu \in \mathbb{R}$, then $\rho = |\mu|$. Thus we can assume that x and y are linearly independent, and $\|x\| = \|y\| = 1$. Moreover, we can identify the space they span, $L = \langle\{x, y\}\rangle$, with $\mathbb{R}^2$, in such a way that $x = (1, 0)$ and $y = (0, 1)$. Let $x : \theta \in [0, 2\pi] \mapsto x(\theta) = (\cos\theta, \sin\theta)/\|(\cos\theta, \sin\theta)\| \in S_L$ be an angle parametrization of the unit sphere S_L in L. Thus, $x = x(0)$ and $y = x(\pi/2)$.

For each point $x(\theta)$, let us consider the interval $[\mu_1(\theta), \mu_2(\theta)]$ of angles that the axis spanned by x forms with the support lines to S_L at $x(\theta)$. Then $x(\theta) \perp_B x(\mu)$ for $\mu \in [\mu_1(\theta), \mu_2(\theta)]$. Since the unit ball is convex, these angles increase monotonically as θ does, i.e., if $\theta_1 < \theta_2$ then $\mu_2(\theta_1) \leq \mu_1(\theta_2)$; and $\mu_2(\theta_1) = \mu_1(\theta_2)$ if and only if $[x(\theta_1), x(\theta_2)] \subset S_L$. Moreover, for each θ, $\mu_1(\theta) = \mu_2(\theta)$ if and only if $x(\theta)$ is a smooth point. Since $x, -x, y, -y$ belongs to S_L, and S_L is a convex curve, we have also that $\frac{\pi}{4} \leq \mu_1(0) \leq \mu_2(0) \leq \frac{3\pi}{4} \leq \mu_1(\frac{\pi}{2}) \leq \mu_2(\frac{\pi}{2}) \leq \frac{5\pi}{4}$. Thus, the set $\mathcal{C} = \{(\theta, \mu) : \theta \in [0, \frac{\pi}{2}], \mu \in [\mu_1(\theta), \mu_2(\theta)]\}$ is a continuous curve in $\mathbb{R}^2$ that goes from $(0, \mu_1(0))$ to $(\frac{\pi}{2}, \mu_2(\frac{\pi}{2}))$. Then, $\mathcal{C}$ cuts the line that goes through the points $(0, \pi)$ and $(\frac{\pi}{2}, \frac{\pi}{2})$ at a unique point $(\theta_0, \mu_0) = (\theta_0, \pi - \theta_0)$, with $0 < \theta_0 < \frac{\pi}{2}, \mu_1(\theta_0) \leq \mu_0 \leq \mu_2(\theta_0)$. Therefore, $x(\theta_0) \perp_B x(\pi - \theta_0)$, which implies (recall that Birkhoff orthogonality is homogeneous) that $x + \rho y \perp_B x - \rho y$, with $\rho = \tan\theta_0$. The uniqueness of ρ follows from the uniqueness of θ_0. ☐

Fig. 4.3 $x + \rho y \perp_B x - \rho y$

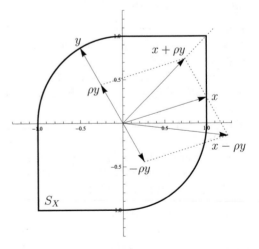

Theorem 4.5.16 *If $x, y \in X \setminus \{0\}$ are such that $x + \rho y \perp_B x - \rho y$, with $\rho > 0$, then*

$$\frac{1}{3}\frac{\|x\|}{\|y\|} \le \rho \le 3\frac{\|x\|}{\|y\|}.$$

Moreover, if x and y are linearly independent, then $\rho = 3\frac{\|x\|}{\|y\|}$ if and only if the unit sphere S_L of the subspace $L = \langle\{x, y\}\rangle$, spanned by x and y, is the parallelogram with vertices $\pm\frac{x}{\|x\|}$ and $\pm\frac{1}{2}\left(\frac{x}{\|x\|} + 3\frac{y}{\|y\|}\right)$. If $\rho = \frac{1}{3}\frac{\|x\|}{\|y\|}$, then an analogous result follows by interchanging the roles of x and y (Fig. 4.4).

Proof Let $\rho > 0$ be such that $x + \rho y \perp_B x - \rho y$. Then $\|x + \rho y + \lambda(x - \rho y)\| \ge \|x + \rho y\|$ for every $\lambda \in \mathbb{R}$. By taking $\lambda = 1$, it follows that $2\|x\| \ge \|x + \rho y\| \ge \rho\|y\| - \|x\|$, which leads to the upper bound. Similarly, by taking $\lambda = -1$, the lower bound follows.

Assume now that $x + 3\frac{\|x\|}{\|y\|}y \perp_B x - 3\frac{\|x\|}{\|y\|}y$. To simplify, let $u = \frac{1}{2}\left(\frac{x}{\|x\|} + 3\frac{y}{\|y\|}\right)$. Then

$$\left\|\frac{x}{\|x\|} + 3\frac{y}{\|y\|} + \lambda\left(\frac{x}{\|x\|} - 3\frac{y}{\|y\|}\right)\right\| \ge 2\|u\|$$

holds for every $\lambda \in \mathbb{R}$. By taking $\lambda = 1$, it follows that $2 \ge 2\|u\| \ge 2$, and then $u \in S_L$. Moreover, $\frac{y}{\|y\|} = \frac{1}{3}\left(\frac{-x}{\|x\|}\right) + \frac{2}{3}u$, which implies that $\left[\frac{-x}{\|x\|}, u\right] \subset S_L$. Furthermore, by taking $\lambda = \frac{1}{2}$, it follows that $1 = \|u\| \le \left\|\frac{1}{2}\frac{x}{\|x\|} + \frac{1}{2}u\right\| \le 1$, and then, $\left[\frac{x}{\|x\|}, u\right] \subset S_L$.

Conversely, assume that S_L is the parallelogram with vertices $\pm\frac{x}{\|x\|}$ and $\pm u$. Then, the function

$$f(\lambda) := \left\|\frac{x}{\|x\|} + 3\frac{y}{\|y\|} + \lambda\left(\frac{x}{\|x\|} - 3\frac{y}{\|y\|}\right)\right\| = 2\left\|\lambda\frac{x}{\|x\|} + (1 - \lambda)u\right\|$$

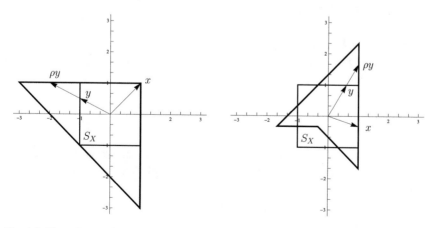

Fig. 4.4 These figures show, for a fixed x, the set of points ρy, such that $y \in S_X$ and $x + \rho y \perp_B x - \rho y$

satisfies $f(\lambda) = 2$ for every $\lambda \in [0, 1]$. Since f is convex, it follows that $f(\lambda) \geq 2 = \|2u\|$ for every $\lambda \in \mathbb{R}$, which implies $x + 3\frac{\|x\|}{\|y\|}y \perp_\mathrm{B} x - 3\frac{\|x\|}{\|y\|}y$.

Finally, note that $x + \frac{1}{3}\frac{\|x\|}{\|y\|}y \perp_\mathrm{B} x - \frac{1}{3}\frac{\|x\|}{\|y\|}y$ if and only if $y + 3\frac{\|y\|}{\|x\|}x \perp_\mathrm{B} y - 3\frac{\|y\|}{\|x\|}x$. $\qquad\square$

From Theorem 4.5.15 it follows that in any normed space X we can consider the map

$$S_X \times S_X \longrightarrow \mathbb{R}^+,$$
$$(x, y) \mapsto \rho(x, y)$$

where $\rho = \rho(x, y)$ is the only scalar that satisfies $x + \rho y \perp_\mathrm{B} x - \rho y$. Since Birkhoff orthogonality has the continuity property, and $\rho(x, y)$ is bounded and univocally determined, simple compactness arguments show that such a map is continuous (with the product topology) in $S_X \times S_X$. Note also that $\rho(x, y) = \rho(y, x)^{-1}$. Moreover, since $S_X \times S_X$ is connected, it follows from Theorem 4.5.16 that the set

$$A_X = \{\rho(x, y) : x, y \in S_X\}$$

is an interval in $\mathbb{R}^+$ of the form $[\frac{1}{a}, a]$, with $1 \leq a \leq 3$. From Theorem 4.5.16 it follows that this interval is maximum, i.e., $A_X = [\frac{1}{3}, 3]$ if and only if the unit sphere S_X has a section through the origin that is a parallelogram. On the other hand, this interval is minimum, i.e., $A_X = \{1\}$, only in inner product spaces: just note that the property $[x, y \in S_X \Rightarrow x + y \perp_\mathrm{B} x - y]$ is equivalent to the property $[x, y \in X, x \perp_\mathrm{I} y \Rightarrow x \perp_\mathrm{B} y]$, which is (see Sect. 4.8) characteristic of inner product spaces. Furthermore, the identity $A_X = \{1\}$ is equivalent to the (at first) weaker identity $\{\rho(x, y) : x, y \in S_X, x \perp_\mathrm{B} y\} = \{1\}$, because the property $[x, y \in S_X, x \perp_\mathrm{B} y \Rightarrow x + y \perp_\mathrm{B} x - y]$ also characterizes inner product spaces [34]. Moreover, simple continuity arguments show (by considering, e.g., $\ell_p(\mathbb{R}^2)$, $1 \leq p \leq 2$) that for any $a \in [1, 3]$ there exists a normed space X such that $A_X = [\frac{1}{a}, a]$.

The above gives, in particular, the next theorem.

Theorem 4.5.17 *A normed space X is an inner product space if and only if for any* $x, y \in X \setminus \{0\}$,

$$x + \frac{\|x\|}{\|y\|}y \perp_\mathrm{B} x - \frac{\|x\|}{\|y\|}y.$$

The fact that an orthogonality has the property of existence of orthogonal diagonals led Benítez and Del Río [40, 62] to weaken the Parallelogram Law characterization of inner product spaces.

Corollary 4.5.18 *A normed space X is an inner product space if and only if*

$$x, y \in X, \ x \perp_B y \quad \Rightarrow \quad \|x + y\|^2 + \|x - y\|^2 \approx 2(\|x\|^2 + \|y\|^2),$$

where "$\approx$" means "$\leq$" or "$\geq$".

Proof It is known (see, e.g., Day [60, p.154]) that the following property characterizes inner product spaces:

$$\forall x, y \in S_X, \ \exists \lambda, \mu \neq 0 \text{ such that } \|\lambda x + \mu y\|^2 + \|\lambda x - \mu y\|^2 \approx 2(\lambda^2 + \mu^2).$$

Now, let $x, y \in S_X$. From Theorem 4.5.15 there exists $\rho > 0$ such that $x + \rho y \perp_B x - \rho y$. From the hypothesis we have

$$\|x + \rho y + x - \rho y\|^2 + \|x + \rho y - x + \rho y\|^2 \approx 2(\|x + \rho y\|^2 + \|x - \rho y\|^2),$$

and then $\|x + \rho y\|^2 + \|x - \rho y\|^2 \approx 2(1 + \rho^2)$. $\qquad\qquad\qquad\qquad\qquad\square$

4.5.5 Birkhoff Orthogonality and Hyperplanes

By considering the closed hyperplane $H = \{x \in X : f(x) = 0\}$, Theorem 4.5.3 can be rewritten.

Theorem 4.5.19 *For any $x \in X$, there exists a closed and homogeneous hyperplane $H \subset X$ such that $x \perp_B H$.*

Proof Just consider Theorem 4.5.2. $\qquad\qquad\qquad\qquad\qquad\qquad\qquad\qquad\square$

Furthermore, for any $x \in X \setminus \{0\}$, the hyperplane H in Theorem 4.5.19 is unique if and only if X is smooth.

It is worthwhile to mention here that, for Banach spaces, the similar property

For any closed and homogeneous hyperplane $H \subset X$ there exists $x \in X \setminus \{0\}$ such that $x \perp_B H$.

is equivalent to the reflexivity of X [89]. Therefore, in Minkowski spaces the above property also holds.

Moreover, Marchaud [104] relaxed the hypothesis in Blaschke's characterization (see p. 107):

Marchaud's characterization. *A centrally symmetric convex body $S \subset \mathbb{R}^3$ is an ellipsoid if and only if the set of contact points between S and any circumscribed cylinder contains a plane section of S.*

The next theorem follows directly from this characterization, while the Kakutani characterization leads to Theorem 4.5.21. The proofs of both theorems can also be found in [87].

Theorem 4.5.20 *A normed space X, with* $\dim X \geq 3$, *is an inner product space if and only if for any $x \in X$ there exists a closed and homogeneous hyperplane $H \subset X$ such that $H \perp_B x$.*

Theorem 4.5.21 *A normed space X, with* $\dim X \geq 3$,[1] *is an inner product space if and only if for any closed and homogeneous hyperplane $H \subset X$ there exists $x \in X \setminus \{0\}$ such that $H \perp_B x$.*

4.6 Isosceles Orthogonality

Definition 4.6.1 Two vectors x and y are said to be *isosceles orthogonal* $(x \perp_I y)$ if

$$\|x - y\| = \|x + y\|.$$

Isosceles orthogonality was introduced by James in [86]. Aside from having the properties of non-degeneracy, simplification and continuity, isosceles orthogonality is symmetric. Thus, we do not need to distinguish between properties on the left and on the right, as in the case of Birkhoff orthogonality. In contrast, since isosceles orthogonality is not (in general) homogeneous, we must distinguish between existence and α-existence properties. Nonetheless, note that trivially $x \perp_I y$ implies $x \perp_I -y$. Observe also that if $x \perp_I y$, then $\|x + y\| \geq \max\{\|x\|, \|y\|\}$.

4.6.1 Existence and Uniqueness

Theorem 4.6.2 *Isosceles orthogonality has the existence property, i.e., for any $x, y \in X$, linearly independent, and any $\rho > 0$, there exists $z = \alpha x + \beta y$, with $\alpha, \beta \in \mathbb{R}$, $\beta > 0$, such that $\|z\| = \rho$ and $x \perp_I z$.*

Proof For $\theta \in [0, \pi]$, let $z(\theta) = \rho \frac{(\cos\theta)x + (\sin\theta)y}{\|(\cos\theta)x + (\sin\theta)y\|}$. The continuous function $f : \theta \in [0, \pi] \mapsto f(\theta) = \|x + z(\theta)\| - \|x - z(\theta)\|$ satisfies $f(0) = -f(\pi)$. Therefore, there exists $\theta_0 \in (0, \pi)$ such that $f(\theta_0) = 0$, i.e., $x \perp_I z(\theta_0)$. □

It follows from Theorem 4.6.2 that for any $x \in S_X$, we can find in any two-dimensional subspace $L \subset X$ that contains x, a vector $y \in S_X$ such that $x \perp_I y$. The next theorem shows where the other vectors in L, that are isosceles orthogonal to x, are located.

---

[1] If $\dim X = +\infty$, it is necessary to assume that X is complete.

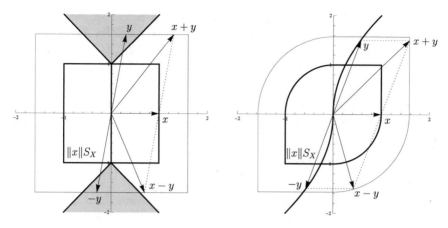

Fig. 4.5 Set of points $y \in X$, such that $x \perp_{\mathrm{I}} y$

Theorem 4.6.3 ([3]) *Let* $x, y \in S_x$ *be such that* $x \perp_{\mathrm{I}} y$. *If* $x \perp_{\mathrm{I}} \alpha x + \beta y$, $\alpha, \beta \in \mathbb{R}$, *then*

$$|\alpha| \leq \begin{cases} \min\left\{|\beta|(\|x+y\|-1), \frac{(1-|\beta|)(2-\|x+y\|)}{\|x+y\|}\right\} & \text{if } |\beta| \leq 1, \\ \max\left\{|\beta|(2-\|x+y\|)-1, \frac{(|\beta|-1)(2-\|x+y\|)}{\|x+y\|}\right\} & \text{if } |\beta| \geq 1, \end{cases}$$

and the estimations are sharp.

The inequalities in Theorem 4.6.3 show right away that if $x, y \in S_X$, $x \perp_{\mathrm{I}} y$ and $x \perp_{\mathrm{I}} \alpha x + y$ then $\alpha = 0$. The next corollary completes this result (see Fig. 4.5).

Corollary 4.6.4 *Let* $x, y \in S_X$. *If* $x \perp_{\mathrm{I}} \alpha_1 x + \beta y$ *and* $x \perp_{\mathrm{I}} \alpha_2 x + \beta y$, *with* $|\beta| \leq 1$, *then* $\alpha_1 = \alpha_2$.

Proof We can assume that, without loss of generality, $\alpha_2 \leq \alpha_1$, $\alpha_1 \geq 0$ and $\beta > 0$. Let us show first that then $\alpha_1 \leq 1$. Let $\gamma = \|(\alpha_1 - 1)x + \beta y\| = \|(\alpha_1 + 1)x + \beta y\|$. Then, $(\alpha_1+1)\gamma = \|(\alpha_1-1)[(\alpha_1+1)x+\beta y]+2\beta y\| \leq |\alpha_1-1|\gamma+2\beta \leq |\alpha_1-1|\gamma+2$. Thus $(\alpha_1+1-|\alpha_1-1|)\gamma \leq 2$. Moreover, $\alpha_1+1 = \|(\alpha_1+1)x+\beta y-\beta y\| \leq \gamma+1$. Therefore, $(\alpha_1 + 1 - |\alpha_1 - 1|)\alpha_1 \leq (\alpha_1 + 1 - |\alpha_1 - 1|)\gamma \leq 2$, leading to $\alpha_1 \leq 1$.

Assume now that $\alpha_2 < \alpha_1$. Then, $0, \alpha_2 + 1$ and $\alpha_1 - 1$ are in the interior of the interval $[\alpha_2 - 1, \alpha_1 + 1]$. Since the function $f(\lambda) = \|\lambda x + \beta y\|$ is convex, and $f(\alpha_1-1) = f(\alpha_1+1)$ and $f(\alpha_2-1) = f(\alpha_2+1)$, it follows that $f(\lambda) = f(0) = \beta$ for $\lambda \in [\alpha_2 - 1, \alpha_1 + 1]$. In particular we have $\|\alpha_1 x + \beta y\| = \|\alpha_2 x + \beta y\| = \beta$. Moreover, for $i = 1, 2$, $\alpha_i - \beta$ and $\alpha_i + \beta$ are in $[\alpha_2 - 1, \alpha_1 + 1]$, and thus $\beta x \perp_{\mathrm{I}} \alpha_i x + \beta y$. Then,

$$x \perp_{\mathrm{I}} \frac{\alpha_1 x + \beta y}{\|\alpha_1 x + \beta y\|}, \quad x \perp_{\mathrm{I}} \left(\frac{\alpha_2 - \alpha_1}{\beta}\right) x + \frac{\alpha_1 x + \beta y}{\|\alpha_1 x + \beta y\|}$$

and, from Theorem 4.6.3, it follows that $\alpha_1 = \alpha_2$, against the hypothesis. $\qquad\square$

Let $x \in X \setminus \{0\}$. The next theorem shows that for any two-dimensional subspace L that contains x, and any $\rho \leq \|x\|$, there is only one (except for the sign) vector $y \in L$ such that $\|y\| = \rho$ and $x \perp_I y$. On the other hand, Theorem 4.6.6 shows that if we do not restrict the size of ρ, then the uniqueness of y characterizes strictly convex spaces.

Theorem 4.6.5 ([3]) *Let $x, y \in S_X$ be linearly independent. If $x \perp_I \alpha_1 x + \beta_1 y$ and $x \perp_I \alpha_2 x + \beta_2 y$, with $\|\alpha_1 x + \beta_1 y\| = \|\alpha_2 x + \beta_2 y\| \leq 1$, then $(\alpha_1, \beta_1) = \pm(\alpha_2, \beta_2)$.*

Theorem 4.6.6 *A normed space X is strictly convex if and only if isosceles orthogonality has the uniqueness property.*

Proof Assume that X is not strictly convex. Let $u, v \in S_X$ be linearly independent and such that $\|u + v\| = 2$. Take $x = u - v$ and $y = u + v$. The convex function $f(\lambda) = \|y + \lambda x\|$ satisfies $f(-1) = f(0) = f(1) = 2$. Then $f(\lambda) = 2$ for any $\lambda \in [-1, 1]$. Let $z_\mu = \mu x + 2y$. Then, for $\mu \in [-1, 1]$, $\|z_\mu\| = 2f\left(\frac{\mu}{2}\right) = 4$, and
$$\|x + z_\mu\| = 2f\left(\tfrac{\mu+1}{2}\right) = 4 = 2f\left(\tfrac{\mu-1}{2}\right) = \|x - z_\mu\|,$$ i.e., $x \perp_I z_\mu$, against the uniqueness property.

Conversely, assume that X is strictly convex. Let $x, y \in X$ be linearly independent, and $z_1 = \alpha_1 x + \beta_1 y$, $z_2 = \alpha_2 x + \beta_2 y$, with $\beta_1 > 0$, $\beta_2 > 0$, $\|z_1\| = \|z_2\|$, such that $x \perp_I z_1$ and $x \perp_I z_2$. We can assume that $\alpha_2 \beta_1 - \alpha_1 \beta_2 \geq 0$, and (by Theorem 4.6.5) that $\|x\| < \|z_1\|$. Then

$$(\alpha_2 \beta_1 - \alpha_1 \beta_2)\|x\| = \|\beta_1 z_2 - \beta_2 z_1\| \geq |\beta_1 - \beta_2|\|z_1\| \geq |\beta_1 - \beta_2|\|x\|, \qquad (4.2)$$

from which it follows that $\frac{\alpha_1-1}{\beta_1} \leq \frac{\alpha_2-1}{\beta_2}$ and $\frac{\alpha_1+1}{\beta_1} \leq \frac{\alpha_2-1}{\beta_2}$. Let $f(\lambda) = \|y + \lambda x\|$, $\forall \lambda \in \mathbb{R}$. Then, $f\left(\frac{\alpha_1-1}{\beta_1}\right) = f\left(\frac{\alpha_1+1}{\beta_1}\right)$ and $f\left(\frac{\alpha_2-1}{\beta_2}\right) = f\left(\frac{\alpha_2+1}{\beta_2}\right)$. By Lemma 4.5.9, $f(\lambda)$ cannot be constant in any interval. Therefore, $\frac{\alpha_1-1}{\beta_1} = \frac{\alpha_2-1}{\beta_2}$ and $\frac{\alpha_1+1}{\beta_1} = \frac{\alpha_2+1}{\beta_2}$, which gives $\frac{\alpha_1}{\beta_1} = \frac{\alpha_2}{\beta_2}$. By (4.2), $z_1 = z_2$. $\qquad \square$

4.6.2 α-Existence and α-Uniqueness

The fact that isosceles orthogonality has the α-existence property (that Roberts orthogonality lacks) motivated James [86] to consider this orthogonality in 1945.

Theorem 4.6.7 *For any $x, y \in X$, there exists $\alpha \in \mathbb{R}$ such that $x \perp_I \alpha x + y$.*

Proof Let $f(\lambda) = \|(\lambda + 1)x + y\| - \|(\lambda - 1)x + y\|$, $\lambda \in \mathbb{R}$. Then

$$\lim_{\lambda \to +\infty} f(\lambda) = \lim_{\lambda \to +\infty} \left((\lambda - 1)\|x + \tfrac{1}{\lambda+1}y\| + 2\|x + \tfrac{1}{\lambda+1}y\| - \|(\lambda - 1)x + y\| \right),$$

and, since

$$\left| (\lambda - 1) \|x + \tfrac{1}{\lambda+1}y\| - \|(\lambda - 1)x + y\| \right|$$

$$\leq \left\| (\lambda - 1)(x + \tfrac{1}{\lambda+1}y) - ((\lambda - 1)x + y) \right\| = \tfrac{2}{\lambda+1}\|y\|,$$

it follows that $\lim_{\lambda \to +\infty} f(\lambda) = 2\|x\|$. Moreover, since this limit does not depend on y, it follows that $\lim_{\lambda \to -\infty} f(\lambda) = \lim_{\lambda \to +\infty} f(-\lambda) = -2\|x\|$. Therefore, there must exist $\alpha \in \mathbb{R}$ such that $f(\alpha) = 0$, i.e., $x \perp_I \alpha x + y$. □

Theorem 4.6.8 *If* $x \perp_I \alpha x + y$, $x \neq 0$, *then*

$$|\alpha| \leq \begin{cases} \dfrac{\|y\|}{\|x\|} & \text{if } \|y\| \leq \|x\|, \\[2mm] \dfrac{2\|y\| - \|x\|}{\|x\|} & \text{if } \|y\| \geq \|x\|. \end{cases}$$

If x *and* y *are linearly independent, then the bound is attained only if* X *is not strictly convex. Moreover, if* $\|y\| > \|x\|$, *then the bound is attained only if the unit sphere in the subspace spanned by* x *and* y *is a parallelogram.*

Proof We can assume, without loss of generality, that $\alpha \geq 0$. To simplify the notation, let $z_\alpha^+ = (\alpha + 1)x + y$, $z_\alpha^- = (\alpha - 1)x + y$. Assume that $\|y\| \leq \|x\|$. Then

$$\|z_\alpha^-\| = \|z_\alpha^+\| \geq (\alpha + 1)\|x\| - \|y\| \geq \alpha\|x\| \geq \alpha\|y\|$$

$$= \frac{\alpha}{2}\left\| (\alpha + 1)z_\alpha^- - (\alpha - 1)z_\alpha^+ \right\| \geq \frac{\alpha}{2}(\alpha + 1 - |\alpha - 1|)\|z_\alpha^+\|.$$

Therefore, $2 \geq \alpha(\alpha + 1 - |\alpha - 1|)$, which implies $\alpha \leq 1$. Moreover, $(\alpha + 1)\|x\| \leq \|z_\alpha^+\| + \|y\| = \|z_\alpha^-\| + \|y\| \leq (1 - \alpha)\|x\| + 2\|y\|$, which gives $\alpha \leq \|y\|/\|x\|$. On the other hand, assume that $\|x\| \leq \|y\|$. If $\alpha \leq 1$, then $\alpha \leq 1 \leq (2\|y\| - \|x\|)/\|x\|$. Moreover, if $\alpha \geq 1$, then

$$\|y\| = \frac{1}{2}\left\| (\alpha + 1)z_\alpha^- - (\alpha - 1)z_\alpha^+ \right\|$$

$$\geq \frac{1}{2}(\alpha + 1 - \alpha + 1)\|z_\alpha^+\| \geq (\alpha + 1)\|x\| - \|y\|,$$

from which it follows that $\alpha \leq (2\|y\| - \|x\|)/\|x\|$.

Assume now that x and y are linearly independent, $\|y\| \leq \|x\|$, and $x \perp_I \alpha x + y$ with $\alpha = \|y\|/\|x\|$. Then it follows from the above that $\|z_\alpha^+\| = \|x\|$, which implies that

$$\frac{x}{\|x\|} = \frac{\|x\|}{\|x\| + \|y\|} \frac{z_\alpha^+}{\|z_\alpha^+\|} + \frac{\|y\|}{\|x\| + \|y\|} \frac{-y}{\|y\|},$$

and then the segment $\left[z_\alpha^+ / \|z_\alpha^+\|, -y/\|y\| \right]$ is included in S_X.

Finally, assume that x and y are linearly independent, $\|y\| > \|x\|$, and $x \perp_I \alpha x + y$ with $\alpha = (2\|y\| - \|x\|)/\|x\|$. Then

$$\frac{\|y\|}{\|x\|}\|z_\alpha^-\| = \left\| \frac{\|y\| - \|x\|}{\|x\|} z_\alpha^+ + y \right\| \le \frac{\|y\| - \|x\|}{\|x\|}\|z_\alpha^+\| + \|y\|,$$

from which it follows that $\|z_\alpha^+\| \le \|y\|$. Moreover, $2\|y\| = \|z_\alpha^+ - y\| \le \|z_\alpha^+\| + \|y\|$. Therefore, $\|z_\alpha^+\| = \|y\|$. Then

$$\frac{z_\alpha^-}{\|z_\alpha^-\|} = \frac{\|y\| - \|x\|}{\|y\|}\frac{z_\alpha^+}{\|z_\alpha^+\|} + \frac{\|x\|}{\|y\|}\frac{y}{\|y\|} \quad \text{and} \quad \frac{x}{\|x\|} = \frac{1}{2}\frac{z_\alpha^+}{\|z_\alpha^+\|} - \frac{1}{2}\frac{y}{\|y\|}$$

which implies that the unit sphere in the subspace spanned by x and y is the parallelogram with vertices $\pm y/\|y\|$ and $\pm z_\alpha^+/\|z_\alpha^+\|$. $\square$

The following remarks complete Theorem 4.6.8.

Remarks 4.6.9

(a) Assume that X is not strictly convex, i.e., that there exist $u, v \in S_X$ such that $[u, v] \subset S_X$. Take $x = \beta u + (1 - \beta)v$, $y = \frac{\beta-1}{\beta}v$, with $\frac{1}{2} \le \beta \le 1$. Then $\|x\| = 1 \ge \|y\|$, and for $\alpha = \frac{\|y\|}{\|x\|}$ we have $(\alpha + 1)x + y = u$ and $(\alpha - 1)x + y = (2\beta - 1)(-u) + (2 - 2\beta)(-v) \in S_X$, which implies $x \perp_I \alpha x + y$.
(b) Assume that the parallelogram with vertices $\pm u$, $\pm v$ belongs to S_X. Take $x = (v - u)/2$, $y = \lambda u$, with $\lambda \ge 1$. Then $\|x\| = 1 \le \lambda = \|y\|$, and for $\alpha = 2\lambda - 1$ we have $(\alpha + 1)x + y = \lambda v$, and $(\alpha - 1)x + y = \lambda(\frac{\lambda-1}{\lambda}v + \frac{1}{\lambda}u) \in \lambda S_x$. Then, $x \perp_I \alpha x + y$.

Theorem 4.6.10 *If $x \perp_I \alpha_1 x + y$ and $x \perp_I \alpha_2 x + y$, with $\alpha_1 < \alpha_2$, then $x \perp_I \alpha x + y$ for $\alpha_1 \le \alpha \le \alpha_2$.*

Proof The convex function $f(\lambda) = \|\lambda x + y\|$ satisfies $f(\alpha_1 - 1) = f(\alpha_1 + 1)$, $f(\alpha_2 - 1) = f(\alpha_2 + 1)$, with $\alpha_1 - 1 < \alpha_2 - 1$ and $\alpha_1 + 1 < \alpha_2 + 1$. Hence it must be constant in $[\alpha_1 - 1, \alpha_2 + 1]$, which implies that $x \perp_I \alpha x + y$ for $\alpha_1 \le \alpha \le \alpha_2$. $\square$

From Corollary 4.6.4 it follows that if $\|y\| \le \|x\| \ne 0$, then the scalar α in Theorem 4.6.7 is unique. Moreover, as Kapoor and Prasad [93] showed in the next theorem, isosceles orthogonality is α-unique if and only if X is strictly convex. Thus (recall Theorem 4.6.6), uniqueness and α-uniqueness are equivalent properties for isosceles orthogonality.

Theorem 4.6.11 *A normed space X is strictly convex if and only if isosceles orthogonality has the α-uniqueness property.*

Proof Let $x, y \in X \setminus \{0\}$ and $\alpha_1 < \alpha_2$ be such that $x \perp_I \alpha_1 x + y$, $x \perp_I \alpha_2 x + y$. Then the function $f(\lambda) = \|\lambda x + y\|$ is constant in the interval $[\alpha_1 - 1, \alpha_2 + 1]$, which implies that X is not strictly convex (Lemma 4.5.9). Conversely, assume that X is not strictly convex. Take $u, v \in S_X$, $u \ne v$, such that $\|u + v\| = 2$. Let $x = \frac{u-v}{2}$,

$y = u + v$, and $f(\lambda) = \|y + \lambda x\|$, $\lambda \in \mathbb{R}$. Then $f(-2) = f(0) = f(2) = 2$, which implies $f(\lambda) = 2$ for any $\lambda \in [-2, 2]$. Hence, $f(\alpha - 1) = f(\alpha + 1)$ (i.e., $x \perp_I \alpha x + y$) for any $\alpha \in [-1, 1]$. $\square$

4.6.3 Homogeneity

One of the main reasons that makes isosceles orthogonality interesting is that, as James [86] proved, it is homogeneous only in inner product spaces. This property (Theorem 4.6.13) follows easily from the following strong characterization by Ficken [74].

Theorem 4.6.12 *A normed space X is an inner product space if and only if*

$$x, y \in X, \ \|x\| = \|y\| \quad \Rightarrow \quad \|\alpha x + y\| = \|x + \alpha y\|, \ \textit{for all } \alpha \in \mathbb{R}.$$

Proof The necessity is evident. To see the sufficiency, we begin by showing that from the hypothesis it follows that X is strictly convex. Assume that there exist $u, v \in S_X$, $u \neq v$, such that the segment $[u, v]$ is a maximal segment contained in S_X. Take $x = (u+v)/2$, $y = -v$. Then $\|x\| = \|y\|$ and $\|3x+y\| = 2\|\frac{3}{4}u+\frac{1}{4}v\| = 2$. Then $2 = \|x + 3y\| = 2\|\frac{1}{4}u - \frac{5}{4}v\|$, which implies that $[u, v]$ is not maximal.

Now, to see that X is an inner product space, we will show that the parallelogram equality

$$\|x + y\|^2 + \|x - y\|^2 = 2(\|x\|^2 + \|y\|^2)$$

holds for any $x, y \in X$. For this, we can assume that x and y are linearly independent and that $\|x\| \leq \|y\|$. Let us consider the convex function

$$f(\lambda) = \left\| 2\|y\|^2 x + \lambda y \right\|, \quad \lambda \in \mathbb{R},$$

and let

$$\lambda_1 = \|x - y\|^2 - 2\|y\|^2, \qquad \lambda_2 = \|x - y\|^2 - 2\|x\|^2,$$

$$\lambda_3 = 2\|x\|^2 - \|x + y\|^2, \qquad \lambda_4 = 2\|y\|^2 - \|x + y\|^2.$$

From the hypothesis, we get the following identities:

$$f(\lambda_1) = \|y\|\|x - y\|^2 \left\| \frac{2\|y\|}{\|x - y\|} \left(\frac{x - y}{\|x - y\|} \right) + \frac{y}{\|y\|} \right\|$$

$$= \|y\|\|x - y\|^2 \left\| \frac{x - y}{\|x - y\|} + \frac{2\|y\|}{\|x - y\|} \left(\frac{y}{\|y\|} \right) \right\|$$

$$= \|y\|\|x - y\|\|x + y\|,$$

$$f(\lambda_2) = 2\|x\|\|y\|^2 \left\| \frac{x}{\|x\|} + \frac{\|x-y\|^2 - 2\|x\|^2}{2\|x\|\|y\|} \left(\frac{y}{\|y\|}\right) \right\|$$

$$= 2\|x\|\|y\|^2 \left\| \frac{\|x-y\|^2 - 2\|x\|^2}{2\|x\|\|y\|} \left(\frac{x}{\|x\|}\right) + \frac{y}{\|y\|} \right\|$$

$$= 2\|x\|\|y\|\|x-y\| \left\| \frac{\|x-y\|}{2\|x\|} \left(\frac{x}{\|x\|}\right) - \frac{x-y}{\|x-y\|} \right\|$$

$$= 2\|x\|\|y\|\|x-y\| \left\| \frac{x}{\|x\|} - \frac{\|x-y\|}{2\|x\|} \left(\frac{x-y}{\|x-y\|}\right) \right\|$$

$$= \|y\|\|x-y\|\|x+y\|,$$

and, similarly, $f(\lambda_3) = f(\lambda_4) = \|y\|\|x-y\|\|x+y\|$. Thus, $f(\lambda)$ takes the same value at the four points λ_i. If $\|x\| < \|y\|$, then $\lambda_1 < \lambda_2$ and $\lambda_3 < \lambda_4$. Since X is strictly convex, f cannot be constant in any interval (Lemma 4.5.9). Therefore, $\lambda_1 = \lambda_3$ and $\lambda_2 = \lambda_4$, and the parallelogram equality holds. If $\|x\| = \|y\|$, then, applying the same argument to the points x and $y_n = (1 + \frac{1}{n})y$, $n \in \mathbb{N}$, it follows that the parallelogram equality holds for x and y_n, and making $n \to +\infty$, for x and y. □

Theorem 4.6.13 *A normed space X is an inner product space if and only if isosceles orthogonality is homogeneous.*

Proof The necessity is evident. To prove the sufficiency we apply Theorem 4.6.12. Assume that isosceles orthogonality is homogeneous. Let $x, y \in X$ be such that $\|x\| = \|y\|$. Then $x + y \perp_I x - y$. For $\alpha \neq -1$, let $\lambda = (\alpha - 1)/(\alpha + 1)$. Then $x + y \perp_I \lambda(x - y)$, i.e., $\|\alpha x + y\| = \|x + \alpha y\|$. □

The next two theorems reduce the hypothesis above. The first one is just a reformulation of Theorem 4.4.2 and gives a slight improvement. The second one, by Lorch [102], reduces the hypothesis in Theorem 4.6.13 significantly.

Theorem 4.6.14 *A normed space X is an inner product space if and only if for any two-dimensional subspace $P \subset X$ and any $x \in P$ there exists $y \in P$ such that $x \perp_I \lambda y$ for every $\lambda \in \mathbb{R}$.*

Theorem 4.6.15 *A normed space X is an inner product space if and only if there exists $\alpha \neq 0, \pm 1$ such that*

$$x, y \in X, \ x \perp_I y \quad \Rightarrow \quad x \perp_I \alpha y.$$

Proof Assume that the above implication holds. We can consider, without loss of generality, that $\alpha > 1$. Then we will see that isosceles orthogonality is homogeneous. As in the proof of Theorem 4.6.12, we begin by showing that X is strictly convex. Let $u, v \in S_X, u \neq v$, be such that $[u, v]$ is a maximal segment in S_X. Let $0 < \beta < 1/(1 + \alpha)$, $y = \beta(u - v)$ and $x = u - y$. Then $\|x + y\| = 1$ and $\|x - y\| = \|(1 - 2\beta)u + 2\beta v\| = 1$, i.e., $x \perp_I y$. Then $x \perp_I \alpha y$, which means $\|x - \alpha y\| = \|(1 - (1 + \alpha)\beta)u + (1 + \alpha)\beta v\| = 1 = \|x + \alpha y\| = \|(1 - (1 - \alpha)\beta)u + (1 - \alpha)\beta v\|$. Since $(1 - \alpha)\beta < 0$, it follows that $[u, v]$ is not maximal.

Next, let us show that $x \perp_I y$ implies $x \perp_B y$ and $y \perp_B x$. Let $x, y \in X \setminus \{0\}$ be such that $x \perp_I y$. From the hypothesis it follows that for all $n \in \mathbb{N}$, $\|x + \alpha^{-n} y\| = \|x - \alpha^{-n} y\|$, and since the function $f(\lambda) = \|x + \lambda y\|$, is convex, this implies that $f(\lambda)$ attains its infimum at 0, which means $x \perp_B y$. Since isosceles orthogonality is symmetric, we get also $y \perp_B x$.

Finally, let $x, y \in X \setminus \{0\}$, $x \perp_I y$, and $\lambda \in \mathbb{R}$. Then $\lambda y \perp_B x$. Assume that $x \not\perp_I \lambda y$. Due to the existence property, there exists $y' = \alpha x + \beta y$, with $\beta > 0$, $\|y'\| = \|y\|$, such that $x \perp_I \lambda y'$. Then $\lambda y' \perp_B x$, which contradicts the strict convexity of X (recall Theorem 4.5.10).   □

Remarks 4.6.16

(a) It follows easily from the above that the hypothesis in Theorem 4.6.12 can also be reduced. Namely, a normed space X is an inner product space if and only if there exists $\alpha \neq 0, \pm 1$ such that

$$x, y \in X, \|x\| = \|y\| \quad \Rightarrow \quad \|\alpha x + y\| = \|x + \alpha y\|.$$

(b) Moreover, another characterization of inner product spaces based on a property weaker than homogeneity of isosceles orthogonality is the following one: a normed space X is an inner product space if and only if there exists $\delta > 0$ such that

$$x, y \in S_X, \ x \perp_I y, \ 0 < \lambda < \delta \quad \Rightarrow \quad x \perp_I \lambda y.$$

Furthermore, δ may depend on x, y. Similar arguments as those used in the proof of Theorem 4.6.15 show that the above property implies the following:

$$x, y \in S_X, \ x \perp_I y \quad \Rightarrow \quad x \perp_B y,$$

which characterizes inner product spaces (see Theorem 4.8.21).

4.6.4 Additivity

The original proof of the next theorem was given by James [86].

Theorem 4.6.17 *A normed space X is an inner product space if and only if isosceles orthogonality is additive.*

Proof Necessity is evident. Assume that isosceles orthogonality is additive. Then

$$x, y \in X, \ x \perp_I y \quad \Rightarrow \quad x \perp_I 2y,$$

which implies (Theorem 4.6.15) that X is an inner product space. $\qquad\square$

4.6.5 Orthogonal Diagonals

The next theorem follows right away from the definition of isosceles orthogonality.

Theorem 4.6.18 *Let X be a normed space. For any $x, y \in X \setminus \{0\}$ there exists a unique $\rho > 0$ such that $x + \rho y \perp_I x - \rho y$. Moreover, $\rho = \|x\|/\|y\|$.*

Corollary 4.9.9 will show that Corollary 4.5.18 is also valid for isosceles orthogonality.

4.6.6 Isosceles Orthogonality and Hyperplanes

Lemma 4.6.19 *If $x, y \in X$ and $x \perp_I \lambda y$ for every $\lambda \in \mathbb{R}$, then $x \perp_B y$.*

Proof Assume that $x \not\perp_B y$, and let δ be such that $\|x + \delta y\| < \|x\|$. Since $\lambda \mapsto \|x + \lambda y\|$ is convex, necessarily $\|x - \delta y\| > \|x\| > \|x + \delta y\|$, and $x \not\perp_I \delta y$. $\qquad\square$

Theorem 4.6.20 *A normed space X is an inner product space if and only if any of the following properties holds:*[2]

(a) *For any $x \in X$ there exists a closed and homogeneous hyperplane $H \subset X$ such that $x \perp_I H$;*
(b) *For any closed and homogeneous hyperplane $H \subset X$ there exists $x \in X \setminus \{0\}$ such that $x \perp_I H$.*

Proof

(a) If X is an inner product space, just consider $H = \{x\}^\perp$. Conversely, assume that (a) holds. We consider two cases:

   CASE 1: Assume that $\dim X \geq 3$. Take $x \in X$, and let H be a closed and homogeneous hyperplane such that $x \perp_I H$. Then $y \perp_I \lambda x$ for every $y \in H$ and $\lambda \in \mathbb{R}$. From Lemma 4.6.19 it follows that $y \perp_B x$. Thus, $H \perp_B x$. By Theorem 4.5.20, X is an inner product space.

---

[2] If $\dim X = +\infty$, in property (b) it is necessary to assume that X is complete.

CASE 2: Assume that dim $X = 2$. First we show that X is strictly convex.
Assume that there exist $u, v \in S_X$, $u \neq v$, such that $[u, v]$ is a
maximal segment in S_X. Let $x = \frac{3}{4}u + \frac{1}{4}v$, $y = u - v$, and let H be
a line through the origin such that $x \perp_I H$. Then, $x \perp_B H$, which
implies that $H = \langle\{y\}\rangle$. Hence $x \perp_I \frac{3}{4}y$, which implies $1 = \|v\| =$
$\|x - \frac{3}{4}y\| = \|x + \frac{3}{4}y\| = \|\frac{3}{2}u - \frac{1}{2}v\|$, and $[u, v]$ is not maximal.
Therefore, X is strictly convex.

Now, we will see that isosceles orthogonality is homogeneous, and that then
(Theorem 4.6.13) X is an inner product space. Let $x, y \in X \setminus \{0\}$, $x \perp_I y$. Let
H be a line through the origin such that $x \perp_I H$. By Theorem 4.6.6, $y \in H$,
and then $x \perp_I \lambda y$ for every $\lambda \in \mathbb{R}$.

(b) This statement follows completely similarly to the above, by considering
Theorem 4.5.21 in Case 1.   □

When dim $X \geq 3$, the key in the proof of Theorem 4.6.20 lies in the fact that if
x is isosceles orthogonal to the whole hyperplane H, then $H \perp_B x$. From the more
restrictive hypothesis in the next theorem we cannot get the same conclusion, which
forces a different proof. Moreover, note that the hypothesis in this theorem holds in
any two-dimensional normed space.

Theorem 4.6.21 *A normed space X, dim $X \geq 3$, is an inner product space if and
only if for any $x \in S_X$ there exists a closed and homogeneous hyperplane $H \subset X$
such that $x \perp_I H \cap S_X$.*

Proof If X is an inner product space, just consider $H = \{x\}^\perp$. Conversely, let
$x \in S_X$, and let $H_x \subset X$ be a closed and homogeneous hyperplane such that $x \perp_I$
$H_x \cap S_X$. From Theorem 4.6.5, it follows that if $y \in S_X$ and $x \perp_I y$, then $y \in H_x$.
Thus, if $x, y, z \in S_X$, $x \perp_I y$, $x \perp_I z$, then $(y+z)/\|y+z\| \in H_x$, and therefore $x \perp_I$
$(y + z)/\|y + z\|$. This means that Singer orthogonality is additive. A very technical
proof (see Pei-Kee Lin [99]) shows that if dim $X \geq 3$, then Singer orthogonality is
additive only in inner product spaces.   □

Property (b) in Theorem 4.6.20 leads naturally to the following question.

Question 4.6.22 *Let X be a Banach space, dim $X \geq 3$, such that for any closed
and homogeneous hyperplane $H \subset X$ there exists $x \in S_X$ such that $x \perp_I H_x \cap S_X$.
Is X necessarily a Hilbert space?*

4.7 Pythagorean Orthogonality

Definition 4.7.1 Two vectors x and y are said to be *Pythagorean orthogonal*
$(x \perp_{P_-} y)$ if

$$\|x - y\|^2 = \|x\|^2 + \|y\|^2.$$

Pythagorean orthogonality was also introduced by James (see [86]), and it has the properties of non-degeneracy, simplification, continuity and symmetry, but it is not (in general) homogeneous. Thus we must, as with isosceles orthogonality, distinguish between existence and α-existence properties. In the literature, the definition of Pythagorean orthogonality also appears as

$$x \perp_{P_+} y \quad :\Leftrightarrow \quad \|x + y\|^2 = \|x\|^2 + \|y\|^2.$$

Since $x \perp_{P_+} y$ if and only if $x \perp_{P_-} -y$, the properties for P_+-orthogonality follow from those for P_--orthogonality.

4.7.1 Existence and Uniqueness

Theorem 4.7.2 *Pythagorean orthogonality has the existence property, i.e., for any $x, y \in X$, linearly independent, and any $\rho > 0$, there exists $z = \alpha x + \beta y$, with $\alpha, \beta \in \mathbb{R}$, $\beta > 0$, such that $\|z\| = \rho$ and $x \perp_P z$.*

Proof For $\theta \in [0, \pi]$, let $z(\theta) = \rho \frac{(\cos\theta)x+(\sin\theta)y}{\|(\cos\theta)x+(\sin\theta)y\|}$. The continuous function $g :$ $\theta \in [0, \pi] \mapsto g(\theta) = \|x - z(\theta)\|^2 - \|x\|^2 - \rho^2$ satisfies $g(0) = -2\rho\|x\| \leq 0 \leq 2\rho\|x\| = g(\pi)$. Therefore, there exists $\theta_0 \in (0, \pi)$ such that $g(\theta_0) = 0$, i.e., $x \perp_P z(\theta_0)$. $\square$

Theorem 4.7.3 *A normed space X is strictly convex if and only if Pythagorean orthogonality has the uniqueness property.*

Proof Assume that Pythagorean orthogonality has not the uniqueness property, i.e., that there exist $x, y \in X$, linearly independent, $\alpha, \alpha' \in \mathbb{R}$, $\beta > 0$, $\beta' > 0$, such that $z = \alpha x + \beta y$ and $z' = \alpha' x + \beta' y$ satisfy $z \neq z'$, $\|z\| = \|z'\|$, $x \perp_P z$ and $x \perp_P z'$. Let us show that then X is not strictly convex. We consider the convex function $f(\lambda) = \|\lambda_0 z + \lambda(z' - z)\|$, where $\lambda_0 = \beta(1 - \alpha') - \beta'(1 - \alpha)$. Let $\lambda_1 = 0$, $\lambda_2 = \beta$, and $\lambda_3 = \beta + \alpha\beta' - \beta\alpha'$. Then $f(\lambda_0) = \|\lambda_0 z'\| = \|\lambda_0 z\| = f(\lambda_1)$, and $f(\lambda_2) = \|(\alpha'\beta - \alpha\beta')(x - z)\| = \|(\alpha'\beta - \alpha\beta')(x - z')\| = f(\lambda_3)$. Since $z \neq z'$, it follows that $\lambda_0 \neq \lambda_1$ and $\lambda_2 \neq \lambda_3$. Moreover, $\lambda_0 < \lambda_3$ and $\lambda_1 < \lambda_2$. Since f is convex, it must be constant in the interval $[\min\{\lambda_0, \lambda_1\}, \max\{\lambda_2, \lambda_3\}]$, which implies that X is not strictly convex (Lemma 4.5.9).

Conversely, assume that X is not strictly convex. Let $u, v \in S_X$, $u \neq v$, such that $[u, v] \subset S_X$. Let L be the two-dimensional subspace spanned by u and v. We can identify L with $\mathbb{R}^2$ in such a way that $(0, 1) \in S_L$, $u = (1, \gamma)$, $v = (1, -\gamma)$, with $0 < \gamma \leq 1$. For any $\rho > 0$, it holds that $0 < \sqrt{1 + \rho^2} - \rho < 1$, and then there exists $\delta(\rho) > 0$ such that $x_\rho := \left(\sqrt{1 + \rho^2} - \rho, \delta(\rho)\right) \in S_L$. Moreover, since $(0, 1)$ and $-v$ belong to S_L, it follows from the convexity of B_L that

$$\delta(\rho) \leq \left(1 - \gamma\right)\left(\sqrt{1 + \rho^2} - \rho\right) + 1. \tag{4.3}$$

Straightforward computations show that there exists $\bar{\rho} > 0$ such that

$$(1 - \gamma)(\sqrt{1 + \bar{\rho}^2} - \bar{\rho}) + 1 \leq \gamma \sqrt{1 + \bar{\rho}^2}. \tag{4.4}$$

Let $z_1 = (-\bar{\rho}, 0)$ and $z_2 = (-\bar{\rho}, \bar{\rho}\gamma)$. Then, $\|z_1\| = \|z_2\| = \bar{\rho}$. By taking

$$\mu_1 = \frac{\delta(\bar{\rho}) + \gamma \sqrt{1 + \bar{\rho}^2}}{2\gamma \sqrt{1 + \bar{\rho}^2}}, \qquad \mu_2 = \frac{\delta(\bar{\rho}) + \gamma(\sqrt{1 + \bar{\rho}^2} - \bar{\rho})}{2\gamma \sqrt{1 + \bar{\rho}^2}}$$

we have $x_{\bar{\rho}} - z_i = \sqrt{1 + \bar{\rho}^2}(\mu_i u + (1 - \mu_i)v)$, $i = 1, 2$. Moreover, from (4.3) and (4.4) it follows that $0 < \mu_i \leq 1$, $i = 1, 2$. Then, for $i = 1, 2$, $\|x_{\bar{\rho}} - z_i\| = \sqrt{1 + \bar{\rho}^2}$, i.e., $x_{\bar{\rho}} \perp_{\mathrm{P}_-} z_i$, from which it follows that Pythagorean orthogonality has not uniqueness property. □

4.7.2   α-Existence and α-Uniqueness

The α-existence of Pythagorean orthogonality was proved by James [86].

Theorem 4.7.4 *For any $x, y \in X$ there exists $\alpha \in \mathbb{R}$ such that $x \perp_{\mathrm{P}_-} \alpha x + y$.*

Proof We can assume that $x \neq 0$. Let us consider the function

$$f(\alpha) = \|x\|^2 + \|\alpha x + y\|^2 - \|(\alpha - 1)x + y\|^2, \quad \alpha \in \mathbb{R}.$$

We must show that there exists $\bar{\alpha} \in \mathbb{R}$ such that $f(\bar{\alpha}) = 0$. It follows easily that

$$\begin{aligned}
f(\alpha) = \|x\|^2 &+ \left(\frac{2\alpha - 1}{\alpha^2}\right) \|\alpha x + y\|^2 \\
&+ \left(\left\|(\alpha - 1)x + \left(\frac{\alpha - 1}{\alpha}\right)y\right\| - \|(\alpha - 1)x + y\|\right) \\
&\cdot \left(\left\|(\alpha - 1)x + \left(\frac{\alpha - 1}{\alpha}\right)y\right\| + \|(\alpha - 1)x + y\|\right).
\end{aligned} \tag{4.5}$$

Assume that $\alpha > 1$. Then, we have the following bounds for the terms in (4.5):

$$\left\|(\alpha - 1)x + \left(\frac{\alpha - 1}{\alpha}\right)y\right\| - \|(\alpha - 1)x + y\| \geq -\frac{\|y\|}{\alpha}$$

and

$$\left\|(\alpha - 1)x + \left(\frac{\alpha - 1}{\alpha}\right)y\right\| + \|(\alpha - 1)x + y\| \leq 2(\alpha - 1)\|x\| + \left(\frac{2\alpha - 1}{\alpha}\right)\|y\|$$

from which it follows that

$$f(\alpha) \geq \|x\|^2 + \left(\frac{2\alpha - 1}{\alpha^2}\right)(\alpha\|x\| - \|y\|)^2$$

$$- \frac{\|y\|}{\alpha}\left(2(\alpha - 1)\|x\| + \left(\frac{2\alpha - 1}{\alpha}\right)\|y\|\right)$$

$$= 2\|x\|\left(\alpha\|x\| - 3\|y\| + \frac{2}{\alpha}\|y\|\right).$$

This shows that taking $\alpha_1 > 1$ sufficiently large we get $f(\alpha_1) > 0$. On the other hand, assume that $\alpha < 0$. In this case, we get the following bounds:

$$\left\|(\alpha - 1)x + \left(\frac{\alpha - 1}{\alpha}\right)y\right\| - \|(\alpha - 1)x + y\| \leq -\frac{\|y\|}{\alpha}$$

and

$$\left\|(\alpha - 1)x + \left(\frac{\alpha - 1}{\alpha}\right)y\right\| + \|(\alpha - 1)x + y\| \leq 2(1 - \alpha)\|x\| + \left(\frac{2\alpha - 1}{\alpha}\right)\|y\|.$$

Then it follows from (4.5) that

$$f(\alpha) \leq \|x\|^2 + \left(\frac{2\alpha - 1}{\alpha^2}\right)(\alpha\|x\| + \|y\|)^2$$

$$- \frac{\|y\|}{\alpha}\left(2(1 - \alpha)\|x\| + \left(\frac{2\alpha - 1}{\alpha}\right)\|y\|\right)$$

$$= 2\|x\|\left(\alpha\|x\| + 3\|y\| - \frac{2\|y\|}{\alpha}\right),$$

which implies that there exists $\alpha_2 < 0$, sufficiently small, such that $f(\alpha_2) < 0$. Since $f(\alpha)$ is continuous, it follows that there exists $\bar{\alpha} \in [\alpha_2, \alpha_1]$ such that $f(\bar{\alpha}) = 0$, as we wished to prove.   □

The α-uniqueness property of Pythagorean orthogonality was studied by Kapoor and Prasad [93].

Theorem 4.7.5 *Pythagorean orthogonality has the α-uniqueness property in every normed space X.*

Proof We prove it by contradiction. Let $x, y \in X \setminus \{0\}$, and assume that there exist $\alpha, \alpha' \in \mathbb{R}, \alpha < \alpha'$, such that $x \perp_P \alpha x + y$ and $x \perp_P \alpha' x + y$.

Let us consider the strictly convex function $f : t \in \mathbb{R} \mapsto f(t) = \|y + tx\|^2$. It satisfies $f(\alpha - 1) = f(\alpha) + \|x\|^2$ and $f(\alpha' - 1) = f(\alpha') + \|x\|^2$. Then,

$$f(\alpha' - 1) + f(\alpha) = f(\alpha - 1) + f(\alpha')   \tag{4.6}$$

Assume that $\alpha' - \alpha > 1$, and let $\mu = \frac{1}{\alpha' - \alpha}$. Since $0 < \mu < 1$, $\alpha' - 1 = \mu\alpha + (1 - \mu)\alpha'$, and $\alpha = \mu(\alpha' - 1) + (1 - \mu)(\alpha - 1)$, it follows that $f(\alpha' - 1) < \mu f(\alpha) + (1 - \mu)f(\alpha')$ and $f(\alpha) < \mu f(\alpha' - 1) + (1 - \mu)f(\alpha - 1)$. By adding both inequalities we get a contradiction with (4.6). Moreover, if $\bar{\mu} := \alpha' - \alpha < 1$, then $\alpha = \bar{\mu}(\alpha' - 1) + (1 - \bar{\mu})\alpha'$ and $\alpha' - 1 = (1 - \bar{\mu})(\alpha - 1) + \bar{\mu}\alpha$. Therefore, $f(\alpha) < \bar{\mu}f(\alpha' - 1) + (1 - \bar{\mu})f(\alpha')$ and $f(\alpha' - 1) < (1 - \bar{\mu})f(\alpha - 1) + \bar{\mu}f(\alpha)$. By adding these inequalities, we get a new contradiction. Finally, if $\alpha' - \alpha = 1$ then $\alpha = \frac{1}{2}(\alpha - 1) + \frac{1}{2}\alpha'$, which implies $f(\alpha) < \frac{1}{2}f(\alpha - 1) + \frac{1}{2}f(\alpha') = \frac{1}{2}f(\alpha' - 1) + \frac{1}{2}f(\alpha) = f(\alpha)$, also absurd.   □

4.7.3 Homogeneity

As it occurs with isosceles orthogonality, Pythagorean orthogonality is homogeneous only in inner product spaces. This was proved by James [86] as a direct consequence of the parallelogram law characterization.

Theorem 4.7.6 *A normed space X is an inner product space if and only if Pythagorean orthogonality is homogeneous.*

Proof The necessity is evident. To prove the sufficiency we will see that any $x, y \in X$ satisfy the parallelogram law. By Theorem 4.7.4, there exists $\alpha \in \mathbb{R}$ such that $x \perp_P \alpha x + y$. Since we are assuming that Pythagorean orthogonality is homogeneous, we have $\|\lambda x - (\alpha x + y)\|^2 = \lambda^2 \|x\|^2 + \|\alpha x + y\|^2$ for every $\lambda \in \mathbb{R}$. Taking $\lambda = \alpha$ it follows that $\|y\|^2 = \alpha^2 \|x\|^2 + \|\alpha x + y\|^2$, and taking $\lambda = \alpha \pm 1$, we get $\|x \mp y\|^2 = (\alpha \pm 1)^2 \|x\|^2 + \|\alpha x + y\|^2 = \alpha^2 \|x\|^2 + \|\alpha x + y\|^2 + (1 \pm 2\alpha)\|x\|^2 = \|y\|^2 + (1 \pm 2\alpha)\|x\|^2$. Therefore,

$$\|x + y\|^2 + \|x - y\|^2 = 2(\|x\|^2 + \|y\|^2),$$

as we wished to prove.   □

With the next theorem, Kapoor and Prasad [93] strongly reduced the hypothesis in the above one.

Theorem 4.7.7 *A normed space X is an inner product space if and only if there exists $\alpha \neq 0, 1$ such that*

$$x, y \in X, \; x \perp_P y \quad \Rightarrow \quad x \perp_P \alpha y.$$

Proof The necessity is evident. With respect to sufficiency, the case $\alpha = -1$ will be studied in Theorem 4.8.7. In the other cases, by considering (if necessary) $1/\alpha$ or α^2, we can assume, without loss of generality, that $0 < \alpha < 1$.

Assume that $x, y \in X$ are such that $x \perp_{\mathrm{P_-}} y$. Then, for any $n \in \mathbb{N}$, $x \perp_{\mathrm{P_-}} \alpha^n y$, which implies that

$$\left(\|x - \alpha^n y\| + \|x\| \right) \left(\frac{\|x - \alpha^n y\| - \|x\|}{-\alpha^n} \right) = -\alpha^n \|y\|^2.$$

Taking the limit as n goes to infinity, it follows that $2\|x\| N_-(x, y) = 0$, and then (Theorem 4.5.6) that $x \perp_{\mathrm{B}} y$. Thus, Pythagorean orthogonality implies Birkhoff orthogonality. We will see in Theorem 4.8.3 that this occurs only in inner product spaces.   □

Remark 4.7.8 Another characterization of inner product spaces based on a property weaker than homogeneity of Pythagorean orthogonality is the following one: a normed space X is an inner product space if and only if there exists $\delta > 0$ such that

$$x, y \in S_X, \ x \perp_{\mathrm{P_-}} y, \ |\lambda| < \delta \quad \Rightarrow \quad x \perp_{\mathrm{P_-}} \lambda y.$$

Furthermore, δ may depend on x, y. Note that as the map $\lambda \in \mathbb{R} \mapsto \|x + \lambda y\|$ is convex, if $x \perp_{\mathrm{P_-}} \lambda y$ for $|\lambda| < \delta(x, y)$, then $\|x + \lambda y\| \geq 1$ for every $\lambda \in \mathbb{R}$, i.e., $x \perp_{\mathrm{B}} y$. Thus, the above property implies the following one:

$$x, y \in S_X, \ x \perp_{\mathrm{P_-}} y, \quad \Rightarrow \quad x \perp_{\mathrm{B}} y.$$

We will see (Theorem 4.8.20) that this characterizes inner product spaces.

4.7.4 Additivity

From Theorem 4.7.7 the next statement follows directly.

Theorem 4.7.9 *A normed space X is an inner product space if and only if Pythagorean orthogonality is additive.*

4.7.5 Orthogonal Diagonals

Existence, uniqueness and other properties of orthogonal diagonals for Pythagorean orthogonality were studied by Benítez [36].

Theorem 4.7.10 *Let X be a normed space. For any $x, y \in X \setminus \{0\}$ there exists a unique $\rho > 0$ such that $x + \rho y \perp_{\mathrm{P_-}} x - \rho y$. Moreover,*

$$\frac{1}{\sqrt{2}} \frac{\|x\|}{\|y\|} \leq \rho \leq \frac{1}{\sqrt{2} - 1} \frac{\|x\|}{\|y\|}.$$

Proof Let $x, y \in X \setminus \{0\}$, and let $f(\lambda) := \|\lambda x + y\|^2 + \|\lambda x - y\|^2 - 4\|y\|^2$. Then, $x + \rho y \perp_P x - \rho y$ if and only if $f(1/\rho) = 0$. Since $f(0) = -2\|y\|^2 < 0$ and $f(\lambda) \geq 2(\lambda\|x\| - \|y\|)^2 - 4\|y\|^2 = 2(\lambda^2\|x\|^2 - 2\lambda\|x\|\|y\| - \|y\|^2) > 0$ for λ large enough, the continuity and convexity of f imply the existence of a unique $\lambda > 0$ such that $f(\lambda) = 0$.

Assume now that $x + \rho y \perp_P x - \rho y$, with $\rho > 0$. Then, $4\rho^2\|y\|^2 = \|x + \rho y\|^2 + \|x - \rho y\|^2 \leq 2(\|x\| + \rho\|y\|)^2$, from which it follows that

$$\left(\rho - \frac{1}{\sqrt{2}-1}\frac{\|x\|}{\|y\|}\right)\left(\rho + \frac{1}{\sqrt{2}+1}\frac{\|x\|}{\|y\|}\right) \leq 0,$$

and the right bound is obtained. On the other hand, the convex function $\lambda \in \mathbb{R} \mapsto g(\lambda) = \|x + \lambda\rho y\|^2 + \|x - \lambda\rho y\|^2 - 4\rho^2\|y\|^2$ satisfies $g(1) = g(-1) = 0$. Therefore, $g(0) \leq 0$, from which the left bound follows. □

Remark 4.7.11

(a) The bounds in Theorem 4.7.10 are sharp. Consider the space $X = (\mathbb{R}^2, \|\cdot\|_\infty)$, $\|(x_1, x_2)\|_\infty = \max\{|x_1|, |x_2|\}$. For $x = (1, 1)$, $y = (1, -1)$ and $\rho = (\sqrt{2} - 1)^{-1}$ we have $x + \rho y \perp_P x - \rho y$. Moreover, taking $\bar{x} = (1, 0)$, $\bar{y} = (0, 1)$ and $\bar{\rho} = (\sqrt{2})^{-1}$, we have $\bar{x} + \bar{\rho}\bar{y} \perp_P \bar{x} - \bar{\rho}\bar{y}$.

(b) The right bound in Theorem 4.7.10 is attained only in special situations. Consider $\rho = \frac{1}{\sqrt{2}-1}\frac{\|x\|}{\|y\|}$, and assume that $x + \rho y \perp_P x - \rho y$. Clearly, if x and y are linearly dependent, then $x = \pm\rho y$. On the other hand, let $L = \langle\{x, y\}\rangle$ be the subspace spanned by x and y. Let us show that then S_L is the parallelogram with vertices $\pm\frac{x}{\|x\|}, \pm\frac{y}{\|y\|}$. From the hypothesis it follows that

$$4(3 + 2\sqrt{2})\|x\|^2 = 4\rho^2\|y\|^2 = \|x + \rho y\|^2 + \|x - \rho y\|^2$$
$$\leq 2(\|x\| + \rho\|y\|)^2 = 4(3 + 2\sqrt{2})\|x\|^2,$$

which implies that $\|x + \rho y\| = \|x - \rho y\| = \|x\| + \rho\|y\|$. Let $\alpha = \frac{\|x\|}{\|x\|+\rho\|y\|}$. Then $0 < \alpha < 1$ and $\left\|\alpha\frac{x}{\|x\|} + (1 - \alpha)\frac{\pm y}{\|y\|}\right\| = \frac{\|x \pm \rho y\|}{\|x\|+\rho\|y\|} = 1$. Therefore, the segments $\left[\frac{x}{\|x\|}, \frac{\pm y}{\|y\|}\right]$ are contained in S_L.

(c) It is easy to give examples that show that attaining the left bound in Theorem 4.7.10 does not determine the shape of the sphere S_L defined in (b).

The next theorem is entirely analogous to Theorem 4.5.17.

Theorem 4.7.12 *A normed space X is an inner product space if and only if for any $x, y \in X \setminus \{0\}$,*

$$x + \frac{\|x\|}{\|y\|}y \perp_P x - \frac{\|x\|}{\|y\|}y.$$

Proof The necessity is evident. Conversely, it follows from the hypothesis that if $x, y \in S_X$, then $\|x + y\|^2 + \|x - y\|^2 = 4$, which is a characteristic property of inner product spaces given by Day [59].   □

Similar to what we did with Birkhoff orthogonality, we can consider the map $(x, y) \in X \setminus \{0\} \times X \setminus \{0\} \mapsto \rho(x, y)$, such that $x + \rho(x, y)y \perp_P x - \rho(x, y)y$. Like there, it is easy to see that this map is continuous (with the product topology). Since $S_X \times S_X$ is a connected set, it follows that the set

$$A_X = \{\rho(x, y) : \|x\| = \|y\|\} = \{\rho(x, y) : x, y \in S_X\}$$

is a subinterval contained in $[(\sqrt{2})^{-1}, (\sqrt{2} - 1)^{-1}]$. Theorem 4.7.12 says that X is an inner product space if and only if $A_X = \{1\}$. The next theorem improves this result.

Theorem 4.7.13 *A normed space X is an inner product space if and only if $1 \notin$ Int A_X.*

Proof The necessity is evident. To prove the sufficiency, let us assume that $A_X = [a, b]$, and $1 \leq a$. Let $x, y \in S_X$. Then $1 \leq \rho(x, y)$, and since the convex function $f(\lambda) = \|\lambda x + y\|^2 + \|\lambda x - y\|^2$ satisfies $f(0) = 2$ and $f((\rho(x, y)^{-1}) = 4$, it follows that $f(1) \geq 4$. Therefore, for $x, y \in S_X$, the inequality $\|x + y\|^2 + \|x - y\|^2 \geq 4$ holds, which is a characteristic property of inner product spaces, see [59] and [131]. If $b \leq 1$, we get the inequality with "$\leq$", which is also characteristic for inner product spaces.   □

The next example shows that computing $\rho(x, y)$ is not easy, even with simple norms.

Example 4.7.14 Let us consider the space $X = (\mathbb{R}^2, \|\cdot\|_\infty)$. Cumbersome computations show that for $x, y \in S_X$, $\rho(x, y)$ takes the following values.

Case 1. Let $x = (1, \alpha)$, $y = (1, \beta)$, with $\alpha, \beta \in [-1, 1]$. Then

$$\rho(x, y) = \frac{1 - \alpha\beta + \sqrt{4 + 3\alpha^2 - 2\alpha\beta - \beta^2}}{3 - \beta^2}.$$

Case 2. Let $x = (1, \alpha)$, $y = (\beta, 1)$, with $\alpha, \beta \in [-1, 1]$. The functions

$$\beta_1(\alpha) = \frac{1 - (1 - \alpha)\sqrt{3 - 4\alpha + 2\alpha^2}}{2 - 2\alpha + \alpha^2}, \quad \beta_2(\alpha) = \frac{-1 + (1 + \alpha)\sqrt{3 + 4\alpha + 2\alpha^2}}{2 + 2\alpha + \alpha^2}$$

define three regions in the set $[-1, 1] \times [-1, 1]$. Then

$$\rho(x, y) = \begin{cases} \dfrac{\alpha - \beta + \sqrt{3 + 4\alpha^2 - 2\alpha\beta - \alpha^2\beta^2}}{3 - \beta^2}, & \text{if } \beta \leq \beta_1(\alpha), \\[3ex] \dfrac{1}{\sqrt{2 - \beta^2}}, & \text{if } \beta_1(\alpha) \leq \beta \leq \beta_2(\alpha), \\[3ex] \dfrac{\beta - \alpha + \sqrt{3 + 4\alpha^2 - 2\alpha\beta - \alpha^2\beta^2}}{3 - \beta^2}, & \text{if } \beta \geq \beta_2(\alpha). \end{cases}$$

The value of $\rho(x, y)$ for other possible locations of vectors x and y can be deduced from the above ones. Moreover, from the values of $\rho(x, y)$, it is easily checked that, effectively, $A_X = [(\sqrt{2})^{-1}, (\sqrt{2} - 1)^{-1}]$.

Corollary 4.9.9 will show that Corollary 4.5.18 is also valid for Pythagorean orthogonality.

4.7.6   Pythagorean Orthogonality and Hyperplanes

If H is an homogeneous hyperplane and $x \perp_{P_-} H$, then $x \perp_I H$. This shows that the next theorem follows directly from Theorem 4.6.20. Nevertheless, we include here a proof not based on Theorems 4.5.20 and 4.5.21 (recall that they are used in Theorem 4.6.20).

Theorem 4.7.15 *A normed space X is an inner product space if and only if any of the following properties holds:*[3]

(a) *For any $x \in X$ there exists a closed and homogeneous hyperplane $H \subset X$ such that $x \perp_{P_-} H$.*
(b) *For any closed and homogeneous hyperplane $H \subset X$ there exists $x \in X \setminus \{0\}$ such that $x \perp_{P_-} H$.*

Proof

(a) If X is an inner product space, just consider $H = \{x\}^\perp$. Conversely, assume that (a) holds. Let $x, y \in X \setminus \{0\}$ be such that $x \perp_{P_-} y$. Let $H \subset X$ be a closed homogeneous hyperplane such that $x \perp_{P_-} H$, and let $z \in H \cap \langle\{x, y\}\rangle$. Since $x \notin H$, we can assume that $z = \alpha x + y$, with $\alpha \in \mathbb{R}$. The α-uniqueness (Theorem 4.7.5) implies that $\alpha = 0$, and then $y \in H$. Therefore $x \perp_{P_-} \lambda y$ for all $\lambda \in \mathbb{R}$. Theorem 4.7.6 completes the proof.
(b) If X is an inner product space, just consider $x \in H^\perp \setminus \{0\}$. Conversely, assume that (b) holds. We will see that then (a) holds. Let $x \in X$. By Theorem 4.5.19

---

[3] If dim $X = +\infty$, in property (b) it is necessary to assume that X is complete.

there exists a closed homogeneous hyperplane H such that $x \perp_B H$. Let $z \in X \setminus \{0\}$ be such that $z \perp_{P_-} H$. Then $x = \lambda z + h$, with $\lambda \in \mathbb{R}$ and $h \in H$. Hence, $\lambda^2 \|z\|^2 = \|x - h\|^2 \geq \|x\|^2 = \|\lambda z + h\|^2 = \lambda^2 \|z\|^2 + \|h\|^2$, which implies $h = 0$, and then $x \perp_{P_-} H$. □

We also have trivially that if H is a homogeneous hyperplane and $x \perp_{P_-} H \cap S_X$, then $x \perp_I H \cap S_X$. Thus, Theorem 4.6.21 gives the next theorem (previously proved in [1] and [43]).

Theorem 4.7.16 *A normed space X, dim $X \geq 3$, is an inner product space if and only if for any $x \in S_X$ there exists a closed and homogeneous hyperplane $H \subset X$ such that $x \perp_{P_-} H \cap S_X$.*

As with Theorem 4.6.21, the above theorem is not true if dim $X = 2$. A simple counterexample is the space $\mathbb{R}^2$ endowed with a norm whose unit sphere is a regular hexagon. Also as with isosceles orthogonality, the next question follows naturally.

Question 4.7.17 *Let X be a Banach space, dim $X \geq 3$, such that for any closed and homogeneous hyperplane $H \subset X$ there exists $x \in S_X$ such that $x \perp_{P_-} H_x \cap S_X$. Is X necessarily a Hilbert space?*

4.8 Relations Between Birkhoff, Isosceles, and Pythagorean Orthogonality

As we know, all the orthogonalities defined for a normed space and considered here coincide in the case of an inner product space. In fact, this is the origin of their definitions. But nevertheless, as the following theorems will show, they are essentially different for non-inner product spaces.

Basically, the results in this section are divided into two families. In the first one we consider properties of the type $[x, y \in X, \ x \perp_F y \Rightarrow x \perp_G y]$, whereas in the second we restrict the vectors x, y to the unit sphere. Obviously, some results of the first family are trivial consequence of the corresponding ones in the second. But we include them here because they are interesting (mainly historically) by themselves.

Clearly, a normed space X, dim $X \geq 2$, is an inner product space if and only if any two-dimensional subspace is an inner product space. Thus, to prove that a property of the type $[x, y \in X (\text{or } S_X), \ x \perp_F y \Rightarrow x \perp_G y]$ characterizes inner product spaces we can assume that dim $X = 2$, i.e., $X = (\mathbb{R}^2, \| \cdot \|)$.

4.8.1 Implications over the Whole Space

The original proofs of Theorem 4.8.2 and Theorem 4.8.3 were given by Ohira [120].

Lemma 4.8.1 *Assume that for every $u, v \in S_X$, $u + v \perp_B u - v$. Then X is strictly convex.*[4]

Proof Assume that X is not strictly convex, and let $[u, v] \subset S_X$ be a maximal segment in the unit sphere. Let $w = u + t(u - v)$, with $0 < t < 1$. Then $\|w\| > 1$. Let $\rho = (1 + \|w\|)/\|w\|$ and $\mu = (1 + t)/(1 + \|w\|)$. From the identity $v + \frac{w}{\|w\|} = \rho(\mu u + (1 - \mu)v)$ it follows that $\|v + \frac{w}{\|w\|}\| = \rho$. Moreover,

$$\left\| v + \frac{w}{\|w\|} + \left(\frac{t - \|w\|}{t + \|w\|}\right)\left(v - \frac{w}{\|w\|}\right) \right\| = \left\| \frac{2(t + 1)}{t + \|w\|} u \right\| = \frac{2(t + 1)}{t + \|w\|} < \rho,$$

against the assumption $v + \frac{w}{\|w\|} \perp_B v - \frac{w}{\|w\|}$. $\qquad \square$

Theorem 4.8.2 *A normed space X is an inner product space if and only if any of the following properties holds:*

(a) *If $x, y \in X$ and $x \perp_B y$, then $x \perp_I y$.*

(b) *If $x, y \in X$ and $x \perp_I y$, then $x \perp_B y$.*

Proof

(a) Let us show first that if (a) holds, then X is strictly convex. Assume that X is not strictly convex, and let $[u, v]$ be a maximal segment contained in S_X. Let $x = u$, $y = u - v$. Then $x \perp_B y$, which implies $x \perp_I y$. Therefore $\|2u - v\| = 1$, and $u = \frac{1}{2}(2u - v) + \frac{1}{2}v$, and then $[u, v]$ is not maximal. Now, let $x, y \in X$ be such that $x \perp_I y$. By Theorem 4.5.4 there exists $\alpha \in \mathbb{R}$ such that $x \perp_B \alpha x + y$. Then, $x \perp_I \alpha x + y$, and by Theorem 4.6.11, $\alpha = 0$. Thus, Birkhoff and isosceles orthogonalities are equivalent, and then isosceles orthogonality is homogeneous, which implies (Theorem 4.6.13) that X is an inner product space.

(b) Let $u, v \in S_X$. Since $u + v \perp_I u - v$, we have $u + v \perp_B u - v$. By Lemma 4.8.1, X is strictly convex. Let $x, y \in X$, $x \perp_B y$. By Theorem 4.6.7, there exists $\alpha \in \mathbb{R}$ such that $\alpha y + x \perp_I y$. Then $\alpha y + x \perp_B y$, and since X is strictly convex, from Theorem 4.5.10 it follows that $\alpha = 0$. Therefore, Birkhoff and isosceles orthogonalities are equivalent. $\qquad \square$

Theorem 4.8.3 *A normed space X is an inner product space if and only if any of the following properties holds:*[5]

(a) *If $x, y \in X$ and $x \perp_B y$, then $x \perp_{P_-} y$.*

(b) *If $x, y \in X$ and $x \perp_{P_-} y$, then $x \perp_B y$.*

---

[4] In fact, this property characterizes inner product spaces (see the footnote on page 142).

[5] Similarly if P_- is replaced by P_+.

Proof

(a) Let $x, y \in X$, $x \perp_{P_-} y$, and let $\alpha \in \mathbb{R}$ be such that $x \perp_B \alpha x + y$. Then $x \perp_{P_-} \alpha x + y$, and since Pythagorean orthogonality has the uniqueness property, $\alpha = 0$. Thus, Pythagorean and Birkhoff orthogonalities are equivalent, and Pythagorean orthogonality is homogeneous.

(b) Let us show first that if (b) holds, then X is strictly convex. Assume that $u, v \in S_X$, $u \neq v$, are such that $\|u + v\| = 2$. Take $x = \frac{1}{2}(u+v)$ and $y = \frac{1}{2}(v-u)$. Let $\alpha \in \mathbb{R}$ be such that $x \perp_{P_-} \alpha x + y$. Then $x \perp_B \alpha x + y$. Since $[u, v] \subset S_X$, we have $\alpha = 0$, but then $1 = \|u\|^2 = \|x - y\|^2 = \|x\|^2 + \|y\|^2 = 1 + \|y\|^2$, which is absurd. The proof is finished similarly to the case (b) in Theorem 4.8.2. $\square$

To prove Theorem 4.8.6 we use the characterization of inner product spaces stated on Theorem 4.8.5. It is known as *"rhombus equality"*, and it was given by Day [59] as a debilitation of the parallelogram equality. Day gave two proofs of this characterization: the first one is complex and based on trigonometrical arguments; the second one (that we include here) is a nice proof based on an unpublished result from Löwner.

Lemma 4.8.4 (Löwner Ellipse) *Let C be a symmetric closed convex curve in the plane. There is a unique ellipse of minimal area circumscribed about C. Moreover, this ellipse touches C in at least four different points (two and their symmetric images).*

Proof Let us assume that C is centered at the origin. Let E be a minimal area ellipse circumscribed about C. We consider several steps:

E is centered at the origin: Without loss of generality we can assume that E is centered at the point $(0, t)$, i.e., $E = \{(x, y) : ax^2 + by^2 - 2atx = 1 - at^2\}$. For every $(x, y) \in C$, $ax^2 + by^2 - 2atx \leq 1 - at^2$. If $(x, y) \in C$, then $(-x, -y) \in C$. Thus, the ellipse $E' = \{(x, y) : ax^2 + by^2 = 1 - at^2\}$ circumscribes C, but its area is $A(E') = \frac{\pi\sqrt{1-at^2}}{\sqrt{ab}} < \frac{\pi}{\sqrt{ab}} = A(E)$.

E is unique: Assume that there are two circumscribed ellipses of minimal area. With an affine transformation we can assume that they are $E = \{(x, y) : x^2 + y^2 = 1\}$ and $E' = \{(x, y) : ax^2 + by^2 = 1\}$. Since $\pi = A(E) = A(E') = \frac{\pi}{\sqrt{ab}}$, it follows that $ab = 1$. If $(x, y) \in C$, then $x^2 + y^2 \leq 1$ and $ax^2 + by^2 \leq 1$, which implies that the ellipse $E'' = \{(x, y) : \left(\frac{a+1}{2}\right)x^2 + \left(\frac{b+1}{2}\right)y^2 = 1\}$ also circumscribes C, but $A(E'') = \frac{2\pi}{\sqrt{(a+1)(b+1)}} = \frac{2\pi}{\sqrt{2+a+\frac{1}{a}}} < \pi$, except if $a = 1$.

E touches C in at least two points (and their symmetric images): We can assume that $E = \{(x, y) : x^2 + y^2 = 1\}$ and that $(1, 0) \in E \cap C$. For $\varepsilon > 0$, let $E_\varepsilon = \{(x, y) : \left(\frac{1}{1+\varepsilon}\right)x^2 + (1+\varepsilon)y^2 = 1\}$ (see Fig. 4.6). Since $A(E_\varepsilon) = A(E) = \pi$, and E is unique, E_ε cannot circumscribe C. Let $p_\varepsilon \in C$ be a point outside E_ε and inside E. Take $\varepsilon_n \to 0$. Then there exists $p \in C$ such that $p_{\varepsilon_n} \to p$. Since p_{ε_n} is outside E_{ε_n} and inside E, it follows that $p \in E$. Finally, let us show that

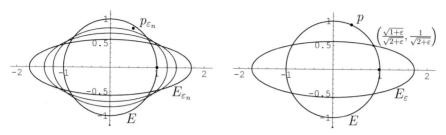

Fig. 4.6 Proof of Lemma 4.8.4

$p = (p_1, p_2) \neq \pm(1, 0)$. The second coordinate of points in $E \cap E_\varepsilon$ is $\frac{\pm 1}{\sqrt{2+\varepsilon}}$, and then $|p_2| \geq \frac{1}{\sqrt{2}}$. □

Theorem 4.8.5 (Rhombus Equality) *A normed space X is an inner product space if and only if for every $u, v \in S_X$, the equality*

$$\|u + v\|^2 + \|u - v\|^2 = 4$$

holds.

Proof Let E be the minimal area ellipse circumscribed about S_X. By Lemma 4.8.4 there exist two independent points $u, v \in S_X \cap E$. Let $\| \cdot \|_E$ be the norm defined by E. Then

$$4 = \|u + v\|_E^2 + \|u - v\|_E^2 \leq \|u + v\|^2 + \|u - v\|^2 = 4,$$

which implies that $\|u + v\|_E = \|u + v\|$ and $\|u - v\|_E = \|u - v\|$, and then $\frac{u \pm v}{\|u \pm v\|} \in S_X \cap E$. Pursuing with the same process (take in the next step, i.e., $u_1 = \frac{u+v}{\|u+v\|}$, $v_1 = v$) we get that S_X and E meet in a dense set of points, and then $S_X = E$. □

Theorem 4.8.6 *A normed space X is an inner product space if and only if any of the following properties holds:*[6]

(a) *If $x, y \in X$ and $x \perp_I y$, then $x \perp_{P_-} y$.*
(b) *If $x, y \in X$ and $x \perp_{P_-} y$, then $x \perp_I y$.*

Proof

(a) For $u, v \in S_X$ we have $u + v \perp_I u - v$, and then $u + v \perp_{P_-} u - v$, from which it follows that $\|u + v\|^2 + \|u - v\|^2 = 4$. By Theorem 4.8.5, X is an inner product space.

(b) Let us show first that property (b) implies that X is strictly convex. Assume that there exist $u, v \in S_X$, $u \neq v$, such that $[u, v] \subset S_X$. Let $x = \frac{1}{2}(u + v)$, and let

---

[6] Similarly if P_- is replaced by P_+.

$\rho > 0$ be small enough such that for all $z \in \rho S_X$, the rays $[0, x \pm z\rangle$ cut $[u, v]$. By Theorem 4.7.2 there exists $z \in \rho S_X$ such that $x \perp_{P_-} z$. From (b) it follows that also $x \perp_{P_-} -z$. Then $\|x + z\| = \|x - z\|$, which implies that $x + z \in S_X$. But this is absurd because $1 = \|x + z\|^2 = \|x\|^2 + \|z\|^2 = 1 + \rho^2$.

Now, let $u, v \in S_X$, $x = u + v$ and $y = u - v$. Then $x \perp_I y$. Let $z \in X$ be such that $\|z\| = \|y\|$ and $x \perp_{P_-} z$. Then $x \perp_I z$. By Theorem 4.6.6, it follows that $y = z$, and then, $4 = \|u + v\|^2 + \|u - v\|^2$. Again, by Theorem 4.8.5, X is an inner product space. $\square$

Finally, we can consider also the relation between P_- and P_+ orthogonalities.

Theorem 4.8.7 *A normed space X is an inner product space if and only if any of the following properties holds:*

(a) *If $x, y \in X$ and $x \perp_{P_-} y$, then $x \perp_{P_+} y$.*
(b) *If $x, y \in X$ and $x \perp_{P_+} y$, then $x \perp_{P_-} y$.*

Proof If $x \perp_{P_\mp} y$ implies $x \perp_{P_\pm} y$, then also $x \perp_{P_\mp} y$ implies $x \perp_I y$. Thus, the proof follows from Theorem 4.8.6. $\square$

4.8.2 The Joly Construction

Before going on with properties of the type $[u, v \in S_X, \; u \perp_F v \Rightarrow u \perp_G v]$, we will deal with a kind of curves defined by Joly [91], closely related to this topic.

For given vectors $x = (x_1, x_2)$, $y = (y_1, y_2)$ in the plane, let us consider them oriented according to the sign of $x \wedge y := x_1 y_2 - x_2 y_1$.

Lemma 4.8.8 (Joly's Construction) *Let S_X be the unit sphere[7] of a norm in the plane. The sets*

$$S_B = \{u + v : u, v \in S_X, \; u \perp_B v, \; u \wedge v > 0\},$$

$$S'_B = \{u + v : u, v \in S_X, \; u \perp_B v, \; u \wedge v < 0\}$$

are rectifiable Jordan curves enclosing two times the area enclosed by S_X (Fig. 4.7).

Proof First, we will parametrize S_B. Let $u_0, v_0 \in S_X$ be such that $u_0 \perp_B v_0$, $u_0 \wedge v_0 > 0$ and in the arc in S_X from u_0 to v_0 there is no additional vector w such that $u_0 \perp_B w$. Take angle parameterizations of S_X, $u(\theta), v(\theta), 0 \leq \theta \leq 2\pi$, such that $u_0 = u(0)$, $v_0 = v(0)$ and $u(\theta) \wedge v(\theta) > 0$. Due to the existence properties of Birkhoff orthogonality, we know that for each $\theta \in [0, 2\pi]$ there exists (at least) one $\mu \in [0, 2\pi]$ such that $u(\theta) \perp_B v(\mu)$, i.e., $u(\theta) + v(\mu) \in S_B$. Similarly, for each μ there exists (at least) one θ with the same roles. But, since S_X is not necessarily

---

[7] Joly [91] proved his result for gauges.

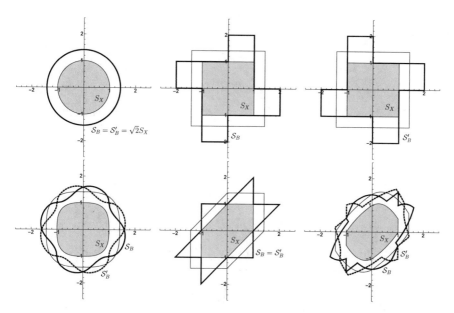

Fig. 4.7 Sets $\mathcal{S}_B$ and $\mathcal{S}'_B$ for different norms

strictly convex nor smooth, the above correspondence can be not one-to-one. Thus, neither θ nor μ is enough to parametrize $\mathcal{S}_B$. But simple geometric arguments show that if we consider the set $M = \{(\theta, \mu) \in [0, 2\pi) \times [0, 2\pi) : u(\theta) \perp_{\mathrm{B}} v(\mu),\ u(\theta) \wedge v(\mu) > 0\}$, then the map $(\theta, \mu) \in M \mapsto v = \theta + \mu \in [0, 4\pi)$ is a homeomorphism such that the two inverse components, $v \mapsto \theta$ and $v \mapsto \mu$, are increasing functions. This allows to parametrize $\mathcal{S}_B$ with the continuous function of bounded variation

$$v \in [0, 4\pi] \mapsto s(v) := u(\theta) + v(\mu) \in \mathcal{S}_B,$$

where $s(4\pi) = u_0 + v_0$.

The area enclosed by $\mathcal{S}_B$, $A(\mathcal{S}_B)$, is then given by the Riemann–Stieltjes integral

$$
\begin{aligned}
A(\mathcal{S}_B) &= \frac{1}{2} \int_{\mathcal{S}_B} s \wedge ds \\
&= \frac{1}{2} \int_{\mathcal{S}_B} u \wedge du + \frac{1}{2} \int_{\mathcal{S}_B} v \wedge dv + \frac{1}{2} \int_{\mathcal{S}_B} u \wedge dv + \frac{1}{2} \int_{\mathcal{S}_B} v \wedge du \\
&= 2A(\mathcal{S}_X) + \int_{\mathcal{S}_B} v \wedge du,
\end{aligned}
$$

where the last equality follows by recalling that $u(\theta)$ and $v(\mu)$ are parameterizations of $\mathcal{S}_X$, and applying integration by parts in the sum at the right. Thus, it only remains to show that $\int_{\mathcal{S}_B} v \wedge du = 0$. But this follows by considering that for each partition

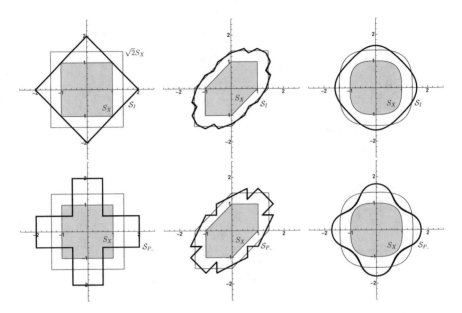

Fig. 4.8 Sets $\mathcal{S}_I$ and $\mathcal{S}_{P_-}$ for different norms

$0 = \nu_0 < \nu_1 < \cdots < \nu_n = 4\pi$ there correspond partitions $0 = \theta_0 < \theta_1 < \cdots < \theta_n = 2\pi$, $0 = \mu_0 < \mu_1 < \cdots < \mu_n = 2\pi$, and there exist μ'_k, $k = 1, \ldots, n$, such that $\mu_{k-1} \leq \mu'_k \leq \mu_k$, and $v(\mu'_k) \wedge \big(u(\theta_k) - u(\theta_{k-1})\big) = 0$.

The corresponding results for $\mathcal{S}'_B$ follow in the same way. $\qquad\square$

In a way similar to the above, it can be proved that the results in Lemma 4.8.8 hold also with isosceles and Pythagorean orthogonalities [5].[8] Since these orthogonalities are symmetric, we can remove the assumption on the orientation.

Lemma 4.8.9 *Let S_X be the unit sphere of a norm in the plane. The sets*

$$\mathcal{S}_I = \{u + v : u, v \in S_X, \ u \perp_I v\},$$

$$\mathcal{S}_{P_-} = \{u + v : u, v \in S_X, \ u \perp_{P_-} v\}$$

are rectifiable Jordan curves enclosing two times the area enclosed by S_X (see Fig. 4.8).

We can also consider the corresponding curve for P_+-orthogonality that, as is easily seen, coincides with $\sqrt{2}S_X$, i.e.,

---

[8] In [5] it is also proved that the corresponding curve for area orthogonality has the same property. But nevertheless, this is not a general property of all orthogonalities. If we consider the Carlsson orthogonality, $x \perp_C y :\Leftrightarrow 5\|x - y\|^2 = \|x - 2y\|^2 + \|y - 2x\|^2$, then $A(\mathcal{S}_C) \approx 2.3381 A(S_X)$.

$$\mathcal{S}_{P_+} = \{u + v : u, v \in S_X, \ u \perp_{P_+} v\} = \sqrt{2}S_X.$$

Theorem 4.8.10 *Let S_X be the unit sphere of a norm in the plane, and let $\mathcal{S}_B$ and $\mathcal{S}'_B$ be the curves defined in Lemma 4.8.8. The following properties are equivalent:*

(a) *Birkhoff orthogonality is symmetric.*
(b) $\mathcal{S}_B = \mathcal{S}'_B.$[9]

Proof If Birkhoff orthogonality is symmetric, then it trivially follows that $\mathcal{S}_B = \mathcal{S}'_B$. Conversely, assume that $\mathcal{S}_B = \mathcal{S}'_B$. Let $u, v \in S_X$, $u \perp_B v$, $u \wedge v > 0$. We will see that $v \perp_B u$. Since $\mathcal{S}_B = \mathcal{S}'_B$, there exist $u', v' \in S_X$, $u' \perp_B v'$, $u' \wedge v' < 0$, and $u + v = u' + v'$. Let $\alpha, \beta \in \mathbb{R}$ be such that $u' = \alpha u + \beta v$, and then $v' = (1 - \alpha)u + (1 - \beta)v$. Since $u \perp_B v$ and $u' \perp_B v'$, it follows that $0 \leq \alpha \leq 1$, and from $u \wedge v > 0$ and $u' \wedge v' < 0$ we get that $\alpha < \beta$. Moreover, $\beta = \|u' - \alpha u\| \leq 1 + \alpha$. Since $u' \perp_B v'$, it follows that $\beta - \alpha = \|(\alpha - 1)u' + \alpha v'\| \geq 1 - \alpha$, and then $\beta \geq 1$. From $\|v'\| = 1$ it follows that $\beta - \alpha \geq 1$. Therefore, $\beta - \alpha = 1$, which means that v belongs to the segment $[-u, u']$. Since $(-u) \perp_B v$ and $(-u) \wedge v < 0$, there exist $\bar{u}, \bar{v} \in S_X$, $\bar{u} \perp_B \bar{v}$, $\bar{u} \wedge \bar{v} > 0$, such that $v - u = \bar{u} + \bar{v}$. The same arguments as those used above lead to v being in the segment $[u, \bar{u}]$. But this makes no sense, except if $v = u'$ or $v = \bar{u}$. In both cases it follows that $v \perp_B u$. □

Theorem 4.8.11 *Let S_X be the unit sphere of a norm in the plane, and let $\mathcal{S}_B$ and $\mathcal{S}_I$ be the curves defined in Lemma 4.8.8 and Lemma 4.8.9, respectively. Then the following properties are equivalent:*[10]

(a) $\mathcal{S}_B = \mathcal{S}_I.$
(b) $\mathcal{S}'_B = \mathcal{S}_I.$
(c) *If $u, v \in S_X$ and $u \perp_B v$, then $u \perp_I v$.*
(d) *If $u, v \in S_X$ and $u \perp_I v$, then $u \perp_B v$.*

Proof Property (c) (respectively, (d)) implies that $\mathcal{S}_B \subset \mathcal{S}_I$ (respectively, $\mathcal{S}_I \subset \mathcal{S}_B$). Since these curves are Jordan curves, both properties imply that $\mathcal{S}_B = \mathcal{S}_I$. To prove that (a) implies (c) and (d), we will see that if $u, v, u', v' \in S_X$ are such that $u \perp_B v$, $u' \perp_I v'$, $u \wedge v > 0$, $u' \wedge v' > 0$ and $u + v = u' + v'$, then $u = u'$ and $v = v'$.

Let $\alpha, \beta \in \mathbb{R}$ be such that $u' = \alpha u + \beta v$. Then $v' = (1 - \alpha)u + (1 - \beta)v$. Since $u \perp_B v$, it follows that $0 \leq \alpha \leq 1$, and since $\mathrm{sign}(u \wedge v) = \mathrm{sign}(u' \wedge v')$, it follows that $\alpha > \beta$. Now, let us consider two cases:

Assume that $\beta \leq 0$. Then, $\alpha - \beta = \|\alpha v' + (\alpha - 1)u'\| \leq 1$, and $1 - \beta = \|\beta v' + (\alpha - \beta)u\| \leq -\beta + \alpha - \beta$. Therefore, $\alpha - \beta = 1$, and then $v = \frac{1-\alpha}{1-\beta}(-u) + \frac{\alpha-\beta}{1-\beta}v'$, which implies that the points v, $-u$ and v' are in a segment contained in S_X. Then

---

$\frac{1}{2}(v' - u') = \frac{2\alpha-1}{2}(-u) + \frac{1-2\beta}{2}v \in S_X$, and therefore, $2 = \|v' - u'\| = \|u' + v'\| = \|u + v\| = \left\|\frac{1}{1-\beta}v' + \frac{\alpha-\beta}{1-\beta}u\right\| \leq \frac{2}{1-\beta}$, which implies $\beta = 0$ and $\alpha = 1$.

Assume that $\beta > 0$. Since $\|u'\| = \|v'\| = 1$, it follows that $\alpha + \beta = 1$, which implies that the points u, v, u' and v' are in a segment contained in S_X. Therefore, $2 = \|u' + v'\| = \|v' - u'\| = \|(2\alpha - 1)(v - u)\| \leq 2|2\alpha - 1|$. Hence $|2\alpha - 1| \geq 1$, that jointly with $\alpha > \beta > 0$ gives $\alpha = 1$, and therefore $\beta = 0$, a contradiction.

Similarly with $\mathcal{S}'_B$. $\square$

Question 4.8.12 *Does some analogue of Theorem 4.8.11 hold for other orthogonalities?*

In a way similar to the above one, the following can be proved (see [5]):

(i) $\mathcal{S}_B = \mathcal{S}_{P_-}$ if and only if $[u, v \in S_X, u \perp_B v, u \wedge v > 0 \Rightarrow u \perp_{P_-} v]$.
(ii) $\mathcal{S}_B = \mathcal{S}_{P_+}$ if and only if $[u, v \in S_X, u \perp_B v, u \wedge v > 0 \Rightarrow u \perp_{P_+} v]$.
(iii) $\mathcal{S}_{P_-} = \mathcal{S}_{P_+}$ if and only if $[u, v \in S_X, u \perp_{P_\mp} v \Rightarrow u \perp_{P_\pm} v]$.

The corresponding results with $\mathcal{S}'_B$ also hold.

Question 4.8.13 *Can the condition $u \wedge v > 0$ be removed in the properties (i) and (ii)?*

Theorem 4.8.20 will show that a normed space X is an inner product space if and only if the property $[u, v \in S_X, u \perp_B v \Rightarrow \|u + v\| = \sqrt{2}]$ holds. This, jointly with the above results, implies that if any three of the curves $\mathcal{S}_B, \mathcal{S}'_B, \mathcal{S}_{P_-}$ and $\mathcal{S}_{P_+}$ coincide, then S_X is an ellipse, i.e., X is an inner product space. Nevertheless, the curves $\mathcal{S}_{P_-}$ and $\mathcal{S}_{P_+}$ can coincide without S_X being an ellipse: e.g., if S_X is a regular octagon, then $\mathcal{S}_I = \mathcal{S}_{P_-} = \mathcal{S}_{P_+}$.

Question 4.8.14 *If $\mathcal{S}_K = \mathcal{S}_L$ with $K \in \{B, B'\}$ and $L \in \{P_-, P_+\}$, is S_X and ellipse?*

The Rectangular Constant

Joly [91] defined the *rectangular constant* of a normed space X as

$$\mu(X) = \sup\left\{\frac{\|x\| + \|y\|}{\|x + y\|} : x, y \in X, x \perp_B y\right\},$$

and he proved that for any normed space X the inequalities

$$\sqrt{2} \leq \mu(X) \leq 3 \tag{4.7}$$

hold. The right bound follows by having into account that if $x \perp_B y$, then

$$\frac{\|x\| + \|y\|}{\|x + y\|} \leq \frac{2\|x\| + \|x + y\|}{\|x + y\|} \leq 3.$$

Moreover, $\mu(X) = \sup\{\mu(X_2) : X_2 \subset X, \dim X_2 = 2\}$. Thus, to get the left bound, we can assume that $\dim X = 2$. Since $A(\mathcal{S}_B) = 2$, necessarily there exist $u, v \in S_X$ such that $u \perp_B v$ and $\|u + v\| \geq \sqrt{2}$, which leads to the left bound.

Both bounds in (4.7) are sharp. To attain the right bound, consider, e.g., the space $X = (\mathbb{R}^2, \|\cdot\|_\infty)$, $x = (1, -1)$ and $y = (0, 2)$. In fact, $\mu(X) = 3$ if and only if X is not uniformly non-square[11] [34]. Moreover, if X is an inner product space and $x \perp_B y$, then $\|x + y\|^2 = \|x\|^2 + \|y\|^2$, which easily leads to $\mu(X) = \sqrt{2}$. Does this identity hold in other spaces? For a while we pay attention to this important question.

Assume that $\mu(X) = \sqrt{2}$. Then, $\mu(X_2) = \sqrt{2}$ for any two-dimensional subspace $X_2 \subset X$. This implies that if $u, v \in S_{X_2}$, $u \perp_B v$, then $\|u + v\| \geq \sqrt{2}$. Since $A(\mathcal{S}_B) = A(\mathcal{S}_B') = 2A(S_{X_2})$, it follows that if $u, v \in S_{X_2}$, $u \perp_B v$, then $\|u + v\| = \sqrt{2}$. Therefore, $\mathcal{S}_B = \sqrt{2} S_{X_2} = \mathcal{S}_B'$, and by Theorem 4.8.10, Birkhoff orthogonality is symmetric in X_2, and, a fortiori, in X. The reciprocal result is not true: i.e., symmetry of Birkhoff orthogonality in X does not imply $\mu(X) = \sqrt{2}$. In fact, we can only get (trivially) that $\mu(X) \leq 2$; and this bound is attained, e.g., in the space $\mathbb{R}^2$ endowed with the norm whose unit sphere is affine to a regular hexagon [79]. Actually, if $\dim X = 2$ and Birkhoff orthogonality is symmetric, then $\mu(X) = 2$ only in such spaces [34].

From the above it follows (recall Theorem 4.5.11) that if $\dim X \geq 3$ and $\mu(X) = \sqrt{2}$, then X is an inner product space. Joly [91] conjectured that the same is true if $\dim X = 2$, and this was solved affirmatively by Del Río and Benítez [63]. Thus, we have

Theorem 4.8.15 *A normed space X is an inner product space if and only if $\mu(X) = \sqrt{2}$.*

The proof of the above theorem, when $\dim X = 2$, is not trivial and involves the following lemmas that are of interest by themselves.

For $x = (x_1, x_2)$, $y = (y_1, y_2)$ vectors in $\mathbb{R}^2$, denote by $A(x, y)$ the area of the parallelogram they determine, i.e.,

$$A(x, y) = |x_1 y_2 - x_2 y_1|.$$

Lemma 4.8.16 *Let $\dim X = 2$. For $x, y \in S_X$,*

$$x \perp_B y \quad \Leftrightarrow \quad A(x, y) = \sup_{z \in S_X} A(z, y).$$

Proof It is enough to note that for $\alpha, \beta \in \mathbb{R}$ the identity

---

[11] A Banach space X is said to be *uniformly non-square* if there exists $\delta \in (0, 1)$ such that for any $x, y \in S_X$, either $\|x + y\| \leq 2(1 - \delta)$ or $\|x - y\| \leq 2(1 - \delta)$.

$$\|\alpha x + \beta y\| = \frac{|\alpha| A(x, y)}{A\left(\frac{\alpha x + \beta y}{\|\alpha x + \beta y\|}, y\right)}$$

holds.   □

Lemma 4.8.17 ([63]) *Let* dim $X = 2$. *Birkhoff orthogonality is symmetric in X if and only if the area of the parallelograms determined by vectors* $x, y \in S_X, x \perp_B y$, *is constant.*

Lemma 4.8.18 ([63]) *Let* dim $X = 2$ *and* $\mu(X) = \sqrt{2}$.

(i) *If* $u, v \in S_X$ *and* $u \perp_B v$, *then* $u + v \perp_B u - v$.[12]
(ii) *If* $u, v, u', v' \in S_X$ *are such that* $u \perp_B v$, $u' \perp_B v'$ *and* $(u \wedge v)(u' \wedge v') > 0$, *then* $A_q(u, u') = A_q(v, v')$, *where* $A_q(x, y)$, *for* $x, y \in X \setminus \{0\}$, *denotes the area of the quadrant* $\{\alpha x + \beta y : \alpha, \beta \in \mathbb{R}^+, \|\alpha x + \beta y\| \leq 1\}$.

Sketch of the Proof of Theorem 4.8.15 After an affine transformation, it can be assumed that $(1, 0), (0, 1) \in S_X$ and $(1, 0) \perp_B (0, 1)$. For $0 \leq \theta \leq 2\pi$, let $x(\theta) = (x_1(\theta), x_2(\theta))$ be the vector in S_X that makes an angle θ with $(1, 0)$, measured in the positive sense. Let $y(\theta) = (y_1(\theta), y_2(\theta))$ be the unique (by Lemma 4.8.18 (ii)) vector in S_X such that $x(\theta) \perp_B y(\theta)$ and $x(\theta) \wedge y(\theta) > 0$.

The key of the proof is to demonstrate that the equalities

$$\int_0^\theta x_i \, dx_i + \int_0^\theta y_i \, dy_i = 0, \quad i = 1, 2, \tag{4.8}$$

hold for every $0 \leq \theta < 2\pi$, because in such a case, $x_i^2(\theta) + y_i^2(\theta) = 1, i = 1, 2$; and since from Lemma 4.8.17 (recall that $\mu(X) = \sqrt{2}$ leads to the symmetry of Birkhoff orthogonality) it follows that $x_1(\theta)y_2(\theta) - x_2(\theta)y_1(\theta) = 1$, we obtain $\left(x_1(\theta) - y_2(\theta)\right)^2 + \left(x_2(\theta) + y_1(\theta)\right)^2 = 0$. This implies that S_X is the circle $x_1^2 + x_2^2 = 1$.

The proof of (4.8) is very cumbersome and uses again Lemma 4.8.18.   □

Considering only unit vectors in $\mu(X)$ leads to the constant

$$\mu'(X) = \sup\left\{\frac{2}{\|x + y\|} : x, y \in S_X, x \perp_B y\right\},$$

introduced by Baronti [34]. Obviously, $\mu'(X) \leq \mu(X)$, but now

$$\sqrt{2} \leq \mu'(X) \leq 2.$$

---

[12] Some years after the paper from Del Río–Benítez [63] had been published, Baronti [34] showed that this property characterizes inner product spaces.

The left bound is obtained as in (4.7), and the right one follows trivially. As with $\mu(X)$, it follows that if $\mu'(X) = \sqrt{2}$, then Birkhoff orthogonality is symmetric. Therefore, if $\dim X \geq 3$, then $\mu'(X) = \sqrt{2}$ if and only if X is an inner product space.

Assume now that $\dim X = 2$ and $\mu'(X) = \sqrt{2}$. If $x, y \in S_X$, $x \perp_B y$, then $\|x + y\| \geq \sqrt{2}$. Therefore (recall that $A(S_B) = 2A(S_X)$),

$$x, y \in S_X, \ x \perp_B y \quad \Rightarrow \quad \|x + y\| = \sqrt{2}. \tag{4.9}$$

In [40] it is proved that property (4.9) implies property (i) in Lemma 4.8.18. Moreover, looking inside the proof of property (ii), it can be seen that the hypothesis actually used is $\mu'(X) = \sqrt{2}$, instead of the (at first) stronger $\mu(X) = \sqrt{2}$. All the above leads to the following theorem.

Theorem 4.8.19 *A normed space X is an inner product space if and only if $\mu'(X) = \sqrt{2}$.*

4.8.3 Implications over the Unit Sphere

As a consequence of Theorem 4.8.19, we get the following two theorems that improve the results in Theorem 4.8.3 and Theorem 4.8.2, respectively.

Theorem 4.8.20 *A normed space X is an inner product space if and only if any of the following properties holds:*[13]

(a) *If $x, y \in S_X$ and $x \perp_B y$, then $x \perp_{P_+} y$.*
(b) *If $x, y \in S_X$ and $x \perp_{P_+} y$, then $x \perp_B y$.*

Proof The necessity is evident. Let us prove the sufficiency.

(a) The proof follows directly from Theorem 4.8.19.
(b) We can assume, without loss of generality, that $\dim X = 2$. Let $x, y \in S_X$, $x \perp_{P_+} y$. Then, $x \perp_B y$, and since Pythagorean orthogonalities are symmetric, we have also $y \perp_B x$. Therefore, $S_{P_+} \subset S_B$ and $S_{P_+} \subset S'_B$. Since these sets are Jordan curves, it follows that $S_B = S'_B = S_{P_+} = \sqrt{2}S_X$. Again Theorem 4.8.19 completes the proof. □

Theorem 4.8.21 *A normed space X is an inner product space if and only if any of the following properties holds:*

(a) *If $x, y \in S_X$ and $x \perp_B y$, then $x \perp_I y$.*
(b) *If $x, y \in S_X$ and $x \perp_I y$, then $x \perp_B y$.*

---

[13] Similarly if P_+ is replaced by P_-.

Proof The necessity is evident. To prove the sufficiency, we can assume, without loss of generality, that $\dim X = 2$. From Theorem 4.8.11 it follows that any of the properties (a) and (b) implies that Birkhoff and isosceles orthogonalities are equivalent over the unit sphere, and therefore Birkhoff orthogonality is symmetric. Let $x, y \in S_X$, $x \perp_B y$. Since $\|x + y\| = \|x - y\|$, we have $\frac{x+y}{\|x+y\|} \perp_I \frac{x-y}{\|x-y\|}$ and, therefore, $\frac{x+y}{\|x+y\|} \perp_B \frac{x-y}{\|x-y\|}$. From Lemma 4.8.17, it follows that

$$A(x, y) = A\left(\frac{x+y}{\|x+y\|}, \frac{x-y}{\|x-y\|}\right) = \frac{2A(x, y)}{\|x+y\|^2},$$

and then $\|x + y\| = \sqrt{2}$. Theorem 4.8.20 completes the proof. □

Remark 4.8.22 In contrast to what happens in Theorem 4.8.6 and Theorem 4.8.7, the properties $[x, y \in S_X, \ x \perp_K y \ \Rightarrow \ x \perp_L y]$, with $K, L \in \{I, P_+, P_-\}$, $K \neq L$, do not characterize inner product spaces. In the space $\mathbb{R}^2$ endowed with a norm whose unit sphere is a regular octagon, the isosceles, Pythagorean + and Pythagorean − orthogonalities coincide over the unit sphere.[14] We do not know what happens if the dimension of the space is greater than or equal to three.

At the end of this section we mention some further contributions treating the rectangular modulus which is strongly related to Joly's rectangular constant, based on Birkhoff orthogonality and yielding characterizations of inner product spaces, uniform rotundity, and uniform smoothness. These are [132, 133], and [134]. In the latter paper also results of the following type are shown: if X_1 and X_2 are isomorphic Banach spaces where X_1 is close to X_2 in the Banach–Mazur distance, then the corresponding rectangular moduli are also close.

4.9 Other Orthogonalities

In this section we collect the properties listed in Sect. 4.3 of the remaining orthogonalities defined in Sect. 4.2. Only simple proofs (or sketches) of some results are included; the rest can be found in the provided references.

4.9.1 Carlsson Orthogonalities

Definition 4.9.1 Let n be a positive integer, and $a_i, b_i, c_i, i = 1, \dots, n$, be numbers satisfying

$$\sum_{i=1}^{n} a_i b_i^2 = \sum_{i=1}^{n} a_i c_i^2 = 0 \quad \text{and} \quad \sum_{i=1}^{n} a_i b_i c_i = 1.$$

---

[14] Also coincide with area orthogonality.

A vector x is said to be *Carlsson orthogonal to* a vector y ($x \perp_C y$) if

$$\sum_{i=1}^{n} a_i \|b_i x + c_i y\|^2 = 0.$$

Note that the condition $\sum_{i=1}^{n} a_i b_i c_i = 1$ in Definition 4.9.1 can be changed into $\sum_{i=1}^{n} a_i b_i c_i \neq 0$.

Symmetry

Some Carlsson orthogonalities are symmetric, like isosceles, Pythagorean and (a) orthogonalities, and some others are not. For example, the Carlsson orthogonality $x \perp_{2I} y :\Leftrightarrow \|x + 2y\| = \|x - 2y\|$ is not symmetric in $(\mathbb{R}^2, \|\ \|_\infty)$: if $x = (1, \frac{1}{2})$ and $y = (-\frac{3}{4}, 1)$, then $x \perp_{2I} y$, but $y \not\perp_{2I} x$. It was conjectured in [6] that a Carlsson orthogonality is "trivially" symmetric (isosceles, Pythagorean, ...) or it is symmetric only in inner product spaces. Such is the case, e.g., of the above orthogonality and, in general, of any λI-orthogonality ($\lambda \neq \pm 1$): if $\perp_{\lambda I}$ is symmetric then $x \perp_I y$ implies $x \perp_I \lambda^2 y$, which is a characteristic property of inner product spaces (Theorem 4.6.15). Similarly, from Theorem 4.7.7 it follows that symmetry of λP-orthogonality is also characteristic for inner product spaces.

It was proved in [19] that if an (a, b)-orthogonality is symmetric and additive in X, then X is an inner product space. In light of Theorem 4.9.5, the symmetry hypothesis is superfluous. We ask the following:

Question 4.9.2 *Does symmetry alone imply the same result?*

Existence and Uniqueness Properties

Concerning the existence of Carlsson orthogonalities, we have:

Theorem 4.9.3 *All Carlsson orthogonalities have the existence and α-existence properties in any normed space.*

Proof Carlsson [50] proved the α-existence property on the right. He followed a similar proof given by James [86] with respect to isosceles orthogonality: For $x, y \in X$, consider the function $f(t) = \sum_{i=1}^{n} a_i \|b_i x + c_i (tx + y)\|^2$. Then $x \perp_C \alpha x + y$ if and only if $f(\alpha) = 0$. Recall that $\sum_{i=1}^{n} a_i c_i^2 = 0$. Since

$$\frac{f(t)}{t} = \frac{1}{t} \sum_{c_i \neq 0} a_i \left[\left\| \left(t + \frac{b_i}{c_i} \right) c_i x + c_i y \right\|^2 - \|t c_i x + c_i y\|^2 \right] + \frac{\|x\|^2}{t} \sum_{c_i = 0} a_i b_i^2$$

and having in mind [50, Lemma 2.2] that

$$\frac{1}{t} \left[\|(t + a)x + y\|^2 - \|tx + y\|^2 \right] \xrightarrow[t \to \pm\infty]{} 2a \|x\|^2,$$

it follows that $\lim_{t \to \pm\infty} \frac{f(t)}{t} = 2\|x\|^2$. Therefore, $f(t)$ must take positive and negative values. Since it is continuous, it must be null at some point. Since the roles of the scalars b_i and c_i are similar, the same proof gives the α-existence in the left.

The existence (on the right and on the left) property follows as in the cases of isosceles and Pythagorean orthogonalities. For given $x, y \in X$, $\rho > 0$, consider $z(\theta) \in \rho S$ as defined in the proof of Theorem 4.6.2. Let $g(\theta) = \sum_{i=1}^{n} a_i \|b_i x + c_i z(\theta)\|^2$. Then $g(0) = 2\rho \|x\| = -g(\pi)$. Therefore, there exists $0 < \theta_0 < \pi$ such that $g(\theta_0) = 0$, i.e., $x \perp_C z(\theta_0)$. $\square$

With respect to uniqueness, only partial results are known. On the one hand, recall that the following properties are equivalent:

- X is strictly convex;
- isosceles orthogonality is unique (Theorem 4.6.6);
- isosceles orthogonality is α-unique (Theorem 4.6.11);
- Pythagorean orthogonality is unique (Theorem 4.7.3).

On the other hand, Pythagorean orthogonality is α-unique in every normed space (Theorem 4.7.5). Therefore, in view of that, it seems that Carlsson orthogonalities do not have a common behavior regarding α-uniqueness.

Question 4.9.4 *Is the uniqueness of any Carlsson orthogonality equivalent to strict convexity?*

Following Carlsson [50], a C-orthogonality is said to have property (H) in X if

$$x, y \in X, \quad x \perp_C y \quad \Rightarrow \quad \lim_{\substack{m \to +\infty \\ m \in \mathbb{N}}} \frac{1}{m} \sum_{i=1}^{n} a_i \|m b_i x + c_i y\|^2 = 0. \tag{H}$$

A normed space X with a C-orthogonality having property (H) has also the following properties [50]:

- C-orthogonality is α-unique, and α depends continuously on x and y;
- the norm is Gâteaux differentiable;
- C-orthogonality and B-orthogonality are equivalent;
- B-orthogonality is symmetric.

Homogeneity and Additivity
We know that isosceles and Pythagorean orthogonalities are homogeneous (Theorems 4.6.13 and 4.7.6) or additive (Theorems 4.6.17 and 4.7.9) only in inner product spaces. We will see that the same occurs with any Carlsson orthogonality.

First, note that if C-orthogonality is homogeneous or additive on the left in X then it has property (H) above, and then Birkhoff orthogonality is symmetric. But if $\dim X \geq 3$, Birkhoff orthogonality is symmetric only in inner product spaces (Theorem 4.5.11). Therefore, homogeneity or additivity on the left (and on the right, by interchanging the roles of b_i and c_i) of any Carlsson orthogonality characterizes inner product spaces of dimension greater than or equal to three. In addition, Carlsson [50] has shown, with a very involved proof based on functional equations, that the same result is true in two-dimensional spaces. Therefore,

Theorem 4.9.5 *Let X be a normed space. The following properties are equivalent:*

(i)   *X is a inner product space.*
(ii) *A Carlsson orthogonality is homogeneous in X.*
(iii) *A Carlsson orthogonality is left (or right) additive in X.*

For some particular cases the proof of the above result is not so hard. For example, if λI-orthogonality (resp. λP-orthogonality) is homogeneous (resp. additive), then I-orthogonality (resp. P_+-orthogonality) is also homogeneous (resp. additive), and then the space is an inner product space.

Let us show now how the a-orthogonality case ($a \neq 0$) was solved with simple tools in [65]. Since $x \perp_a y$ is equivalent to $x \perp_{\frac{1}{a}} y$, we can consider $|a| < 1$. For $n \in \mathbb{N}$, let us consider the property

$$P_n : x, y \in X, \ x \perp_a y \ \Rightarrow \ \left(1 + a^{2^n}\right)\|x - y\|^2 = \left\|x - a^{2^{n-1}}y\right\|^2 + \left\|y - a^{2^{n-1}}x\right\|^2.$$

We shall see by induction that if a-orthogonality is homogeneous then P_n holds for every $n \in \mathbb{N}$. First, P_1 holds in any case. Then, assume that a-orthogonality is homogeneous and that P_n holds. Let $\beta = a^{2^{n-1}}$. Then, for $x, y \in X$, $x \perp_a y$, we have $\beta x \perp_a y$ and $x \perp_a \beta y$. Since P_n holds, we get

$$(1 + \beta^2)\|\beta x - y\|^2 = \|\beta x - \beta y\|^2 + \|y - \beta^2 x\|^2$$

and

$$(1 + \beta^2)\|x - \beta y\|^2 = \|x - \beta^2 y\|^2 + \|\beta y - \beta x\|^2.$$

Adding both identities, and having in mind that P_n holds, it follows that $2\beta^2\|x - y\|^2 + \|x - \beta^2 y\|^2 + \|y - \beta^2 x\|^2 = (1 + \beta^2)\left(\|\beta x - y\|^2 + \|x - \beta y\|^2\right) = (1 + \beta^2)^2\|x - y\|^2$, and then $(1 + \beta^4)\|x - y\|^2 = \|x - \beta^2 y\|^2 + \|y - \beta^2 x\|^2$, which means that P_{n+1} holds.

Therefore, since P_n holds for every n, we got that $x \perp_a y$ implies $x \perp_{P_-} y$. Let us see that the converse is also true. So, assume that $x \perp_{P_-} y$. Since a-orthogonality has the α-existence property, there exists $\alpha \in \mathbb{R}$ such that $x \perp_a \alpha x + y$. From the homogeneity of a-orthogonality and the fact that a-orthogonality implies P_--orthogonality it follows that $\gamma x \perp_{P_-} \alpha x + y$ for any $\gamma \in \mathbb{R}$. In particular, $\alpha x \perp_{P_-} \alpha x + y$ and $(1 + \alpha)x \perp_{P_-} \alpha x + y$, that jointly with $x \perp_{P_-} y$ leads to the identity $\|x\|^2 = (1 + 2\alpha)\|x\|^2$. Hence $\alpha = 0$, i.e., $x \perp_a y$. Therefore, both orthogonalities are equivalent, and then P_--orthogonality is homogeneous.

Assume now that a-orthogonality is additive and let $x \perp_a y$. Following the ideas of James [86, Theorem 5.3], from the additivity and symmetry we get that $x \perp_a \frac{m}{n} y$ for any $m, n \in \mathbb{N}$, and then, by continuity, $x \perp_a \lambda y$ for any $\lambda \in \mathbb{R}^+$. Moreover, from the α-existence property we get $\alpha \in \mathbb{R}$ such that $x \perp_a \alpha x - y$. Then, by additivity, $x \perp_a \alpha x$, which implies $\alpha = 0$. Thus, we get homogeneity.

In the next theorem by Boussouis [44] the homogeneity hypothesis in Theorem 4.9.5 is weakened.

Theorem 4.9.6 *If for every two-dimensional subspace $L \subset X$ and any $x \in L$ there exists $y \in L \setminus \{0\}$ such that $x \perp_C \lambda y$ for every $\lambda \in \mathbb{R}$, then X is an inner product space.*

In the proof of the above result the next interesting lemma was used.

Lemma 4.9.7 *If $x, y \in X$ are such that $x \perp_C \lambda y$ for every $\lambda \in \mathbb{R}$, then $x \perp_B y$.*

Orthogonal Diagonals

Regarding the existence of orthogonal diagonals in the sense of Carlsson orthogonalities, we have:

Theorem 4.9.8 *Carlsson orthogonalities have the property of existence of orthogonal diagonals.*

Proof This result was proved by Del Río [62]. Similarly to the case of Pythagorean orthogonality, just consider the continuous function $f(\lambda) = \sum_{i=1}^{n} a_i \|(b_i + c_i)x + \lambda(b_i - c_i)y\|^2$. Then $x + \rho y \perp_C x - \rho y$ if and only if $f(\rho) = 0$. The existence of such $\rho > 0$ follows from the fact that $f(0) = 2\|x\|^2 > 0$ and $\lim_{\lambda \to +\infty} \frac{f(\lambda)}{\lambda^2} = -2\|y\|^2 < 0$. ∎

With respect to uniqueness of orthogonal diagonals, little is known. Except for the cases of isosceles and Pythagorean orthogonalities, there are almost no results in the literature, and the problem (in general) seems to be not trivial. In the particular case of a-orthogonality the uniqueness follows easily: $x + \rho y \perp_a x - \rho y$ if and only if $g(1/\rho) = 0$, where $g(\lambda) = \|\lambda(1 - a)x + (1 + a)y\|^2 + \|\lambda(1 - a)x - (1 + a)y\|^2 - 4(1 + a^2)\|y\|^2$. Since g is symmetric and convex, it has a unique positive root. Moreover, if $|a| < 1$ then X is an inner product space if and only if $\rho = \|x\|/\|y\|$ (see [47]).

Nevertheless, to extend the above proof of uniqueness of orthogonal diagonals to any Carlsson orthogonality seems to be not easy because the analogous function g is, in general, the difference of two convex functions.

As stated in [18, p. 50], the proof of Corollary 4.5.18 is also valid for the following corollary.

Corollary 4.9.9 *A normed space X is an inner product space if and only if*

$$x, y \in X, \ x \perp_C y \quad \Rightarrow \quad \|x + y\|^2 + \|x - y\|^2 \approx 2(\|x\|^2 + \|y\|^2),$$

where "$\approx$" means "$\leq$" or "$\geq$".

Relations of Carlsson Orthogonalities with Other Orthogonalities

Obviously, by Theorem 4.9.5 Carlsson orthogonality is equivalent to a homogeneous orthogonality only in inner product spaces. Moreover, a long process developed in the papers [8, 44, 64] and [46] gives rise to the following more interesting results.

Theorem 4.9.10 *Let X be a normed space and let $\perp \in \{\perp_B, \perp_S, \perp_D, \perp_A\}$. The following properties are equivalent:*

(i) *X is an inner product space.*
(ii) *If $x, y \in X$ and $x \perp y$, then $x \perp_C y$.*
(iii) *If $x, y \in X$ and $x \perp_C y$, then $x \perp y$.*

4.9.2 Boussouis Orthogonalities

Definition 4.9.11 Let (Ω, μ) be a positive measure space and a, b, c be μ-measurable functions from Ω to $\mathbb{R}$ such that $a \neq 0$, μ-a.e., ab^2 and ac^2 are μ-integrable,

$$\int_\Omega ab^2\, d\mu = \int_\Omega ac^2\, d\mu = 0, \quad \text{and} \quad \int_\Omega abc\, d\mu = 1.$$

A vector x is said to be *Boussouis orthogonal to* another vector y ($x \perp_M y$) if

$$\int_\Omega a(\omega)\|b(\omega)x + c(\omega)y\|^2\, d\mu(\omega) = 0$$

Boussouis orthogonalities [47, 48] generalize Carlsson orthogonalities: if, for $n \geq 2$, we take $\Omega = \{1, 2, \ldots, n\}$, μ the counting measure, and $a(\omega), b(\omega), c(\omega)$, $\omega = 1, \ldots, n$, such that

$$\sum_{\omega \in \Omega} a(\omega)b(\omega)^2 = \sum_{\omega \in \Omega} a(\omega)c(\omega)^2 = 0, \quad \sum_{\omega \in \Omega} a(\omega)b(\omega)c(\omega) = 1,$$

then we get Carlsson orthogonality.

As in the case of Carlsson orthogonality, the roles of $b(\omega)$ and $c(\omega)$ are symmetric. Thus, all that happens to the left will also happen to the right.

The proof of the next theorem is similar to the case of Carlsson orthogonality [47, 48].

Theorem 4.9.12 *Boussouis orthogonalities have the existence, α-existence and existence of diagonals properties in any normed space.*

With respect to uniqueness and symmetry, nothing more than what is known about Carlsson orthogonalities appears in the literature.

Theorem 4.9.13 *A normed space X is an inner product space if and only if Boussouis orthogonality is homogeneous.*

The proof of the above theorem is long and nontrivial [47, 48]. It is proved that the homogeneity of Boussouis orthogonality implies that Roberts orthogonality has the existence property (Theorem 4.4.2).

As in the cases of isosceles (Remark 4.6.16.a) and Pythagorean (Remark 4.7.8) orthogonalities the hypothesis in Theorem 4.9.13 can be slightly weakened:

X is an inner product space if and only if for any $x, y \in X$, $x \perp_M y$, there exists $\delta(x, y) > 0$ such that $x \perp_M \lambda y$ for every $0 < \lambda < \delta(x, y)$.

As corollary of Theorem 4.9.13, the following is obtained.

Theorem 4.9.14 *A normed space X is an inner product space if and only if Boussouis orthogonality is additive.*

The next results can be found in [47] and [48].

Theorem 4.9.15 *Let X be a normed space and let $\perp \in \{\perp_B, \perp_S, \perp_A, \perp_D, \perp_{DP}\}$. Then the following properties are equivalent:*

(i)   *X is an inner product space.*
(ii) *If $x, y \in X$ and $x \perp y$, then $x \perp_M y$.*
(iii) *If $x, y \in X$ and $x \perp_M y$, then $x \perp y$.*

In fact, the result in Theorem 4.9.15(ii) proved in [47] is more general because it is proved also that X is an inner product space if and only if a UM-orthogonality implies a M-orthogonality, where both orthogonalities can be defined with different parameters.

4.9.3   Singer Orthogonality

Definition 4.9.16 Two vectors x and y are said to be *Singer orthogonal* $(x \perp_S y)$ if

$$\left\| \frac{x}{\|x\|} + \frac{y}{\|y\|} \right\| = \left\| \frac{x}{\|x\|} - \frac{y}{\|y\|} \right\|.$$

Singer orthogonality [135] was the first orthogonality defined by restricting a known orthogonality to the unit sphere: $x \perp_S y \Leftrightarrow \frac{x}{\|x\|} \perp_I \frac{y}{\|y\|}$.

It is clear that Singer orthogonality is symmetric and homogeneous.

Theorem 4.9.17 *Singer orthogonality has the existence and the α-existence properties in any normed space.*

Proof Since Singer orthogonality is homogeneous, existence and α-existence are equivalent properties. Nevertheless, we will give independent proofs of both properties.

To prove the existence, let $x, y \in X$, linearly independent, and $\rho > 0$. Let $\bar{x} = \frac{x}{\rho\|x\|}$, $\bar{y} = y$ and $\bar{\rho} = \frac{1}{\rho}$. Since isosceles orthogonality has the existence property, there exists $\bar{\alpha}, \bar{\beta} \in \mathbb{R}$, $\bar{\beta} > 0$, such that $\bar{x} \perp_I \bar{\alpha}\bar{x} + \bar{\beta}\bar{y}$ and $\|\bar{\alpha}\bar{x} + \bar{\beta}\bar{y}\| = \bar{\rho}$. Then, taking $\alpha = \frac{\rho\bar{\alpha}}{\|x\|}$ and $\beta = \rho^2\bar{\beta}$, it follows that $x \perp_S \alpha x + \beta y$ and $\|\alpha x + \beta y\| = \rho$.

To prove the α-existence we can assume, without loss of generality, that $x, y \in X$ are linearly independent. The function

$$\lambda \in \mathbb{R} \to f(\lambda) = \left\| \frac{x}{\|x\|} + \frac{x+\lambda y}{\|x+\lambda y\|} \right\| - \left\| \frac{x}{\|x\|} - \frac{x+\lambda y}{\|x+\lambda y\|} \right\|$$

is continuous and satisfies $\lim_{\lambda \to \pm\infty} = \pm 2$, which implies the existence of $\alpha \in \mathbb{R}$ such that $f(\alpha) = 0$, i.e., $x \perp_S \alpha x + y$. $\qquad\square$

Theorem 4.9.18 *Singer orthogonality has the uniqueness and the α-uniqueness properties in any normed space.*

Proof It is enough to prove the α-uniqueness property. Assume that $x, y \in X$, $x \perp_S \alpha_i x + y, i = 1, 2$. Then, $\frac{x}{\|x\|} \perp_I \bar{\alpha}_i \frac{x}{\|x\|} + \beta_i \frac{y}{\|y\|}, i = 1, 2$, where $\bar{\alpha}_i = \frac{\alpha_i\|x\|}{\|\alpha_i x + y\|}$, $\beta_i = \frac{\|y\|}{\|\alpha_i x + y\|}$. From Theorem 4.6.5 it follows that $\|\alpha_1 x + y\| = \|\alpha_2 x + y\|$ and $\alpha_1 = \alpha_2$. $\qquad\square$

The homogeneity and uniqueness of Singer orthogonality implies that it is additive in any two-dimensional space. Nevertheless, the only spaces of dimension greater than two in which Singer orthogonality is additive are the inner product spaces.

Theorem 4.9.19 ([99]) *If* $\dim X \geq 3$, *then Singer orthogonality is additive if and only if X is an inner product space.*

With respect to Singer orthogonal diagonals, the existence and uniqueness were proved in [62] and [1], respectively.

Theorem 4.9.20 *Singer orthogonality has the properties of existence and uniqueness of orthogonal diagonals in any normed space.*

Proof Let $x, y \in X$. Let us show that there is a unique $\alpha > 0$ such that $x + \alpha y \perp_S x - \alpha y$. Without loss of generality we can assume that x and y are linearly independent. The existence of α follows from the continuity of the function $f : \lambda \in \mathbb{R} \mapsto f(\lambda) = \left\| \frac{x+\lambda y}{\|x+\lambda y\|} + \frac{x-\lambda y}{\|x-\lambda y\|} \right\| - \left\| \frac{x+\lambda y}{\|x+\lambda y\|} - \frac{x-\lambda y}{\|x-\lambda y\|} \right\|$ and the fact that $f(0) = 2$ and $\lim_{\lambda \to +\infty} f(\lambda) = -2$.

Assume now that there exist $0 < \alpha_1 < \alpha_2$ such that $x + \alpha_i y \perp_S x - \alpha_i y$, $i = 1, 2$. Let us consider the parametrization of the unit sphere of the two-dimensional subspace spanned by x and y, namely

$$s : \theta \in [0, 2\pi] \to s(\theta) = \frac{(\cos \theta) \frac{x}{\|x\|} + (\sin \theta) \frac{y}{\|y\|}}{\left\| (\cos \theta) \frac{x}{\|x\|} + (\sin \theta) \frac{y}{\|y\|} \right\|} \in S_X.$$

Then $x = s(0)$ and $y = s(\pi/2)$. Moreover, there exist $0 < \theta_1 < \theta_2 < \frac{\pi}{2} < \bar{\theta}_2 < \bar{\theta}_1 < \pi$ such that $s(\theta_i) = \frac{x + \alpha_i y}{\|x + \alpha_i y\|}$, $s(\bar{\theta}_i) = \frac{\alpha_i y - x}{\|\alpha_i y - x\|}$, $i = 1, 2$. Since $s(\theta_i) \perp_I s(\bar{\theta}_i)$, $i = 1, 2$, then, having in mind (by the Monotonicity Lemma [112, Proposition 31]) that if $0 \le \theta < \theta' < \theta'' \le \pi$, then $\|s(\theta) + s(\theta')\| \ge \|s(\theta) + s(\theta'')\|$ and $\|s(\theta) - s(\theta')\| \le \|s(\theta) - s(\theta'')\|$, it follows that

$$\|s(\theta_1) + s(\bar{\theta}_1)\| \le \|s(\theta_1) + s(\bar{\theta}_2)\| \le \|s(\theta_2) + s(\bar{\theta}_2)\| = \|s(\theta_2) - s(\bar{\theta}_2)\|$$

$$\le \|s(\theta_1) - s(\bar{\theta}_2)\| \le \|s(\theta_1) - s(\bar{\theta}_1)\| = \|s(\theta_1) + s(\bar{\theta}_1)\|,$$

from which we get $s(\theta_1) \perp_I s(\bar{\theta}_2)$. This contradicts Theorem 4.6.5. $\square$

Corollary 4.9.21 ([40, 62]) *A normed space X is an inner product space if and only if*

$$x, y \in X, \ x \perp_S y \quad \Rightarrow \quad \|x + y\|^2 + \|x - y\|^2 \approx 2(\|x\|^2 + \|y\|^2),$$

where "$\approx$" means "$\le$" or "$\ge$".

4.9.4 DP-Orthogonality

Definition 4.9.22 A pair of vectors x and y is said to be *DP-orthogonal* ($x \perp_{DP} y$) if either $\|x\| \|y\| = 0$ or

$$\left\| \frac{x}{\|x\|} - \frac{y}{\|y\|} \right\| = \sqrt{2}.$$

DP-orthogonality [67] is another example of orthogonality defined by restricting a known orthogonality to the unit sphere: $x \perp_{DP} y \Leftrightarrow \frac{x}{\|x\|} \perp_{P_-} \frac{y}{\|y\|}$.

It is clear that DP-orthogonality is symmetric and positive homogeneous ($x \perp_{DP} y \Rightarrow x \perp_{DP} \lambda y$, $\forall \lambda \ge 0$). From Theorem 4.3.1 it follows that existence (resp. uniqueness) in the sphere and α-existence (resp. α-uniqueness) are equivalent properties.

Theorem 4.9.23 ([67]) *DP-orthogonality has the α-existence property.*

Proof It follows from the continuity of the function $f(\alpha) = \left\| \frac{x}{\|x\|} - \frac{\alpha x+y}{\|\alpha x+y\|} \right\|$ and the fact that $\lim_{\alpha \to +\infty} f(\alpha) = 0$ and $\lim_{\alpha \to -\infty} f(\alpha) = 2$. ◻

Theorem 4.9.24 DP-*orthogonality has the* α-*uniqueness property if and only if* P$_-$-*orthogonality has the uniqueness property in the unit sphere, i.e., when* $\rho = \|x\|$ *in the definition of the uniqueness property.*

Proof Assume that DP-orthogonality has the α-uniqueness property. Let $x \perp_{\text{P}_-} \alpha_i x + \beta_i y$, $\beta_i > 0$, $\|\alpha_i x + \beta_i y\| = \|x\|$, $i = 1, 2$. Then $x \perp_{\text{DP}} \frac{\alpha_i}{\beta_i} x + y$, $i = 1, 2$, which implies $\frac{\alpha_1}{\beta_1} = \frac{\alpha_2}{\beta_2}$. Since $\beta_2\|x\| = \|\beta_2\alpha_1 x + \beta_2\beta_1 y\| = \|\beta_1\alpha_2 x + \beta_2\beta_1 y\| = \beta_1\|x\|$, it follows that $\beta_1 = \beta_2$, and then $\alpha_1 = \alpha_2$. Conversely, assume that P$_-$-orthogonality has the uniqueness property in the unit sphere. Let $x, y \in E$, $x \perp_{\text{DP}} \alpha_i x + y$, $i = 1, 2$. By taking for $i = 1, 2$, $\bar{\alpha}_i = \frac{\|x\|\alpha_i}{\|\alpha_i x+y\|}$, $\bar{\beta}_i = \frac{\|x\|}{\|\alpha_i x+y\|}$, we have $x \perp_{\text{P}_-} \bar{\alpha}_i x + \bar{\beta}_i y$, $\bar{\beta}_i > 0$, and $\|\bar{\alpha}_i x + \bar{\beta}_i y\| = \|x\|$. Therefore, $\bar{\alpha}_1 = \bar{\alpha}_2$ and $\bar{\beta}_1 = \bar{\beta}_2$, which implies $\alpha_1 = \alpha_2$. ◻

From above and Theorem 4.7.3 it follows that in any strictly convex space DP-orthogonality has the α-uniqueness property. However, there are non-strictly convex spaces where DP-orthogonality has the α-uniqueness property. This is the case, e.g., if E is $\mathbb{R}^2$ endowed with norms whose spheres are regular octagons.

4.9.5 Diminnie Orthogonality

Definition 4.9.25 Two vectors x and y are said to be *Diminnie orthogonal* ($x \perp_{\text{D}} y$) if

$$\|x, y\| := \sup\{f(x)g(y) - f(y)g(x) : f, g \in S_{X^*}\} = \|x\|\|y\|.$$

Clearly, Diminnie orthogonality [65] is symmetric and homogeneous. Moreover, if $x \perp_{\text{B}} \alpha x + y$, then there exist $\alpha_1 \le \alpha \le \alpha_2$ such that $x \perp_{\text{D}} \alpha_1 x + y$ and $x \perp_{\text{D}} \alpha_2 x + y$, which gives, in particular, the α-existence property and, a fortiori, the existence property. The key in the proof of these results is that

$$x, y \in X, \ x \perp_{\text{B}} y \quad \Rightarrow \quad \|x, y\| \ge \|x\|\|y\|. \tag{4.10}$$

If Diminnie orthogonality is additive then it is unique; and if it is unique then it is equivalent to Birkhoff orthogonality. Therefore, if $\dim X \ge 3$, then additivity and uniqueness properties for Diminnie orthogonality are verified only in inner product spaces. Nevertheless, there are non-inner product spaces of dimension two in which Diminnie orthogonality is additive and unique.

In [9] it is proved that if X is strictly convex, then for every $x, y \in X$, $x \ne 0$, there exist at most two values of $\alpha \in \mathbb{R}$ such that $x \perp_{\text{D}} \alpha x + y$. If X is not strictly convex, then α can range over a closed interval of $\mathbb{R}$, e.g., if X is $\mathbb{R}^2$ endowed with the sup norm.

The key of the results in the above paragraph is the geometric interpretation of Diminnie orthogonality given also in [9] for vectors in a two-dimensional space (recall that an orthogonality is a binary relation): Let X be $\mathbb{R}^2$ endowed with a norm, and let P be a minimum area parallelogram bounding S_X. Then $x, y \in S_X$ are Diminnie orthogonal if, and only if, they are midpoints of two concurrent sides of a parallelogram whose area equals the area of P. Moreover, if u, v are midpoints of two concurrent sides of P, then $u, v \in S_X$, $u \perp_B v$, $v \perp_B u$ and $u \perp_D v$.

Diminnie orthogonality has the existence of orthogonal diagonals property, and they are unique if, and only if, Birkhoff orthogonality is symmetric [9]. As with other orthogonalities, the existence of orthogonal diagonals gives rise to the following result.

Corollary 4.9.26 *A normed space X is an inner product space if and only if*

$$x, y \in X, \ x \perp_D y \quad \Rightarrow \quad \|x + y\|^2 + \|x - y\|^2 \approx 2(\|x\|^2 + \|y\|^2),$$

where "$\approx$" means "$\leq$" or "$\geq$".

Recall that the next result, proved in [47], holds also for Birkhoff orthogonality.[15]

Theorem 4.9.27 *A normed space X is an inner product space if and only if for every $x, y \in S_X$, $x + y \perp_D x - y$.*

With respect to the relations with other orthogonalities, from [9, 44] and [46] we get the next theorem.

Theorem 4.9.28 *Let X be a normed space. Then the following statements hold:*

(a) *If Diminnie orthogonality implies (or is implied by) Birkhoff orthogonality, then both orthogonalities are equivalent.*

(b) *Diminnie orthogonality is equivalent to Birkhoff orthogonality if, and only if, Birkhoff orthogonality is symmetric.*

(c) *Let $\perp_1 \in \{\perp_S, \perp_{DP}, \perp_A\}$ and $\perp_2 \in \{\perp_C, \perp_S, \perp_{DP}, \perp_A\}$. The following properties are equivalent:*

   (i) *X is an inner product space.*

   (ii) *if $x, y \in X$, then $x \perp_D y$ iff $x \perp_1 y$.*

   (iii) *if $x, y \in X$ and $x \perp_D y$, then $x \perp_2 y$.*

   (iv) *if $x, y \in X$ and $x \perp_C y$, then $x \perp_D y$.*

4.9.6 Area Orthogonality

Definition 4.9.29 A pair of vectors x and y is said to be *area orthogonal* ($x \perp_A y$) if either $\|x\|\|y\| = 0$ or x and y are linearly independent and $\pm\frac{x}{\|x\|}$ and $\pm\frac{y}{\|y\|}$

---

[15] See the comments before Theorem 4.5.17.

determine, in the unit ball of the plane generated by them (identified with $\mathbb{R}^2$), four quadrants of equal area.

Area orthogonality [1, 10] is homogeneous, symmetric, and has the properties of existence, uniqueness, existence and uniqueness of orthogonal diagonals. As a consequence of this last property, we have the next result.

Corollary 4.9.30 *A normed space X is an inner product space if and only if*

$$x, y \in X, \ x \perp_A y \quad \Rightarrow \quad \|x + y\|^2 + \|x - y\|^2 \approx 2(\|x\|^2 + \|y\|^2),$$

where "$\approx$" means "$\leq$" or "$\geq$".

From the homogeneity and uniqueness of area orthogonality it follows that it is additive in two-dimensional spaces. It is natural to ask the following:

Question 4.9.31 *What happens when area orthogonality is additive in a space whose dimension is at least three?*

From [10] and [47] we get the following results. Recall that Boussouis orthogonality embraces Carlsson orthogonality.

Theorem 4.9.32 *Let X be a normed space and let $\perp$ denote Birkhoff orthogonality or Boussouis orthogonality. Then the following properties are equivalent:*

(i) *X is an inner product space.*
(ii) *If $x, y \in X$ and $x \perp y$, then $x \perp_A y$.*
(iii) *If $x, y \in X$ and $x \perp_A y$, then $x \perp y$.*
(iv) *If $x, y \in X$ and $x \perp_D y$, then $x \perp_A y$.*

In S_X, the area orthogonality can coincide with Singer orthogonality and with DP-orthogonality without X being an inner product space. For example, this happens if X is $\mathbb{R}^2$ with a regular octagon as unit sphere.

4.9.7   Height Orthogonality

Definition 4.9.33 Two vectors x and y are said to be *height orthogonal* ($x \perp_H y$) if either $\|x\|\|y\| = 0$ or

$$\|x - y\| = \left\| \|y\|\frac{x}{\|x\|} + \|x\|\frac{y}{\|y\|} \right\|.$$

Height orthogonality was introduced in [16], and also studied in [11]. The next results are obtained from these papers. As we recalled at the beginning, this orthogonality is based on the fact that the height onto the hypotenuse in a right triangle divides it into two similar triangles.

Height orthogonality has the properties of symmetry, existence, uniqueness and α-existence. Nevertheless, it is, in general, not α-unique, and the (sharp) bound $|\alpha| \leq \frac{\|y\|}{\|x\|}\left(\frac{1+\sqrt{2}}{2}\right)$ is satisfied. It has also the property of existence of diagonals, and if $x + \delta y \perp_H x - \delta y$ then, $\frac{\|x\|}{\|y\|\sqrt{2}} \leq |\delta| \leq \frac{\|x\|\sqrt{2}}{\|y\|}$, both bounds being sharp. In particular, if the lower bound is attained for some $x, y \in X$, then the subspace of X spanned by $\{x, y\}$ is isometrically isomorphic to $(\mathbb{R}^2, \|\cdot\|_\infty)$. As it occurs with all the orthogonalities that have the existence of diagonals property, we have:

Corollary 4.9.34 *A normed space X is an inner product space if and only if*

$$x, y \in X, \ x \perp_H y \quad \Rightarrow \quad \|x + y\|^2 + \|x - y\|^2 \approx 2(\|x\|^2 + \|y\|^2),$$

where "$\approx$" means "$\leq$" or "$\geq$".

All the orthogonality types studied here, before height orthogonality, have the following behavior with respect to homogeneity property: either they are homogeneous in any normed space or only in inner product spaces. But with height orthogonality this is different. A two-dimensional normed space X is said to have the $\pi/2$-property if it is isometrically isomorphic to $\mathbb{R}^2$ with a norm whose unit sphere is invariant under rotations of $\pi/2$ radians. A normed space is said to have the $\pi/2$-property if any two-dimensional subspace has this property.

Theorem 4.9.35 *Let X be a two dimensional space.*

(i) *If there exist $x, y \in X \setminus \{0\}$ such that $x \perp_H \lambda y$ for every $\lambda \in \mathbb{R}$, then X has the $\pi/2$-property.*

(ii) *If X has the $\pi/2$-property, then height orthogonality is homogeneous.*

Corollary 4.9.36 *A normed space has the $\pi/2$-property if and only if height orthogonality is homogeneous.*

Question 4.9.37 *What happens if $\dim X \geq 3$, and height orthogonality is homogeneous?*

With respect to the relations of height orthogonality to other orthogonalities, the only result we know is the following.

Theorem 4.9.38 *Let X be a normed space. The following properties are equivalent:*

(i) *X is an inner product space.*

(ii) *If $x, y \in X \setminus \{0\}$ and $x \perp_H y$, then $x \perp_{P_+} y$ and $\frac{x}{\|x\|^2} \perp_{P_+} \frac{y}{\|y\|^2}$.*

4.10   A Survey on Further Results

In this final part of our paper we want to give a survey on more recent results treating all types of orthogonalities presented and discussed above. This survey is built on the following fundamental expositions referring to the field (i.e., results derived or discussed in these contributions are almost nowhere repeated here): the exposition [77] of Freese et al., the comprehensive doctoral dissertation [1] as well as the survey papers [6] and [8] of Alonso and Benítez, [38] of Benítez, [14] of Alonso et al., and [51] of Chmieliński. Here also the survey [26] on angles in normed spaces has to be mentioned; its third section contains a lot of material treating angle preserving orthogonality types (recall that an angle function defined in a normed space preserves a certain orthogonality type when it attains the "Euclidean value" $\pi/2$ only for orthogonal pairs of vectors). There are more survey-like papers, e.g. restricted to only one orthogonality type, but extending the view on it to operator orthogonality and possibly also to infinite-dimensional spaces. Examples are the papers [123] and [124] of Paul et al., treating Birkhoff orthogonality of operators. The books [18] of Amir (here mainly the sections 3, 4, 7, 18, and 19) and [84] of Istrățescu (here mainly Chapters 4 and 8) on characterizations of inner product spaces contain also interesting material on orthogonality types in normed linear spaces. And in Chapters 8–12 of Dragomir's monograph [69] on semi-inner products, different orthogonality concepts in normed spaces and their relationships with semi-inner products are presented. The sequence of orthogonality types discussed below follows roughly the sequence given at the beginning of our paper.

Roberts orthogonality was introduced in [127], and in [129] results from this paper were proved again. Singer's paper [135] contains also basic results on Roberts orthogonality. Since some orthogonalities coincide only in inner product spaces, Papini and Wu [122] propose to measure their differences via some suitable constant. In this way it is studied "how far away" Birkhoff orthogonality and isosceles orthogonality are from Roberts orthogonality. These investigations continue in [25], where further geometric constants quantifying the difference between Roberts orthogonality and Birkhoff orthogonality are studied. Roberts orthogonality is characterized via bisectors of two points and also via certain linear transformations, and these two characterizations yield interesting geometric constants suitable for this quantification. Similarly, Mizuguchi [117] studies certain constants in general Banach spaces for measuring differences between Birkhoff, isosceles and Roberts orthogonalities. Zamani and Moslehian [144] introduce two types of approximate Roberts orthogonalities, and they investigate their relations to an analogously defined approximate Birkhoff orthogonality. In addition, the class of linear mappings preserving such approximate Roberts orthogonality is studied. This is continued by the same authors in [145], where besides approximate Roberts orthogonality also an approximate isosceles orthogonality is considered. The approximate Roberts orthogonality set of x with respect to y is studied, i.e., the set of all scalars s such that y is approximately Roberts orthogonal to $x - sy$.

Some geometrical properties (particularly for norms satisfying an approximate parallelogram law) of such sets are presented. The paper [23] contains an interesting characterization of Roberts orthogonality. Namely, let A be a unital C^*-algebra with unit e. The Davis–Wielandt shell of $a \in A$ is defined as the joint numerical radius of the elements a and $a * a$. The authors define and characterize the Roberts orthogonality of $a \in A$ to the unit e in terms of the Davis–Wielandt shell. They prove that, for certain classes of elements of A (for example, the class of isometries) the Roberts orthogonality of a and e is equivalent to the symmetry of the numerical range of a with respect to the origin. The survey [20] also treats Roberts orthogonality, and Birkhoff orthogonality, in normed spaces and C^*-algebras. The authors discuss some basic properties of Roberts orthogonality and present several important results on Birkhoff orthogonality and Roberts orthogonality for a general C^*-algebra and, particularly, the space of all bounded linear operators on a Hilbert space. And in [21] Roberts orthogonality in Hilbert C^*-modules is investigated; such modules in which Birkhoff orthogonality coincides with Roberts orthogonality are described. The Roberts orthogonality of complex double row matrices is studied in [22].

Our second concept is that of *Birkhoff orthogonality*, suggested by Birkhoff and properly studied by James. The name *Birkhoff–James orthogonality*, often alternatively used for this orthogonality, should not be confused with *James orthogonality*, alternatively used for *isosceles orthogonality* (our next orthogonality discussed below). As a starting point regarding basic literature on Birkhoff orthogonality till 2012, we take the exposition [14] of Alonso et al. (which covers in great parts also earlier surveys). Fundamental properties of Birkhoff orthogonality, its connections and differences with other orthogonality concepts, and its applications are comprehensively discussed there. Thus, in our survey here we discuss only results on Birkhoff orthogonality which are not considered in [14] (see therefore also the exposition [87] of James).

We mention now results in the spirit of *elementary geometry in normed planes and spaces*. It is well known that the symmetry of Birkhoff orthogonality in the planar case characterizes Radon curves as norm circles, and comprehensive discussions of these curves are given in Chapter 4 of the monograph [138], in Section 6 of the survey [112], and in the expository paper [109]. Martini and Wu [110] derive theorems on tangent segments of unit norm circles having equal lengths; this yields the notion of arc-length orthogonality, and the Euclidean plane is characterized via a relation between Birkhoff orthogonality and arc-length orthogonality. In [13], Alonso et al. show that a characterization of planar curves of constant width via the total length of perpendicular intersecting chords cannot be extended in the Birkhoff sense to normed planes. Fankhänel [72] studies bunches of metric ellipses and hyperbolas defined in normed planes which are pairwise Birkhoff orthogonal (regarding corresponding tangents at their intersection points). It is well known that a triangle in a strictly convex normed plane has at most one circumcenter, but need not always have one. Väisälä [140] proves that it has no circumcenter if and only if the normal directions (in the sense of left Birkhoff orthogonality) of all its sides are equal. Fankhänel [71] introduces a translation invariant angular measure for normed planes. Combined with Birkhoff orthogonality, this yields a class of interesting norms whose unit circles contain arcs which are pieces of Radon curves. Naszódi et al. [118] introduce an angular measure on the unit circle of a normed plane using Birkhoff orthogonal points there, and they generalize Fankhänel's work using the concept of Auerbach points. Mandal et al. [103] give a purely geometric characterization of polygonal Radon norms among all polygonal norms. The authors of [113] study the group generated by all left reflections in normed planes defined in terms of Birkhoff orthogonality. Considering products of three or two left reflections, characterizations of the Euclidean plane, of Radon planes, and of smooth norms are obtained.

We continue now with more recent results on *Birkhoff orthogonality in classical curve theory and differential geometry*. A planar convex curve is said to be a *Zindler curve* if all its perimeter-halving chords have equal lengths. The authors of [111] characterize Zindler curves in normed planes using Birkhoff orthogonality, and they show that (in contrast to the Euclidean subcase) the convex curves with equal area-halving chords in general differ from Zindler curves. *Legendre curves* are smooth curves in the plane which can have singular points, but still have a well-defined smooth normal vector field. Balestro et al. [27] generalize this notion for normed planes, studying their extended curvature properties, invariance under isometries, evolutes, involutes, and pedal curves. These results show that using an inner product is not necessary for this generalization; only the norm and its established Birkhoff orthogonality are necessary. In the expository paper [28], Balestro, Martini, and Shonoda systematize and complete the list of possible curvature notions of curves in normed planes. They add new results involving characterizations of curves of constant curvature, characterizations of Radon planes and the Euclidean subcase, and they give also analogues to classical statements like the four vertex theorem and the fundamental theorem on planar curves. As applications the curvature behavior of curves of constant width and also new results on notions like evolutes, involutes, and parallel curves are obtained. The concept of orthogonality applied in this paper is the Birkhoff one. And the papers [29–31], and [32], treating analogously the differential geometry of surfaces in three-dimensional normed spaces, are based on the concept of Birkhoff orthogonality between vectors and planes. Results of this type, i.e., on differential geometry in normed spaces based on Birkhoff orthogonality, are clearly also related to *Finsler geometry*. This is nicely discussed by Busemann in [49], see also the introductory parts of [30] and [31].

Chmieliński and Wójcik [53] investigate several orthogonality types in normed spaces, among them also exact and *approximate Birkhoff orthogonality*. Investigating interrelations, they derive characterizations of smooth and semi-smooth norms. This is successfully continued in [141] and [56]. In the latter paper properties and applications of an approximate Birkhoff orthogonality relation are studied, and this approximate orthogonality is then characterized in the class of linear bounded operators on a Hilbert space. In [54] Chmieliński and Wójcik compare two different notions of approximate Birkhoff orthogonality, improve known results about relations between them and show then connections with geometric properties of the respective normed spaces (such as smoothness, strict convexity, and uniform convexity). The notion of approximate symmetry of Birkhoff orthogonality is introduced, and geometric properties related to it are presented, with special emphasis on uniformly convex Banach spaces and their duals. Defining further a Birkhoff symmetry constant for a given Banach space, they are able to measure how far Birkhoff orthogonality is away from being symmetric. In [55] they continue, also with a partial survey, by concentrating on the class of linear operators which approximately preserve or reverse the Birkhoff orthogonality, and in a natural way they add respective stability problems. We should also mention a different notion of approximate Birkhoff orthogonality, introduced in [68]. Panagakou et al. [121] use older results to introduce a Birkhoff ε-orthogonality of vectors in complex normed spaces, studying the rich structure of this concept; they continue with introducing the Birkhoff ε-orthogonality set of vector-valued polynomials in one complex variable (to explore its position in the complex plane) and survey extensions of results on matrix polynomials to these vector-valued polynomials.

We finish this part with some further results on Birkhoff orthogonality. Wu et al. [143] study nonzero linear operators from a Banach space X to X, at least three-dimensional, which preserve Birkhoff orthogonality. Their existence guarantees that X is an inner product space (without the smoothness assumption from older papers). Sain et al. [130] introduce the notion of *Birkhoff strong orthogonality* and prove that for strictly convex spaces this setting is equivalent to usual Birkhoff orthogonality. This setting yields results on Hamel bases and generalized conjugate diameters in strictly convex, smooth spaces of finite dimensions. In the interesting paper [24] the following results on normed spaces X are derived: if X has sufficiently large dimension, then there is a nonzero vector mutually Birkhoff orthogonal to each among a fixed number of given vectors. Further on, if X is nonsmooth, the cardinality of the set of pairwise Birkhoff orthogonal vectors may exceed the dimension of X, but is always bounded from above by a function of the dimension; and any given pair of elements of X can be extended to a finite set, in which all consecutive elements are mutually Birkhoff orthogonal.

What we wrote above, namely that Birkhoff orthogonality is comprehensively presented in the expository paper [14], holds analogously for *isosceles orthogonality*, discussed now. Using isosceles orthogonality, Ji and Wu [90] introduce a constant $D(X)$ for arbitrary real normed spaces X to quantify the difference between Birkhoff and isosceles orthogonalities; an interesting upper bound on $D(X)$ is obtained in [98]. Similarly, Papini and Wu [122] introduce a constant for an analogous purpose, but referring to isosceles and Roberts orthogonality, and they study also the relations of this constant to the classical modulus of convexity. Mizuguchi [117] continues this, deriving and summarizing results on several constants in general Banach spaces, and quantifying relations between Birkhoff, isosceles, and Roberts orthogonality; also in the paper [25] these three types are compared via constants. In particular, Mizuguchi [116] investigates the difference between isosceles and Birkhoff orthogonalities for Radon planes. We mention here also the paper [35], where Hilbert spaces are characterized via a new parameter. Continuing Chmielínski and Wójcik [52], Zamani and Moslehian [145] study some type of approximate isosceles orthogonality and the mappings preserving it. Jahn [85] studies properties of binary relations for approximate Birkhoff orthogonality and isosceles orthogonality even for gauges (i.e., for general convex distance functions, without the symmetry axiom).

We come now to *properties of isosceles orthogonality*. Based on the classical paper [86] of James who showed that isosceles orthogonality is homogeneous only for inner product spaces, Hao and Wu [82] study homogeneous directions of isosceles orthogonality that also other spaces can have. They investigate relations to isometric reflections and characterize Hilbert spaces by the property that the interior of such a direction set in the unit sphere is non-empty. To measure the non-homogeneity of isosceles orthogonality in a normed space X, a suitable constant (which is related to the known James constant $J(X)$) is introduced. Also Komuro et al. [96] deal with the James constant. It is known that for a Hilbert space $J(X) = \sqrt{2}$ holds, but that (in two dimensions only) also non-Hilbert spaces exist that reach this value. Constructing a remarkable family of two-dimensional spaces X with $J(X) = \sqrt{2}$, they succeed to give even a characterization of such planes, and this construction gives also a negative answer to a conjecture on 4-covering numbers of convex sets. In the review of [96] (see the Mathematical Review of MR3653924) it is stated that a further consequence of these results is a characterization of those normed planes where isosceles orthogonality implies (or is equivalent to) Pythagorean orthogonality. Inspired by the concept of homogeneous directions of isosceles orthogonality from [82] (see above), He and Wang [83] introduce the more general notion of almost homogeneous direction of isosceles orthogonality; they prove that, surprisingly, these two concepts are equivalent, and they derive also a nice characterization of Hilbert spaces. And in [57] the homogeneity of isosceles orthogonality is used (together with angular distance equalities) to get new characterizations of inner product spaces.

As a generalization of Pythagorean and isosceles orthogonality, the notion of *Carlsson orthogonality* was introduced in [50]; it is also discussed in the surveys [6] and [38]. Boussouis [44] proves that a normed space is an inner product space if and only if several orthogonality types (like, e.g., those of Birkhoff and Singer) imply Carlsson orthogonality. The Birkhoff case is also contained in [64]. Boussouis [45] proves that Carlsson orthogonality implies Birkhoff orthogonality if and only if the normed space under consideration is an inner product space. We refer here also to the *Boussouis orthogonalities* introduced in [47] and [48], that generalize Carlsson orthogonalities and are discussed in an subsection above. And in [95] the authors derive results on properties of a Carlsson orthogonality; relations to Gateaux differentiability of the norm are investigated, too. Dehghani and Zamani [61] give new characterizations of real inner product spaces by studying also the relations between Birkhoff orthogonality and Carlsson orthogonality and its subcases, in particular Hermite–Hadamard isosceles orthogonality and Hermite–Hadamard Pythagorean orthogonality (for the latter see also [70]).

Like isosceles orthogonality, *Pythagorean orthogonality* is a special case of Carlsson orthogonality, see [50]. Properties of it (like uniqueness) and related characterizations of inner product spaces were studied already by James [86]; see also Kapoor and Prasad [93]. Continuing investigations of Menger, Wilson,

Valentine, and Wayment on the homogeneity of Pythagorean orthogonality and its equivalence to "homogeneity" of the Euclidean law of cosines for right angles, Martin and Valentine [106] remove the right-angle restriction and extend the concept to metric spaces. This yields characterizations of inner product spaces and new results on normed spaces, treating also Wilson's definition of angles. Comparing properties that follow from Birkhoff, Pythagorean, and isosceles orthogonality, the authors of [77] give several characterizations of inner product spaces. Again based on considerations of James [86] and others, Diminnie and Andalafte [66] study the so-called α-orthogonality as a common generalization of isosceles ($\alpha = -1$) and Pythagorean ($\alpha = 0$) orthogonality in a normed space, and they verify that the homogeneity or additivity of this orthogonality type implies that the space is an inner product space; see also Perfect's paper [125]. Following Sullivan [136], Liu [100] studies L^p-orthogonality which for $p = 2$ yields Pythagorean orthogonality. He shows that various additional conditions on L^p-orthogonality imply that the respective Banach space is uniformly convex. Fernández [73] studies generalizations of isosceles and Pythagorean orthogonality for generalized normed spaces which are derived from generalized inner products. Freese and Andalafte [75] study metrized versions of orthogonality types and show that a complete, convex, externally convex metric space in which the metric Pythagorean orthogonality is homogeneous is a real inner product space; see also [76] for related results. Alsina and Tomás [15] use functionals satisfying conditions which are more general than usual axioms of inner products and consider some weak versions of orthogonality relations in real normed spaces; in this way they find new characterizations of inner product spaces for isosceles and Pythagorean orthogonalities. They also confirm that this does not work for Birkhoff orthogonality. As we say above, Dragomir and Kikianty [70] introduce (using integral means, and coming from the concept of Pythagorean orthogonality) the so-called Hermite–Hadamard Pythagorean orthogonality, which is additive and homogeneous only in inner product spaces. These investigations are continued in [61], where the relations to Birkhoff orthogonality are clarified. Classes of linear mappings that preserve this orthogonality are obtained, yielding again characterizations of inner product spaces. Considering the conditional Cauchy functional equation $f(x + y) = f(x) + f(y)$ defined on a normed space of at least three dimensions, Szabó [137] studies the subcase of pairs x, y that are Pythagorean orthogonal (in the sense of P_+-orthogonality), and he clarifies that any odd Pythagorean orthogonally additive mapping satisfies the above equation for any pair x, y (the even case is not clarified.) Ban and Gal [33] introduce and investigate defects of (also Pythagorean) orthogonality and present applications to Fourier series and approximation theory.

We discuss now some results on *Pythagorean orthogonality in elementary geometry* of normed planes. Alsina et al. [17] generalize a property of isosceles trapezoids, getting a characterization of inner product spaces and yielding a definition of a new orthogonality type, which generalizes many known orthogonalities, e.g. also the Pythagorean, the Birkhoff and the isosceles one. It is "empty" for two-dimensional spaces, but the authors present interesting three-dimensional examples (different to inner product spaces). In [81] height vectors of triangles and their orthogonalities to sides are used to characterize inner product spaces; this is done for Pythagorean, Birkhoff, and isosceles orthogonality. Similarly, in [139] analogues of circumcenters of triangles are used for the same purpose and regarding the same types of orthogonalities. Martini and Spirova [107] introduce the notions of perfect parallelograms and of golden rectangles in strictly convex normed planes. Based on the close relation of the first notion to Pythagorean orthogonality, it is shown that the concept of golden rectangles can be generalized suitably only for this orthogonality type. Wu et al. [142] introduce the so-called circle-uniqueness of Pythagorean orthogonality and use it to give a characterization of strict convexity of normed spaces. They get further results on relations between circle-uniqueness of Pythagorean orthogonality and the shape of the unit sphere.

The concept of *Singer orthogonality* was introduced in [135], and Alonso and Benítez discussed it in their surveys [6, 8] and [38]. The authors of [101] and [2] continue the list of characterizations of inner product spaces given there. Benítez conjectured in [39] that if for dimensions at least three Singer orthogonality is additive, then the respective normed space is an inner product space. This is confirmed by Lin [99]. Boussouis [44] confirms that a normed space is an inner product space if and only if Singer orthogonality implies Carlsson orthogonality. The authors of [58] present two new concepts of orthogonality; the second of them is equivalent to Singer orthogonality, and it helps to see Singer orthogonality in a new way. In [115] it is shown how Singer orthogonality and isosceles orthogonality can be defined in quasi-inner product spaces via best approximations. And in [128] a new orthogonality type based on some angular distance in normed spaces is proposed, which generalizes Singer and isosceles orthogonalities to a large extent. Further important properties of this orthogonality are derived, and it is proved that a real normed space is an inner product space if and only if this orthogonality is either homogeneous or additive.

We continue with the concept of *Diminnie orthogonality* which was introduced by Diminnie in [65], compared there with Birkhoff orthogonality and also used to characterize inner product spaces. This is continued in [9], where interesting relationships to Birkhoff, Pythagorean, and isosceles orthogonality are studied. Boussouis [46] proves that if Carlsson orthogonality implies Diminnie orthogonality, then the corresponding normed linear space is an inner product space (the converse implication is given in [44]). Rätz [126] derives the following characterization of inner product spaces: For a normed space X, a functional $f : X \to \mathbb{R}$ is called an orthogonally additive mapping if $f(x + y) = f(x) + f(y)$ for all $x, y \in X$, where x is Diminnie orthogonal to y. It is shown that X is an inner product space if and only if not each orthogonally additive functional is additive.

We finish this survey by mentioning some orthogonalities which are not so widespread in the literature. The notion of *area orthogonality* was introduced by Alonso in 1984 (see Sect. 4.9.6). In 1995 Alsina, Guijarro and Tomás introduced the *height orthogonality* (see Sect. 4.9.7). Laugwitz [97] considers a generalized inner product in normed spaces X. For dimensions at least three, Kapoor and Prasad [94] alternatively prove his result that the corresponding *G-orthogonality* (defined via Gateaux derivatives) is symmetric if and only if X is an inner product space, and they also provide new proofs for further related results with the help of this orthogonality notion. Guijarro and Tomás [80] use perpendicular bisectors and a new orthogonality type based on them (and being not symmetric) to characterize the Euclidean plane among all normed planes. Now let $[p_1, q_1]$, $[p_2, q_2]$ be two chords of the unit circle of a normed plane. As introduced in [108], the first segment is said to be *chordal-orthogonal* to the second if the affine hull of $q_2, -p_2$ is parallel to that of p_1, q_1. The authors prove that a normed plane is Euclidean if and only if this chordal-orthogonality is symmetric, and further results (e.g., related to correspondingly defined orthocenters) are also derived. Without any metric, Alonso and Spirova [12] introduce the notion of *affine orthogonality* with respect to a given convex body, and they use this concept to characterize sets of constant width (and also other special types of convex figures) in strictly convex and smooth normed planes. Two further concepts of orthogonality in normed spaces are introduced in [114] and [119], and they are used to give criteria for properties of the respective norms, such as smoothness and strict convexity.

References

1. J. Alonso, Ortogonalidad en espacios normados, in *Publicaciones de la Sección de Matemáticas, Universidad de Extremadura [Publications of the Mathematics Section of the University of Extremadura]*, vol. 4 (Universidad de Extremadura, Facultad de Ciencias, Departamento de Matemáticas, Badajoz, 1984). Dissertation, Universidad de Extremadura, Cáceres, 1984
2. J. Alonso, Some results on Singer orthogonality and characterizations of inner product spaces. Arch. Math. **61**(2), 177–182 (1993). https://doi.org/10.1007/BF01207467
3. J. Alonso, Uniqueness properties of isosceles orthogonality in normed linear spaces. Ann. Sci. Math. Québec **18**(1), 25–38 (1994)
4. J. Alonso, Some properties of Birkhoff and isosceles orthogonality in normed linear spaces, in *Inner Product Spaces and Applications*. Pitman Research Notes in Mathematics Series, vol. 376 (Longman, Harlow, 1997), pp. 1–11
5. J. Alonso, C. Benítez, The Joly's construction: a common property of some generalized orthogonalities. Bull. Soc. Math. Belg. Sér. B **39**(3), 277–285 (1987)
6. J. Alonso, C. Benítez, Orthogonality in normed linear spaces: a survey. I. Main properties. Extracta Math. **3**(1), 1–15 (1988)
7. J. Alonso, C. Benítez, Some characteristic and noncharacteristic properties of inner product spaces. J. Approx. Theory **55**(3), 318–325 (1988). https://doi.org/10.1016/0021-9045(88)90098-6
8. J. Alonso, C. Benítez, Orthogonality in normed linear spaces: a survey. II. Relations between main orthogonalities. Extracta Math. **4**(3), 121–131 (1989)
9. J. Alonso, C. Benítez, Complements on Diminnie orthogonality. Math. Nachr. **165**, 99–106 (1994). https://doi.org/10.1002/mana.19941650108
10. J. Alonso, C. Benítez, Area orthogonality in normed linear spaces. Arch. Math. **68**(1), 70–76 (1997). https://doi.org/10.1007/PL00000397
11. J. Alonso, M.L. Soriano, On height orthogonality in normed linear spaces. Rocky Mountain J. Math. **29**(4), 1167–1183 (1999). https://doi.org/10.1216/rmjm/1181070401
12. J. Alonso, M. Spirova, Characterization of different classes of convex bodies via orthogonality. Bull. Belg. Math. Soc. Simon Stevin **18**(4), 707–721 (2011). http://projecteuclid.org/getRecord?id=euclid.bbms/1320763132

13. J. Alonso, H. Martini, Z. Mustafaev, On orthogonal chords in normed planes. Rocky Mountain J. Math. **41**(1), 23–35 (2011). https://doi.org/10.1216/RMJ-2011-41-1-23
14. J. Alonso, H. Martini, S. Wu, On Birkhoff orthogonality and isosceles orthogonality in normed linear spaces. Aequationes Math. **83**(1–2), 153–189 (2012). https://doi.org/10.1007/s00010-011-0092-z
15. C. Alsina, M.S. Tomás, On some orthogonality relations in real normed spaces and characterizations of inner products. Boll. Unione Mat. Ital. Sez. B Artic. Ric. Mat. (8) **10**(3), 513–520 (2007)
16. C. Alsina, P. Guijarro, M.S. Tomás, A characterization of inner product spaces based on orthogonal relations related to height's theorem. Rocky Mountain J. Math. **25**(3), 843–849 (1995). https://doi.org/10.1216/rmjm/1181072190
17. C. Alsina, P. Cruells, M.S. Tomàs, Isosceles trapezoids, norms and inner products. Arch. Math. **72**(3), 233–240 (1999). https://doi.org/10.1007/s000130050327
18. D. Amir, Characterizations of inner product spaces, in *Operator Theory: Advances and Applications*, vol. 20 (Birkhäuser Verlag, Basel, 1986)
19. E.Z. Andalafte, C.R. Diminnie, R.W. Freese, (α, β)-orthogonality and a characterization of inner product spaces. Math. Jpn. **30**(3), 341–349 (1985)
20. L. Arambašić, R. Rajić, On Birkhoff-James and Roberts orthogonality. Spec. Matrices **6**, 229–236 (2018). https://doi.org/10.1515/spma-2018-0018
21. L. Arambašić, R. Rajić, Another characterization of orthogonality in Hilbert C^*-modules. Math. Inequal. Appl. **22**(4), 1421–1426 (2019). https://doi.org/10.7153/mia-2019-22-99
22. L. Arambašić, R. Rajić, Roberts orthogonality for 2×2 complex matrices. Acta Math. Hungar. **157**(1), 220–228 (2019). https://doi.org/10.1007/s10474-018-0870-3
23. L. Arambašić, T. Berić, R. Rajić, Roberts orthogonality and Davis-Wielandt shell. Linear Algebra Appl. **539**, 1–13 (2018). https://doi.org/10.1016/j.laa.2017.10.023
24. L. Arambašić, A. Guterman, B. Kuzma, R. Rajić, S. Zhilina, Symmetrized Birkhoff-James orthogonality in arbitrary normed spaces. J. Math. Anal. Appl. **502**(1), 125203 (2021). https://doi.org/10.1016/j.jmaa.2021.125203
25. V. Balestro, H. Martini, R. Teixeira, Geometric constants for quantifying the difference between orthogonality types. Ann. Funct. Anal. **7**(4), 656–671 (2016). https://doi.org/10.1215/20088752-3661053
26. V. Balestro, A.G. Horváth, H. Martini, R. Teixeira, Angles in normed spaces. Aequationes Math. **91**(2), 201–236 (2017). https://doi.org/10.1007/s00010-016-0445-8
27. V. Balestro, H. Martini, R. Teixeira, On Legendre curves in normed planes. Pacific J. Math. **297**(1), 1–27 (2018). https://doi.org/10.2140/pjm.2018.297.1
28. V. Balestro, H. Martini, E. Shonoda, Concepts of curvatures in normed planes. Expo. Math. **37**(4), 347–381 (2019). https://doi.org/10.1016/j.exmath.2018.04.002
29. V. Balestro, H. Martini, R. Teixeira, Surface immersions in normed spaces from the affine point of view. Geom. Dedicata **201**, 21–31 (2019). https://doi.org/10.1007/s10711-018-0380-z
30. V. Balestro, H. Martini, R. Teixeira, Differential geometry of immersed surfaces in three-dimensional normed spaces. Abh. Math. Semin. Univ. Hambg. **90**(1), 111–134 (2020). https://doi.org/10.1007/s12188-020-00219-7
31. V. Balestro, H. Martini, R. Teixeira, On curvature of surfaces immersed in normed spaces. Monatsh. Math. **192**(2), 291–309 (2020). https://doi.org/10.1007/s00605-020-01394-8
32. V. Balestro, H. Martini, R. Teixeira, Some topics in differential geometry of normed spaces. Adv. Geom. **21**(1), 109–118 (2021). https://doi.org/10.1515/advgeom-2020-0001
33. A.I. Ban, S.G. Gal, On the defect of orthogonality in real normed linear spaces. Bull. Math. Soc. Sci. Math. Roumanie (N.S.) **44(92)**(4), 331–343 (2001)
34. M. Baronti, Su alcuni parametri degli spazi normati. Raporti Scientifico Dell'Instituto di Matematica **48**, 945–957 (1980)
35. M. Baronti, C. Franchetti, The isosceles orthogonality and a new 2-dimensional parameter in real normed spaces. Aequationes Math. **89**(3), 673–683 (2015). https://doi.org/10.1007/s00010-014-0255-9

36. C. Benítez, *A Property of Some Orthogonalities in Normed Spaces (Spanish)* (Actas de las Primeras Jornadas Matemáticas Hispano-Lusitanas, Madrid, 1973), pp. 55–62
37. C. Benítez, A property of Birkhoff orthogonality, and a characterization of pre-Hilbert spaces (Spanish). Collect. Math. **26**(3), 211–218 (1975)
38. C. Benítez, Orthogonality in normed linear spaces: a classification of the different concepts and some open problems. Rev. Mat. Univ. Complut. Madrid **2**(suppl.), 53–57 (1989). Congress on Functional Analysis (Madrid, 1988)
39. C. Benítez, A note on certain orthogonality in normed linear spaces. Math. Nachr. **153**, 7–8 (1991). https://doi.org/10.1002/mana.19911530102
40. C. Benítez, M. del Rio, Characterization of inner product spaces through rectangle and square inequalities. Rev. Roumaine Math. Pures Appl. **29**(7), 543–546 (1984)
41. G. Birkhoff, Orthogonality in linear metric spaces. Duke Math. J. **1**(2), 169–172 (1935). https://doi.org/10.1215/S0012-7094-35-00115-6
42. W. Blaschke, *Kreis und Kugel.* (Veit u. Co., Leipzig, 1916)
43. Á.P. Bosznay, On a problem concerning orthogonality in normed linear spaces. Studia Sci. Math. Hungar. **26**(1), 63–65 (1991)
44. B. Boussouis, Caractérisation d'un espace préhilbertien au moyen des orthogonalités généralisées. Extracta Math. **7**(1), 20–24 (1992)
45. B. Boussouis, Relations entre l'orthogonalité de Birkhoff-James et l'orthogonalité de Carlsson. Ann. Sci. Math. Québec **17**(2), 139–143 (1993)
46. B. Boussouis, Comparaison de l'orthogonalité de Diminnie aux orthogonalités de type Carlsson. Extracta Math. **9**(3), 173–180 (1994)
47. B. Boussouis, Orthogonalité et caractérisation des espaces préhilbertiens. Ph.D. thesis, Université Sidi Mohamed Ben Abdellah, Fès, Morocco (1995)
48. B. Boussouis, Orthogonalité et caractérisation des espaces préhilbertiens. Ann. Sci. Math. Québec **24**(1), 1–17 (2000)
49. H. Busemann, The geometry of Finsler spaces. Bull. Amer. Math. Soc. **56**, 5–16 (1950). https://doi.org/10.1090/S0002-9904-1950-09332-X
50. S.O. Carlsson, Orthogonality in normed linear spaces. Ark. Mat. **4**, 297–318 (1962)
51. J. Chmieliński, Orthogonality preserving property and its Ulam stability, in *Functional Equations in Mathematical Analysis.* Springer Optimization and Its Applications, vol. 52 (Springer, New York, 2012), pp. 33–58. https://doi.org/10.1007/978-1-4614-0055-4_4
52. J. Chmieliński, P. Wójcik, Isosceles-orthogonality preserving property and its stability. Nonlinear Anal. **72**(3–4), 1445–1453 (2010). https://doi.org/10.1016/j.na.2009.08.028
53. J. Chmieliński, P. Wójcik, ρ-orthogonality and its preservation—revisited, in *Recent Developments in Functional Equations and Inequalities.* Banach Center Publications, vol. 99 (Institute of Mathematics of the Polish Academy of Sciences, Warsaw, 2013), pp. 17–30. https://doi.org/10.4064/bc99-0-2
54. J. Chmieliński, P. Wójcik, Approximate symmetry of Birkhoff orthogonality. J. Math. Anal. Appl. **461**(1), 625–640 (2018). https://doi.org/10.1016/j.jmaa.2018.01.031
55. J. Chmieliński, P. Wójcik, Birkhoff-James orthogonality reversing property and its stability, in *Ulam Type Stability* (Springer, Cham, 2019), pp. 57–71
56. J. Chmieliński, T. Stypuła, P. Wójcik, Approximate orthogonality in normed spaces and its applications. Linear Algebra Appl. **531**, 305–317 (2017). https://doi.org/10.1016/j.laa.2017.06.001
57. F. Dadipour, F. Sadeghi, A. Salemi, Characterizations of inner product spaces involving homogeneity of isosceles orthogonality. Arch. Math. **104**(5), 431–439 (2015). https://doi.org/10.1007/s00013-015-0762-5
58. F. Dadipour, F. Sadeghi, A. Salemi, An orthogonality in normed linear spaces based on angular distance inequality. Aequationes Math. **90**(2), 281–297 (2016). https://doi.org/10.1007/s00010-014-0333-z
59. M.M. Day, Some characterizations of inner-product spaces. Trans. Amer. Math. Soc. **62**, 320–337 (1947)

60. M.M. Day, *Normed Linear Spaces*, 3rd edn. Ergebnisse der Mathematik und ihrer Grenzgebiete, Band 21 (Springer, New York, 1973)
61. M. Dehghani, A. Zamani, Characterization of real inner product spaces by Hermite-Hadamard type orthogonalities. J. Math. Anal. Appl. **479**(1), 1364–1382 (2019). https://doi.org/10.1016/j.jmaa.2019.07.002
62. M. del Río, Ortogonalidad en espacios normados y caracterización de espacios prehilbertianos, in *Publicaciones del Departamento de Análisis Matemático: serie B*, vol. 14 (Universidad de Santiago de Compostela, Santiago de Compostela, 1975)
63. M. del Río, C. Benítez, The rectangular constant for two-dimensional spaces. J. Approx. Theory **19**(1), 15–21 (1977)
64. J. Desbiens, Une nouvelle caractérisation des espaces de Hilbert. Ann. Sci. Math. Québec **14**(1), 17–22 (1990)
65. C.R. Diminnie, A new orthogonality relation for normed linear spaces. Math. Nachr. **114**, 197–203 (1983). https://doi.org/10.1002/mana.19831140115
66. C.R. Diminnie, R.W. Freese, E.Z. Andalafte, An extension of Pythagorean and isosceles orthogonality and a characterization of inner-product spaces. J. Approx. Theory **39**(4), 295–298 (1983). https://doi.org/10.1016/0021-9045(83)90073-4
67. C.R. Diminnie, E.Z. Andalafte, R.W. Freese, Angles in normed linear spaces and a characterization of real inner product spaces. Math. Nachr. **129**, 197–204 (1986). https://doi.org/10.1002/mana.19861290118
68. S.S. Dragomir, On approximation of continuous linear functionals in normed linear spaces. An. Univ. Timişoara Ser. Ştiinţ. Mat. **29**(1), 51–58 (1991)
69. S.S. Dragomir, *Semi-Inner Products and Applications* (Nova Science Publishers, Inc., Hauppauge, 2004)
70. S.S. Dragomir, E. Kikianty, Orthogonality connected with integral means and characterizations of inner product spaces. J. Geom. **98**(1–2), 33–49 (2010). https://doi.org/10.1007/s00022-010-0048-9
71. A. Fankhänel, On angular measures in Minkowski planes. Beitr. Algebra Geom. **52**(2), 335–342 (2011). https://doi.org/10.1007/s13366-011-0011-4
72. A. Fankhänel, On conics in Minkowski planes. Extracta Math. **27**(1), 13–29 (2012)
73. R. Fernández, Orthogonality in generalized normed spaces. Stochastica **12**(2–3), 141–148 (1988)
74. F.A. Ficken, Note on the existence of scalar products in normed linear spaces. Ann. Math. (2) **45**, 362–366 (1944)
75. R. Freese, E. Andalafte, Metrizations of orthogonality and characterizations of inner product spaces. J. Geom. **39**(1–2), 28–37 (1990). https://doi.org/10.1007/BF01222138
76. R. Freese, E. Andalafte, Weak additivity of metric Pythagorean orthogonality. J. Geom. **54**(1–2), 44–49 (1995). https://doi.org/10.1007/BF01222851
77. R.W. Freese, C.R. Diminnie, E.Z. Andalafte, A study of generalized orthogonality relations in normed linear spaces. Math. Nachr. **122**, 197–204 (1985). https://doi.org/10.1002/mana.19851220120
78. S. Gähler, Lineare 2-normierte Räume. Math. Nachr. **28**, 1–43 (1964)
79. N. Gastinel, J.L. Joly, Condition numbers and general projection method. Linear Algebra Appl. **3**, 185–224 (1970)
80. P. Guijarro, M.S. Tomás, Perpendicular bisectors and orthogonality. Arch. Math. **69**(6), 491–496 (1997). https://doi.org/10.1007/s000130050151
81. P. Guijarro, M.S. Tomás, Characterizations of inner product spaces by geometrical properties of the heights in a triangle. Arch. Math. **73**(1), 64–72 (1999). https://doi.org/10.1007/s000130050021
82. C. Hao, S. Wu, Homogeneity of isosceles orthogonality and related inequalities. J. Inequal. Appl. **2011**, 84 (2011). https://doi.org/10.1186/1029-242X-2011-84
83. C. He, D. Wang, A remark on the homogeneity of isosceles orthogonality. J. Funct. Spaces **3**, Art. ID 876015 (2014). https://doi.org/10.1155/2014/876015

84. V.I. Istrățescu, Inner product structures, in *Mathematics and Its Applications*, vol. 25 (D. Reidel Publishing Co., Dordrecht, 1987). https://doi.org/10.1007/978-94-009-3713-0. Theory and applications
85. T. Jahn, Orthogonality in generalized Minkowski spaces. J. Convex Anal. **26**(1), 49–76 (2019)
86. R.C. James, Orthogonality in normed linear spaces. Duke Math. J. **12**, 291–302 (1945)
87. R.C. James, Inner product in normed linear spaces. Bull. Amer. Math. Soc. **53**, 559–566 (1947)
88. R.C. James, Orthogonality and linear functionals in normed linear spaces. Trans. Amer. Math. Soc. **61**, 265–292 (1947)
89. R.C. James, Reflexivity and the sup of linear functionals, in *Proceedings of the International Symposium on Partial Differential Equations and the Geometry of Normed Linear Spaces (Jerusalem, 1972)*, vol. 13 (1972), pp. 289–300
90. D. Ji, S. Wu, Quantitative characterization of the difference between Birkhoff orthogonality and isosceles orthogonality. J. Math. Anal. Appl. **323**(1), 1–7 (2006). https://doi.org/10.1016/j.jmaa.2005.10.004
91. J.L. Joly, Caractérisations d'espaces hilbertiens au moyen de la constante rectangle. J. Approx. Theory **2**, 301–311 (1969)
92. S. Kakutani, Some characterizations of Euclidean space. Jpn. J. Math. **16**, 93–97 (1939)
93. O.P. Kapoor, J. Prasad, Orthogonality and characterizations of inner product spaces. Bull. Austral. Math. Soc. **19**(3), 403–416 (1978). https://doi.org/10.1017/S0004972700008947
94. O.P. Kapoor, J. Prasad, On characterizations of inner-product spaces. Publ. Inst. Math. **35**(49), 173–177 (1984)
95. E. Kikianty, S.S. Dragomir, On Carlsson type orthogonality and characterization of inner product spaces. Filomat **26**(4), 859–870 (2012). https://doi.org/10.2298/FIL1204859K
96. N. Komuro, K.S. Saito, R. Tanaka, On the class of Banach spaces with James constant $\sqrt{2}$, III. Math. Inequal. Appl. **20**(3), 865–887 (2017). https://doi.org/10.7153/mia-20-55
97. D. Laugwitz, On characterizations of inner product spaces. Proc. Amer. Math. Soc. **50**, 184–188 (1975). https://doi.org/10.2307/2040537
98. J.H. Li, B. Ling, S.Y. Liu, A new upper bound of geometric constant $D(X)$. J. Inequal. Appl. **9**, Paper No. 203 (2017). https://doi.org/10.1186/s13660-017-1474-0
99. P.K. Lin, A remark on the Singer-orthogonality in normed linear spaces. Math. Nachr. **160**, 325–328 (1993). https://doi.org/10.1002/mana.3211600116
100. Z. Liu, L^p-orthogonality in Banach spaces. J. Math. Res. Exposition **4**(4), 31–35 (1984)
101. Z. Liu, Y.D. Zhuang, Singer orthogonality and characterizations of inner product spaces. Arch. Math. **55**(6), 588–594 (1990). https://doi.org/10.1007/BF01191695
102. E.R. Lorch, On certain implications which characterize Hilbert space. Ann. Math. (2) **49**, 523–532 (1948)
103. K. Mandal, D. Sain, K. Paul, A geometric characterization of polygonal Radon planes. J. Convex Anal. **26**(4), 1113–1123 (2019)
104. A. Marchaud, Un théorème sur les corps convexes. Ann. Sci. École Norm. Sup. (3) **76**, 283–304 (1959)
105. G. Marino, P. Pietramala, A note about inner products on Banach spaces. Boll. Un. Mat. Ital. A (7) **1**(3), 425–427 (1987)
106. C.F. Martin, J.E. Valentine, Angles in metric and normed linear spaces. Colloq. Math. **34**(2), 209–217 (1975/76). https://doi.org/10.4064/cm-34-2-209-217
107. H. Martini, M. Spirova, Golden rectangles in normed planes. Mitt. Math. Ges. Hamburg **29**, 125–134 (2010)
108. H. Martini, M. Spirova, A new type of orthogonality for normed planes. Czechoslovak Math. J. **60(135)**(2), 339–349 (2010). https://doi.org/10.1007/s10587-010-0039-x
109. H. Martini, K.J. Swanepoel, Antinorms and Radon curves. Aequationes Math. **72**(1–2), 110–138 (2006). https://doi.org/10.1007/s00010-006-2825-y
110. H. Martini, S. Wu, Tangent segments and orthogonality types in normed planes. J. Geom. **99**(1–2), 89–100 (2010). https://doi.org/10.1007/s00022-011-0063-5

111. H. Martini, S. Wu, On Zindler curves in normed planes. Canad. Math. Bull. **55**(4), 767–773 (2012). https://doi.org/10.4153/CMB-2011-112-x
112. H. Martini, K.J. Swanepoel, G. Weiß, The geometry of Minkowski spaces—a survey. I. Expo. Math. **19**(2), 97–142 (2001). https://doi.org/10.1016/S0723-0869(01)80025-6
113. H. Martini, M. Spirova, K. Strambach, Geometric algebra of strictly convex Minkowski planes. Aequationes Math. **88**(1–2), 49–66 (2014). https://doi.org/10.1007/s00010-013-0204-z
114. P.M. Miličić, Sur la g-orthogonalité dans des espaces normés. Mat. Vesnik **39**(3), 325–334 (1987)
115. P.M. Miličić, Singer orthogonality and James orthogonality in the so-called quasi-inner product space. Math. Morav. **15**(1), 49–52 (2011). https://doi.org/10.5937/matmor1101049m
116. H. Mizuguchi, The differences between Birkhoff and isosceles orthogonalities in Radon planes. Extracta Math. **32**(2), 173–208 (2017)
117. H. Mizuguchi, Measurement of the difference between two orthogonality types in Banach spaces. J. Nonlinear Convex Anal. **19**(9), 1579–1586 (2018)
118. M. Naszódi, V. Prokaj, K. Swanepoel, Angular measures and Birkhoff orthogonality in Minkowski planes. Aequationes Math. **94**(5), 969–977 (2020). https://doi.org/10.1007/s00010-020-00715-4
119. M. Nur, H. Gunawan, A new orthogonality and angle in a normed space. Aequationes Math. **93**(3), 547–555 (2019). https://doi.org/10.1007/s00010-018-0582-3
120. K. Ōhira, On some characterizations of abstract Euclidean spaces by properties of orthogonality. Kumamoto J. Sci. Ser. A. **1**(1), 23–26 (1952)
121. V. Panagakou, P. Psarrakos, N. Yannakakis, Birkhoff-James ε-orthogonality sets of vectors and vector-valued polynomials. J. Math. Anal. Appl. **454**(1), 59–78 (2017). https://doi.org/10.1016/j.jmaa.2017.04.033
122. P.L. Papini, S. Wu, Measurements of differences between orthogonality types. J. Math. Anal. Appl. **397**(1), 285–291 (2013). https://doi.org/10.1016/j.jmaa.2012.07.059
123. K. Paul, D. Sain, Birkhoff-James orthogonality and its application in the study of geometry of Banach space, in *Advanced Topics in Mathematical Analysis* (CRC Press, Boca Raton, 2019), pp. 245–284
124. K. Paul, D. Sain, P. Ghosh, Symmetry of Birkhoff-James orthogonality of bounded linear operators, in *Ulam Type Stability* (Springer, Cham, 2019), pp. 331–344
125. H. Perfect, Pythagorean orthogonality in a normed linear space. Proc. Edinburgh Math. Soc. (2) **9**, 168–169 (1958)
126. J. Rätz, Characterization of inner product spaces by means of orthogonally additive mappings. Aequationes Math. **58**(1–2), 111–117 (1999). https://doi.org/10.1007/s000100050098. Dedicated to János Aczél on the occasion of his 75th birthday
127. B.D. Roberts, On the geometry of abstract vector spaces. Tohoku Math. J. **39**, 42–59 (1934)
128. J. Rooin, S. Rajabi, M.S. Moslehian, p-angular distance orthogonality. Aequationes Math. **94**(1), 103–121 (2020). https://doi.org/10.1007/s00010-019-00664-7
129. F.B. Saidi, A characterisation of Hilbert spaces via orthogonality and proximinality. Bull. Austral. Math. Soc. **71**(1), 107–112 (2005). https://doi.org/10.1017/S0004972700038053
130. D. Sain, K. Paul, K. Jha, Strictly convex space: strong orthogonality and conjugate diameters. J. Convex Anal. **22**(4), 1215–1225 (2015)
131. I.J. Schoenberg, A remark on M. M. Day's characterization of inner-product spaces and a conjecture of L. M. Blumenthal. Proc. Amer. Math. Soc. **3**, 961–964 (1952)
132. I. Şerb, Rectangular modulus and geometric properties of normed spaces. Math. Pannon. **10**(2), 211–222 (1999)
133. I. Şerb, Rectangular modulus, Birkhoff orthogonality and characterizations of inner product spaces. Comment. Math. Univ. Carolin. **40**(1), 107–119 (1999)
134. I. Şerb, Geometric properties of normed spaces and estimates for rectangular modulus. Math. Pannon. **12**(1), 27–38 (2001)
135. I. Singer, Angles abstraits et fonctions trigonométriques dans les espaces de Banach (Romanian). Acad. R. P. Romî ne. Bul. Şti. Secţ. Şti. Mat. Fiz. **9**, 29–42 (1957)

136. F.E. Sullivan, Structure of real L^p spaces. J. Math. Anal. Appl. **32**, 621–629 (1970). https://doi.org/10.1016/0022-247X(70)90285-4
137. G. Szabó, Pythagorean orthogonality and additive mappings. Aequationes Math. **53**(1–2), 108–126 (1997). https://doi.org/10.1007/BF02215968
138. A.C. Thompson, Minkowski geometry, *Encyclopedia of Mathematics and Its Applications*, vol. 63 (Cambridge University Press, Cambridge, 1996)
139. M.S. Tomás, Circumcenters in real normed spaces. Boll. Unione Mat. Ital. Sez. B Artic. Ric. Mat. (8) **8**(2), 421–430 (2005)
140. J. Väisälä, Observations on circumcenters in normed planes. Beitr. Algebra Geom. **58**(3), 607–615 (2017). https://doi.org/10.1007/s13366-017-0338-6
141. P. Wójcik, Characterizations of smooth spaces by approximate orthogonalities. Aequationes Math. **89**(4), 1189–1194 (2015). https://doi.org/10.1007/s00010-014-0293-3
142. S. Wu, X. Dong, D. Wang, Circle-uniqueness of Pythagorean orthogonality in normed linear spaces. J. Funct. Spaces 4, Art. ID 634842 (2014). https://doi.org/10.1155/2014/634842
143. S. Wu, C. He, G. Yang, Orthogonalities, linear operators, and characterization of inner product spaces. Aequationes Math. **91**(5), 969–978 (2017). https://doi.org/10.1007/s00010-017-0494-7
144. A. Zamani, M.S. Moslehian, Approximate Roberts orthogonality. Aequationes Math. **89**(3), 529–541 (2015). https://doi.org/10.1007/s00010-013-0233-7
145. A. Zamani, M.S. Moslehian, Approximate Roberts orthogonality sets and (δ, ε)-(a,b)-isosceles-orthogonality preserving mappings. Aequationes Math. **90**(3), 647–659 (2016). https://doi.org/10.1007/s00010-015-0383-x

Chapter 5
Convex Bodies: Mixed Volumes and Inequalities

Ivan Izmestiev

Abstract We give a brief introduction into the theory of mixed volumes of convex bodies and discuss the inequalities involving volumes and mixed volumes: the Brunn–Minkowski, the Alexandrov–Fenchel, and the two Minkowski inequalities. Along the way we discuss the Steiner formula and the integral-geometric formulas, namely the proportionality of the average width to the total mean curvature and the formulas of Cauchy and Crofton. We also pay attention to the interplay between the discrete and the smooth, that is between convex polyhedra and convex hypersurfaces. The connections between the second Minkowski inequality, the Wirtinger inequality, and the spectrum of the Laplacian lead to the definition of a discrete spherical Laplacian enjoying spectral properties similar to its smooth counterpart.

Keywords Convex body · Steiner formula · Mixed volume · Alexandrov–Fenchel inequality · Minkowski inequalities · Brunn–Minkowski inequality

1991 Mathematics Subject Classification 52A39, 52A40

5.1 Introduction

This survey is based on the notes to a minicourse given at a CIMPA school in Varanasi in December 2019. The author thanks the organizers of that school for the opportunity to read the course and for the hospitality.

Supported by the Swiss National Science Foundation grants 169391 and 179133.

I. Izmestiev (✉)
TU Wien, Institute of Discrete Mathematics and Geometry, Vienna, Austria
e-mail: izmestiev@dmg.tuwien.ac.at

© The Author(s), under exclusive license to Springer Nature Switzerland AG 2022
A. Papadopoulos (ed.), *Surveys in Geometry I*,
https://doi.org/10.1007/978-3-030-86695-2_5

A reader wishing to learn more about convex geometry or to fill in the omitted technical details is advised to consult [1, 5, 6, 13].

5.2 Basic Notions

5.2.1 Convex Bodies

Definition 5.2.1 A subset K of $\mathbb{R}^d$ is called *convex* if for any $p, q \in K$ the segment

$$[p, q] = \{\lambda p + (1 - \lambda)q \mid 0 \le \lambda \le 1\}$$

is contained in K.

By induction one can prove that a convex set K contains any *convex combination*

$$\lambda_1 p_1 + \cdots + \lambda_n p_n, \quad \lambda_i \ge 0, \quad \sum_i \lambda_i = 1$$

of any of its points.

Definition 5.2.2 A convex set $K \subset \mathbb{R}^d$ is called a *convex body* if it is compact and has a non-empty interior.

The non-empty interior requirement is not very important. One defines the *dimension* of a convex set as the dimension of its affine hull. Then every compact convex set is a convex body within its affine hull. For example, a triangle in the plane is a convex body; a triangle in space is not, but it is a convex body within the plane it spans.

Every convex body in $\mathbb{R}^d$ is homeomorphic to the d-ball: it can be isotopically shrunk to a closed ε-neighborhood of an interior point. Accordingly, the boundary of a convex body is homeomorphic to the sphere.

Definition 5.2.3 A *convex polytope* is the convex hull (that is the set of all convex combinations) of finitely many points.

A *convex polyhedron* is the intersection of finitely many closed half-spaces.

Theorem 5.2.4 (Weyl–Minkowski) *Every convex polytope is a convex polyhedron. Every bounded convex polyhedron is a convex polytope.*

For a proof, see [1].

Definition 5.2.5 A *smooth convex body* is a convex body whose boundary is a smooth hypersurface.

The global convexity (in the sense of Definition 5.2.1) is equivalent to the non-negativity of the curvature (extrinsic or intrinsic), see [10, 12].

A general convex body is neither polyhedral nor smooth. Take for example a Cantor set in the circle and form its convex hull.

Bounded convex sets in $\mathbb{R}^d$ are measurable, even in the Jordan sense. We denote by $\mathrm{vol}_d(K)$ the d-measure of $K \subset \mathbb{R}^d$, but often omit the lower index.

5.2.2 Steiner Formula

For a convex body $K \subset \mathbb{R}^d$ and $r > 0$ consider the closed r-neighborhood of K:

$$K_r = \{x \in \mathbb{R}^d \mid \mathrm{dist}(x, K) \leq r\}.$$

This set is sometimes called the parallel body.

Theorem 5.2.6 *The volume of K_r is a polynomial in r of degree d:*

$$\mathrm{vol}_d(K_r) = \sum_{k=0}^{d} a_k(K) r^k.$$

We will prove this theorem in Sect. 5.3.3. The above formula is called the *Steiner formula*. Steiner [16] proved it only for polytopes and smooth bodies in $\mathbb{R}^3$, providing explicit formulas for the coefficients of the polynomials.

Theorem 5.2.7 (Steiner) *For every convex polytope $K \subset \mathbb{R}^3$ one has*

$$\mathrm{vol}(K_r) = \mathrm{vol}(K) + r \cdot \mathrm{area}(\partial K) + r^2 \cdot \frac{1}{2} \sum_e \theta_e \ell_e + r^3 \cdot \frac{1}{3} \sum_v \theta_v. \qquad (5.1)$$

Here the first sum is over all edges with ℓ_e the edge length and θ_e the exterior angle at e (the angle between the normals to the faces adjacent to e). The second sum is over all vertices, where θ_v is the exterior angle at v.

For every smooth convex body $K \subset \mathbb{R}^3$ one has

$$\mathrm{vol}(K_r) = \mathrm{vol}(K)$$
$$+ r \cdot \mathrm{area}(\partial K) + r^2 \cdot \frac{1}{2} \int_{\partial K} (\kappa_1 + \kappa_2) \, d\mathrm{area} + r^3 \cdot \frac{1}{3} \int_{\partial K} \kappa_1 \kappa_2 \, d\mathrm{area}. \qquad (5.2)$$

Here κ_1 and κ_2 are the principal curvatures of ∂K.

Proof The r-neighborhood of a convex polytope decomposes into several building blocks: the polytope itself, the prisms over the faces, the cylindrical wedges along the edges, and ball sectors at the vertices. See Fig. 5.1.

The volumes of the prisms sum up to the linear term in Eq. (5.1), the volumes of the wedges to the quadratic term, and the volumes of the sectors to the cubical term.

Fig. 5.1 Building blocks of
an r-neighborhood of a
convex polytope

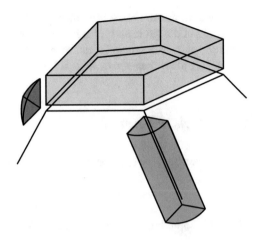

Let us now deal with the smooth case. A local parametrization $\sigma : U \rightarrow \mathbb{R}^3$
of ∂K gives rise to a local parametrization of the boundary of the r-neighborhood
of K:

$$\Sigma : U \times [0, r] \rightarrow \mathbb{R}^3, \quad \Sigma(u, v, t) = \sigma(u, v) + tv,$$

where v is the outward unit normal field. This implies that the area element on the
t-equidistant surface is related to the area element on ∂K as

$$|\Sigma_u \times \Sigma_v| = |(\sigma_u + tv_u) \times (\sigma_v + tv_v)|$$
$$= \det(E + tS)|\sigma_u \times \sigma_v| = (1 + t\kappa_1)(1 + t\kappa_2)|\sigma_u \times \sigma_v|.$$

This leads to

$$\operatorname{vol}(K_r) = \operatorname{vol}(K) + \int_{\partial K} \int_0^r (1 + t\kappa_1)(1 + t\kappa_2)\, dt\, d\text{area}$$

$$= \operatorname{vol}(K) + \int_{\partial K} \left(r + \frac{r^2}{2}(\kappa_1 + \kappa_2) + \frac{r^3}{3}\kappa_1\kappa_2 \right) d\text{area}$$

and proves Eq. (5.2).   □

Let us look at the cubical term in Eqs. (5.1) and (5.2). By the Gauss-Bonnet
theorem, $\int_{\partial K} \kappa_1\kappa_2\, d\text{area} = 4\pi$, so that the coefficient at r^3 in (5.2) equals $\frac{4\pi}{3}$. The
coefficient at r^3 in (5.1) is also $\frac{4\pi}{3}$, since the ball sectors fit together to form a ball
of radius r whose volume is $\frac{4\pi}{3}r^3$. This fact can be seen as an analog of the Gauss-

Bonnet formula for convex polytopes. (Note that the Gauss-Bonnet theorem was not yet properly stated in 1840 when Steiner made his discovery.)

To summarize, we have in both cases

$$\text{vol}(K_r) = \text{vol}(K) + r\,\text{area}(\partial K) + r^2 \cdot S(K) + r^3 \cdot \frac{4\pi}{3},$$

where

$$S(K) = \begin{cases} \frac{1}{2}\sum_e \theta_e \ell_e, & \text{if } K \text{ is a polytope} \\ \frac{1}{2}\int_{\partial K}(\kappa_1 + \kappa_2)\,d\text{area}, & \text{if } K \text{ has smooth boundary.} \end{cases} \qquad (5.3)$$

This explains why one often calls the expression $\frac{1}{2}\sum_e \theta_e \ell_e$ the *total mean curvature* of a polytope. In fact, the function S is continuous on the space of convex bodies (more about this in Sect. 5.3.3). Thus, when a convex body K with smooth boundary is approximated by a sequence of convex polytopes K_i, their discrete total mean curvature converges to the integral of the mean curvature over the boundary of K:

$$K_i \to K \Rightarrow S(K_i) \to S(K).$$

Moreover, the function S has a unique continuous extension to the space of all convex bodies. (This follows from certain topological properties of this space.) It is thus possible to define the total mean curvature of the boundary of any convex body so that it naturally extends the integral of the mean curvature for smooth surfaces.

On the boundary of every convex surface there is a *mean curvature measure*. The total mass of this measure is the total mean curvature. For details see [13].

Theorem 5.2.7 and its proof can be easily extended to $\mathbb{R}^d$. In the discrete case, the coefficient of the Steiner polynomial at r^k is the sum of volumes of $(d-k)$-dimensional faces multiplied by their exterior angles. In the smooth case it is the integral of the elementary symmetric polynomial of degree $k-1$ of the principal curvatures.

5.2.3   Integral Geometry

Here is a seemingly unrelated question: What is the average width of the unit cube? That is, we want to compute

$$w(Q) := \frac{1}{4\pi}\int_{\mathbb{S}^2} |\text{pr}_v(Q)|\,d\text{area},$$

where for a unit vector $v \in \mathbb{S}^2$ we denote by $\text{pr}_v(Q)$ the projection of the unit cube Q to the line spanned by v, and by $|\text{pr}_v(Q)|$ the length of this projection.

The minimum width is obviously 1 (the vector v is parallel to one of the edges), the maximum width is $\sqrt{3}$ (the vector v is parallel to one of the big diagonals), so the answer should be some number between 1 and 1.732. The reader is invited to test his/her intuition: which of the following answers is correct?

$$A: \quad \frac{1+\sqrt{3}}{2} \qquad B: \quad \sqrt{2} \qquad C: \quad \frac{3}{2} \qquad D: \quad \frac{\pi}{2}$$

The correct answer can be derived from the following theorem.

Theorem 5.2.8 *The average width of a convex body is proportional to its total mean curvature. Namely, for every convex body K one has*

$$w(K) = \frac{1}{2\pi} S(K),$$

where S is the function on the space of convex bodies discussed at the end of the previous section.

From Eq. (5.3) we compute

$$S(K) = \frac{1}{2} \cdot 12 \cdot 1 \cdot \frac{\pi}{2} = 3\pi,$$

which implies that the answer C is correct.

Remark 5.2.9 If you forget the proportionality factor in Theorem 5.2.8, then you can recover it by taking for K the unit ball (provided you remember the area of the unit sphere).

Corollary 5.2.10 *If $K \subset L$, then $S(K) \leq S(L)$.*

For smooth bodies this means that the integral of the mean curvature is monotone with respect to inclusion.

The other coefficients in the Steiner formula have a similar meaning, and this for any dimension of the ambient space.

Theorem 5.2.11 *Let $K \subset \mathbb{R}^d$ be a compact convex set. Then the coefficient at r^{d-k} of the Steiner polynomial of K is proportional to the average volume of projections of K to k-dimensional linear subspaces of $\mathbb{R}^d$.*

More exactly, if the Steiner polynomial is written as

$$\mathrm{vol}(K_r) = \sum_{k=0}^{d} \binom{d}{k} W_k(K) r^k,$$

then one has

$$W_{d-k}(K) = \frac{\beta_d}{\beta_k} \int\limits_{\mathrm{Gr}_k(\mathbb{R}^d)} \mathrm{vol}_k(\mathrm{pr}_A(K))\, d\mu.$$

Here β_k is the volume of the k-dimensional ball of unit radius, $\mathrm{Gr}_k(\mathbb{R}^d)$ is the set of all k-dimensional linear subspaces of $\mathbb{R}^d$ (the Grassmanian), the integration is done with respect to the unique rotation-invariant measure of total mass 1, and pr_A denotes the orthogonal projection to A.

The quantities $W_k(K)$ are called the *quermassintegrals* of K. (In German, "quer" means "across" and "mass" means "measure".)

A special case of the above theorem are the formulas of Crofton and Cauchy, which deal with curves in the plane and surfaces in the space, respectively.

Theorem 5.2.12

(1) *(Crofton formula) The length of a closed convex curve is half the integral of its projection length:*

$$\ell(\gamma) = \frac{1}{2} \int\limits_{\mathbb{S}^1} |\mathrm{pr}_v(\gamma)|\, dv.$$

(2) *(Cauchy formula) The area of a closed convex surface is*

$$\mathrm{area}(\partial K) = \frac{1}{2\pi} \int\limits_{\mathbb{S}^2} \mathrm{area}(\mathrm{pr}_v(K))\, d\mathrm{area}.$$

Proof We prove the Crofton formula, and do it for polygons only. The length of a polygon is the sum of the lengths of its edges:

$$\ell(\gamma) = \sum_e |e|.$$

On the other hand, since the projection of a convex polygon is 2-to-1 almost everywhere, one has

$$|\mathrm{pr}_v(\gamma)| = \frac{1}{2} \sum_e |\mathrm{pr}_v(e)|.$$

Thus it suffices to show that for every segment e one has

$$|e| = \frac{1}{4} \int\limits_{\mathbb{S}^1} |\mathrm{pr}_v e|\, dv.$$

The length of the projection depends on the angle between the segment and the projection direction:

$$|\operatorname{pr}_v(e)| = |e| \cdot |\cos(v, e)|$$

Thus it suffices to integrate the cosine function:

$$\int_{\mathbb{S}^1} |\operatorname{pr}_v(e)|\, dv = |e| \cdot \int_0^{2\pi} |\cos x|\, dx = 4|e|,$$

and the Crofton formula is proved.

The Cauchy formula for convex polytopes is similarly reduced to projecting a single face of the polytope. This implies that the Cauchy formula holds for some proportionality constant independent of the polytope. This constant can either be computed by some trick or extracted from considering the special case of the sphere (which can be arbitrarily closely approximated by convex polytopes). □

5.3   Technical Tools

5.3.1   The Support Function

Let K be a convex body in $\mathbb{R}^d$.

Definition 5.3.1 A hyperplane $H \subset \mathbb{R}^d$ is called a *support hyperplane* of K if $K \cap H \neq \emptyset$ and K is contained in one of the halfspaces bounded by H.

For a smooth convex body, support hyperplanes are tangent hyperplanes. In particular, through every boundary point of a smooth convex body there is a unique support hyperplane. In the non-smooth case through every boundary point there is at least one support hyperplane (this is one of the so called separation theorems), but this hyperplane is not necessarily unique (think of a vertex of a polytope).

Lemma 5.3.2 *For every non-zero vector $v \in \mathbb{R}^d$ there is exactly one support hyperplane with outward normal v (meaning that v points away from K).*

Proof The linear functional $x \mapsto \langle v, x \rangle$ attains its maximum on K. □

Definition 5.3.3 The *support function* of K is defined as

$$h_K : \mathbb{R}^d \to \mathbb{R}, \quad h_K(v) = \max\{\langle v, x \rangle \mid x \in K\}.$$

If $\|v\| = 1$, then the value $h_K(v)$ is the signed distance from the origin to the support hyperplane of K with outward unit normal v. The sign is positive if and

Fig. 5.2 The support
function of a convex body

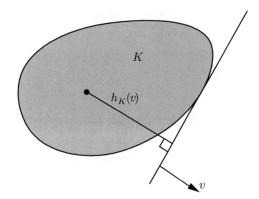

only if the origin and the body lie on the same side of the support hyperplane. See
Fig. 5.2.

Example 5.3.4 The support function of a point is linear: if $K = \{p\}$, then $h_K(v) = \langle v, p \rangle$.

The support function of a polytope is piecewise linear, because the maximum of $\langle v, x \rangle$ is attained when x is one of the vertices of the polytope. The linearity domains are the normal cones of vertices (cones spanned by the outward normals to the faces adjacent to a vertex).

Theorem 5.3.5 *The support function of a convex body is positively homogeneous and convex:*

$$h(\lambda v) = \lambda h(v) \text{ for } \lambda \geq 0$$

$$h(v + w) \leq h(v) + h(w)$$

Conversely, every positively homogeneous convex function is the support function of some convex body.

Proof The homogeneity is obvious, and the convexity follows from taking the maxima in $\langle v + w, x \rangle = \langle v, x \rangle + \langle w, x \rangle$.

For the opposite direction, given h, define

$$K = \{x \in \mathbb{R}^d \mid \langle v, x \rangle \leq h(v)\}.$$

Then surely $h_K \leq h$. One can prove that the convexity of h implies the equality $h_K = h$.   □

Remark 5.3.6 The Minkowski functional of a convex body L is defined as

$$d_L(v) = \min\{\lambda \mid \lambda v \in L\}.$$

If the interior of K contains the origin, then $h_K = d_{K^\circ}$, where K° is the *polar body* of K defined as

$$K^\circ = \{y \in \mathbb{R}^d \mid \langle x, y \rangle \leq 1\}.$$

5.3.2 The Minkowski Sum and Scaling of Convex Bodies

Definition 5.3.7 The *Minkowski sum* of $K, L \subset \mathbb{R}^d$ is

$$K + L := \{p + q \mid p \in K, q \in L\}.$$

Example 5.3.8 The Minkowski sum $K + \{p\}$ is the translate of K by p.

It is not hard to show that if K and L are both convex, then $K + L$ is also convex.

Definition 5.3.9 For $\lambda \geq 0$ denote

$$\lambda K = \{\lambda p \mid p \in K\}.$$

Thus λK is the image of K under scaling/homothety with center at the origin and coefficient λ. It is possible to define λK for $\lambda < 0$, but we will never use this. A nice property of positive scalings is that they satisfy

$$\lambda_1 K + \lambda_2 K = (\lambda_1 + \lambda_2)K, \quad \text{for } \lambda_1, \lambda_2 \geq 0.$$

Example 5.3.10 If we denote by B the unit ball in $\mathbb{R}^d$ centered at the origin, then rB is the ball of radius r centered at the origin. The r-neighborhood of K can be written as a Minkowski sum:

$$K_r = K + rB.$$

Theorem 5.3.11 *The support function of the Minkowski sum is the sum of the support functions:*

$$h_{K+L} = h_K + h_L.$$

Besides, one has

$$h_{\lambda K} = \lambda h_K \quad \text{for } \lambda \geq 0.$$

The proof is straightforward.

5.3.3 The Hausdorff Distance

Definition 5.3.12 Let $K, L \subset \mathbb{R}^d$ be two compact sets. The *Hausdorff distance* between K and L is defined as

$$\mathrm{dist}(K, L) = \max \left\{ \max_{p \in K} \mathrm{dist}(p, L), \max_{q \in L} \mathrm{dist}(q, K) \right\}$$

In other words, we take the point of K the most distant from L and the point of L the most distant from K and compare the two distances.

Lemma 5.3.13 *The Hausdorff distance can equivalently be defined as*

$$\mathrm{dist}(K, L) = \min\{r \mid K \subset L_r, L \subset K_r\}.$$

Proof One has

$$\mathrm{dist}(p, L) \leq r \text{ for all } p \in K \Leftrightarrow K \subset L_r,$$

and similarly exchanging K and L. This implies the lemma.   □

Corollary 5.3.14 *The Hausdorff distance between K and L is the uniform distance between the restrictions of their support functions to $\mathbb{S}^{d-1}$:*

$$\mathrm{dist}(K, L) = \left\| h_K \big|_{\mathbb{S}^{d-1}} - h_L \big|_{\mathbb{S}^{d-1}} \right\|_\infty := \max\{ |h_K(v) - h_L(v)| \mid v \in \mathbb{S}^{d-1} \}$$

Proof One has

$$K \subset L_r \Leftrightarrow h_K \leq h_{L_r} \Leftrightarrow h_K \leq h_L + r h_B.$$

The function h_B is identical to 1 when restricted to the unit sphere, thus

$$K \subset L_r \Leftrightarrow h_K \big|_{\mathbb{S}^{d-1}} - h_L \big|_{\mathbb{S}^{d-1}} \leq r.$$

Exchanging K and L leads to the statement.   □

Theorem 5.3.15 (Blaschke Selection Theorem) *Let K_i be a sequence of convex bodies in $\mathbb{R}^d$ such that all K_i are contained in the same ball. Then this sequence contains a subsequence which converges in the Hausdorff metric to some convex body.*

One can derive this from the Arzela–Ascoli theorem by restricting the support functions to the unit sphere (so that the Hausdorff metric becomes the uniform metric, see above) and noticing that the restricted support functions of uniformly bounded convex bodies are uniformly Lipschitz. For a direct proof, see [13].

Theorem 5.3.16 *For every convex body K and every $\varepsilon > 0$ there is a convex polytope P such that* $\mathrm{dist}(P, K) < \varepsilon$.

Proof Take the convex hull of an ε-net in K. □

Theorem 5.3.17 *The volume function is continuous on the space of convex bodies.*

Sketch of Proof Let K be a convex body. Without loss of generality, the interior of K contains 0. Then there is $\rho > 0$ such that $2\rho B \subset K$. Let L be another convex body such that $\mathrm{dist}(K, L) < \delta$. If $\delta < \rho$, then $\rho B \subset L$ and one has

$$K \subset L_\delta = L + \delta B \subset L + \frac{\delta}{\rho} L = \left(1 + \frac{\delta}{\rho}\right) L$$

and similarly $L \subset (1 + \frac{\delta}{\rho})K$. This implies

$$\mathrm{vol}\, K \leq \left(1 + \frac{\delta}{\rho}\right)^d \mathrm{vol}\, L$$

(and the same inequality with K and L exchanged). Now, for every $\varepsilon > 0$ the number δ can be chosen so small that the above inequalities imply $|\,\mathrm{vol}\, K - \mathrm{vol}\, L\,| < \varepsilon$. □

We can now prove the Steiner formula in full generality: the volume of the r-neighborhood of any convex body is a polynomial of degree d in r.

Proof of Theorem 5.2.6 If K is a polytope, then the theorem follows from the decomposition of K_r into parts similarly to what we have done in dimension 3.

For an arbitrary convex body K choose a sequence of polytopes P^i converging to K in the Hausdorff metric. Since $\mathrm{dist}(K_r, P_r^i) = \mathrm{dist}(K, P^i)$, we also have $P_r^i \to K_r$ for every $r \geq 0$. Thus we have

$$\mathrm{vol}(K_r) = \lim_{i \to \infty} \left(\sum_{k=0}^{d} a_k(P^i) r^k \right).$$

It remains to note that a pointwise limit of polynomials of bounded degree is again a polynomial, and its coefficients are the limits of the coefficients of the sequence. □

5.3.4 The Support Function and the Volume

The intersection of a convex body K with one of its support hyperplanes is called a *face* of K. Faces of polytopes are polytopes of smaller dimension. Indeed, a face of a polyhedron is obviously a polyhedron, and Theorem 5.2.4 says that polytopes and compact polyhedra are the same. Faces of the maximum possible dimension $d-1$ are called *facets* of the polytope.

Let $F_1, \ldots, F_n$ be all facets of a polytope P, and let $v_1, \ldots, v_n$ be their outward unit normals. Then the numbers

$$h_i := h_P(v_i)$$

are called the *support numbers* of P. These are just the signed distances from 0 to the hyperplanes spanned by the F_i.

Theorem 5.3.18 *The volume of a convex polytope P is given by the formula*

$$\mathrm{vol}_d(P) = \frac{1}{d} \sum_{i=1}^{n} h_i \, \mathrm{vol}_{d-1}(F_i).$$

The volume of a smooth convex body K is given by the formula

$$\mathrm{vol}_d(K) = \frac{1}{d} \int_{\partial K} h_K \, d\mathrm{area}.$$

(Strictly speaking, this is the integral of $h_K \circ \Gamma$, where $\Gamma \colon \partial K \to \mathbb{S}^{d-1}$ is the Gauss map.)

Proof First assume that the origin lies inside the polytope. Subdivide a polytope into pyramids with apex at the origin and facets as the bases. The volume of the i-th pyramid is $\frac{1}{d} h_i \, \mathrm{vol}_{d-1}(F_i)$, and the formula follows. If the origin does not lie inside the polytope, then the polytope is a "signed union" of pyramids; at the same time $\frac{1}{d} h_i \, \mathrm{vol}_{d-1}(F_i)$ gives the signed volume.

In the smooth case consider the function $\|x\|^2$ on $\mathbb{R}^d$. Due to $\Delta \|x\|^2 = 2d$ one has

$$2d \cdot \mathrm{vol}(K) = \int_K \Delta(\|x\|^2)\, dx$$

$$= \int_{\partial K} \langle \mathrm{grad}(\|x\|^2), N \rangle \, d\mathrm{area} = \int_{\partial K} \langle 2x, N \rangle = 2 \int_{\partial K} h \, d\mathrm{area}.$$

□

Theorem 5.3.19 *For every convex polytope P one has*

$$\sum_{i=1}^{n} \mathrm{vol}_{d-1}(F_i) v_i = 0.$$

For every smooth convex body K one has

$$\int_{\partial K} v \, d\text{area} = 0,$$

where v is the field of outward unit normals.

Proof Translate the polytope P by an arbitrary vector w. This does not change its volume. Facets of $P + w$ are $F_i + w$, and they have the same outward normals as the facets of P. Let us compute the support numbers h_i^x of $P + w$.

$$h_i^w = h_{P+w}(v_i) = h_P(v_i) + h_{\{w\}}(v_i) = h_i + \langle v_i, w \rangle.$$

It follows that

$$\text{vol}_d(P + w) = \frac{1}{d} \sum_{i=1}^{n} (h_i + \langle v_i, w \rangle) \, \text{vol}_{d-1}(F_i) = \text{vol}_d(P) + \frac{1}{d} \sum_{i=1}^{n} \langle v_i, w \rangle \, \text{vol}_{d-1}(F_i),$$

and therefore

$$0 = \sum_{i=1}^{n} \langle v_i, w \rangle \, \text{vol}_{d-1}(F_i) = \left\langle \sum_{i=1}^{n} \text{vol}_{d-1}(F_i) v_i, w \right\rangle.$$

Since this equality holds for all w, it follows that the vector $\sum_{i=1}^{n} \text{vol}_{d-1}(F_i) v_i$ vanishes.

In the smooth case consider an arbitrary linear function $f(x) = \langle w, x \rangle$ on $\mathbb{R}^d$ and apply the divergence theorem again. The Laplacian of a linear function is zero, therefore we have

$$0 = \int_K \Delta f \, dx = \int_{\partial K} \langle w, N \rangle \, d\text{area} = \left\langle w, \int_{\partial K} N \, d\text{area} \right\rangle$$

Since the vector w is arbitrary, it follows that the integral vanishes.   □

5.4   Mixed Volumes and Inequalities

5.4.1   Definition and Examples

Let $K_1, \ldots, K_n$ be convex bodies in $\mathbb{R}^d$. Operations described in Sect. 5.3.2 allow to form linear combinations with arbitrary nonnegative coefficients: $\lambda_1 K_1 + \cdots + \lambda_n K_n$.

Theorem 5.4.1 *The volume of a nonnegative combination of convex bodies in $\mathbb{R}^d$ is a homogeneous polynomial of degree d in the coefficients of the combination:*

$$\mathrm{vol}(\lambda_1 K_1 + \cdots + \lambda_n K_n) = \sum_{i_1,\ldots,i_d=1}^{n} a_{i_1\ldots i_d}\lambda_{i_1}\cdots\lambda_{i_d} \qquad (5.4)$$

Example 5.4.2 Consider the rectangles $K_1 = [0, a_1] \times [0, b_1]$ and $K_2 = [0, a_2] \times [0, b_2]$. One has

$$\mathrm{vol}(\lambda K_1 + \mu K_2) = (\lambda a_1 + \mu a_2)(\lambda b_1 + \mu b_2) = a_1 b_1 \lambda^2 + (a_1 b_2 + a_2 b_1)\lambda\mu + a_2 b_2 \mu^2.$$

Here we have taken $n = d = 2$. You can check what happens for $n = 2, d = 3$ or $n = 3, d = 2$.

Proof It suffices to consider the case when all K_i are polytopes. For general bodies the result follows by approximation similarly to the proof of the Steiner formula.

The proof for polytopes is by induction on the dimension d. The base $d = 1$ is obvious.

For brevity denote $K = \lambda_1 K_1 + \cdots + \lambda_n K_n$. By Theorem 5.3.18 one has

$$\mathrm{vol}_d(K) = \frac{1}{d}\sum_{j=1}^{m} h_K(v_j)\,\mathrm{vol}_{d-1}(F_K(v_j)),$$

where the sum is over all outward unit normals to the facets of K, and $F_K(v_j)$ denotes the facet with the normal $\dot{v}_j$. Let us express $h_K(v_j)$ and $\mathrm{vol}_{d-1}(F_j)$ in terms of λ_i and K_i. By Theorem 5.3.11 one has

$$h_K(v_j) = \lambda_1 h_{K_1}(v_j) + \cdots + \lambda_n h_{K_n}(v_j).$$

It is not hard to prove that

$$F_K(v_j) = \lambda_1 F_{K_1}(v_j) + \cdots + \lambda_n F_{K_n}(v_j),$$

where $F_{K_i}(v_j)$ is the intersection of K_i with its support hyperplane that has the outward normal v_j. (Observe that $F_{K_i}(v_j)$ is not necessarily a facet of K_i, it can be a face of a lower dimension.)

By the induction hypothesis $\mathrm{vol}_{d-1}(F_K(v_j))$ is a degree $d - 1$ homogeneous polynomial in λ_i. It follows that $\mathrm{vol}_d(K)$ is a homogeneous degree d polynomial in λ_i. □

The coefficients $a_{i_1\ldots i_d}$ in (5.4) are uniquely defined if one requires that they are symmetric with respect to the permutation of indices:

$$a_{i_1\ldots i_d} = a_{i_{\sigma(1)}\ldots i_{\sigma(d)}}.$$

Definition 5.4.3 Under the symmetry assumption the coefficients $a_{i_1 \dots i_d}$ in (5.4) are called the *mixed volumes*. The coefficient $a_{i_1 \dots i_d}$ is denoted by $\mathrm{vol}(K_{i_1}, \dots, K_{i_d})$.

It is convenient to think of the formula (5.4) as an analog of the multinomial expansion:

$$(\lambda_1 K_1 + \dots + \lambda_n K_n)^d = \sum_{i_1,\dots,i_d=1}^{n} \lambda_{i_1} \cdots \lambda_{i_d} K_{i_1} \cdots K_{i_d}.$$

That is, the mixed volume $\mathrm{vol}(K_{i_1}, \dots, K_{i_d})$ can be seen as a "product" of polytopes $K_{i_1}, \dots, K_{i_d}$, and this product is required to be commutative.

Theorem 5.4.4 *Mixed volumes have the following properties.*

(1) $\mathrm{vol}(K, \dots, K) = \mathrm{vol}(K)$.
(2) $\mathrm{vol}(K_1 + L_1, K_2, \dots, K_d) = \mathrm{vol}(K_1, K_2, \dots, K_d) + \mathrm{vol}(L_1, K_2, \dots, K_d)$,
    $\mathrm{vol}(\lambda K_1, K_2, \dots, K_d) = \lambda \, \mathrm{vol}(K_1, K_2, \dots, K_d)$ *for* $\lambda \geq 0$.
(3) *If* $K_1 \subset L_1$, *then* $\mathrm{vol}(K_1, K_2, \dots, K_d) \leq \mathrm{vol}(L_1, K_2, \dots, K_d)$. *In particular* $\mathrm{vol}(K_1, K_2, \dots, K_d) > 0$ *for any convex bodies* $K_1, \dots, K_d$.

The first two items follow directly from the definition. The monotonicity with respect to inclusion requires more work, see the end of this section. The positivity follows from the monotonicity because every convex body contains a ball.

In the special case $n = 2$ the formula (5.4) reads

$$\mathrm{vol}(\lambda K + \mu L) = \sum_{k=0}^{d} \binom{d}{k} \mathrm{vol}(K, \dots, K, L, \dots, L) \lambda^k \mu^{d-k},$$

where we have used the symmetry of the mixed volume for sorting K and L.

In view of the above example the Steiner formula becomes a special case of Theorem 5.4.1:

$$\mathrm{vol}(K_r) = \mathrm{vol}(K + rB) = \sum_{k=0}^{d} \binom{d}{k} \mathrm{vol}(\underbrace{K, \dots, K}_{d-k}, \underbrace{B, \dots B}_{k}) r^k.$$

Thus the quermassintegrals are the mixed volumes of a body with the ball:

$$W_k(K) = \mathrm{vol}([d-k]K, [k]B).$$

(We use numbers in square brackets to indicate the multiplicity of the argument.)

Theorem 5.4.5 *For any two convex bodies one has*

$$\mathrm{vol}_d(K, \dots, K, L) = \frac{1}{d} \sum_j h_L(v_j) \, \mathrm{vol}_{d-1}(F_K(v_j)).$$

Here the sum is taken over the normals to the facets of K.

This formula looks very natural, but there is more behind it than it seems. By using the arguments from the proof of Theorem 5.4.1 one looks at

$$\text{vol}_d(K + \lambda L) = \frac{1}{d} \sum_j h_{K+\lambda L} \, \text{vol}_{d-1}(F_{K+\lambda L}(v_j))$$

and comparing the linear term in λ one gets

$$\text{vol}_d([d-1]K, L) = \frac{1}{d^2} \Bigg(\sum_j h_L(v_j) \, \text{vol}_{d-1}(F_K(v_j))$$

$$+ (d-1) \sum_j h_K(v_j) \, \text{vol}_{d-1}([d-2]F_K(v_j), F_L(v_j)) \Bigg).$$

Thus Theorem 5.4.5 is equivalent to the statement that the two sums on the right hand side of the last formula are equal. This is not obvious.

Sketch of the proof By definition of mixed volumes one has

$$\text{vol}(K + \lambda L) = \text{vol}(K) + d \cdot \lambda \, \text{vol}([d-1]K, L) + o(\lambda).$$

Therefore

$$\text{vol}([d-1]K, L) = \frac{1}{d} \lim_{\lambda \to 0} \frac{\text{vol}(K + \lambda L) - \text{vol}(K)}{\lambda}.$$

Let K be a convex polytope. Then the body $K + \lambda L$ is the union of several pieces: K itself, then prisms over the facets of K with the heights equal to $h_L(v_j)$, then wedges at the faces of codimension 2, and so on. The total volume of the prisms is linear in λ and yields the desired formula. □

Remark 5.4.6 The last theorem can be generalized to

$$\text{vol}_d(K_1, \dots, K_d) = \frac{1}{d} \sum_j h_{K_1}(v_j) \, \text{vol}_{d-1}(F_{K_2}(v_j), \dots, F_{K_d}(v_j)).$$

This formula implies the monotonicity of the mixed volume under inclusion, the third property from Theorem 5.4.4.

5.4.2 Alexandrov–Fenchel and Minkowski Inequalities

The Alexandrov–Fenchel inequality is a very general and powerful inequality for mixed volumes. In this section we will show how it implies some of the other geometric inequalities.

Theorem 5.4.7 (Alexandrov–Fenchel inequality) *For any convex bodies*

$$K, L, M_1, \ldots, M_{d-2} \subset \mathbb{R}^d$$

one has

$$\mathrm{vol}(K, L, M_1, \ldots, M_{d-2})^2 \geq \mathrm{vol}(K, K, M_1, \ldots, M_{d-2}) \, \mathrm{vol}(L, L, M_1, \ldots, M_{d-2}).$$

For a proof see [13] or [14]. The equality cases are not known in general. For polytopes, they were obtained only recently, see [15].

Theorem 5.4.8 (Second Minkowski Inequality) *For any convex bodies* $K, L \subset \mathbb{R}^d$ *one has*

$$\mathrm{vol}(K, \ldots, K, L)^2 \geq \mathrm{vol}(K) \, \mathrm{vol}(K, \ldots, K, L, L).$$

This is clearly a special case of the Alexandrov–Fenchel inequality. The equality case is known [4] but its description is a bit long.

Theorem 5.4.9 (First Minkowski Inequality) *For any convex bodies* $K, L \subset \mathbb{R}^d$ *one has*

$$\mathrm{vol}(K, \ldots, K, L)^d \geq \mathrm{vol}(K)^{d-1} \, \mathrm{vol}(L).$$

Equality holds if and only if K *and* L *are homothetic.*

Let us show that the first Minkowski inequality (without the description of equality cases) follows from the Alexandrov–Fenchel inequality.

Lemma 5.4.10 *The sequence* $V_0, \ldots, V_d$ *defined as*

$$V_k = \mathrm{vol}([d-k]K, [k]L)$$

is logarithmically concave, that is

$$V_k^2 \geq V_{k-1} V_{k+1} \quad \textit{for all } 1 \leq k \leq d-1.$$

Proof In the Alexandrov–Fenchel inequality put

$$M_i = \begin{cases} K, & i = 1, \ldots, d - k - 1, \\ L, & i = d - k, \ldots, d - 2. \end{cases}$$

□

Alexandrov–Fenchel Implies First Minkowski By Lemma 5.4.10 the sequence $\log V_k$ satisfies

$$\log V_k \geq \frac{\log V_{k-1} + \log V_{k+1}}{2}.$$

This implies

$$\log V_1 \geq \frac{(d-1)\log V_0 + \log V_d}{d},$$

which is equivalent to the first Minkowski inequality.   □

An interesting special case is $L = B$, the unit ball.
From two ways of writing the Steiner formula

$$\mathrm{vol}(K_r) = \mathrm{vol}(K) + r \cdot \mathrm{area}(\partial K) + r^2 \cdot S(K) + r^3 \cdot \frac{4\pi}{3}$$

$$= W_0(K) + 3W_1(K) \cdot r + 3W_2(K) \cdot r^2 + W_3(K) \cdot r^3$$

one obtains

$$W_0(K) = \mathrm{vol}(K), \quad W_1(K) = \frac{\mathrm{area}(\partial K)}{3}, \quad W_2(K) = \frac{S(K)}{3}, \quad W_3(K) = \frac{4\pi}{3}.$$

The first Minkowski inequality thus says

$$\mathrm{area}(\partial K)^3 \geq \frac{4\pi}{3} \mathrm{vol}(K)^2,$$

which is the isoperimetric inequality in $\mathbb{R}^3$. The equality condition in the first Minkowski inequality ensures that the ball is the only maximizer of the volume while the surface area is fixed. A generalization to higher dimensions is straightforward.

Another instance of the first Minkowski inequality is

$$\mathrm{vol}(K, B, B)^3 \geq \mathrm{vol}(B)^2 \, \mathrm{vol}(K),$$

which can be rewritten as

$$\left(\frac{w(K)}{2}\right)^3 \geq \frac{3}{4\pi}\,\mathrm{vol}(K),$$

where $w(K)$ denotes the average width of K. This is known as the Urysohn inequality.

5.5 The Brunn–Minkowski Inequality and Symmetrization

The Brunn–Minkowski inequality is another inequality for volumes. It provides an alternative approach to the first Minkowski inequality.

5.5.1 The Brunn–Minkowski Inequality

Theorem 5.5.1 (Brunn–Minkowski Inequality) *Let $K_0, K_1 \subset \mathbb{R}^d$ be compact convex subsets of $\mathbb{R}^d$. Consider the continuous family of convex sets*

$$K_\lambda = (1-\lambda)K_0 + \lambda K_1, \quad \lambda \in [0,1].$$

Then the function $\mathrm{vol}_d(K_\lambda)^{\frac{1}{d}}$ is concave.

Besides, $\mathrm{vol}_d(K_\lambda)^{\frac{1}{d}}$ is linear if and only if either both K_0 and K_1 have dimension smaller than d and are contained in parallel hyperplanes or they are homothetic.

Example 5.5.2 Let $d = 2$ and let K_0 and K_1 be two segments in the plane. If they are parallel, then all their convex combinations are parallel segments and all have 2-volume equal to 0. Here the first equality condition is fulfilled.

Let K_0 be the $[-1, 1]$ segment in the x-axis, and K_1 be the $[-1, 1]$ segment in the y-axis. Then $(1-\lambda)K_0 + \lambda K_1$ is the $[-(1-\lambda), 1-\lambda] \times [-\lambda, \lambda]$ rectangle, and one has

$$\sqrt{\mathrm{vol}_2(K_\lambda)} = 2\sqrt{\lambda(1-\lambda)} = \sqrt{1 - (1-2\lambda)^2}.$$

Thus the graph of $\sqrt{\mathrm{vol}_2(K_\lambda)}$ is an upper half of an ellipse, which is concave.

The Brunn–Minkowski inequality holds for non-convex sets as well. For its numerous connections to other inequalities see [8].

5.5.2 Brunn–Minkowski Implies Minkowski

Let us prove Theorems 5.4.9 and 5.4.8 (without the equality cases) under the assumption of validity of Theorem 5.5.1.

Consider the function

$$f(\lambda) = \mathrm{vol}_d(K_\lambda)^{\frac{1}{d}} - (1 - \lambda)\,\mathrm{vol}_d(K_0)^{\frac{1}{d}} - \lambda\,\mathrm{vol}_d(K_1)^{\frac{1}{d}}.$$

It differs from $\mathrm{vol}_d(K_\lambda)^{\frac{1}{d}}$ by a linear function, thus by the Brunn–Minkowski theorem it is concave. Besides, one has $f(0) = f(1) = 0$.

This implies first that $f'(0) \geq 0$, with the same equality condition as in the Brunn–Minkowski inequality. By a patient calculation of $f'(0)$ one shows

$$f'(0) = \frac{1}{d}\,\mathrm{vol}_d(K_0)^{\frac{1-d}{d}}\,\mathrm{vol}_d(K_0, \ldots, K_0, K_1) - \mathrm{vol}_d(K_1)^{\frac{1}{d}},$$

so that $f'(0) \geq 0$ is equivalent to

$$\mathrm{vol}_d(K_0, \ldots, K_0, K_1)^d \geq \mathrm{vol}_d(K_0)^{d-1}\,\mathrm{vol}_d(K_1),$$

which is the first Minkowski inequality.

The concavity of $f(\lambda)$ also implies $f''(0) \leq 0$. By another careful calculation one shows

$$f''(0) = -(d-1)V_0^{\frac{2d-1}{d}}(V_1^2 - V_0 V_2),$$

so that $f''(0) \leq 0$ implies $V_1^2 \geq V_0 V_2$, which is the second Minkowski inequality.

5.5.3 Steiner Symmetrization and Schwartz Rounding

In order to prove the Brunn–Minkowski inequality, we reformulate it as follows. In the space $\mathbb{R}^{d+1}$ lift the body K_1 to the hyperplane $x_{d+1} = 1$ while the body K_0 remains in the coordinate hyperplane $x_{d+1} = 0$. Form the convex hull $\widetilde{K}$ of $K_0 \cup K_1$. Then one can show that $\widetilde{K} \cap \{x_{d+1} = \lambda\} = K_\lambda \times \{\lambda\}$. See Fig. 5.3.

Now replace each body K_λ by the ball B_λ of the same volume in the same hyperplane $x_{d+1} = \lambda$ centered at the point $(0, \lambda)$. The Brunn–Minkowski inequality is equivalent to the statement that the radius of B_λ is a concave function of λ, in other words that the resulting "body of revolution" is convex. Besides, this body is a cylinder or a truncated cone if and only if K_0 and K_1 are homothetic.

The transformation of $\widetilde{K}$ into a "body of revolution" described above is called the *Schwartz rounding*. We will prove that the result of Schwartz rounding is convex

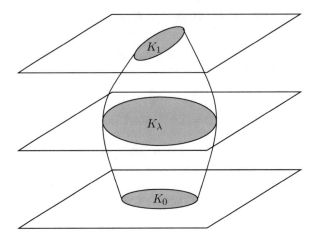

Fig. 5.3 To the proof of the Brunn–Minkowski inequality

by realizing it as a process with infinitely many steps, where each step preserves convexity.

Definition 5.5.3 Let $K \subset \mathbb{R}^d$ be a convex body, and $H \subset \mathbb{R}^d$ be a hyperplane. The *Steiner symmetral* of K about H is obtained as follows. Draw all lines perpendicular to H; they intersect K in segments; move each segment along its line so that its midpoint goes to H.

We denote the Steiner symmetral of K about H by $\mathrm{St}_H(K)$.

From the definition it is immediate that the Steiner symmetrization preserves the volume.

Lemma 5.5.4 *The Steiner symmetral of a convex body is convex.*

The proof is a simple geometric exercise.

Take d hyperplanes $H_1, H_2, \ldots H_d$ in $\mathbb{R}^{d+1}$, all passing through the coordinate axis x_{d+1} and such that their normals span $\mathbb{R}^d$ and form angles incommensurable with π. Denote by St_k the Steiner symmetrization about H_k. We apply to the body $\widetilde{K}$ the sequence of symmetrizations $\mathrm{St}_d \circ \cdots \circ \mathrm{St}_2 \circ \mathrm{St}_1$, then repeat it over and over again.

Lemma 5.5.5 *The sequence of convex bodies* $(\mathrm{St}_d \circ \cdots \circ \mathrm{St}_2 \circ \mathrm{St}_1)^n (\widetilde{K})$ *converges to a "body of revolution", that is to a subset of $\mathbb{R}^{d+1}$ such that all hyperplane sections $\{x_{d+1} = \lambda\}$ are balls centered at $(0, \lambda)$.*

Sketch of Proof The $(d + 1)$-st dimension plays no role in the process, so that it suffices to prove the following statement. If $H_1, \ldots, H_d$ are hyperplanes in $\mathbb{R}^d$ whose normals span $\mathbb{R}^d$ and have angles incommensurable with π, then the consecutive symmetrization described above applied to a body $K \subset \mathbb{R}^d$ converges to a ball.

Denote $(\text{St}_d \circ \cdots \circ \text{St}_2 \circ \text{St}_1)^n(K)$ by K_n. By the Blaschke selection theorem there is a subsequence K_{i_n} which converges to a body K'. How can we conclude that K' is a ball? We will exhibit a function on the space of convex bodies which decreases with every symmetrization step and has a unique minimum on the set of bodies of fixed volumes.

Definition 5.5.6 The *moment of inertia* of K with respect to $p \in \mathbb{R}^d$ is defined as

$$\mu(K) = \int_K \|x - p\|^2 \, dx.$$

One can prove the following properties of the moment of inertia.

(1) The moment of inertia with respect to p is the sum of moments of inertia with respect to d pairwise perpendicular hyperplanes passing through p. Here the moment of inertia with respect to H is

$$\mu_H(K) = \int_K \text{dist}^2(x, H) \, dx.$$

(2) The Steiner symmetrization about H decreases the moment of inertia, and does it strictly if the body is not symmetric about H.

Property 2 implies that among the bodies of a given volume the ball is the unique minimizer of the moment of inertia. Indeed, the existence of minimizer follows from the Blaschke selection theorem, and the minimizer must be symmetric in every direction.

Now, back to our converging subsequence $K_{i_n} \to K'$, the body K' must be symmetric about H_1. Otherwise $\mu(\text{St}_1(K')) < \mu(K')$ which contradicts the fact that $(\text{St}_n \circ \cdots \circ \text{St}_1)(K') = K'$ (see Property 2 above). Similarly one can show that $\text{St}_1(K')$ is symmetric with respect to H_2. But since $\text{St}_1(K') = K'$, this means that K' is symmetric with respect to H_2. Continuing in the same manner, one shows that K' is symmetric with respect to all H_k.

From the spanning property of H_k and the irrationality of the angles between them one can conclude that K' is a ball.   □

The above proof does not discuss the equality condition: the Schwartz rounding is a truncated cone if and only if the K_0 and K_1 are homothetic. I do not know how to do this but I believe that this is doable. Blaschke gave an argument for $d = 2$ in [3], but it seems to be specific for this dimension.

5.5.4 A Couple of Other Inequalities

Theorem 5.5.7 (Blaschke-Santalo Inequality) *Let* $K \subset \mathbb{R}^d$ *be a convex body symmetric with respect to* 0. *Then*

$$\text{vol}(K) \cdot \text{vol}(K^\circ) \le \beta_d^2,$$

where K° *is the polar body, and* β_d *is the volume of the unit ball.*
Equality occurs if and only if K *is an ellipsoid.*

The proof is based on the following lemma.

Lemma 5.5.8 *If* K *is symmetric with respect to* 0, *then for every hyperplane* H *through* 0 *one has*

$$\text{vol}((\text{St}_H K)^\circ) \ge \text{vol}(K^\circ).$$

It allows to prove the Blaschke-Santalo inequality with the help of Schwartz rounding.

The lower bound for the same product of volumes is still an open problem. This is due to the fact that the minimizers are non-symmetric and non-unique.

Conjecture 5.5.9 (Mahler) *Let* $K \subset \mathbb{R}^d$ *be a convex body symmetric with respect to* 0. *Then*

$$\text{vol}(K) \cdot \text{vol}(K^\circ) \ge \frac{4^d}{d!}$$

The equality is assumed to be attained for the so-called Hanner polytopes: iterations of products and convex hulls of 0-*symmetric segments on the coordinate lines.*

5.6 Isoperimetric Inequality and Laplacian Eigenvalues

5.6.1 Normal Parametrization of Convex Plane Curves

A convex body $K \subset \mathbb{R}^d$ is called *strictly convex* if every support hyperplane of K has only one point in common with K. The boundary of a strictly convex body has a *normal parametrization*: this is the map

$$\gamma: \mathbb{S}^{d-1} \to \partial K,$$

which associates to $v \in \mathbb{S}^{d-1}$ the unique common point of ∂K and the support hyperplane with the outward unit normal v. If the boundary of K is smooth, then γ is the inverse of the Gauss map.

Let us express the map γ in terms of the support function $h\colon \mathbb{S}^{d-1} \to \mathbb{R}$. In this section we do it for $d = 2$.

Theorem 5.6.1 *The normal parametrization of a smooth strictly convex curve in $\mathbb{R}^2$ is given by*

$$\gamma(v) = h(v) \cdot v + h'(v) \cdot Jv,$$

where $Jv \in \mathbb{R}^2$ is the image of rotation of v by $90°$.

Proof Let $\gamma(v) = a(v)v + b(v)Jv$. Since (v, Jv) is an orthonormal basis, one has

$$a(v) = \langle \gamma(v), v \rangle = h(v).$$

By differentiating $h = \langle \gamma(v), v \rangle$ one obtains

$$h'(v) = \langle \gamma'(v), v \rangle + \langle \gamma(v), v' \rangle = \langle \gamma(v), Jv \rangle = b(v).$$

$\square$

Lemma 5.6.2 *The velocity vector of the normal parametrization of a smooth strictly convex curve has norm*

$$\|\gamma'(v)\| = h + h''.$$

Proof By differentiating the formula from Theorem 5.6.1 one obtains

$$\gamma' = h'v + hv' + h''Jv + h'(Jv)' = (h + h'')v',$$

which implies $\|\gamma'\| = |h + h''|$. Since v' and γ' both look in the same direction, $h + h''$ is non-negative. $\square$

Theorem 5.6.3 *The length of a smooth strictly convex curve can be computed from its support function by the following formula:*

$$\ell(\gamma) = \int_{\mathbb{S}^1} h(v)\, dv.$$

Proof By Lemma 5.6.2 one has

$$\ell(\gamma) = \int_{\mathbb{S}^1} h(v)\, dv = \int_{\mathbb{S}^1} (h + h'')\, dv = \int_{\mathbb{S}^1} h\, dv.$$

$\square$

Observe that this theorem is equivalent to the Crofton formula for smooth convex curves, see Theorem 5.2.12 since $|\operatorname{pr}_v(\gamma)| = h(v) + h(-v)$.

Theorem 5.6.4 *The area of a strictly convex figure with smooth boundary can be computed from its support function as follows:*

$$\operatorname{area}(K) = \frac{1}{2}\int_{\mathbb{S}^1}(h^2 - (h')^2)\,dv.$$

Proof One has

$$\operatorname{area}(K) = \frac{1}{2}\int_\gamma h\circ\gamma^{-1}\,d\ell(\gamma)$$

(the figure K is made of infinitely thin triangles with height h and base $d\ell$). By Lemma 5.6.2 one has $d\ell(\gamma) = (h + h'')dv$, which implies

$$\operatorname{area}(K) = \frac{1}{2}\int_{\mathbb{S}^1} h(h + h'')\,dv = \frac{1}{2}\int_{\mathbb{S}^1}(h^2 - (h')^2)\,dv.$$

□

5.6.2 The Isoperimetric Inequality in the Plane

The eigenfunctions of the Laplacian $f \mapsto -f''$ on $\mathbb{S}^1$ are 1, $\cos nt$, $\sin nt$. Thus the spectrum of the Laplacian on $\mathbb{S}^1$ is $\{n^2 \mid n \in \mathbb{Z}\}$.

Theorem 5.6.5 (Wirtinger Inequality) *Let $f: \mathbb{R} \to \mathbb{R}$ be a 2π-periodic smooth function with zero average:*

$$\int_0^{2\pi} f\,dt = 0.$$

Then one has

$$\int_0^{2\pi} f^2\,dt \le \int_0^{2\pi}(f')^2\,dt.$$

Besides, equality occurs if and only if f is constant.

Proof The assumption $\int_0^{2\pi} f\,dt = 0$ can be written as

$$\langle f, 1\rangle_{L^2} = 0.$$

That is, f is orthogonal to the kernel of the Laplacian. From our knowledge of the spectrum of Δ we conclude that

$$\langle f, f'' \rangle \le -\langle f, f \rangle.$$

Now do integration by parts:

$$\int_0^{2\pi} (f')^2 \, dt = -\int_0^{2\pi} f f'' \, dt \ge \int_0^{2\pi} f^2 \, dt.$$

□

Theorem 5.6.6 (Isoperimetric Inequality) *For every strictly convex figure K with smooth boundary the following inequality between its area and perimeter holds:*

$$\ell(\partial K)^2 \ge 4\pi \, \mathrm{area}(K).$$

Sketch of Proof Let $h \colon [0, 2\pi] \to \mathbb{R}$ be the support function of K. Denote by c its average:

$$c = \frac{1}{2\pi} \int_0^{2\pi} h \, dt.$$

By the formulas from the previous section one has

$$\ell(\partial K) = \int_0^{2\pi} h(t) \, dt = 4\pi^2 c^2,$$

$$\mathrm{area}(K) = \frac{1}{2} \int_0^{2\pi} (h^2 - (h')^2) \, dt = \frac{1}{2} \int_0^{2\pi} \left((f + c)^2 - (f')^2 \right) dt$$

$$= \frac{1}{2} \int_0^{2\pi} f^2 \, dt + c \int_0^{2\pi} f \, dt + \pi c^2 - \frac{1}{2} \int_0^{2\pi} (f')^2 \, dt \le \pi c^2 = \frac{\ell(\partial K)}{4\pi}.$$

□

5.6.3 Higher Dimensions

Let us state higher-dimensional analogs of the formulas from the previous subsections. Let K be a strictly convex body in $\mathbb{R}^d$ with smooth boundary. Denote the normal parametrization of ∂K by $\gamma \colon \mathbb{S}^{d-1} \to \partial K$.

Theorem 5.6.7 *The normal parametrization of the boundary of a smooth strictly convex body is given by*

$$\gamma(v) = h(v) \cdot v + \mathrm{grad}^{\mathbb{S}} h,$$

where $\mathrm{grad}^{\mathbb{S}} h$ *denotes the gradient of the restriction of* h *to* $\mathbb{S}^{d-1}$.

The proof is similar to that of Theorem 5.6.1.

Let $D^2 h(v)$ be the Hessian matrix of h at v (now h viewed as a function on $\mathbb{R}^d$), that is, the matrix of second partial derivatives of h. Due to the homogeneity of h, the vector v is an eigenvector of $D^2 h(v)$. Thus its orthogonal complement $v^\perp$ is invariant under the linear operator $D^2 h(v)$. It can be shown that

$$D^2 h(v)\big|_{v^\perp} = \mathrm{Hess}^{\mathbb{S}} h(v),$$

where $\mathrm{Hess}^{\mathbb{S}} h(v) \colon T_v \mathbb{S}^{d-1} \to T_v \mathbb{S}^{d-1}$ is the *Hessian operator* defined as

$$\mathrm{Hess}^{\mathbb{S}} h(X) = \nabla_X^{\mathbb{S}}(\mathrm{grad}^{\mathbb{S}} h).$$

Theorem 5.6.8 *The volume of a smooth strictly convex body is given by*

$$\mathrm{vol}(K) = \frac{1}{d} \int_{\mathbb{S}^{d-1}} h(v) \det\left(D^2 h(v)\big|_{v^\perp} \right) d\mathrm{vol}_{\mathbb{S}^{d-1}}$$

$$= \frac{1}{d} \int_{\mathbb{S}^{d-1}} h \det(h \cdot \mathrm{id} + \mathrm{Hess}^{\mathbb{S}} h) \, d\mathrm{vol}_{\mathbb{S}^{d-1}}.$$

Proof Use the formula

$$\mathrm{vol}(K) = \frac{1}{d} \int_{\partial K} h \circ \gamma^{-1} \, d\mathrm{vol}_{\partial K},$$

and move the integration to $\mathbb{S}^{d-1}$. The integrand will be multiplied by the determinant of the Jacobian of γ. This Jacobian can be computed by differentiating the formula from Theorem 5.6.7 and is equal to $h \cdot \mathrm{id} + \mathrm{Hess}^{\mathbb{S}} h$. □

Let us write the quermassintegrals of K in terms of the support function. One has

$$\mathrm{vol}(B + tK) = \mathrm{vol}(B) + dt W_{d-1}(K) + \frac{d(d-1)}{2} t^2 W_{d-2} K) + \cdots$$

On the other hand, one can substitute the support function $1 + th$ into the formula of Theorem 5.6.8 and expand the result in the powers of t. Doing some integrations by parts (a smooth analog of Theorem 5.4.5 and its generalization in Remark 5.4.6) one arrives at the formulas

$$W_{d-1}(K) = \frac{1}{d} \int_{\mathbb{S}^{d-1}} h \, d\mathrm{vol}_{\mathbb{S}^{d-1}}$$

$$W_{d-2}(K) = \frac{1}{d} \int_{\mathbb{S}^{d-1}} h \left(h - \frac{\Delta^{\mathbb{S}} h}{d-1} \right) d\mathrm{vol}_{\mathbb{S}^{d-1}},$$

where $\Delta^{\mathbb{S}} h = - \mathrm{tr}\, \mathrm{Hess}^{\mathbb{S}} h$.

Recall that the spectrum of the Laplacian on $\mathbb{S}^{d-1}$ is

$$\{k(k+d-2) \mid k \geq 0\}$$

(the eigenfunctions are the restrictions of homogeneous harmonic polynomials in d variables). Important for us is that the smallest nonzero eigenvalue is $d-1$. If we repeat our argument from the case $d=2$, then we get not the isoperimetric inequality but the following special case of the second Minkowski inequality.

Theorem 5.6.9 *For every convex body K one has $(W_{d-1})^2 \geq W_d W_{d-2}$.*

Remark 5.6.10 Hilbert gave a proof of the general case of the second Minkowski inequality based on similar arguments, estimating the spectrum of a certain differential operator. See [5], and for a modern exposition [14].

5.7 A New Take on the Mixed Volumes

5.7.1 *Strongly Isomorphic Polytopes and Volume Polynomials*

Take n unit vectors $v_1, \ldots, v_n \in \mathbb{R}^d$ and n real numbers $h_1, \ldots, h_n$. Then the solution set of the system of linear inequalities

$$P(v, h) := \{x \in \mathbb{R}^d \mid \langle v_i, x \rangle \leq h_i\} \tag{5.5}$$

is a convex polyhedron. This polyhedron is compact if and only if every vector in $\mathbb{R}^d$ is a non-negative linear combination of vectors v_i. Every facet of $P(v, h)$ has one of the vectors v_i as the outward unit normal, but it is not necessarily true that every v_i is the normal of some facet. Indeed, one of the inequalities in (5.5) may be a consequence of the others.

For a given collection of v_i, let us look at only those $h \in \mathbb{R}^n$, for which the polytope $P(v, h)$ has n facets. All these polytopes have pairwise parallel facets, but their combinatorics may be different. Think of an octahedron whose faces are translated by small distances: each of the vertices of the octahedron splits in two vertices and can do this in two different ways.

Definition 5.7.1 Two polytopes are called *strongly isomorphic* if they have pairwise parallel facets and have the same combinatorics (that is, if a set of facets of one polytope has a common point, then the corresponding facets of the other polytope also have a common point).

Choose h^0 so that $P(v, h^0)$ has n facets and denote by $C \subset \mathbb{R}^n$ the set of all h such that the polytope $P(v, h)$ is strongly isomorphic to $P(v, h^0)$. If one chooses h^0 so that $P(v, h^0)$ is simple, that is, at every vertex exactly d facets meet, then the set C is open.

Remark 5.7.2 The set $C \subset \mathbb{R}^n$ is a convex polyhedral cone called the type cone of a given strong isomorphy class. The space of all $h \in \mathbb{R}^n$ for which the system (5.5) contains no redundant inequalities is also a convex polyhedral cone. Its subdivision into type cones is related to the so-called secondary polytope, see [7, 9, 11].

Theorem 5.7.3 *The volume of a polytope $P(v, h)$ for $h \in C$ is a homogeneous degree d polynomial in the variables $h_1, \ldots, h_n$:*

$$\mathrm{vol}(P(v, h)) = V_C(h).$$

It should be stressed that the polynomial V_C depends on the choice of the domain C. In particular, a monomial $h_{i_1} \cdots h_{i_d}$ is present in V_C if and only if the intersection of the facets with normals $v_{i_1}, \ldots, v_{i_d}$ is non-empty (we assume that the polytopes in C are simple).

Example 5.7.4 For $d = 2$ the combinatorics is always the same. The 2-volume can be computed as

$$\mathrm{vol}(P(v, h)) = \frac{1}{2} \sum_{i=1}^n h_i \ell_i(h),$$

where $\ell_i(h)$ is the length of the side of $P(v, h)$ with the outward normal v_i. It is easy to see that ℓ_i is a linear combination of h_{i-1}, h_i and h_{i+1} (if the vectors v_i are ordered accordingly to their direction), the coefficients are certain trigonometric functions of the angles between v_{i-1} and v_i and between v_i and v_{i+1}.

Every homogeneous polynomial of degree d gives rise to a unique symmetric multilinear form, which we denote by the same symbol. That is, V_C satisfies and is uniquely determined by the properties

$$V_C(h, \ldots, h) = V_C(h),$$

$$V_C(h^1, \ldots, h^d) = V_C(h^{\sigma(1)}, \ldots, h^{\sigma(d)}),$$

$$V_C(\ldots, \lambda h^i + \mu g^i, \ldots) = \lambda_i V_C(\ldots, h_i, \ldots) + \mu V_C(\ldots, g_i, \ldots).$$

This leads to an alternative definition of the mixed volume for strongly isomorphic polytopes.

Definition 5.7.5 Let $P(v, h^i)$, $i = 1, \ldots, d$ be strongly isomorphic polytopes. Their mixed volume is defined as

$$\mathrm{vol}(P(v, h^1), \ldots, P(v, h^d)) := V_C(h^1, \ldots, h^d).$$

Example 5.7.6 In the $d = 2$ case one has

$$\mathrm{vol}(P(v, h^1), P(v, h^2)) = \frac{1}{2} \sum_{i=1}^{n} h_i^1 \ell_i(h^2) = \frac{1}{2} \sum_{i=1}^{n} h_i^2 \ell_i(h^1).$$

In the last equation, symmetry in h^1 and h^2 is not straightforward, but it follows from $\frac{\partial \, \mathrm{vol}}{\partial h_i} = \ell_i$, which can be proved geometrically, compare Theorem 5.4.5.

5.7.2 Alexandrov–Fenchel Inequality Revisited

Let $h^1, \ldots, h^{d-2} \in C \subset \mathbb{R}^n$ be the support parameters of $d-2$ strongly isomorphic polytopes. Consider the symmetric bilinear form

$$q(x, y) := V_C(x, y, h^1, \ldots, h^{d-2}).$$

(If $x, y \in C$, then this is the mixed volume of some polytopes, but in principle one can substitute any $x, y \in \mathbb{R}^n$.)

Recall that any quadratic form has a so-called *signature* (n_+, n_-, n_0): the number of positive, negative, respectively zero coefficients after diagonalization. The number n_+ is also the maximum dimension of a subspace of $\mathbb{R}^n$ on which the quadratic form is positive definite. For the quadratic form q one has $n_+ \geq 1$ because for any $h \in C$ the number

$$q(h, h) = V_C(h, h, h^1, \ldots, h^{d-2})$$

is the mixed volume of convex polytopes, which is positive.

Theorem 5.7.7 *The quadratic form q has $n_+ = 1$.*

Let us show that this theorem is a reformulation of the Alexandrov–Fenchel inequalities.

Equivalence of Theorems 5.7.7 and 5.4.7 It can be shown that any $K, L, M_1, \ldots, M_{d-2}$ can be simultaneously approximated by strongly isomorphic convex polytopes. This reduces the Alexandrov–Fenchel inequality to the case when $M_i = P(v, h^i)$, $h^i \in C$ for some type cone C.

Take any $h, g \in C$. Assume that $n_+ = 1$. Then on the one hand $q(h, h) > 0$, on the other hand the restriction of q to the linear space span$\{h, g\}$ is not positively definite. It follows that

$$\det \begin{pmatrix} q(h, h) & q(h, g) \\ q(h, g) & q(g, g) \end{pmatrix} \le 0$$

But this is exactly the Alexandrov–Fenchel inequality.

In the opposite direction, if all determinants as above are non-positive, then $\mathbb{R}^n$ contains no 2-dimensional subspaces on which the form q would be positive definite. This implies that $n_+ = 1$. $\quad\square$

If all M_i are the same, then the nullspace dimension n_0 is equal to d (the nullspace comes from the support parameters of points). This is a consequence of the equality case in the second Minkowski inequality.

In particular, the area quadratic form from Example 5.7.6 has signature $(1, 2, n-3)$. This allows to see the space of n-gons with fixed directions of sides as an $(n-3)$-dimensional hyperbolic polyhedron, see [2]. Their construction is generalized to higher dimensions in [7].

5.7.3 The Discrete Spherical Laplacian

The formula

$$W_{d-2}(P) = \frac{1}{d} \sum_{i=1}^n h_i \, \mathrm{vol}_{d-1}(F_i(h, [d-2]\mathbf{1}))$$

(here $\mathbf{1}$ is the vector $(1, \ldots, 1) \in \mathbb{R}^n$) holds for all vectors $h \in \mathbb{R}^n$ such that the polytopes $P(v, h)$ and $P(v, \mathbf{1})$ are strongly isomorphic. It is a discrete analog of the formula

$$W_{d-2}(K) = \frac{1}{d} \int_{\mathbb{S}^{d-1}} h\left(h - \frac{\Delta^{\mathbb{S}} h}{d-1}\right) d\mathrm{dvol}_{\mathbb{S}^{d-1}}$$

from Sect. 5.6.3. For any $v_1, \ldots, v_n$ the polytope $P(v, \mathbf{1})$ has n facets and is circumscribed about the unit sphere.

The function $W_{d-2}(P)$ is a quadratic form in h:

$$W_{d-2}(P) = q(h, h).$$

By analogy with the smooth case it is natural to see it as the discrete counterpart of the quadratic form

$$\left\langle h, h - \frac{\Delta^{\mathbb{S}} h}{d-1} \right\rangle_{L^2}.$$

This leads to the notion of a discrete spherical Laplacian associated with any set of points $v_1, \ldots, v_n \in \mathbb{S}^{d-1}$ such that any open hemisphere contains at least one of these points.

By the argument from the previous section, the inequality $(W_{d-1})^2 \geq W_d W_{d-2}$ implies that the form q has $n_+ = 1$. This, in turn, implies that the smallest positive eigenvalue of the discrete spherical Laplacian is $d - 1$ in a perfect analogy to the smooth case.

References

1. A. Barvinok, *A Course in Convexity*, volume 54 of Graduate Studies in Mathematics (American Mathematical Society, Providence, RI, 2002)
2. C. Bavard, É. Ghys, Polygones du plan et polyèdres hyperboliques. Geom. Dedicata **43**(2), 207–224 (1992)
3. W. Blaschke, *Kreis und Kugel*, 2te Aufl. (Walter de Gruyter & Co., Berlin, 1956)
4. G. Bol, Beweis einer Vermutung von H. Minkowski. Abh. Math. Sem. Hansischen Univ. **15**, 37–56 (1943)
5. T. Bonnesen, W. Fenchel, *Theory of Convex Bodies* (BCS Associates, Moscow, ID, 1987). Translated from the German and edited by L. Boron, C. Christenson and B. Smith
6. G. Ewald, *Combinatorial Convexity and Algebraic Geometry*, volume 168 of Graduate Texts in Mathematics (Springer, New York, 1996)
7. F. Fillastre, I. Izmestiev, Shapes of polyhedra, mixed volumes and hyperbolic geometry. Mathematika **63**(1), 124–183 (2017)
8. R. Gardner, The Brunn-Minkowski inequality. Bull. Am. Math. Soc. **39**, 355–405 (2002)
9. I.M. Gelfand, M.M. Kapranov, A.V. Zelevinsky, *Discriminants, Resultants, and Multidimensional Determinants*. Mathematics: Theory & Applications (Birkhäuser Boston Inc., Boston, MA, 1994)
10. J. Hadamard, Sur certaines propriétés des trajectoires en dynamique. J. Math. Pures Appl. **3**, 331–387 (1897)
11. P. McMullen, Representations of polytopes and polyhedral sets. Geom. Dedicata **2**, 83–99 (1973)
12. R. Sacksteder, On hypersurfaces with no negative sectional curvatures. Amer. J. Math. **82**, 609–630 (1960)
13. R. Schneider, *Convex Bodies: The Brunn-Minkowski Theory*, volume 151 of Encyclopedia of Mathematics and its Applications, expanded edition (Cambridge University Press, Cambridge, 2014)
14. Y. Shenfeld, R. van Handel, Mixed volumes and the Bochner method. Proc. Am. Math. Soc. **147**(12), 5385–5402 (2019)
15. Y. Shenfeld, R. van Handel, The extremals of the Alexandrov-Fenchel inequality for convex polytopes (2020). https://arxiv.org/abs/2011.04059
16. J. Steiner, Über parallele Flächen, in *Monatsbericht der Akademie der Wissenschaften zu Berlin* (1840), pp. 114–118

Chapter 6
Compactness and Finiteness Results for Gromov-Hyperbolic Spaces

(After Joint Works by Gérard Besson, Gilles Courtois, Sylvestre Gallot, and Andrea Sambusetti)

Gérard Besson and Gilles Courtois

Abstract This is a series of lectures on Bishop–Gromov's type inequalities adapted to metric spaces. We consider the case of Gromov-hyperbolic spaces and draw consequences of these inequalities such as compactness and finiteness Theorems. This course is intended to be elementary in the sense that the necessary background is described in detail.

Keywords Gromov-hyperbolic space · Bishop–Gromov inequalities · Comparison theorems · Entropy · Ricci curvature · Compactness theorems in metric spaces · Finiteness theorems i metric spaces · Gromov–Hausdorff topology

AMS Codes 51K10, 53C23, 53C21, 53E20, 57K30

6.1 Introduction

This text is the transcript of a series of lectures given by the authors during a CIMPA school (see here). It was held in Varanasi, India from December 5 till December 15, 2019; it was organised by Bankteshwar Tiwari and Athanase Papadopoulos.

The theme is the Geometry of (some) compact Metric Spaces. Why should we care about metric spaces? In dimension $n > 24$, the only known examples of manifolds with a Riemannian metric of positive sectional curvature also do carry a symmetric metric. *Grosso modo* these closed positively curved symmetric Riemannian manifolds are constructed using Lie groups and somehow, in large

G. Besson
Institut Fourier, Université de Grenoble, Saint Martin d'Hères, France
e-mail: g.besson@univ-grenoble-alpes.fr

G. Courtois (✉)
Institut de Mathématiques de Jussieu – Paris Rive Gauche (IMJ-PRG), Paris, France
e-mail: courtoisgls@gmail.com

© The Author(s), under exclusive license to Springer Nature Switzerland AG 2022
A. Papadopoulos (ed.), *Surveys in Geometry I*,
https://doi.org/10.1007/978-3-030-86695-2_6

dimensions, they are the only examples we know. Concerning negative sectional curvature on closed manifolds of dimension greater than 2 the situation is a bit better since there are several series of examples, and some sporadic ones, of such manifolds not carrying symmetric metrics. Altogether, they constitute a sparse and statistically not significant sample.

On the other hand if one considers metric spaces, we get an incredible flexibility for constructing examples and we are left with an impressive zoology. Following Felix Klein's program we get fascinating properties of the groups acting on these spaces. However, the difficulty is now to give a meaning to positive or negative curvature. Great advances have recently been made in this direction. Several good notions of *synthetic curvature* have been introduced. Among them A. D. Aleksandrov's approach for a synthetic version of upper or lower bounds on sectional curvature for metric spaces has been fundamental; for details and references the reader is referred to the survey [1] and the book [3]. Recently synthetic versions of the Ricci curvature has been introduced by various authors and the reader could check the seminal article [6] and the book [28] for a detailed introduction and lots of references. In these lectures we explore the negative curvature side and the key notion, which was introduced by M. Gromov, and which is now called *Gromov-hyperbolicity* (see [17, 19]).

These lectures intend to be a guide for reading the articles [4, 5], the joint works mentioned in the subtitle. In the two first lectures we give the basic definitions and we progress towards compactness and finiteness results (see Theorems 6.4.6, 6.7.4, 6.8.18, 6.10.3, 6.11.8). One key step is the proof of a version of the celebrated Bishop–Gromov inequality (see Theorems 6.2.8, 6.3.2, 6.5.3, 6.5.10). The original Bishop–Gromov inequality has had a tremendous influence on the Riemannian Geometry of manifolds and we do hope that our version will have some impact.

We are happy to thank the organisers of this school: Athanase Papadopoulos and particularly Bankteshwar Tiwari and all the local organisers who made our stay a really exceptional experience.

6.2 Lecture 1

This series of lectures is based on the preprints [4, 5] which contain all the details of the statements that follow. The main character is a metric space that we will usually denote by (X, d). For the sake of simplicity we will always assume that the distance is symmetric that is,

$$\forall x, y \in X, \quad d(x, y) = d(y, x).$$

This includes Riemannian and Finsler manifolds endowed with the distance coming from the Riemannian or symmetric Finsler metric.

Eventually we wish to study various families, denoted by $\mathcal{M}$, of metric spaces satisfying extra assumptions, endowed with the (pointed) Gromov–Hausdorff distance between metric spaces, denoted by d_{GH}. We then aim at proving that the metric space

$$(\mathcal{M}, d_{GH})$$

is compact, where $\mathcal{M} = \{(X, d);$ metric space with extra assumptions$\}$. It is the core of this course to make precise and minimal these assumptions. The reader is supposed to be familiar with the Gromov–Hausdorff distance between metric spaces and, if necessary, is referred to [20] Definition 3.4 page 72 or [9] page 70.

6.2.1   Some Assumptions

A metric space (X, d) is endowed with the topology induced by the distance function d. We always assume that (X, d) is proper, *i.e.* that closed metric balls are compact. Notice that this excludes infinite dimensional Banach spaces.

6.2.2   Geodesics

A continuous curve (Fig. 6.1),

$$c : [0, 1] \longrightarrow X , \quad c(0) = x \text{ and } c(1) = y ,$$

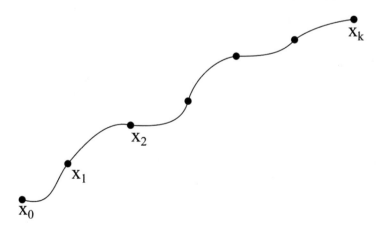

Fig. 6.1 A curve

is said to be rectifiable if

$$\sup \left\{ \sum_{i=0}^{k-1} d(x_i, x_{i+1}); \ 0 = x_0 \leq x_1 \leq \cdots \leq x_k = 1, \ \text{subdivision of } [0,1] \right\} < +\infty.$$

This number is then called the *length* of the curve c, denoted by length(c) (see [9], Definition 1.18, page 12). A metric space (X, d) is called a *length space* if $\forall x, y \in X$ there is a rectifiable curve between x and y, and:

$$d(x, y) = \inf \left\{ \text{length}(c); \ c(0) = x, \ c(1) = y \right\}.$$

A *geodesic segment* is a map

$$c : I \longrightarrow X,$$

for $I \subset \mathbf{R}$ an interval, which is an isometric embedding, *i.e.*

$$\forall s, t \in I, \quad d(c(s), c(t)) = |t - s|.$$

A length space (X, d) is said to be a *geodesic space* if any two points in X can be joined by at least one geodesic segment.

Remarks 6.2.1

(i) Spaces with few or no rectifiable curves are of some interest. They, for example, appear as boundary at infinity of negatively curved manifolds when the curvature is not constant.

(ii) An easy but important example is a metric graph. We could, for example, decide that the edges are isometric to $[0, 1]$.

On Fig. 6.2 there are two geodesics between v_0 and v_6.

Fig. 6.2 metric graph

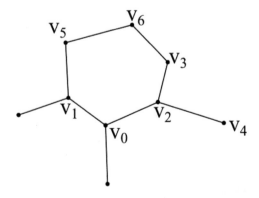

6.2.3   *Gromov-Hyperbolic Spaces*

The main references that we use and suggest are [11] and [9]. The main idea in this subsection is not to use curvature to prove comparison theorems on triangles, à *la* Toponogov, but to use triangle comparison to define a weak notion of sectional curvature bounded above.

We consider a proper geodesic space X and a triangle $\Delta \subset X$. A triangle is the union of three points, a, b and c, and a choice of a (minimising) geodesic between each couple of them. We call α (resp. β, γ) the length of the side opposite to a (resp. b, c) (Fig. 6.3).

Because α, β and γ satisfy the triangle inequalities there exists a Euclidean triangle $\bar{\Delta} \subset \mathbf{R}^2$ with the same side lengths. We call it the *comparison triangle* of Δ.

Exercise 6.2.2 *Prove the existence of $\bar{\Delta}$.*

Now $\bar{\Delta} \subset \mathbf{R}^2$ has an inscribed circle and we get a map $\bar{f} : \bar{\Delta} \longrightarrow T^*$, onto a tripod which is an isometry when restricted to each side of $\bar{\Delta}$, see Fig. 6.4 below.

Exercises 6.2.3

1. Prove the equalities on the segments of $\bar{\Delta}$ shown in the Fig. 6.4.
2. Prove that, for any triangle, the associated tripod is unique.

There is also an obvious isometry between the sides of Δ and $\bar{\Delta}$ of same length. Together with $\bar{f}$ this combines into a map

$$f_\Delta : \Delta \longrightarrow T^*,$$

which is an isometry between the segment $[a, b] \subset \Delta$ (resp. $[b, c]$, $[c, a]$) and the segment $[a^*, b^*] \subset T^*$ (resp. $[b^*, c^*]$, $[c^*, a^*]$). Now let $\delta \geq 0$ be a real number, we get the

Fig. 6.3 $\Delta \subset X$ and $\bar{\Delta} \subset \mathbf{R}^2$

Fig. 6.4 $\bar{\Delta} \subset \mathbf{R}^2$ and tripod T^*

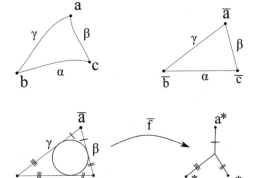

Fig. 6.5 Tree, $\delta = 0$

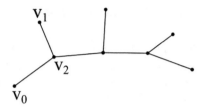

Definition 6.2.4 The space (X, d) is said to be δ-*hyperbolic* (in the sense of Gromov) if for all triangles Δ the following property holds. For all $x, y \in \Delta$, with $f_\Delta(x) = f_\Delta(y)$ we have,

$$d(x, y) \le \delta.$$

Such triangles are said to be δ-*thin*.

Notice that there could exist several triangles based on three points a, b and c, and the above definition requires that all of them are δ-thin (Fig. 6.5).

Examples 6.2.5

1. Obviously, R^2 endowed with the Euclidean distance is never δ-hyperbolic.
2. On the other hand a tree is 0-hyperbolic; indeed, all triangles are tripods.

Exercise 6.2.6 *Prove that the hyperbolic plane is hyperbolic in the sense of Gromov and compute δ.*

Note that δ-hyperbolicity characterizes large triangles since, by compactness, for small ones there always exist a δ satisfying the definition (we recall that the space X is proper). The interpretation of Definition 6.2.4 is that large triangles look like δ-thickened tripods. Looking at the hyperbolic plane with a metric of constant curvature equal to $\kappa < 0$ we can check that, somehow, δ^{-2} behaves like $|\kappa|$. In particular an upper bound on δ can be viewed as a weak and synthetic version of an upper bound on κ; indeed, $\delta \le C \simeq \kappa \le -C^{-2}$.

6.2.4 Measures

We endow the metric space (X, d) with a Borel measure that we assume to be non-negative and non-trivial (*i.e* not identically zero). At this stage let us give the three main examples.

(a) For a discrete set, possibly infinite, say $\{p_1, p_2, \cdots\}$ we shall consider the counting measure that is,

$$\mu = \sum_1^{+\infty} \delta_{p_k},$$

where δ_p is the Dirac mass at the point p. In the sequel the discrete set will always be the discrete orbit of a group action.

(b) For a graph we shall consider the 1-dimensional Lebesgue measure supported on the edges which make them of length 1.

(c) Finally, if (X, d) is a Riemannian manifold (M, d_g) (where d_g is the distance induced by the Riemannian metric) the Riemannian measure, noted dv_g, is the most natural one.

6.2.5 Groups Acting by Isometries

Let $\Gamma \subset \mathrm{Isom}(X, d)$ be a subgroup of the isometry group of the metric space (X, d). We say that the action is proper if for some (hence for all) $x \in X$ we have,

$$\forall R > 0 \quad \#\{\gamma;\, d(x, \gamma x) \leq R\} < +\infty.$$

This is equivalent to Γ being discrete in $\mathrm{Isom}(X, d)$ for the compact-open topology. We shall also consider groups Γ which are finitely generated. We pick a finite symmetric generating set,

$$\Sigma = \{\gamma_1, \cdots, \gamma_n, \gamma_1^{-1}, \cdots, \gamma_n^{-1}\}.$$

To this data we associate a graph, the *Cayley graph* of the group, denoted by $\mathcal{G}(\Gamma, \Sigma)$. We turn this graph into a metric space, as before, by deciding that the edges are isometric to $[0, 1]$. The distance then defined on $\mathcal{G}$ is noted $d_{\mathcal{G}}$. Now, if the metric space $(\mathcal{G}(\Gamma, \Sigma), d_{\mathcal{G}})$ is δ-hyperbolic for some δ we say that the group is hyperbolic in the sense of Gromov. Notice that δ then depends on Σ.

We consider the counting measure on the orbits of the action of Γ on the space X, namely

$$\mu_x^{\Gamma} = \sum_{\gamma \in \Gamma} \delta_{\gamma x} \quad \text{that is,} \quad \mu_x^{\Gamma}(B_X(x, R)) = \#\{\gamma;\, d(x, \gamma x) < R\}.$$

Here $B_X(x, R) \subset X$ is a ball of radius R around the point $x \in X$. Finally we shall consider co-compact actions, that is, actions for which the topological space X/Γ is compact. When (X, d) is a metric space on which a discrete group Γ acts by isometries we shall call $\bar{d}$ the distance on X/Γ.

6.2.6 Entropy of (X, d, μ)

For a metric space with a measure as described in Sect. 6.2.4 we define the following well-known quantity,

$$\mathrm{Ent}(X, d, \mu) = \lim_{R \to +\infty} \frac{1}{R} \log(\mu(B(x, R))).$$

The name "Entropy" comes from the dynamical nature of this invariant in the case of Riemannian manifolds.

Exercise 6.2.7

1. *Prove that it does not depend on x,*
2. *and that the* $\underline{\lim}$ *is indeed a limit when the action is co-compact.*

If $\mu = \mu_y^\Gamma$ for some $y \in X$, by abuse of language, we shall denote its entropy by $\mathrm{Ent}(X, d)$.

6.2.7 Bishop–Gromov's Inequality

This is a major comparison theorem that has been our guideline in [4, 5]. Let us recall the Riemannian case. We consider a complete Riemannian manifold (M, g) whose Ricci curvature satisfies,

$$\mathrm{Ric} \geq (n - 1)\kappa g, \quad \text{for } \kappa \in \mathbf{R}.$$

We can now compare the volumes of the balls around a point $m \in M$ with the volumes of balls of the same radius in a simply-connected Riemannian manifold of constant sectional curvature equal to κ denoted by M_κ. More precisely, let $R > 0$ be a real number, the classical Bishop's inequality is:

$$\mathrm{Vol}(B_M(m, R)) \leq \mathrm{Vol}(B_{M_\kappa}(R)).$$

Let us define the function

$$\varphi(R) = \frac{\mathrm{Vol}(B_M(m, R))}{\mathrm{Vol}(B_{M_\kappa}(R))},$$

Bishop's Inequality can then be rewritten: for any $R > 0$, $\varphi(R) \leq 1$. The improvement by Gromov is then that the function φ is non increasing, that is, if $0 < r \leq R$,

$$\varphi(R) = \frac{\mathrm{Vol}(B_M(m, R))}{\mathrm{Vol}(B_{M_\kappa}(R))} \leq \frac{\mathrm{Vol}(B_M(m, r))}{\mathrm{Vol}(B_{M_\kappa}(r))} = \varphi(r).$$

Notice that volumes of balls in M_κ are explicitly computable.

An interesting and obvious consequence is the following. If (M, g) is a Riemannian manifold as above, say with $\kappa \leq 0$, Bishop's Inequality yields,

$$\mathrm{Ent}(M, d_g, dv_g) \leq (n - 1)|\kappa| = \mathrm{Ent}(M_\kappa).$$

Here, by abuse of language, $\mathrm{Ent}(M_\kappa)$ is computed with the distance and the volume element coming from the constant curvature metric. When $\kappa > 0$ both M and M_κ are closed manifolds, their entropies are equal to zero and the above inequality is trivial. On the contrary it is particularly interesting when (M, g) is the universal cover of a closed Riemannian manifold and g is the pulled-back metric on M and $\kappa \leq 0$.

The conclusion that is our guideline is that any upper bound on the entropy could play the role of a (very) weak and asymptotic version of a lower bound on the Ricci curvature. Although the Ricci curvature may not be defined on a metric space, the entropy exists for a metric space with mild assumptions (the one we described). We then wish to study the consequences of this weak and synthetic "Ricci curvature bounded below" condition.

6.2.8 Main Theorem

We now state the main theorem of the first part of this course.

Theorem 6.2.8 *Let (X, d) be a δ-hyperbolic metric space, $\Gamma \subset \mathrm{Isom}(X, d)$ a discrete group (acting properly) such that*

$$\mathrm{diam}(X/\Gamma, \bar{d}) \leq D \quad and \quad \mathrm{Ent}(X, d) \leq H.$$

Then, for every $x \in X$ and $10(D + 2\delta) \leq r \leq R < +\infty$,

$$\frac{\#\{\gamma;\ d(x, \gamma x) < R\}}{\#\{\gamma;\ d(x, \gamma x) < r\}} = \frac{\mu_x^\Gamma(B_X(x, R))}{\mu_x^\Gamma(B_X(x, r))} \leq 3\left(\frac{R}{r}\right)^{25/4} e^{6HR}.$$

Remarks 6.2.9

(i) This inequality does not give any information about small balls, which is natural since neither the δ-hyperbolicity nor the assumption on the entropy "see" the geometry at small scale.

(ii) Furthermore it is by no means optimal.

(iii) However it is a strong Bishop–Gromov inequality obtained without any pointwise assumption on the curvature, which may not exist in this context. Only large scale assumptions are required.

6.3 Lecture 2

In this lecture we intend to sketch the proof of Theorem 6.2.8 which will be restated below.

Fig. 6.6 Quadrangle

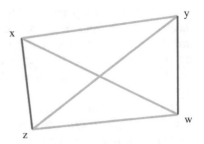

6.3.1 Quadrangle Inequality for δ-Hyperbolic Metric Spaces

We need another characterisation of δ-hyperbolicity. Let us consider a proper geodesic metric space (X, d) which is furthermore δ-hyperbolic and let x, y, z and w be four points in X. If we choose a geodesic between any two of them we get a quadrangle with its diagonals as shown on Fig. 6.6 below.

The quadrangle inequality is the following,

$$d(x, z) + d(y, w) \leq \text{Max} \left\{ d(x, y) + d(z, w), d(x, w) + d(y, z) \right\} + 2\delta .$$

Exercise 6.3.1 *Prove that this inequality is equivalent to hyperbolicity in the sense of Gromov (see [9] p. 410).*

6.3.2 A Simple Version of Theorem 6.2.8

Let us now state a weaker but simpler version of Theorem 6.2.8 which we prove below. The Main theorem follows from it by iteration, see the details in [5].

Theorem 6.3.2 *Let (X, d) be a δ-hyperbolic metric space and $\Gamma \subset \text{Isom}(X, d)$ a discrete group acting properly such that*

$$\text{diam}(X / \Gamma, \bar{d}) \leq D \quad \text{and} \quad \text{Ent}(X, d) \leq H.$$

Then, for every $x \in X$ and $R \geq 12(D + 2\delta)$,

$$\frac{\mu_x^\Gamma (B(x, R))}{\mu_x^\Gamma (B(x, \frac{5}{6}R))} \leq 1 + 2e^{HR} .$$

We start with several (not too) technical lemmas.

6.3.2.1 Easy Lemma

Let μ be a Borel measure on X, finite on compact sets, and $\Omega \subset X$ a measurable set. We then have,

$$\int_\Omega \mu\big(B(y,r)\big)d\mu(y) = \int_X \mu\big(B(z,r) \cap \Omega\big)d\mu(z).$$

Exercise 6.3.3 *Prove this inequality. Hint: this is an easy consequence of Fubini's Theorem and of the symmetry of the distance d.*

6.3.2.2 More Technical Lemmas

The main lemma of this proof, which translates the δ-hyperbolicity, is the following.

Lemma 6.3.4 *Let (X,d) be a δ-hyperbolic metric space. $\forall R, R' \in (0,+\infty)$, $\forall x, x' \in X$ such that $d(x,x') < R + R'$, $\exists y \in X$ such that,*

$$B(x,R) \cap B(x',R') \subset B(y,r),$$

with $r = \mathrm{Min}\left\{R, R', \frac{1}{2}(R + R' - d(x,x')) + 2\delta\right\}$.

On Fig. 6.7, we have $R = R'$ and $d(x,y) = d(y,x') = \frac{1}{2}d(x,x')$. Hence $d(x,x') < 2R$ and $d(y, \partial B(x,R)) = R - \frac{d(x,x')}{2} = \frac{1}{2}(2R - d(x,x'))$. The proof of Lemma 6.3.4 relies on the quadrangle inequality (see [5], Lemma 8.4).

Exercises 6.3.5

1. *Show that this is not true in $\mathbf{R}^2$ ($\delta = +\infty$).*
2. *Show that this is trivially true on a tree ($\delta = 0$).*

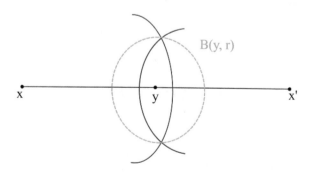

Fig. 6.7 Intersection of balls

We now assume that (X, d) satisfies the hypotheses of Theorem 6.3.2. We also assume that μ is Γ-invariant. We then have,

Lemma 6.3.6 *For every R and R' such that $8\delta \leq R' \leq R$,*

$$\int_{B(x,R)\setminus B(x,R-\frac{1}{2}R')} \mu\big(B(x', R')\big)d\mu(x') \leq \mu\Big(B(x, R + R') \setminus B(x, R - \frac{1}{2}R')\Big)$$

$$\times \mu\Big(B(x, \frac{3}{4}R' + 2\delta + D)\Big).$$

Proof of Lemma 6.3.6 We shall only need, for the main theorem, to consider the measure $\mu = \mu_x^\Gamma$; however the proof of this lemma is valid when μ satisfies the above hypotheses.

We have,

$$I := \int_{B(x,R)\setminus B(x,R-\frac{1}{2}R')} \mu\big(B(x', R')\big)d\mu(x')$$

$$= \int_{\big(B(x,R+R')\setminus B(x,R-\frac{1}{2}R')\big)\cap B(x,R)} \mu\big(B(x', R')\big)d\mu(x').$$

By lemma 6.3.2.1 we get,

$$I = \int_{B(x,R+R')\setminus B(x,R-\frac{1}{2}R')} \mu\big(B(x', R') \cap B(x, R)\big)d\mu(x').$$

Let $x' \in B(x, R + R') \setminus B(x, R - \frac{1}{2}R')$ (as in Fig. 6.8), we have

$$R - R' + 4\delta \leq R - R' + \frac{1}{2}R' = R - \frac{1}{2}R' \leq d(x, x') < R + R'.$$

We can then apply Lemma 6.3.4 to get that there exists y such that,

$$B(x', R') \cap B(x, R) \subset B(y, r),$$

with

$$r = \text{Min}\Big\{R, R', \frac{1}{2}(R+R'-d(x, x'))+2\delta\Big\} = \frac{1}{2}(R+R'-d(x, x'))+2\delta \leq \frac{3}{4}R'+2\delta.$$

Fig. 6.8 Main argument

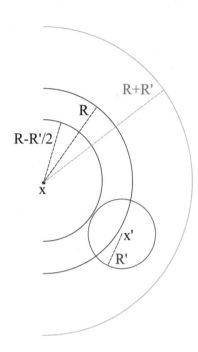

It only remains to include the ball $B(y, r)$ into a ball centred at x. Since the group acts co-compactly with a quotient whose diameter is bounded above by D, by assumption, there exists $\gamma \in \Gamma$ such that $d(y, \gamma x) \leq D$. Hence,

$$\mu\big(B(x', R') \cap B(x, R)\big) \leq \mu\Big(B(y, \frac{3}{4}R' + 2\delta)\Big) \leq \mu\Big(B(\gamma x, \frac{3}{4}R' + D + 2\delta)\Big)$$

$$= \mu\Big(B(x, \frac{3}{4}R' + D + 2\delta)\Big),$$

the last equality coming from the invariance of the measure. We finally get the desired result,

$$I \leq \mu\Big(B(x, \frac{3}{4}R' + D + 2\delta)\Big) \int_{\big(B(x, R+R') \setminus B(x, R-\frac{1}{2}R')\big)} d\mu(x').$$

$\square$

Starting now we specify the measure μ to be μ_x^{Γ}. We however keep the notation μ for the sake of simplicity.

Lemma 6.3.7 *Under the same hypotheses, $\forall x \in X$, $\forall R, R'$ such that $12(D + 2\delta) \leq R' \leq R$, we have*

$$\frac{\mu\big(B(x, R')\big)}{\mu\big(B(x, 5R'/6)\big)} \leq \frac{\mu\big(B(x, R + R')\big) - \mu\big(B(x, R - R'/2)\big)}{\mu\big(B(x, R)\big) - \mu\big(B(x, R - R'/2)\big)}.$$

Exercise 6.3.8 *Show that these ratios are well-defined, i.e. that the denominators are not zero. Hint: the measure is μ_x^Γ hence it counts the number of points of the orbit Γx in the balls, in particular it takes integer values. Now the fact that the denominator on the left hand side is positive is obvious (explain why); for the right hand side it requires a short proof.*

Proof of Lemma 6.3.7 We recall that $\mu = \mu_x^\Gamma$, hence, if $y \in \Gamma x$,

$$\mu\big(B(y, R')\big) = \mu\big(B(x, R')\big),$$

by invariance of μ under the action of Γ. We then get

$$\mu\big(B(x, R')\big)\Big(\mu\big(B(x, R)\big) - \mu\big(B(x, R - R'/2)\big)\Big)$$
$$= \mu\big(B(x, R')\big)\mu\big(B(x, R) \setminus B(x, R - R'/2)\big)$$
$$= \int_{B(x,R)\setminus B(x,R-R'/2)} \mu\big((B(x, R')\big)d\mu(y)$$
$$= \int_{B(x,R)\setminus B(x,R-R'/2)} \mu\big((B(y, R')\big)d\mu(y).$$

The last equality comes from the fact that the only y's weighted by the measure are the points in Γx. For simplicity we shall denote by N the quantity

$$N = \mu\big(B(x, R')\big)\Big(\mu\big(B(x, R)\big) - \mu\big(B(x, R - R'/2)\big)\Big).$$

Thanks to Lemma 6.3.6 we then obtain,

$$N \le \mu\big(B(x, R + R') \setminus B(x, R - R'/2)\big)\mu\big(B(x, 3R'/4 + 2\delta + D)\big)$$
$$\le \mu\big(B(x, R + R') \setminus B(x, R - R'/2)\big)\mu\big(B(x, 3R'/4 + R'/12)\big)$$
$$= \mu\big(B(x, R + R') \setminus B(x, R - R'/2)\big)\mu\big(B(x, 5R'/6)\big).$$

Which proves Lemma 6.3.7. □

6.3.2.3 Proof of Theorem 6.3.2

We now proceed to the proof of Theorem 6.3.2.

Proof of Theorem 6.3.2 Once more we recall that $\mu = \mu_x^\Gamma$ for some $x \in X$. We choose $R' \ge 12(D + 2\delta)$ and define $R = kR'/2$ for k a positive integer. From Lemma 6.3.7 we obtain

$$\frac{\mu\big(B(x, R')\big)}{\mu\big(B(x, 5R'/6)\big)} \le \frac{\mu\big(B(x, \frac{k+2}{2}R')\big) - \mu\big(B(x, \frac{k-1}{2}R')\big)}{\mu\big(B(x, \frac{k}{2}R')\big) - \mu\big(B(x, \frac{k-1}{2}R')\big)}. \tag{6.1}$$

We now set

$$C = \frac{\mu\big(B(x, R')\big)}{\mu\big(B(x, 5R'/6)\big)} - 1 \quad \text{and} \quad a_k = \mu\Big(B\big(x, \frac{k}{2}R'\big)\Big),$$

Equation (6.1) then yields,

$$a_{k+2} - a_{k-1} \geq (C+1)(a_k - a_{k-1}) \implies a_{k+2} - a_k \geq C(a_k - a_{k-1}).$$

We add these inequalities for all integers k between 2 and n to obtain,

$$a_{n+2} + a_{n+1} - a_3 - a_2 \geq C(a_n - a_1),$$

and because the sequence a_k is non decreasing in k

$$2(a_{n+2} - a_1) \geq C(a_n - a_1) \implies a_{n+2} \geq \frac{C}{2}(a_n - a_1) + a_1.$$

Iterating this inequality leads to

$$a_{2n} \geq \left(\frac{C}{2}\right)^{n-1}(a_2 - a_1) + a_1.$$

Exercise 6.3.9 *Show that C and $a_2 - a_1$ are positive (see Exercise 6.3.8).*

Finally, we can now use the second assumption on the metric space (X, d), that is, $\text{Ent}(X, d) \leq H$. More precisely,

$$H \geq \text{Ent}(X, d) = \lim_{n \to +\infty} \left(\frac{1}{nR'} \log \mu\big(B(x, nR')\big)\right)$$

$$= \lim_{n \to +\infty} \frac{1}{nR'} \log(a_{2n}) \geq \frac{1}{R'} \log\left(\frac{C}{2}\right),$$

that is,

$$C \leq 2e^{HR'} \quad \text{and} \quad 1 + C = \frac{\mu\big(B(x, R')\big)}{\mu\big(B(x, 5R'/6)\big)} \leq 1 + 2e^{HR'}.$$

This ends the proof of Theorem 6.3.2. $\qquad\qquad\qquad\qquad\qquad\qquad$ □

6.4 Lecture 3

The theme of this lecture is to illustrate the relations between the geometry and the topology of a space. The idea is that given a topological space Y, the possible metrics on Y which are compatible with the topology of Y have to satisfy constraints and conversely, given a metric on Y, the geometrical properties of this metric provides restrictions on the topology. These questions have been extensively studied in the setting of Riemannian manifolds. Our purpose in this lecture is to consider one example where the metrics under consideration are not necessary Riemannian metrics. One important topological invariant is the fundamental group of Y, which is finitely generated when Y is compact. The abelianisation of the fundamental group of Y is an abelian group called the first homology group with integer coefficients. It is also a topological invariant of Y, simpler than the fundamental group.

6.4.1 Fundamental Group

For "many reasonable" spaces Y, there exist a connected and simply-connected space $\tilde{Y}$ and a group Γ acting on $\tilde{Y}$ freely and properly by homeomorphisms such that $Y = \tilde{Y}/\Gamma$. The group Γ is called the fundamental group of Y and is denoted by $\Gamma = \pi_1 Y$.

- The circle $Y = \mathbf{R}/\mathbf{Z} = S^1$ has fundamental group $\pi_1 Y = \mathbf{Z}$.
- The bouquet of two circles $Y = S^1 \vee S^1 = T/\mathbf{F}_2$, where T is the regular tree with valency 4, has the non abelian free group on 2 generators as fundamental group (Figs. 6.9 and 6.10).

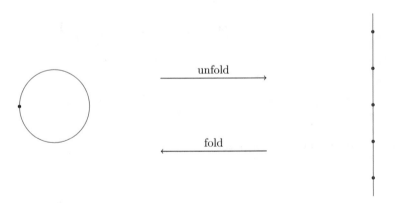

Fig. 6.9 $S^1 = \mathbf{R}/\mathbf{Z}$

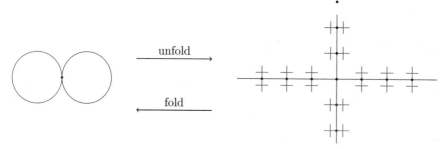

Fig. 6.10 $S^1 \vee S^1 = T/\mathbf{F}_2$

6.4.2 First Homology Group

The fundamental group of a space Y may not be abelian. For example, when Y is a bouquet of two circles, $\pi_1 Y$ is the non abelian free group on 2 generators.

Definition 6.4.1 Let Y be a space with fundamental group $\Gamma = \pi_1 Y$. The first homology group of Y with integral coefficients is defined as the abelianised group $\Gamma/[\Gamma, \Gamma]$. It is denoted by $H_1(Y, \mathbf{Z})$.

Recall that the commutator subgroup $[\Gamma, \Gamma]$ of Γ is the subgroup of Γ generated by the set of "commutators" $\{aba^{-1}b^{-1} \mid a, b \in \Gamma\}$. The homology group $H_1(Y, \mathbf{Z}) = \Gamma/[\Gamma, \Gamma]$ is by definition abelian. When the space Y is compact, $\Gamma = \pi_1 Y$ is finitely generated and so is $H_1(Y, \mathbf{Z})$, therefore by the fundamental Theorem for finitely generated abelian groups,

$$H_1(Y, \mathbf{Z}) \cong \mathbf{Z}^r \times \mathbf{Z}/n_1\mathbf{Z} \times \cdots \times \mathbf{Z}/n_k\mathbf{Z},$$

where r and the n_i's are integers, $r \geq 0$ and $n_i \geq 2$. In this decomposition, the factor $\mathbf{Z}/n_1\mathbf{Z} \times \cdots \times \mathbf{Z}/n_k\mathbf{Z}$ [resp. $\mathbf{Z}^r$] is the torsion [resp. torsion free] part of the homology group $H_1(Y, \mathbf{Z})$. Notice that when $k = 0$, there is no torsion part of $H_1(Y, \mathbf{Z})$ and similarly, there is no torsion free part when $r = 0$.

Definition 6.4.2 Let Y be a compact space and $H_1(Y, \mathbf{Z}) \cong \mathbf{Z}^r \times \mathbf{Z}/n_1\mathbf{Z} \times \cdots \times \mathbf{Z}/n_k\mathbf{Z}$ the first homology group of Y. The first Betti number of Y is defined as $b_1(Y) := r$.

Examples 6.4.3

(1) $H_1(S^1, \mathbf{Z}) \cong \mathbf{Z}$ and $H_1(S^1 \times S^1, \mathbf{Z}) \cong \mathbf{Z}^2$
(2) $H_1(S^1 \vee S^1, \mathbf{Z}) \cong \mathbf{Z}^2$
(3) $H_1(\text{Nil}, \mathbf{Z}) \cong \mathbf{Z}^2$, where $\text{Nil} := Heis(\mathbf{R})/Heis(\mathbf{Z})$ with

$$\text{Heis}(\mathbf{A}) = \left\{ \begin{pmatrix} 1 & x & z \\ 0 & 1 & y \\ 0 & 0 & 1 \end{pmatrix} \middle/ x, y, z \in \mathbf{A} \right\}.$$

6.4.3 Bounding the First Betti Number

The following result is emblematic of the constraints imposed by geometry to topology.

Theorem 6.4.4 ([15, 16, 20]) *There exists a constant $C(\Lambda, n, D) > 0$ such that for every closed n-dimensional Riemannian manifold (M^n, g) with Ricci curvature satisfying $Ric_g \geq -\Lambda g$ and $\mathrm{diam}(M, g) \leq D$, then $b_1(M) \leq C(\Lambda, n, D)$.*

The following examples show that the assumptions on the curvature and the diameter are necessary assumptions in Theorem 6.4.4.

Examples 6.4.5

(1) Let Σ_g be a hyperbolic surface of genus g. The fundamental group Γ of Σ_g has the following presentation, $\Gamma = < a_1, b_1, \cdots, a_g, b_g \mid \Pi_{i=1}^{g}[a_i, b_i] = 1 >$, hence $H_1(\Sigma_g, \mathbf{Z}) = \mathbf{Z}^{2g}$ and the first Betti number of Σ_g is $b_1(\Sigma_g) = 2g$. Notice that the diameter of Σ_g satisfies $D := \mathrm{diam}\,\Sigma_g \geq \log(2g - 1)$. This follows from the fact that the area of a genus g hyperbolic surface satisfies $\mathrm{Vol}\,\Sigma_g = 4\pi(g - 1) \leq \mathrm{Vol}\,B(D) = 2\pi(\cosh D - 1)$, where $B(D)$ is the hyperbolic ball of radius D. In particular, the surfaces Σ_g have constant curvature $\kappa \equiv -1$, diameter and first Betti number tending to infinity with g.

(2) In the preceeding example, the curvature of Σ_g is bounded and the diameter and the Betti number tend to infinity with g. Contracting the hyperbolic metric on Σ_g so that the diameter is equal to 1, the Betti number does not change while the curvature tends to $-\infty$. This second example shows that a lower bound on the curvature is necessary in order to get a bound on the first Betti number (Figs. 6.11 and 6.12).

The next Theorem deals with Gromov hyperbolic metric spaces. We obtain a bound on the first Betti number with an upper bound on the entropy instead of a lower bound on the curvature, The entropy assumption is much weaker than the curvature one.

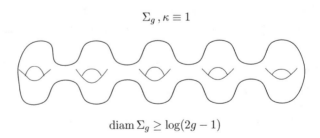

$$\Sigma_g, \kappa \equiv 1$$

$$\mathrm{diam}\,\Sigma_g \geq \log(2g - 1)$$

Fig. 6.11 Large diameter and Betti number

Fig. 6.12 Unbounded curvature and large Betti number

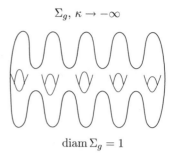

$$\Sigma_g, \ \kappa \to -\infty$$

$$\mathrm{diam}\, \Sigma_g = 1$$

Theorem 6.4.6 *Let* (Y, d) *be a compact, arcwise connected metric space. We suppose that the universal cover* $(\widetilde{Y}, \tilde{d})$ *is* δ-*hyperbolic and that*

$$\mathrm{Ent}(\widetilde{Y}, \tilde{d}) \le H \quad \text{and} \quad \mathrm{diam}(Y, d) \le D.$$

Then

$$b_1(Y) \le 48 \cdot e^{24H(8D+15\delta)}.$$

Proof The proof follows an argument of M. Gromov and relies on a packing argument. Let us denote by Γ the fundamental group of (Y, d). Let us recall that $H_1(Y, \mathbf{Z}) = \Gamma/[\Gamma, \Gamma]$. We first notice that the torsion free part $\mathbf{Z}^r$ of $H_1(Y, \mathbf{Z}) \cong \mathbf{Z}^r \times \mathbf{Z}/n_1\mathbf{Z} \times \cdots \times \mathbf{Z}/n_k\mathbf{Z}$ can be seen as a lattice in $\mathbf{R}^r$. From now on we will assume $r \ge 1$ since Theorem 6.4.6 is trivially true if $r = 0$. We define the two following projection maps, $p_1 : \Gamma \to \Gamma/[\Gamma, \Gamma]$ and $p_2 : \Gamma/[\Gamma, \Gamma] \cong \mathbf{Z}^r \times \mathbf{Z}/n_1\mathbf{Z} \times \cdots \times \mathbf{Z}/n_k\mathbf{Z} \to \mathbf{Z}^r$. We now consider $p := p_2 \circ p_1 : \Gamma \to \mathbf{Z}^r$ and $\phi := \iota \circ p : \Gamma \to \mathbf{R}^r$, where $\iota : \mathbf{Z}^r \hookrightarrow \mathbf{R}^r$ is the natural inclusion.

We now observe that the set $p(S)$ is a generating set of the group $\mathbf{Z}^r$ since p is a surjective morphism. This implies that $\phi(S)$ generates $\mathbf{R}^r$ as a vector space, therefore $r \le \#S$. Hence, we see that the first Betti number $b_1(Y) = r$ of Y can be estimated by the cardinality of any generating set of the fundamental group Γ of Y. Unfortunately, there are no upper bounds of the cardinality of generating sets of Γ; however the following lemma will allow us to replace Γ by a finite index subgroup Γ' with a generating set of bounded cardinality. $\quad\square$

Lemma 6.4.7 *Let* S' *be a finite subset of* Γ *generating a finite index subgroup* $\Gamma' \le \Gamma$. *Then, the set* $\phi(S')$ *generates* $\mathbf{R}^r$ *as a vector space. In particular,* $r \le \#S'$.

Proof Let $V \subset \mathbf{R}^r$ be the vector space generated by $\phi(S')$. We want to show that $V = \mathbf{R}^r$. Since the index of Γ' in Γ is finite, the subset $\{\gamma^k \Gamma'\}_{k \in \mathbf{Z}} \subset \Gamma/\Gamma'$ is finite for every $\gamma \in \Gamma$. Notice that the quotient Γ/Γ' may not be a group and that it is identified with the set of right equivalence classes $\{\gamma\Gamma'\}_{\gamma \in \Gamma}$. Therefore, there exist $k \ne l \in \mathbf{Z}$ such that $\gamma^k \Gamma' = \gamma^l \Gamma'$, hence $\gamma^{k-l} = \gamma' \in \Gamma'$ and $\phi(\gamma) =$

$(k-l)^{-1}\phi(\gamma')$. This implies that V contains the vector space generated by $\phi(S)$, which is $\mathbf{R}^r$.   □

The proof of Theorem 6.4.6 therefore boils down to finding a finite subset S' of Γ generating a finite index subgroup Γ' of Γ such that

$$\#S' \le 4^8 \cdot e^{24H(8D+15\delta)}.$$

In order to construct the set S', we proceed as follows. The idea is that the set S' has to be sufficiently dense to insure that the group Γ' generated by S' satisfies $\#\Gamma/\Gamma' < \infty$ but sufficiently sparse so that its cardinality stays bounded above. This will be a consequence of our Bishop–Gromov estimate Theorem 6.2.8.

We pick up a point $x \in \widetilde{Y}$. Let $R > 0$ and $S' := \{\gamma_i\}_{i \in I}$ be a subset of Γ of maximal cardinality such that

(i) $d(x, \gamma_i(x)) \le 2D + R$
(ii) $d(\gamma_i(x), \gamma_j(x)) \ge R$, for $i \ne j$.

Let Γ' be the subgroup of Γ generated by $S' = \{\gamma_i\}_{i \in I}$.

Claim 6.4.8 $\#\Gamma/\Gamma' < \infty$.

Proof The claim follows from the fact that for every $y \in \widetilde{Y}$, $d(y, \Gamma'x) < R + D$. Indeed, if this is the case, for every $\gamma \in \Gamma$, we have $d(x, \gamma\Gamma'x) = d(\gamma^{-1}x, \Gamma'x) < R + D$, hence for each coset $\gamma\Gamma'$, the orbit $\gamma\Gamma'x$ meets $B(x, R + D)$. Therefore we have

$$\#\Gamma/\Gamma' \le \#\{\gamma \in \Gamma \mid d(x, \gamma x) \le D + R\} < \infty$$

since Γ acts properly discontinuously on $\widetilde{Y}$. It remains to prove that for every $y \in \widetilde{Y}$, $d(y, \Gamma'x) < R + D$, see the Fig. 6.13 below.

Assume by contradiction that this is not the case, then there exists $y \in \widetilde{Y}$ such that $d(y, \Gamma'x) \ge D + R$. Moving y a bit if necessary, we may moreover assume that

$$d(y, \Gamma'x) = d(y, \gamma_0'x) = D + R.$$

Since the diameter of $Y = \widetilde{Y}/\Gamma$ satisfies diam $Y \le D$, we can choose $\alpha \in \Gamma \setminus \Gamma'$ such that $d(y, \alpha x) \le D$. Now, applying twice the triangle inequality gives:

$$R \le d((\gamma_0')^{-1}\alpha x, x) = d(\alpha x, \gamma_0'x) \le 2D + R. \tag{6.2}$$

Moreover, also by the triangle inequality, we have $d(\alpha x, \gamma_0'\gamma_i'x) \ge d(\gamma_0'\gamma_i'x, y) - d(y, \alpha x)$ and since $d(y, \gamma_0'\gamma_i'x) \ge D + R$ and $d(y, \alpha x) \le D$ by assumption, we get

$$d((\gamma_0')^{-1}\alpha x, \gamma_i'x) = d(\alpha x, \gamma_0'\gamma_i'x) \ge R. \tag{6.3}$$

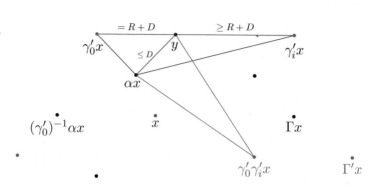

Fig. 6.13 $\#\Gamma/\Gamma' < \infty$

We deduce from (6.2) and (6.3) that the set $\{\gamma_i\}_{i\in I} \cup \{(\gamma'_0)^{-1}\alpha\}\}$ satisfies the above properties (i) and (ii) contradicting the maximality of the cardinality of $\{\gamma_i\}_{i\in I}$ since $\alpha \in \Gamma \setminus \Gamma'$. This ends the proof of the claim.   □

In order to finish the proof of the Theorem 6.4.6 we now bound $\#S'$. Let us consider the counting measure $\mu = \sum_{\gamma\in\Gamma} \delta_{\gamma x}$ of the Γ-orbit of x. Notice that by the properties (i) and (ii) defining S', for every $\gamma'_i \in S'$ we have $B(\gamma'_i x, \frac{R}{2}) \subset B(x, 2D + \frac{3R}{2})$ and the balls $B(\gamma'_i, \frac{R}{2})$ are disjoint. Therefore $\mu(B(x, 2D + \frac{3R}{2})) \geq \#S'\mu(B(x, \frac{R}{2}))$, thus

$$\#S' \leq \frac{\mu(B(x, 2D + \frac{3R}{2}))}{\mu(B(x, \frac{R}{2})}. \tag{6.4}$$

We choose $R = 20(D + 2\delta)$ in (6.4) and we apply Theorem 6.2.8 with (r, R) replaced by $(10(D + 2\delta), 32D + 60\delta)$. This gives

$$\#S' \leq 3\left(\frac{16D + 30\delta}{5D + 10\delta}\right)^{25/4} e^{24H(8D+15\delta)},$$

hence the estimate

$$b_1(Y) \leq \#S' \leq 4^8 \cdot e^{24H(8D+15\delta)}$$

follows. This concludes the proof of Theorem 6.4.6.

6.5 Lecture 4

We now want to extend our version of the Bishop–Gromov inequality to balls of small radii. Using the notation of Lecture 2 we may distinguish the three following cases.

(i) $R \geq r \geq 10(D + 2\delta)$: this is the case described in Sect. 6.3.
(ii) $0 < r \leq R < 10(D + 2\delta)$: this is the case that will be treated in the present section.
(iii) $0 < r < 10(D + 2\delta) \leq R$: this is an easy consequence of the two previous cases and its proof is left to the reader. However for the sake of completeness we state this version at the end of this section.

In the previous sections no local assumption was made on the metric space (X, d). We need now to require another property in order to have some control on the local behaviour.

6.5.1 Busemann Spaces

We call *normal parametrisation* of a geodesic $c : [0, 1] \longrightarrow X$ the parametrisation such that

$$d(c(0), c(t)) = td(c(0), c(1)), \quad \text{for all } t > 0.$$

Definition 6.5.1 A complete, proper, geodesic metric space (X, d) is said to be a *Busemann space* if its metric d is convex, *i.e.* for any two geodesic segments c_1 and c_2 with normal parametrisations,

$$c_1, c_2 : [0, 1] \longrightarrow X \quad \text{such that } c_1(0) = c_2(0),$$

the function $t \mapsto d(c_1(t), c_2(t))$ is convex (Figs. 6.14 and 6.15).

Remarks 6.5.2

1. On every Busemann space any two points are joined by a *unique* geodesic segment.

Fig. 6.14 Convexity in Busemann spaces

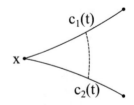

Fig. 6.15 $d(c_1(t), c_2(t))$ is
not convex

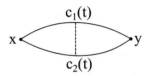

2. Examples of Busemann spaces are given by simply connected manifolds of non positive sectional curvature (Cartan–Hadamard manifolds).
3. Closed geodesic balls or, more generally, closed convex domains in Cartan–Hadamard manifolds are also Busemann spaces.

We will furthermore always assume that the space (X, d) satisfies the *property of extension of geodesics*, that is, for every geodesic segment

$$c : [a, b] \longrightarrow X, \quad a < b$$

there exists $\epsilon > 0$ and a geodesic segment $c' : [a, b + \epsilon] \longrightarrow X$ such that $c'_{|[a,b]} = c$. Notice that this excludes closed convex domains with non-empty boundary in Cartan–Hadamard manifolds. This assumption will be used in the proof of lemma 6.5.8.

We then state the main theorem of this section.

Theorem 6.5.3 *Let (X, d) be a δ-hyperbolic and Busemann metric space, $\Gamma \subset$ Isom(X, d) be a discrete group acting properly such that*

$$\mathrm{diam}(X/\Gamma, \bar{d}) \leq D \quad \text{and} \quad \mathrm{Ent}(X, d) \leq H.$$

Then, for every $x \in X$ and $0 < r \leq R < 12(D + 2\delta)$,

$$\frac{\mu_x^\Gamma(\overline{B}(x, R))}{\mu_x^\Gamma(B(x, r))} \leq C \left(\frac{R}{r}\right)^A \exp\left(18H(11D + 20\delta)\frac{R}{r}\right),$$

here C and A can be explicitly estimated. More precisely $C = \frac{11}{2} \cdot 3^{25/4}$ and $A = 25/4$.

Remark 6.5.4 Notice that on the left hand side, at the numerator, we compute the μ_x^Γ-measure of the closure of the ball $B(x, R)$. This is an improvement since there might be points of the orbit of x on the boundary $\overline{B}(x, R) \setminus B(x, R)$.

The proof of Theorem 6.5.3 relies on arguments of packing and covering big balls by smaller ones.

6.5.2 Packings and Coverings

With the above notation and assumptions, that is, (X, d) is a proper geodesic metric space and $\Gamma \subset \mathrm{Isom}(X)$ is a discrete subgroup, we have the following definitions.

Definition 6.5.5

(i) An *r-packing* of a ball $B(x, R)$ is any family $\left(B(y_j, r)\right)_{j \in J}$ of disjoint balls included in $B(x, R)$. We set

   $\mathrm{Pack}(x, r, R) = $ the maximal number of balls in a r-packing of $B(x, R)$.

(ii) A *$(\Gamma x, r)$-packing* of $B(x, R)$ is any family $\left(B(\gamma_j x, r)\right)_{j \in J}$ of disjoint balls included in $B(x, R)$. We set

   $\mathrm{Pack}_\Gamma(x, r, R) = $ the maximal number of balls in a $(\Gamma x, r)$-packing of $B(x, R)$.

(iii) A *$(\Gamma x, r)$-covering* of $\overline{B}(x, R)$ is any family $\left(B(\gamma_j x, r)\right)_{j \in J}$ of balls such that $\overline{B}(x, R) \cap \Gamma x \subset \cup_{j \in J} B(\gamma_j x, r)$. We set

   $\mathrm{Cover}_\Gamma(x, r, R) = $ the maximal number of balls in a $(\Gamma x, r)$-covring of $\overline{B}(x, R)$.

We then have the following lemma (Fig. 6.16).

Lemma 6.5.6 *Let us assume that Γ acts co-compactly on X and that the quotient space has diameter not greater than $D > 0$, then one has, for every $x \in X$, $0 < r \leq R$,*

(i) $\mathrm{Cover}_\Gamma(x, r, R) \leq \mathrm{Pack}_\Gamma(x, r/2, R + r/2)$.

(ii) $\mathrm{Pack}_\Gamma(x, r, R) \leq \dfrac{\mu_x^\Gamma(B(x, R))}{\mu_x^\Gamma(B(x, r))} \leq \dfrac{\mu_x^\Gamma(\overline{B}(x, R))}{\mu_x^\Gamma(B(x, r))} \leq \mathrm{Cover}_\Gamma(x, r, R)$.

(iii) $\mathrm{Pack}_\Gamma(x, r, R) \leq \mathrm{Pack}(x, r, R) \leq \mathrm{Pack}_\Gamma(x, r - D, R)$, *the last inequality making sense only when $r > D$.*

Exercise 6.5.7 *Prove this lemma.*

Fig. 6.16 $\mathrm{Pack}_\Gamma(x, r, R)$

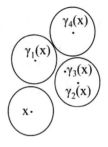

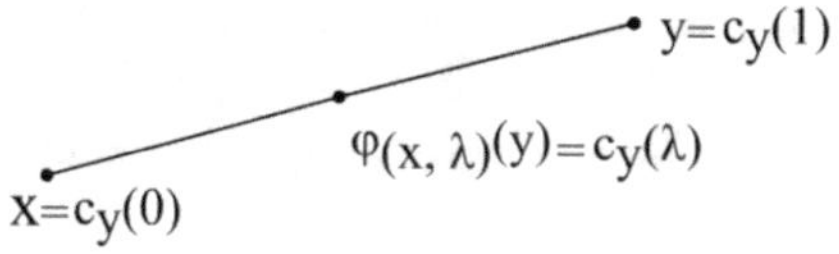

Fig. 6.17 λ-contraction

6.5.3   A Contraction in Busemann Spaces

We now assume that (X, d) is Busemann and has the property of extension of geodesics. The second tool that we shall use essentially depends on the convexity of the distance. We construct a contraction (or a dilation) as follows. Let $x \in X$, $\forall y \in X$ we call $c_y : [0, 1] \longrightarrow X$ the (unique) geodesic normally parametrised joining x to y, that is $c_y(0) = x$ and $c_y(1) = y$ (Fig. 6.17).

Then, $\forall \lambda \in [0, 1]$, we define

$$\varphi_{x,\lambda} : X \longrightarrow X \quad \text{by} \quad \varphi_{x,\lambda}(y) = c_y(\lambda).$$

Lemma 6.5.8 *One has,* $\forall y, y' \in X, \quad d(\varphi_{x,\lambda}(y), \varphi_{x,\lambda}(y')) \leq \lambda d(y, y')$. *As a consequence we get,* $\forall \alpha \geq 1$,

$$\mathrm{Pack}(x, r, R) \leq \mathrm{Pack}(x, \alpha r, \alpha R).$$

Proof The first inequality follows from the convexity of the distance, indeed

$$d(c_y(\lambda), c_{y'}(\lambda)) \leq (1 - \lambda)d(c_y(0), c_{y'}(0)) + \lambda d(c_y(1), c_{y'}(1)) = \lambda d(y, y').$$

The second inequality is left to the reader; notice that it is in the proof of this inequality on the packing numbers that the property of extendable geodesics is needed.   □

We are now ready to prove Theorem 6.5.3.

Proof of Theorem 6.5.3 For the sake of simplicity, let us set

$$I = \frac{\mu_x^\Gamma(\overline{B}(x, R))}{\mu_x^\Gamma(B(x, r))}.$$

Lemma 6.5.6 yields

$$I \leq \mathrm{Cover}_\Gamma(x, r, R) \leq \mathrm{Pack}_\Gamma(x, r/2, R + r/2) \leq \mathrm{Pack}(x, r/2, R + r/2),$$

and Lemmas 6.5.8 and 6.5.6 imply, for $\alpha \geq 1$ and $\alpha(r/2) > D$,

$$I \leq \mathrm{Pack}(x, \alpha r/2, \alpha(R + r/2)) \leq \mathrm{Pack}(x, \alpha(r/2) - D, \alpha(R + r/2)).$$

We now choose α so that $r' = \alpha(r/2) - D = 10(D + 2\delta)$. This clearly implies that $R' = \alpha(R + r/2) \geq 10(D + 2\delta)$ and by the assumption on r (in Theorem 6.5.3) that $\alpha \geq 1$. Finally, we can apply Theorem 6.2.8 and get,

$$I \leq \text{Pack}(x, r', R') \leq \frac{\mu_x^\Gamma(B_X(x, R'))}{\mu_x^\Gamma(B_X(x, r'))} \leq 3\left(\frac{R'}{r'}\right)^{25/4} e^{6HR'}.$$

It then suffices to replace R' and r' by their respective values in terms of R, r, D and δ.   □

Next we prove an inequality relating the Bishop–Gromov ratio associated to a group Γ to the Bishop–Gromov ratio associated to a subgroup $\Gamma' \subset \Gamma$.

Corollary 6.5.9 *Let* (X, d) *be a proper geodesic metric space and* $\Gamma \subset \text{Isom}(X)$ *a discrete subgroup. If* $\Gamma' \subset \Gamma$ *is a subgroup of* Γ *then,* $\forall x \in X$, $\forall r, R$ *such that* $0 < r \leq R$, *one has*

$$\frac{\mu_x^{\Gamma'}(\overline{B}_X(x, R))}{\mu_x^{\Gamma'}(B_X(x, r))} \leq \frac{\mu_x^\Gamma(B_X(x, R + r/2))}{\mu_x^\Gamma(B_X(x, r/2))}.$$

Proof From Lemma 6.5.6 we get

$$\frac{\mu_x^{\Gamma'}(\overline{B}_X(x, R))}{\mu_x^{\Gamma'}(B_X(x, r))} \leq \text{Cover}_{\Gamma'}(x, r, R) \leq \text{Pack}_{\Gamma'}(x, r/2, R + r/2)$$

$$\leq \text{Pack}_\Gamma(x, r/2, R + r/2) \leq \frac{\mu_x^\Gamma(B_X(x, R + r/2))}{\mu_x^\Gamma(B_X(x, r/2))}.$$

The inequality between the two packing numbers follows from the obvious remark that $\Gamma'x \subset \Gamma x$.   □

As announced at the beginning of this section, we end it by stating the intermediate Bishop–Gromov's inequality for μ_x^Γ that is when r is small and R is large. More precisely,

Theorem 6.5.10 *Let* (X, d) *be a* δ*-hyperbolic and Busemann metric space,* $\Gamma \subset \text{Isom}(X, d)$ *be a discrete group acting properly such that*

$$\text{diam}(X/\Gamma, \bar{d}) \leq D \quad \text{and} \quad \text{Ent}(X, d) \leq H.$$

Then, for every $x \in X$ *and* $0 < r < 12(D + 2\delta) \leq R$,

$$\frac{\mu_x^\Gamma(\overline{B}_X(x, R))}{\mu_x^\Gamma(B_X(x, r))} \leq C'\left(\frac{R}{r}\right)^{A'} \exp\left(18H\frac{(11D + 10\delta)^2}{r} + 6HR\right),$$

here C' *and* A' *can be explicitly computed. More precisely,* $C' = \frac{11}{2} \cdot 3^{29/4}$ *and* $A' = 25/4$.

6.6 Lecture 5

In this lecture, we derive from our Bishop–Gromov Theorem a version of the Margulis lemma in the context of cocompact discrete groups of isometries of δ-hyperbolic spaces with bounded entropy. From this "metric Margulis lemma", we deduce estimates on the systole of such actions. Such estimates are classical for manifolds of bounded non positive curvature.

6.6.1 Growth of Groups

Let Γ be a group and S a finite set of generators of Γ. We denote by $(\mathcal{G}(\Gamma, S), d_S)$ the Cayley graph of Γ with the word distance d_S associated to S and for $k \in \mathbf{N}$, $B_S(k)$ the ball of radius k centered at e in the Cayley graph. We also define the counting function $v_{(\Gamma, S)}$ as

$$v_{(\Gamma, S)}(k) := \#B_S(k).$$

A finitely generated group Γ has polynomial growth if for one (hence any) generating set S, there exists a polynomial P_S such that

$$v_{(\Gamma, S)}(k) \le P_S(k).$$

A finitely generated group Γ has exponential growth if for one (hence any) generating set S, there exists $C_S > 0$ such that

$$v_{(\Gamma, S)}(k) \ge e^{C_S k}.$$

Examples 6.6.1

(1) For $\Gamma = \mathbf{Z}^n$ and for every generating set S, we have

$$v_{(\Gamma, S)}(k) \approx k^n.$$

(2) For the non abelian free group $\Gamma = \mathbf{F}_n$ on n generators, $n \ge 2$ and for every generating set S, we have

$$v_{(\Gamma, S)}(k) \approx (2n - 1)^k.$$

(3) Let $\Gamma := \mathrm{Heis}(\mathbf{Z})$ where

$$\mathrm{Heis}(\mathbf{Z}) = \left\{ \begin{pmatrix} 1 & x & z \\ 0 & 1 & y \\ 0 & 0 & 1 \end{pmatrix} \middle/ \; x, y, z \in \mathbf{Z} \right\}.$$

Then, for every generating set S, we have

$$v_{(\Gamma,S)}(k) \approx k^4.$$

(4) More generally, let Γ be a finitely generated nilpotent group, and $\{1\} \trianglelefteq \Gamma_n \trianglelefteq \cdots \trianglelefteq \Gamma_1 = \Gamma$ its lower central series, where $[\Gamma_k, \Gamma] = \Gamma_{k+1}$. Then, for every generating set of Γ, the growth of Γ is given by the formula due to Bass–Guivarc'h,

$$v_{(\Gamma,S)}(k) \approx k^{d(\Gamma)}$$

where $d(\Gamma) = \sum_k k \, \mathrm{rank}(\Gamma_k/\Gamma_{k+1})$. In particular, every finitely generated nilpotent group has polynomial growth, (see [2, 21]).

We now have the following easy lemma, saying that a group (Γ, S) with polynomial growth is "C-doubling" at scale k_n for a sequence k_n tending to infinity, *i.e.* $\frac{v_{(\Gamma,S)}(2k_n)}{v_{(\Gamma,S)}(k_n)} \leq C$, see the regularity Lemma p. 59 in [18] for a more precise statement.

Lemma 6.6.2 *Let us consider a finitely generated group Γ with polynomial growth. Let μ be the counting measure on the Cayley graph $\mathcal{G}(\Gamma, S)$. We claim that there exist $C > 1$ and a sequence of integers k_n tending to infinity such that*

$$\frac{\mu(B_S(2k_n))}{\mu(B_S(k_n))} = \frac{v_{(\Gamma,S)}(2k_n)}{v_{(\Gamma,S)}(k_n)} \leq C.$$

Proof Let us prove this assertion. We assume by contradiction that for every $C > 1$, there exists $N \in \mathbf{N}$ such that, for every $n \geq N$, we have

$$\frac{v_{(\Gamma,S)}(2n)}{v_{(\Gamma,S)}(n)} \geq C.$$

In particular, for every integer k, we have $v_{(\Gamma,S)}(2^k n) \geq v_{(\Gamma,S)}(n) C^k$. Therefore, for every $x \geq n$, we get

$$v_{(\Gamma,S)}(x) \geq \frac{v_{(\Gamma,S)}(n)}{n^{\frac{\log C}{\log 2}}} x^{\frac{\log C}{\log 2}},$$

which contradicts the polynomial growth of Γ since C is arbitrary.   □

Examples 6.6.3

(1) If $\Gamma = \mathbf{Z}^n$ and S its canonical generating set, then,

$$\frac{v_{(\Gamma,S)}(2k)}{v_{(\Gamma,S)}(k)} = 2^n.$$

(2) For Γ nilpotent, it follows directly from Bass–Guivarc'h estimate of the growth that

$$\frac{v_{(\Gamma,S)}(2k)}{v_{(\Gamma,S)}(k)} \approx 2^{d(\Gamma)}.$$

(3) If $\Gamma = \mathbf{F}_n$ is the free group on n generators and S the symmetric set of generators, then

$$\frac{v_{(\Gamma,S)}(2k)}{v_{(\Gamma,S)}(k)} \approx (2n-1)^k.$$

For a finitely generated group Γ, having polynomial growth is a strong constraint. The following theorem, due to Gromov, characterises the groups with polynomial growth.

Theorem 6.6.4 ([18]) *Let Γ be a finite generated group with polynomial growth. Then, Γ is virtually nilpotent.*

According to the lemma 6.6.2, every group with polynomial growth is C-doubling at scale k_n, for some $C > 1$ and some sequence k_n tending to infinity. This property of groups with polynomial growth is a step in the proof of Theorem 6.6.4. With the following observation in mind, the next theorem, due to Breuillard, Green and Tao may be considered as a strong version of Gromov's Theorem. It says roughly that if a group is p-doubling at *one* sufficiently large scale with respect to p, then G is virtually nilpotent.

Theorem 6.6.5 ([8]) *For every $p \in \mathbf{N} \setminus \{0\}$, there exists $N(p) \in \mathbf{N} \setminus \{0\}$ such that the following holds: for every group G and every generating set S of G, if there exists a finite subset $A \subset G$ such that*

1. $S^{N(p)} \subset A$
2. $\frac{\#A \cdot A}{\#A} \leq p$

then, G is virtually nilpotent.

Remark 6.6.6 In the previous theorem, the hypothesis 2 is a p-doubling condition for the set A, while the first one means that the doubling holds at one given scale larger than a universal function $N(p)$ of p.

6.6.2 A Margulis Lemma

Let Γ be a group acting properly discontinuously on a metric space (X, d). For every $x \in X$ and $r > 0$, we set

$$\Gamma_r(x) := \{\gamma \in \Gamma \mid d(x, \gamma x) \leq r\}.$$

When X is simply-connected then Γ is the fundamental group of X/Γ and $\Gamma_r(x)$ is nothing but the subgroup of Γ generated by the "small loops" at p of length less than r, where p is the point on X/Γ having x as a lift. We will denote this subgroup $\Gamma_r(p)$ as well. The following fundamental Theorem due to Margulis says that such groups generated by "small loops" at a point are virtually nilpotent when r is small enough, which gives strong informations on the geometry of neigbourhoods of such points.

Theorem 6.6.7 ([7, 10, 24] Section 37.3) *There exists a positive constant $\varepsilon(n,a)$ such that, for every complete Riemannian manifold M whose sectional curvature satisfies $-a^2 \leq \mathrm{Sect}(M) \leq 0$, for every point $p \in M$ and for every $\varepsilon \leq \varepsilon(n,a)$, the subgroup $\Gamma_\varepsilon(p)$ of $\pi_1(M)$ generated by the loops at p of length less than ε is virtually nilpotent.*

We emphasise that the constant $\varepsilon(n,a)$ depends only on the curvature bound and the dimension. This theorem has been extended to various settings, we do not intend to survey here all of these and refer to [5, 8, 12, 23]. In particular, in [8], the authors deduce from Theorem 6.6.5 a "Ricci gap Theorem", saying that if an n-dimensional Riemannian manifold (M^n, g) with diam $M \leq 1$ satisfies $Ric_g \geq \epsilon(n)$, then the fundamental group of M^n is virtually nilpotent. The purpose of this lecture is to describe how our Bishop–Gromov Theorem 6.2.8 allows to prove a Margulis Lemma in the context of groups acting on metric spaces where the negative curvature assumption is replaced by δ-hyperbolicity and the lower curvature bound by an upper bound on the entropy.

Remark 6.6.8 The following observation is classical, we leave it as an exercise. Let Γ acting properly discontinuously by isometries on a Gromov hyperbolic space (X, d) be such that X/Γ is compact. Then, every virtually nilpotent subgroup of Γ is virtually cyclic (see also [5], Section 8).

Theorem 6.6.5 provides a sufficient condition for a group Γ to be (virtually) nilpotent. This condition is formulated by the p-doubling property of the Cayley graph of Γ at a scale given by the universal function $N(p)$. On the other hand, for a group Γ acting on a δ-hyperbolic space X with entropy bounded by H and diameter of the quotient space X/Γ bounded by D, our Bishop–Gromov Theorem 6.2.8 says that Γ is p-doubling at scale R_0 for $p := p(\delta, D, H)$ and $R_0 := R_0(\delta, D)$. Theorem 6.6.5 then allows to translate the universal function

$$\epsilon(\delta, H, D) := \frac{R_0}{N(p)}$$

into a Margulis constant for such actions.

Theorem 6.6.9 ([4]) *For every $\delta \geq 0$, $H \geq 0$, and $D \geq 0$, there exists $\epsilon_0 := \epsilon_0(\delta, H, D)$ such that the following holds. Let (X, d) be a δ-hyperbolic space, and Γ acting isometrically and properly on X. Assume that $\mathrm{diam}\, X/\Gamma \leq D$ and*

$\text{Ent}(X, d) \leq H$. *Then, for every* $r \leq \epsilon_0$ *and every* $x \in X$, *the group* $\Gamma_r(x)$ *is virtually cyclic.*

Proof Let $N_0 := N(3^{10}e^{300H(D+2\delta)} + 1)$ and $R_0 := 20(D + 2\delta)$, where N is the function defined in Theorem 6.6.5. Denote by ϵ_0 the constant $\frac{R_0}{N_0}$ and choose $0 < r \leq \epsilon_0$.

We consider $G := \Gamma_r(x)$ the subgroup of Γ generated by $S := \{g \in \Gamma \mid d(x, gx) \leq r\}$, and $A \subset G$ where $A = \{g \in G \mid d(x, gx) \leq R_0\}$. We will show that G and A satisfy the hypotheses of Theorem 6.6.5, which will imply that $G = \Gamma_r(x)$ is virtually nilpotent, hence cyclic since X is δ-hyperbolic.

Let us check the hypothesis 2. Let us denote $\mu_x^G = \sum_{g \in G} \delta_{gx}$ the counting measure of the G-orbit of x in X. Notice that $A \cdot A \subset \{g \in G \mid d(x, gx) \leq 2R_0\}$, hence

$$\frac{\#A \cdot A}{\#A} \leq \frac{\mu_x^G(B(x, 2R_0))}{\mu_x^G(B(x, R_0))}.$$

By the Bishop–Gromov inequality for subgroup $G \leq \Gamma$, cf. Corollary 6.5.9, we have

$$\frac{\mu_x^G(B(x, 2R_0))}{\mu_x^G(B(x, R_0))} \leq \frac{\mu_x^\Gamma(B(x, 2R_0 + R_0/2))}{\mu_x^\Gamma(B(x, R_0/2))}.$$

Applying the Bishop–Gromov theorem 6.2.8 for Γ and for (r, R) replaced by $(\frac{R_0}{2}, \frac{5}{2}R_0)$, we get

$$\frac{\mu_x^\Gamma(B(x, 2R_0 + R_0/2))}{\mu_x^\Gamma(B(x, R_0/2))} \leq 3 \cdot 5^{25/4}e^{15HR_0} \leq 3^{10}e^{300H(D+2\delta)}.$$

By the three previous inequalities, we deduce the hypothesis 2 of Theorem 6.6.5

$$\frac{\#A \cdot A}{\#A} \leq p$$

with $p := 3^{10}e^{300H(D+2\delta)}$.

We now check the hypothesis 1. Recall that we have chosen

$$N_0 := N(p + 1) = N(3^{10}e^{300H(D+2\delta)} + 1),$$

$R_0 = 20(D + 2\delta)$ and $0 < r \leq \epsilon_0 = \frac{R_0}{N_0}$.

By definition of the generating set S, for every $g \in S$, we have $d(x, gx) \leq r$. Every $g \in S^{N_0}$ can be written as $g = g_1 g_2 \cdots g_k$ with $k \leq N_0$, hence by the triangle inequality,

$$d(x, gx) \leq kr \leq N_0r \leq R_0,$$

and therefore, $g \in A$, which proves that $S^{N_0} \subset A$ and the hypothesis 1.

By Theorem 6.6.5, we therefore deduce that $\Gamma_r(x)$ is virtually nilpotent. We now notice that, since (X, d) is Gromov hyperbolic and X/Γ is compact, the virtually nilpotent subgroups of Γ are actually virtually cyclic by Remark 6.6.8, which ends the proof of Theorem 6.6.9. $\qquad\qquad\qquad\qquad\qquad\qquad\qquad\qquad\qquad\qquad\qquad$ □

6.6.3 The Thin-Thick Decomposition

This section gives geometric applications of Theorem 6.6.9, These applications are classical in the case of manifolds with bounded non positive curvature. Let Γ a group acting freely and properly discontinuously on (X, d). In this section, we will assume X non compact and Γ torsion free. The systole of this action at $x \in X$ is defined as

Definition 6.6.10 $\mathrm{sys}_\Gamma(x) := \inf_{\gamma \in \Gamma \setminus \{Id\}} d(x, \gamma x)$.

For every $r > 0$, the r-thin part of the action of Γ is the open set defined as

Definition 6.6.11 $X_r := \{x \in X \mid \mathrm{sys}_\Gamma(x) < r\}$.

The r-thick part is simply defined as $X \setminus X_r$. Notice that the thin and thick parts are defined as subsets of X. When (X, d) is Gromov hyperbolic, the action of Γ on (X, d) non elementary and X/Γ compact, then the thin part is not connected, as shown in the next proposition. We recall that Γ is assumed to be torsion free (Fig. 6.18).

Proposition 6.6.12 *Let Γ acting properly discontinuously on a Gromov hyperbolic space (X, d) such that X/Γ is compact, and let $0 < r \le \epsilon_0$, where $\epsilon_0 := \epsilon_0(\delta, H, D)$ is the constant of Theorem 6.6.9. We assume that the action of Γ is non elementary. Then,*

1. *For every $x \in X_r$, there exists a maximal virtually cyclic subgroup $G(x) \subset \Gamma$ containing $\Sigma_r(x) := \{\gamma \in \Gamma \mid d(x, \gamma x) \le r\}$. The map $x \to G(x)$, defined on X_r, is locally constant, hence constant on each connected component of X_r.*
2. *If $X_r \ne \emptyset$, then X_r is not connected.*

Proof Let us prove the first part of the proposition. Let us consider $0 < r \le \epsilon_0$ and $\gamma \in \Sigma_r(x)$ for some $x \in X_r$. Let us assume, to simplify, that γ is a non trivial

Fig. 6.18 r-Thin part

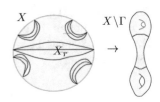

Fig. 6.19 Hyperbolic isometry

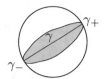

Fig. 6.20 $G(x)$ locally constant

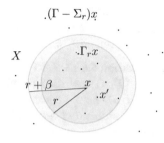

hyperbolic isometry with fixed points $\gamma_\pm$ on the boundary of X (for these notions the reader is referred to [5], Section 8). Notice that every element of $\Gamma_r(x)$ fixes the set $\{\gamma^-, \gamma^+\}$ (Fig. 6.19).

The group

$$G(x) := \{g \in \Gamma \mid g(\{\gamma_-, \gamma_+\}) = \{\gamma_-, \gamma_+\}\},$$

is the unique maximal cyclic subgroup of Γ containing γ. We then have

$$\Sigma_r(x) \subset \Gamma_r(x) \le G(x).$$

Let us show that the map defined on X_r by $x \to G(x)$ is locally constant. Since the action of Γ is proper, there exists $\beta := \beta(x) > 0$ be such that

$$\min_{\gamma \in \Gamma \backslash \Sigma_r(x)} d(x, \gamma x) > r + \beta.$$

Let us consider x' such that $d(x, x') \le \beta/2$. We claim that $\Sigma_r(x') \subset \Sigma_r(x)$. Indeed, take $\gamma \in \Sigma_r(x')$, we then have

$$d(x, \gamma x) \le d(x, x') + d(x', \gamma x') + d(\gamma x', \gamma x) \le r + \beta.$$

Therefore, $G(x') \le G(x)$ hence $G(x') = G(x)$ by maximality. This shows that the map $x \to G(x)$ is locally constant on X_r and concludes the part 1 (Fig. 6.20).

We now prove the part 2 of the proposition. Let us assume, by contradiction, that X_r is non empty and connected. Then, since $G(x)$ is locally constant, $G(x) = G$ does not depend on $x \in X_r$. Notice that X_r is Γ-invariant, $gX_r = X_r$ for every $g \in \Gamma$, and $\Sigma_r(gx) = g\Sigma_r(x)g^{-1}$ for every $x \in X_r$ and $g \in \Gamma$. Given $x \in X$, recall that $G(x) := \{g \in \Gamma \mid g(\{\gamma_-, \gamma_+\}) = \{\gamma_-, \gamma_+\}\}$, where $\{\gamma_-, \gamma_+\}$ are the fixed points of some hyperbolic isometry $\gamma \in \Sigma_r(x)$. Hence for every $g \in \Gamma$, $g\gamma g^{-1} \in \Sigma_r(gx) \subset G(gx) = G = G(x)$. In particular, since the set of fixed points of $g\gamma g^{-1}$ is $\{g\gamma_-, g\gamma_+\}$, we deduce that $\{g\gamma_-, g\gamma_+\} = \{\gamma_-, \gamma_+\}$ for every $g \in \Gamma$, hence $\Gamma = G(x)$ and Γ is elementary, a contradiction. $\qquad\square$

In the following proposition, we show that if a group Γ acts properly on a δ-hyperbolic space (X, d) with bounded entropy and bounded co-diameter, then there exists a point on X with a large systole.

Proposition 6.6.13 *Let Γ act properly discontinuously on a δ-Gromov hyperbolic space (X, d) such that $\mathrm{diam}(X/\Gamma) \leq D$ and $\mathrm{Ent}(X, d) \leq H$ and let $\epsilon_0 := \epsilon_0(\delta, H, D)$ be the constant of Theorem 6.6.9. We assume that the action of Γ is non elementary and that Γ is torsion free. Then, there exists a point $x \in X$ such that $\mathrm{sys}_\Gamma(x) \geq \epsilon_0$.*

Proof By Theorem 6.6.9, there exists $\epsilon_0 := \epsilon_0(\delta, H, D) > 0$ such that for every $0 < r \leq \epsilon_0$ and every $x \in X$, $\Gamma_r(x)$ is cyclic. Now there are two cases. If $X_{\epsilon_0} = \emptyset$, there is nothing to prove. If $X_{\epsilon_0} \neq \emptyset$, then X_r is not connected and since X is connected, we deduce that $X \setminus X_r \neq \emptyset$ and there exists $x \in X$ such that $\mathrm{sys}_\Gamma(x) \geq \epsilon_0$. $\qquad\square$

6.7 Lecture 6

In this lecture we recall the definition of the systole of a proper and geodesic metric space (X, d) on which a discrete subgroup Γ of its isometry group acts. For the sake of simplicity we set $\Gamma^* := \Gamma \setminus \{e\}$, where e is the neutral element of Γ. For the next results we need to exclude "trivial" hyperbolic spaces which we call *elementary*, we recall below the definition.

Definition 6.7.1 A hyperbolic space or group is called *elementary* if its ideal boundary has at most two points.

6.7.1 Systoles

We recall the definition of the systole at a point given in Lecture 5 (see Definition 6.6.10) and we define the global systole of a group action.

Definition 6.7.2 Let (X, d) be as above, we define

(i) $\forall x \in X$, the systole at x is

$$\mathrm{sys}_\Gamma(x) = \inf_{\gamma \in \Gamma^*} \{d(x, \gamma x)\}.$$

(ii) The global systole of the action of Γ is

$$\mathrm{Sys}_\Gamma(X) = \inf_{x \in X} \{\mathrm{sys}_\Gamma(x)\}.$$

Remark 6.7.3 If $x_0 \in X$ is fixed by some element $\gamma_0 \in \Gamma$, *i.e.* if $d(x_0, \gamma_0 x_0) = 0$, then

$$\mathrm{Sys}_\Gamma(X) = \mathrm{sys}_\Gamma(x_0) = 0.$$

We may say that $\gamma_0 \in \mathrm{Isom}(X, d)$ is an elliptic element. In the sequel we shall always assume that the space (X, d) is δ-hyperbolic for some $\delta \geq 0$ in which case an element $\gamma \in \Gamma$ is elliptic if and only if it is a torsion element (see Remark 8.16 in [5]). For the sake of simplicity we shall assume in the sequel that all isometries in Γ are torsion-free. For the general case the reader is refered to [5].

6.7.2   Bounding from Below the Systole

We now intend to prove the following theorem.

Theorem 6.7.4 *Let $\delta \geq 0$ and $H, D > 0$. There exists $s_0 := s_0(\delta, H, D) > 0$ with the following property. Let (X, d) be a non-elementary δ-hyperbolic space which is Busemann and satisfies the property of extension of the geodesics. If $\Gamma \subset \mathrm{Isom}(X, d)$ is discrete, torsion-free and such that,*

$$\mathrm{diam}(X/\Gamma, \bar{d}) \leq D \quad \text{and} \quad \mathrm{Ent}(X, d) \leq H,$$

then,

$$\forall x \in X, \quad \forall \gamma \in \Gamma^*, \quad d(x, \gamma x) > s_0(\delta, H, D).$$

Then,

$$\mathrm{Sys}_\Gamma(X) > s_0(\delta, H, D).$$

To understand this result and its proof one should look at the simplest case, namely the collar lemma for hyperbolic (constant curvature) surfaces summarised in the picture below as well as the separation between thin and thick parts (Fig. 6.21).

Here we have $\mathrm{sys}_\Gamma(x) = \eta \ll 1$, $\mathrm{sys}_\Gamma(y) \geq \epsilon_{\mathrm{Mar}}$, where ϵ_{Mar} denotes the Margulis constant of the surface, and the collar lemma asserts that,

$$d(x, y) \geq C |\log(\eta)|.$$

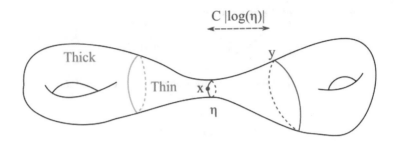

Fig. 6.21 Classical Collar Lemma

Proof of Theorem 6.7.4 For $\gamma \in \Gamma^*$ we define,

$$G = <\gamma> = \{\gamma^k ; k \in \mathbf{Z}\},$$

the cyclic group generated by γ.   □

Lemma 6.7.5 (Collar Lemma) *Let* Γ *act properly on* (X, d) *and let* $R > 0$, $\gamma \in \Gamma^*$ *and* $x \in X$. *If there exists* $y_0 \in X$ *and* $\epsilon > 0$ *such that,*

$$\inf_{k \in \mathbf{N}^*} \{d(y_0, \gamma^k y_0)\} > \epsilon > 0 \quad \text{and} \quad d(x, y_0) \leq R$$

then,

$$\frac{d(x, \gamma x)}{4R} \geq \left(\frac{\mu_{y_0}^{\Gamma}\big(B(y_0, 4R + \epsilon/2)\big)}{\mu_{y_0}^{\Gamma}\big(B(y_0, \epsilon/2)\big)} + 1\right)^{-1}.$$

This is our version of the classical Collar lemma. Notice that the denominator of the left hand side never vanishes. Indeed we always have $\mu_{y_0}^{\Gamma}\big(B(y_0, \epsilon/2)\big) \geq 1$ since y_0 is a point of its own orbit by Γ in $B(y_0, \epsilon/2)$.

Proof of Lemma 6.7.5 By the triangle inequality one has,

$$d(x, \gamma^k x) \leq d(x, \gamma x) + d(\gamma x, \gamma^2 x) + \cdots + d(\gamma^{k-1} x, \gamma^k x) \leq k d(x, \gamma x),$$

so that, if $|k| \leq \left[\dfrac{2R}{d(x, \gamma x)}\right]$ then, $d(x, \gamma^k x) \leq 2R$. We then obtain,

$$2\left[\frac{2R}{d(x, \gamma x)}\right] + 1 \leq \#\{k; \, d(x, \gamma^k x) \leq 2R\} = \mu_x^G\big(\overline{B}(x, 2R)\big).$$

Now,

$$d(y_0, \gamma^k y_0) \leq d(y_0, x) + d(x, \gamma^k x) + d(\gamma^k x, \gamma^k y_0) \leq 2d(y_0, x) + d(x, \gamma^k x),$$

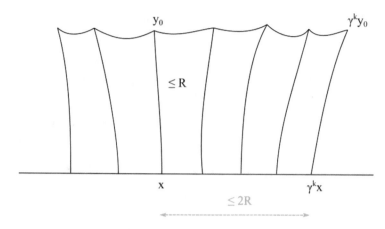

Fig. 6.22 Proof of the Collar lemma

which implies that if $d(x, \gamma^k x) \leq 2R$ then $d(y_0, \gamma^k y_0) \leq 4R$ (Fig. 6.22).
This leads to

$$\mu_x^G\big(\overline{B}(x, 2R)\big) \leq \mu_{y_0}^G\big(\overline{B}(y_0, 4R)\big).$$

These inequalities lead to

$$\frac{4R}{d(x, \gamma x)} - 1 \leq 2\left[\frac{2R}{d(x, \gamma x)}\right] + 1 \leq \frac{\mu_{y_0}^G\big(\overline{B}(y_0, 4R)\big)}{\mu_{y_0}^G\big(\overline{B}(y_0, \epsilon)\big)} \leq \frac{\mu_{y_0}^\Gamma\big(B(y_0, 4R + \epsilon/2)\big)}{\mu_{y_0}^\Gamma\big(B(y_0, \epsilon/2)\big)}.$$

The second inequality comes from the fact that $\mu_{y_0}^G\big(\overline{B}(y_0, \epsilon)\big) = 1$ since, by definition of ϵ, y_0 is the only point of its orbit by G in $\overline{B}(y_0, \epsilon)$. The third inequality comes from Corollary 6.5.9 applied with $\Gamma' = G$. □

We come back to the proof of Theorem 6.7.4. We pick $x \in X$, and $\gamma \in \Gamma^*$. From Proposition 6.6.13 there exists $x_0 \in X$ such that $\forall \alpha \in \Gamma^*$,

$$d(x_0, \alpha x_0) \geq \epsilon_0 = \epsilon_0(\delta, H, D),$$

where $\epsilon_0(\delta, H, D)$ is the constant introduced in Theorem 6.6.9. In particular, $\forall k \neq 0$, $d(x_0, \gamma^k x_0) \geq \epsilon_0 > 0$. However, x_0 may not be close to x as required by Lemma 6.7.5. Now, because of the fact that the action of Γ is co-compact and that the quotient has diameter bounded above by D there exists $\alpha_0 \in \Gamma$ such that $d(x, \alpha_0 x_0) \leq D$. We set $y_0 = \alpha_0 x_0$ and $\forall k \neq 0$,

$$d(y_0, \gamma^k y_0) = d(\alpha_0 x_0, \gamma^k \alpha_0 x_0) = d(x_0, \alpha_0^{-1} \gamma^k \alpha_0 x_0) \geq \epsilon_0 > 0,$$

since $\alpha_0^{-1}\gamma^k\alpha_0 \in \Gamma$. We then apply the previous lemma with $R = D$ and ϵ_0 instead of ϵ and we get,

$$d(x, \gamma x) \geq 4D\left(\frac{\mu_{y_0}^{\Gamma}\big(B(y_0, 4D + \epsilon_0/2)\big)}{\mu_{y_0}^{\Gamma}\big(B(y_0, \epsilon_0/2)\big)} + 1\right)^{-1}.$$

Then it suffices to apply our Bishop–Gromov Inequality to get a lower bound for $d(x, \gamma x)$, for any $x \in X$ and any $\gamma \in \Gamma^*$, in term of δ, H and D and also ϵ_0 which itself depends on δ, H and D. The resulting function of δ, H and D is called s_0. Notice that this is unfortunately not explicit since $\epsilon_0(\delta, H, D)$ is not. This ends the proof of Theorem 6.7.4

6.8 Lecture 7

In this lecture, we study finiteness properties of isomorphism classes of finitely generated groups acting cocompactly on a δ-hyperbolic metric space with bounded entropy. To be more precise, we consider the class of metric spaces (X, d) which satisfy the following assumptions: (X, d) is complete, non elementary δ-hyperbolic and Busemann. The main statement is that the set of isomorphism classes of group Γ which have a proper isometric action on such a space (X, d) with $\mathrm{diam}(X/\Gamma, d) \leq D$ and $\mathrm{Ent}(X, d) \leq H$ has less than $M_0(\delta, H, D)$ elements. There are two steps in the proof. One is to find a generating set S of Γ with bounded cardinality $\#S \leq n_0(\delta, H, D)$. This is a consequence of our Bishop–Gromov Theorem 6.5.3 and the diastole estimate, cf. Lecture 5, Proposition 6.6.13. The second step consists in bounding the number of presentations of the groups Γ. This is a consequence of the fact that Γ inherits the Gromov hyperbolicity of (X, d) and that hyperbolic groups admit bounded presentations.

6.8.1 Marked Groups

In this section, we introduce the notion of "marked groups" which will be useful for formulating the finiteness of isomorphism classes of groups. Given a finitely generated group (Γ, S), where S denotes a symmetric set of generators $S := \{\sigma_1, \cdots, \sigma_k, \sigma_1^{-1}, \cdots, \sigma_k^{-1}\}$, and the non abelian free group on k generators $\mathbf{F}_k := (\mathbf{F}_k, \{e_1, \cdots, e_k, e_1^{-1}, \cdots, e_k^{-1}\})$, we have a morphism

$$\varphi_{\Gamma,S} : \mathbf{F}_k \to \Gamma$$

defined by $\varphi_{\Gamma,S}(e_i^{\pm 1}) = \sigma_i^{\pm 1}$. In particular, we have

$$\Gamma \simeq \mathbf{F}_k/\mathrm{Ker}\,\varphi_{\Gamma,S}.$$

Definition 6.8.1 A marked group is the data of a finitely generated group (Γ, S) with the morphism $\varphi_{\Gamma,S}$. We call $\varphi_{\Gamma,S}$ the marking.

As we are interested in isomorphism classes of groups, we define:

Definition 6.8.2 An isomorphism $\beta : (\Gamma, S) \to (\Gamma', S')$ is an isomorphism $\beta : \Gamma \to \Gamma'$ such that $\beta(\sigma_i) = \sigma_i'$. In particular, $\varphi_{\Gamma',S'} = \beta \circ \varphi_{\Gamma,S}$ and we shall call such a β an isomorphism of marked groups, denoted by $(\Gamma, S) \simeq (\Gamma', S')$.

Remark 6.8.3 Notice that an isomorphism $\beta : (\Gamma, S) \to (\Gamma', S')$ induces an isometry between the Cayley graphs :

$$\bar{\beta} : (\mathcal{G}(\Gamma, S), d_S) \to (\mathcal{G}(\Gamma', S'), d_{S'}).$$

The next classical Proposition states that, when Γ acts by isometries properly discontinuously and cocompactly on (X, d), then Γ is finitely generated and the Gromov hyperbolicity transfers from the space (X, d) to the Cayley graph $(\mathcal{G}(\Gamma, S), d_S)$ for every set S of generators. The main idea is that, under the cocompactness assumption, the Cayley graph of Γ is quasiisometric to (X, d). Gromov hyperbolicity is then a property which is invariant by quasiisometry, [9, 11].

Proposition 6.8.4 *Let (X, d) be a δ-hyperbolic metric space and Γ acting properly discontinuously by isometries on (X, d). Assume that $\mathrm{diam}(X/\Gamma) \leq D$. Let $x \in X$, then $S := S_{6D} = \{\gamma \in \Gamma \,|\, d(x, \gamma x) \leq 6D\}$ is a generating set of Γ and (Γ, S) is δ'-hyperbolic for $\delta' := 8\left(\frac{5\delta}{D} + 4\right)$.*

Proof We only sketch the proof; for details we refer to [4]. Let us choose a point $x \in X$. We first observe that for every $k \in \mathbf{N}^*$, the set

$$S := S_{(k+2)D}(x) = \{\gamma \in \Gamma \,|\, d(x, \gamma x) \leq (k+2)D\}$$

is a generating set of Γ. For $\gamma \in \Gamma$, set $n := \left\lfloor \frac{d(x,\gamma x)}{kD} \right\rfloor + 1$. On a geodesic $[x, \gamma x]$, let us choose $x_0 = x, x_1, \cdots, x_n = \gamma x$ such that $d(x_{i-1}, x_i) = kD$ for $1 \leq i \leq n-1$, and $d(x_{n-1}, x_n) < kD$. By the assumption on the diameter, for every $1 \leq i \leq n$, there is $\gamma_i \in \Gamma$ such that $d(x_i, \gamma_i x) \leq D$. Then, defining $\gamma_0 = Id_X, \gamma_n := \gamma$ and $\sigma_i := \gamma_{i-1}^{-1}\gamma_i$ for $i = 1, \cdots, n$, we see that $\gamma = \sigma_1 \cdots \sigma_n$, and moreover,

$$d(x, \sigma_i x) \leq d(\gamma_{i-1} x, x_{i-1}) + d(x_{i-1}, x_i) + d(x_i, \gamma_i x) \leq (k+2)D.$$

This implies that $S_{(k+2)D} := \{\sigma \in \Gamma \,|\, d(x, \sigma x) \leq (k+2)D\}$ is a generating set of Γ. Choosing $k = 4$, we see that $S := S_{6D}$ is a generating set of Γ, which proves the first part of the proposition (Fig. 6.23). $\square$

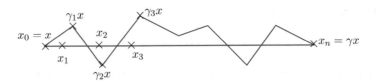

Fig. 6.23 $S_{(k+2)D}$

From now on, we fix $S = S_{6D}$. In order to prove that (Γ, S) is δ'-hyperbolic, we first show that the Cayley graph of (Γ, S) is "quasiisometric" to (X, d) and then, deduce that (Γ, S) is δ'-hyperbolic. Recall that two metric spaces (X, d_X) and (Y, d_Y) are quasiisometric if there exist $a > 0$, $a' > 0$, $b, b' \in \mathbf{R}$ and a map $f : Y \to X$ such that for every $y, y' \in Y$, we have

$$a\, d(y, y') - b \le d(f(y), f(y')) \le a' d(y, y') + b'.$$

Let us define a map

$$f : \mathcal{G}(\Gamma, S) \to X,$$

as follows. The map f sends homothetically the edges $[e, \sigma_i]$ in the Cayley graph onto geodesic segments $[x, \sigma_i x]$ in X, (recall that the edges in the Cayley graph have length equal to 1, hence the homothety factors are the length of the geodesic segments $[x, \sigma_i x]$, that is, the distances $d(x, \sigma_i x)$). We then define f on every edge by $f([\gamma, \gamma \sigma_i]) = \gamma[x, \sigma_i x]$.

The first observation is that the map $f : (\mathcal{G}(\Gamma, S), d_S) \to (X, d)$ is a quasiisometry: for every $s, t \in \mathcal{G}(\Gamma, S)$,

$$4Dd_S(s, t) - 14D \le d(f(s), f(t)) \le 6Dd_S(s, t). \tag{6.5}$$

We skip the proof of (6.5), which is a quite direct consequence of the definition of f, and refer to [4], Lemma 5.7, for details.

The second observation below is a direct consequence of (6.5): given a geodesic segment $[s, t] \subset \mathcal{G}(\Gamma, S)$ between s, t, we have

$$\text{Length}(f[s, t]) \le \frac{3}{2}d(f(s), f(t)) + 21D. \tag{6.6}$$

Indeed, $f([s, t])$ is arcwise geodesic in X; thus, by the second inequality of (6.5), we get $\text{Length} f([s, t]) \le 6Dd_S(s, t)$, and (6.6) follows from the first inequality of (6.5) (Fig. 6.24).

We now recall a fundamental property of δ-hyperbolic spaces which amounts to saying that if a curve c satisfies the property that the length of every subsegment of c is equal to the distance between its end points up to multiplicative and additive constants, then c is at bounded distance from each geodesic joining the end points.

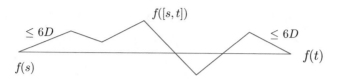

Fig. 6.24 Length $f([s, t])$

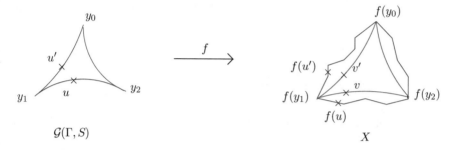

Fig. 6.25 $\mathcal{G}(\Gamma, S)$ is δ'-hyperbolic

Lemma 6.8.5 ([4], Proposition 6.13) *Let (X, d) be a δ-hyperbolic space and $c :$ $[0, T] \to X$ a rectifiable curve such that for every $u, v \in [0, T]$,*

$$\text{Length}(c([u, v]) \leq \lambda d(c(u), c(v)) + C. \tag{6.7}$$

Then, every geodesic in X joining $c(0)$ and $c(T)$ is contained in the C_1-neighborhood of $c([0, T])$ and conversely, $c([0, T])$ is contained in the C_2-neighborhood of each geodesic between $c(0)$ and $c(T)$, where $C_1 := 6\lambda + 2 + \frac{C}{6\lambda + 2}$ and $C_2 := (1 + \lambda)C_1 + \frac{C}{2}$.

We can apply Lemma 6.8.5 to the image $c := f([y, y'])$ by f of geodesic segments $[y, y']$ in $\mathcal{G}(\Gamma, S)$. Let $\bar{c}$ be a geodesic joining $f(y)$ and $f(y')$. By the observation (6.6), the hypothesis of Lemma 6.8.5 is satisfied with $\lambda = \frac{3}{2}$ and $C = 21D$, therefore the curve $c := f([y, y'])$ is contained in the $C_1 := 11\delta + \frac{21D}{11}$-neighborhood of the geodesic $\bar{c}$ and $\bar{c}$ is in the C_2-neighbourhood of c for $C_2 := \frac{55\delta}{2} + \frac{168D}{11}$ (Fig. 6.25).

We are now ready to conclude. Let (y_0, y_1, y_2) be a geodesic triangle in $\mathcal{G}(\Gamma, S)$. We want to show that (y_0, y_1, y_2) is $\frac{\delta'}{4}$-thin, *i.e.* that every point u on one side of (y_0, y_1, y_2) is at distance $\frac{\delta'}{4}$ of the union of the two other sides of the triangle. For every $i \neq j \in \{0, 1, 2\}$, we choose $\bar{c}_{i,j}$ a geodesic joining $f(y_i)$ and $f(y_j)$ and denote by $(f(y_0), f(y_1), f(y_2))$ the geodesic triangle in X whose sides are the $\bar{c}_{i,j}$'s. We can assume that $u \in [y_1, y_2]$. By Lemma 6.8.5 there exists $v \in \bar{c}_{1,2}$ such that $d(f(u), v) \leq C_1$. Since (X, d) is δ-hyperbolic, there exists $v' \in \bar{c}_{0,1} \cup \bar{c}_{0,2}$ such that $d(v, v') \leq \delta$. We can assume that $v' \in \bar{c}_{0,1}$. By Lemma 6.8.5, there exists $f(u') \in f([y_0, y_1])$ such that $d(f(u'), v') \leq C_2$.

By (6.5) we have $d(u, u') \leq \left(\frac{1}{4D} d(f(u), f(u')) + \frac{14D}{4D}\right)$, hence $d(u, u') \leq \frac{1}{4D} \left(d(f(u), v) + d(v, v') + d(v', f(u'))\right) + \frac{14D}{4D}$, and therefore,

$$d(u, u') \leq \frac{1}{4D} (C_1 + \delta + C_2) + \frac{14D}{4D},$$

thus, $d(u, u') \leq \frac{1}{4D} (40\delta + 32D)$. We conclude that $\mathcal{G}(\Gamma, S)$ is $\delta' := 8\left(\frac{5\delta}{D} + 4\right)$-hyperbolic.

6.8.2 Finitely Presented Groups

Let us consider a marked group (Γ, S) with its marking $\varphi_{\Gamma, S} : \mathbf{F}_k \to \Gamma$. We then have $\Gamma \simeq \mathbf{F}_k / \mathrm{Ker}\, \varphi_{\Gamma, S}$.

Definition 6.8.6 A group Γ is finitely presented if there exists a finite generating set S such that $\mathrm{Ker}\, \varphi_{\Gamma, S} = \langle\langle r_1, \cdots, r_n \rangle\rangle$ is finitely generated as a normal subgroup. We write

$$\Gamma = \langle e_1, \cdots, e_k \mid r_1, \cdots, r_n \rangle.$$

Notice that when $\Gamma = \langle e_1, \cdots, e_k \mid r_1, \cdots, r_n \rangle$, then we have $\Gamma \simeq \mathbf{F}_k / \langle\langle r_1, \cdots, r_n \rangle\rangle$.

Examples 6.8.7

1. $\mathbf{Z}^2 = \langle a, b \mid aba^{-1}b^{-1} \rangle$.
2. If M is a genus g surface, $g \geq 2$, then

$$\pi_1(M) = \langle a_1, b_1, \cdots, a_g, b_g \mid \Pi_{i=1}^g [a_i, b_i] \rangle.$$

We set

$$\mathbf{B}_p(3) := \{g \in \mathbf{F}_p \mid d_{S_p}(e, g) = 2, \text{ or } 3\}, \tag{6.8}$$

where d_{S_p} is the distance on the Cayley graph of the non abelian free group $\mathbf{F}_p$ on p generators and $S_p := \{e_1^{\pm 1}, \cdots, e_p^{\pm 1}\}$ the associated generating set.

The next classical proposition says that if (Γ, S) is a δ-hyperbolic group, then there exists a presentation of Γ with relations being words of length less than or equal to 3. Before stating it, let us define, for every δ-hyperbolic group (Γ, S), the "augmented generating set" as follows.

Definition 6.8.8 $\Sigma_S := \{\gamma \in \Gamma \setminus \{e\} \mid d_S(e, \gamma) < 4\delta + 2\}$.

Notice that Σ_S is a generating set of Γ since $S \subset \Sigma_S$.

Proposition 6.8.9 ([11], Theorem 2.5, Chapter 5) *Let (Γ, S) be a δ-hyperbolic group and Σ_S the augmented generating set. We set $p := \#\Sigma_S$ and $\mathcal{R}_{\Gamma,\Sigma_S} := \mathrm{Ker}\,\varphi_{\Gamma,\Sigma_S} \cap \mathbf{B}_p(3)$. Then, $\mathrm{Ker}\,\varphi_{\Gamma,\Sigma_S} = \langle\langle\mathcal{R}_{\Gamma,\Sigma_S}\rangle\rangle$, in particular Γ is finitely presented.*

Remark 6.8.10 By definition of Σ_S, $p := \#\Sigma_S$ is bounded above in term of $\#S$ and δ, hence by the above proposition the same is true for the number of relations $\#\mathcal{R}_{\Gamma,\Sigma_S}$ of the presentation of (Γ, Σ_S). This fact will be used to show that the number of isomorphism classes of marked groups such as (Γ, S) is bounded above by a function depending on $\#S$ and δ.

We now define the family of δ-hyperbolic groups having a generating set with bounded cardinality. In what follows, we fix $\delta > 0$.

Definition 6.8.11 We define MG_k as the class of marked group (Γ, S) such that (Γ, S) is δ-hyperbolic and $\#S \leq k$.

Since we are interested in groups up to isomorphism, we consider the following

Definition 6.8.12 We define G_k as the set of groups Γ such that there exists a generating set S with $(\Gamma, S) \in MG_k$.

Recall that we write $(\Gamma, S) \simeq (\Gamma', S')$ when the marked groups (Γ, S) and (Γ', S') are isomorphic, cf. Definition 6.8.2. We will also write $\Gamma \sim \Gamma'$ when Γ and Γ' are isomorphic. The set of corresponding isomorphism classes are denoted $MG_k/\simeq$ and $G_k/\sim$ respectively.

Proposition 6.8.13 *The set of isomorphism classes $G_k/\sim$ contains less than*

$$M(k) := 2^{N_k} \sum_{p=3}^{N_k} 2^{4p^2(2p-1)}$$

elements, where $N_k := 2k(k-1)^{\lfloor 4\delta+1 \rfloor}$.

Proof We sketch the proof; the details can be found in [4], Proposition 5.9. It follows from the definitions that the map from MG_k onto G_k defined by $(\Gamma, S) \longmapsto \Gamma$ induces a surjective map $MG_k/\simeq \longrightarrow G_k/\sim$, therefore

$$\#(G_k/\sim) \leq \#(MG_k/\simeq). \tag{6.9}$$

Proposition 6.8.13 then amounts to finding a bound of $\#(MG_k/\simeq)$. In order to find such a bound, we will construct a finite to one map from $MG_k/\simeq$ to a finite set.

We first notice that for every $(\Gamma, S) \in MG_k$ the cardinality of the generating set Σ_S introduced in Definition 6.8.8 is bounded above by the number of elements of the ball of radius $\lfloor 4\delta + 2 \rfloor$ of the Cayley graph of the non abelian free group $\mathbf{F}_k$ with respect to the standard set of generators. Therefore,

$$\#\Sigma_S \leq 1 + \sum_{i=1}^{\lfloor 4\delta+2 \rfloor} k(k-1)^{i-1} \leq N_k := 2k(k-1)^{\lfloor 4\delta+1 \rfloor} \tag{6.10}$$

since $1 + X + \cdots + X^n \leq 2X^n - 1$ and by (6.10), the marked group (Γ, Σ_S) then belongs to MG_{N_k}.

We now give several claims which will lead to the construction of a map from $MG_k/\simeq$ to a finite set. We will then sketch the proofs of the claims.

Claim 6.8.14 *The map* $\psi : MG_k \to MG_{N_k}$, *defined by* $\psi((\Gamma, S)) := (\Gamma, \Sigma_S)$, *induces a map* $\bar{\psi} : MG_k/\simeq \to MG_{N_k}/\simeq$.

For every $(\Gamma, \Sigma) \in MG_{N_k}$ we set $p(\Sigma) := \#\Sigma$. Recall that $\mathcal{R}_{\Gamma,\Sigma} = \operatorname{Ker}\varphi_{\Gamma,\Sigma} \cap \mathbf{B}_{p(\Sigma)}(3)$, where $\varphi_{\Gamma,\Sigma}$ is the marking of (Γ, Σ). For $p \in \mathbf{N}$, we denote by $\mathcal{A}(p)$ the set of all subsets of $\mathbf{B}_p(3)$, where $\mathbf{B}_p(3)$ has been defined in (6.8) and we set

$$\mathcal{A} := \cup_{p=3}^{N_k} \{p\} \times \mathcal{A}(p).$$

Claim 6.8.15 *The map* $\Phi : MG_{N_k} \to \mathcal{A}$ *defined by* $\Phi((\Gamma, \Sigma)) := (p(\Sigma), \mathcal{R}_{\Gamma,\Sigma})$ *induces a map* $\bar{\Phi} : MG_{N_k}/\simeq \to \mathcal{A}$. *Moreover,* $\bar{\Phi}_{|\operatorname{Im}\bar{\psi}}$ *is injective, where* $\bar{\Phi}_{|\operatorname{Im}\bar{\psi}}$ *denotes the restriction of* $\bar{\Phi}$ *to* $\operatorname{Im}\bar{\psi} = \bar{\psi}(MG_k/\simeq)$.

By claims 6.8.14 and 6.8.15, the map $\Phi \circ \psi : MG_k \xrightarrow{\psi} MG_{N_k} \xrightarrow{\Phi} \mathcal{A}$ induces a composed map $\bar{\Phi} \circ \bar{\psi}$,

$$MG_k/\simeq \xrightarrow{\bar{\psi}} MG_{N_k}/\simeq \xrightarrow{\bar{\Phi}} \mathcal{A}.$$

We now show that the map $\bar{\Phi} \circ \bar{\psi} : MG_k/\simeq \to \mathcal{A}$ is finite to one.

We observe first that for every $(\Gamma, \Sigma) \in MG_{N_k}$, we have

$$\psi^{-1}(\{(\Gamma, \Sigma)\}) = \{(\Gamma, S) \in MG_k \mid \Sigma_S = \Sigma\}$$

hence,

$$\psi^{-1}(\{(\Gamma, \Sigma)\}) \subset \{(\Gamma, S) \in MG_k \mid S \subset \Sigma\}.$$

Therefore, $\#\psi^{-1}((\Gamma, \Sigma)) \leq 2^{N_k}$.

We now denote by $[(\Gamma, \Sigma)]$ the isomorphism class of $(\Gamma, \Sigma) \in MG_{N_k}$, and we choose some $(\Gamma', S') \in MG_k$ such that

$$[(\Gamma', S')] \in \bar{\psi}^{-1}(\{[(\Gamma, \Sigma)]\}).$$

By definition, $\psi((\Gamma', S')) = (\Gamma', \Sigma_{S'})$ is isomorphic to (Γ, Σ). The isomorphism

$$\beta : (\Gamma', \Sigma_{S'}) \rightarrow (\Gamma, \Sigma)$$

sends (Γ', S') on (Γ, S) for some $S \subset \Sigma$ such that $\Sigma = \Sigma_S$ thus, $(\Gamma, S) \in \psi^{-1}(\{(\Gamma, \Sigma)\})$. We have shown that each isomorphism class in $\bar{\psi}^{-1}(\{[(\Gamma, \Sigma)]\})$ contains a representative (Γ, S) in $\psi^{-1}(\{(\Gamma, \Sigma)\})$, hence, $\#(\bar{\psi}^{-1}(\{[(\Gamma, \Sigma)]\})) \leq \#(\psi^{-1}(\{(\Gamma, \Sigma)\})) \leq 2^{N_k}$.

We deduce from the second part of claim 6.8.15 that the map $\bar{\phi} \circ \psi$ is 2^{N_k}-to-one, hence

$$\#(MG_k/\simeq) \leq 2^{N_k} \#\mathcal{A} = M(k), \tag{6.11}$$

since $\#\mathcal{A} = \sum_{p=3}^{N_k} 2^{4p^2(2p-1)}$.

It remains to sketch the proof of the claims. The claim 6.8.14 is a direct consequence of the fact that an isomorphism between (Γ, S) and (Γ', S) induces an isometry between the corresponding Cayley graphs of (Γ, S) and (Γ', S'). In order to prove the first part of the claim 6.8.15, we consider two isomorphic marked groups (Γ, Σ) and (Γ', Σ') with marking $\varphi_{\Gamma, \Sigma}$ and $\varphi_{\Gamma', \Sigma'}$ respectively. We observe that $\varphi_{\Gamma', \Sigma} = \beta \circ \varphi_{\Gamma, \Sigma}$ where β is the isomorphism between (Γ, Σ) and (Γ', Σ'). This implies that $p(\Sigma) = p(\Sigma')$ and that $\mathrm{Ker}\varphi_{\Gamma, \Sigma} = \mathrm{Ker}\varphi_{\Gamma', \Sigma'}$, therefore $\mathcal{R}_{\Gamma, \Sigma} = \mathcal{R}_{\Gamma', \Sigma'}$ and $\Phi((\Gamma, \Sigma)) = \Phi((\Gamma', \Sigma'))$. This concludes the proof of the claims. Proposition 6.8.13 follows from (6.9) and (6.11). □

6.8.3   Finiteness Theorem

In this section we state a Finiteness Theorem. We will assume that the metric space (X, d) is geodesically complete, non elementary δ-hyperbolic and Busemann. Let $\mathcal{H}(\delta, H, D)$ be the set of torsion free groups Γ which admit a proper and discontinuous action on a metric space satisfying the above properties and such that $\mathrm{diam}\, X/\Gamma \leq D$ and $\mathrm{Ent}(X, d) \leq H$. The set of isomorphism classes of groups in $\mathcal{H}(\delta, H, D)$ is denoted by $\mathcal{H}(\delta, H, D)/\sim$.

A first step toward the finiteness theorem is to show that every group $\Gamma \in \mathcal{H}(\delta, H, D)$ has a generating set with bounded cardinal. We first recall the following property from Lecture 5, Proposition 6.6.13.

Proposition 6.8.16 *Let (X, d) be a non elementary δ-hyperbolic space and a group Γ acting properly by isometries on (X, d). Assume that $\mathrm{diam}(X/\Gamma) \leq D$ and $\mathrm{Ent}(X, d) \leq H$. Then there exists $x_0 \in X$ such that*

$$\mathrm{sys}_\Gamma(x_0) \geq \epsilon_0(\delta, H, D).$$

Theorem 6.8.17 *Given δ, H, D, let $\epsilon_0 := \epsilon_0(\delta, H, D)$ be as in Proposition 6.8.16 and let Γ be a torsion free group acting properly on (X, d), a non elementary δ-hyperbolic Busemann space. Then, there exists $x_0 \in X$ such that*

$$\#S_{6D}(x_0) \le n_0(\delta, H, D),$$

where $n_0(\delta, H, D) := \frac{11}{2} 3^{\frac{25}{4}} \left(\frac{6D}{\epsilon_0}\right)^{\frac{25}{4}} exp\left(18H(11D + 20\delta)\frac{6D}{\epsilon_0}\right)$.

Proof Let $x_0 \in X$ be as in Proposition 6.8.16. In particular, for every $\gamma \in \Gamma \setminus \{Id\}$, we have

$$d(x_0, \gamma x_0) \ge \epsilon_0(\delta, H, D). \tag{6.12}$$

It follows from (6.12) that $\mu_{x_0}^{\Gamma}(B(x_0, \epsilon_0)) = 1$. Applying the Bishop–Gromov Theorem for small radii, Theorem 6.5.3 with (r, R) replaced by $(\epsilon_0, 6D)$, we then deduce

$$\#S_{6D}(x_0) = \frac{\mu_{x_0}^{\Gamma}(B(x_0, 6D))}{\mu_{x_0}^{\Gamma}(B(x_0, \epsilon_0))} \le n_0(\delta, H, D).$$

$\square$

Theorem 6.8.18 $\mathcal{H}(\delta, H, D)/\sim$ *contains less than* $M_0(\delta, H, D) = 2^{N_0}\sum_{p=1}^{N_0} e^{4p^2(p-1)}$ *elements, where* $N_0 = 2n_0(n_0 - 1)^{(\frac{40\delta}{D}+35)}$

Proof Let us choose $x_0 \in X$ such as in Theorem 6.8.17,

$$\#S_{6D}(x_0) \le n_0 := n_0(\delta, H, D),$$

where

$$n_0(\delta, H, D) = \frac{11}{2} 3^{\frac{25}{4}} \left(\frac{6D}{\epsilon_0}\right)^{\frac{25}{4}} exp\left(18H(11D + 20\delta)\frac{6D}{\epsilon_0}\right).$$

By Proposition 6.8.4, $S_{6D}(x_0)$ is a generating set of Γ and $(\Gamma, S_{6D}(x_0))$ is δ'-hyperbolic where $\delta' = 8\left(\frac{5\delta}{D} + 4\right)$ therefore, $(\Gamma, S_{6D}(x_0)) \in MG_{n_0}$, where we have replaced δ by δ' in the definition of MG_{n_0}. Proposition 6.8.13 then implies that there exist less than $M(n_0)$ such groups Γ, where

$$M(n_0) = 2^{N_{n_0}} \sum_{p=3}^{N_{n_0}} 2^{4p^2(2p-1)},$$

and

$$N_{n_0} := 2n_0(n_0 - 1)^{\lfloor 48\delta' + 1\rfloor} = 2n_0(n_0 - 1)^{\lfloor \frac{160\delta}{D} + 129\rfloor},$$

which concludes.   □

6.9   Lecture 8

In this lecture we intend to discuss some homotopical finiteness theorem. We need to define a new property for a proper geodesic metric space (X, d) using triangle comparison as in Fig. 6.3 reproduced below.

6.9.1   CAT(0)-Spaces

We consider, as in Lecture 1, a triangle $\Delta \subset X$ and its comparison triangle $\overline{\Delta} \subset \mathbf{R}^2$ (Fig. 6.26).

We choose a point p on the segment $[b, c]$ and a point $\bar{p}$ on the segment $[\bar{b}, \bar{c}]$ dividing it in the same ratio, that is,

$$d(b, p) = t d(b, c) \text{ and } \bar{d}(\bar{b}, \bar{p}) = t \bar{d}(\bar{b}, \bar{c}) \text{ for } t \in [0, 1].$$

Definition 6.9.1 The space (X, d) is said to be CAT(0) if for every triangle Δ and every such point $p \in \Delta$ we have,

$$d(a, p) \leq \bar{d}(\bar{a}, \bar{p}).$$

Remarks 6.9.2

(i) The interested reader is referred to [9] for a full treatment of the notion of CAT(κ)-spaces for $\kappa \in \mathbf{R}$.

Fig. 6.26 $\Delta \subset X$ and $\overline{\Delta} \subset \mathbf{R}^2$

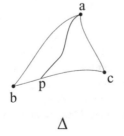

Δ

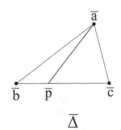

$\overline{\Delta}$

(ii) Contrarily to δ-hyperbolicity, the CAT(0) property imposes some local restrictions.

Exercises 6.9.3

1. Show that CAT(0)-*spaces are Busemann. The converse is not true in general, see* [9] *p.169.*
2. Show that a CAT(-1)-*space is* δ-*hyperbolic and compute* δ.

Notice that in a CAT(0)-space metric balls are contractible. We can also localise this notion, more precisely

Definition 6.9.4 The space (X, d) is said to be *locally* CAT(0) if for all $x \in X$, there exists a neighbourhood U_x of x such that every triangle included in U_x satisfies the CAT(0) property.

Let us recall that we always assume that (X, d) satisfies the *property of extension of geodesics*. We then get

Lemma 6.9.5 (See [9]) *Every locally* CAT(0) *metric space admits a universal cover which is* CAT(0) *and is geodesically complete.*

Let us now choose three real numbers $\delta \geq 0$, $H > 0$ and $D > 0$.

Definition 6.9.6 We consider compact, connected, metric spaces (Y, d) which are locally CAT(0), with diam$(Y, d) \leq D$, and which have a δ-hyperbolic universal cover whose entropy is not greater than H. We denote this space by $\mathcal{M}(\delta, H, D)$.

The following theorem is an easy consequence of Theorem 6.8.18.

Theorem 6.9.7 *With the above notations, we have the following.*

(i) The number of fundamental groups, up to isomorphisms, of elements of $\mathcal{M}(\delta, H, D)$ *is bounded above by* $M_0(\delta, H, D)$.
(ii) The number of elements of $\mathcal{M}(\delta, H, D)$, *up to homotopy equivalence, is bounded above by* $M_0(\delta, H, D)$.

The function $M_0(\delta, H, D)$ is described more precisely in Theorem 6.8.18.

Proof We just have to check that the hypotheses of Theorem 6.8.18 are satisfied.

(i) The fundamental groups of elements of $\mathcal{M}(\delta, H, D)$ are in $\mathcal{H}(\delta, H, D)$, the class defined in Sect. 6.8.3. Notice that, on a CAT(0)-space which is Busemann, the isometries in a fundamental group are torsion-free, see Remark 6.7.3.
(ii) For $(Y, d) \in \mathcal{M}(\delta, H, D)$, its universal cover $\widetilde{Y}$ is contractible, hence it is a $K(\pi, 1)$. This implies that, for $Y, Y' \in \mathcal{M}(\delta, H, D)$, having an isomorphism between $\pi_1(Y)$ and $\pi_1(Y')$ is equivalent to having an homotopy equivalence between Y and Y'. We recall that a homotopy equivalence between Y and Y' is equivalent to having two continuous maps $f : Y \longrightarrow Y'$ and $g : Y' \longrightarrow Y$ such that $g \circ f$ is homotopic to the identity map on Y and $f \circ g$ is homotopic to the identity map on Y'. $\qquad\square$

Remark 6.9.8 Theorem 6.9.7 is certainly true when the hypothesis "locally CAT(0)" is replaced by the hypothesis "locally Busemann". However, we have not yet done the proof.

6.9.2   Towards a Compactness Result

The aim of the next step is to endow $\mathcal{M}(\delta, H, D)$ with a distance function, d_{GH} and show that the metric space $\left(\mathcal{M}(\delta, H, D), d_{GH}\right)$ is compact. Let us first recall a few facts about the Gromov–Hausdorff distance between metric spaces.

6.9.2.1   Gromov–Hausdorff Distance Between Metric Spaces

Let (X, d) be a general metric space and $A, B \subset X$ two compact subsets. We define the Hausdorff distance $d_{X,H}(A, B)$ between subsets A and B of X by,

$$d_{X,H}(A, B) = \inf\left\{\epsilon > 0;\ A \subset B_\epsilon\ \text{and}\ B \subset A_\epsilon\right\},$$

where $A_\epsilon = \left\{x \in X;\ d(x, A) \leq \epsilon\right\}$.

Now let (X, d_X) and (Y, d_Y) be two compact metric spaces. We consider metric spaces (Z, d_Z) such that there exist two isometric embeddings,

$$i : X \longrightarrow Z \ \text{and}\ j : Y \longrightarrow Z$$

and define

$$d_{GH}(X, Y) = \inf_{Z,i,j} \left\{d_{Z,H}\left(i(X), j(Y)\right)\right\}.$$

Here $d_{Z,H}$ is the Hausdorff distance between compact subsets of Z.

Exercise 6.9.9 *Prove that such spaces Z always exist and that d_{GH} is a distance.*

The strength of this definition is clear in the following proposition

Proposition 6.9.10 *The set $\mathcal{M}$ of compact length spaces (modulo isometries) endowed with d_{GH} is complete.*

Remarks 6.9.11

(1) Notice that $\mathcal{M}$ is a *set*.
(2) This proposition is left to the reader. Completeness in this situation is tricky but not difficult.

Obviously this definition is not really convenient since we have to know all possible metric spaces Z. We thus turn to the standard more effective alternative approach.

Definitions 6.9.12 An ϵ-approximation between two metric spaces (X, d_X) and (Y, d_Y) is given by two maps $\varphi : X \to Y$ and $\psi : Y \to X$ which are ϵ-almost reciprocal, i.e. such that $d_Y\big(y, \varphi \circ \psi(y)\big) < \epsilon$ and $d_X\big(x, \psi \circ \varphi(x)\big) < \epsilon$ for all $(x, y) \in X \times Y$ and which furthermore satisfy:

$$\forall x, x' \in X \quad |d_Y\big(\varphi(x), \varphi(x')\big) - d_X(x, x')| < \epsilon \, ;$$

$$\forall y, y' \in Y \quad |d_X\big(\psi(y), \psi(y')\big) - d_Y(y, y')| < \epsilon \, .$$

A sequence $\big((X_n, d_n)\big)_{n \in \mathbf{N}}$ is said to "converge in the Gromov–Hausdorff sense to (X, d)" if there exists a sequence of positive numbers $(\epsilon_n)_{n \in \mathbf{N}}$ approaching zero and, for every n, an ϵ_n-approximation between (X, d) and (X_n, d_n).

Remarks 6.9.13

1. This notion of convergence is equivalent to convergence under the Gromov–Hausdorff distance defined before up to multiplicative factors of ϵ_n.
2. The reader may also check yet another definition closest to the one given by Gromov, in [9], page 72.

We now define the notion of ϵ-net.

Definition 6.9.14 An ϵ-net $Z \subset X$ of a compact metric space (X, d_X) is a finite set of points $\{x_i\}_{i \in I}$ of X such that $\bigcup_{i \in I} B(x_i, \epsilon) = X$.

Proposition 6.9.15 *Let $D > 0$ and $N : (0, 1) \longrightarrow \mathbf{N}$ an integer valued map. We denote by $\mathcal{Met}(D, N)$ the set of all isometry classes of compact metric spaces satisfying*

1. $\mathrm{diam}(M) \le D$,
2. $\forall \epsilon, \exists Z$ a finite ϵ-net such that $\#Z \le N(\epsilon)$.

Then, $\mathcal{Met}(D, N)$ is compact for the distance d_{GH}.

This result is classical and not difficult, just a bit boring to describe. We then get the following theorem.

Theorem 6.9.16 ([20], Theorem 5.3, p. 275) *Let $\mathcal{Man}(n, D)$ be the set of isometry classes of closed Riemannian manifolds (M, g) of dimension n such that $\mathrm{Ric}(g) \ge -(n-1)$ and diameter not greater than D. Then, $\mathcal{Man}(n, D)$ is relatively compact in the space of all isometry classes of compact metrics spaces.*

Remark 6.9.17 The proof consists in showing that $\mathcal{Man}(n, D) \subset \mathcal{Met}(D, N)$. However it is not closed, which explains that the conclusion is only "relatively compact".

Proof We need to show that, for every $\epsilon > 0$, there exists an ϵ-net with a uniformly bounded cardinality, independently of $(M, g) \in \mathcal{Man}(n, D)$.

Now, for $(M, g) \in \mathcal{M}an(n, D)$, let $\epsilon > 0$ and choose $Z \subset M$ to be a maximal subset satisfying that if $z_1, z_2 \in Z$, $z_1 \neq z_2$ then $d_g(z_1, z_2) > \epsilon$, where d_g denotes the distance in M induced by the metric g. This is a maximal packing by balls of radius $\epsilon/2$. Notice that there might be several subset Z and we choose one. The balls $B(z, \epsilon/2) \subset M$, for $z \in Z$ are disjoints, hence, for $m \in M$,

$$\sum_{z \in Z} \mathrm{Vol}\left(B(z, \epsilon/2)\right) \leq \mathrm{Vol}(M) = \mathrm{Vol}\left(B(m, D)\right).$$

As a consequence,

$$\#Z \leq \frac{\mathrm{Vol}(M)}{\inf\limits_{z \in Z}\left\{\mathrm{Vol}\left(B(z, \epsilon/2)\right)\right\}} = \frac{\mathrm{Vol}\left(B(z_0, D)\right)}{\mathrm{Vol}\left(B(z_0, \epsilon/2)\right)} = \frac{\mathrm{Vol}_{-1}(D)}{\mathrm{Vol}_{-1}(\epsilon/2)} := N(\epsilon).$$

Here $\mathrm{Vol}_{-1}(r)$ denotes the volume of a ball of radius r in a simply connected real hyperbolic space of constant sectional curvature equal to -1. This ends the proof of Theorem 6.9.16. □

We finish this lecture by stating one of the main theorems of [4].

Theorem 6.9.18 (see [4]) *The metric space $\left(\mathcal{M}(\delta, D, H), d_{GH}\right)$ is compact.*

Remarks 6.9.19

1. Notice that in the last theorem, contrarily to Theorem 6.9.16, there is no reference to any kind of dimension of the metric spaces in $\mathcal{M}(\delta, D, H)$. However the core of the proof will show that the bound on the dimension is hidden in the bound on H.
2. This result is certainly true with the hypothesis "Busemann" instead of "CAT(0)".

6.10 Lecture 9

In the first part of this lecture, we state a compactness Theorem for the set of isometry classes of locally CAT(0) metric spaces with bounded diameter and universal cover δ-hyperbolic with bounded entropy. When the metric spaces are assumed to be topological manifolds, this leads to a topological finiteness Theorem. In the second part of the lecture, we come back to the growth of groups and raise a few questions that we will discuss in the last lecture.

6.10.1 Compactness

We define $\mathcal{M} := \{(Y, d) \mid (Y, d) \text{ compact metric space}\}$. In this section, for every metric space (Y, d), we will denote by (X, d) its universal cover. In particular, we will not distinguish the corresponding distances. We denote by $\mathcal{M}(\delta, H)$ the set of connected, compact, locally CAT(0) metric spaces satisfying the property of extension of local geodesics (see Sect. 6.5.1), which is not a circle nor a point and such that its universal cover (X, d) is δ-hyperbolic and satisfies $\text{Ent}(X, d) \leq H$. We also denote by $\mathcal{M}(\delta, H, D)$ the subset of $\mathcal{M}(\delta, H)$ of metric spaces (Y, d) such that $\text{diam}(Y, d) \leq D$ and $\mathcal{M}(\delta, H, D)/ \sim$ the set of isometry classes of metric spaces in $\mathcal{M}(\delta, H, D)$.

Theorem 6.10.1 *The set* $\mathcal{M}(\delta, H)/ \sim$ *is complete for the Gromov–Hausdorff topology and so is* $\mathcal{M}(\delta, H, D)/ \sim$.

Proof We just sketch the proof. The details can be found in [4], Theorem 4.5. We consider a Cauchy sequence (Y_n, d_n) in $\mathcal{M}(\delta, H)/ \sim$. Let Γ_n denote the fundamental group of Y_n. The following Lemma, due to G. Reviron is the main tool in the proof.   □

Lemma 6.10.2 ([26]) *Let* (Y_n, d_n) *be a Cauchy sequence of compact metric spaces with universal cover* (X_n, d_n) *and fundamental group* Γ_n. *Assume that the global systole of* Γ_n *is uniformly bounded below,* $\text{sys}_{\Gamma_n}(X_n) \geq s_0 > 0$. *Then,* (Y_n, d_n) *converges for the Gromov–Hausdorff topology to a compact metric space* (Y, d). *The limit space* (Y, d) *has a universal cover* (X, d) *satisfying* $\text{Ent}(X, d) = \lim \text{Ent}(X_n, d_n)$. *Moreover, the fundamental groups of* Y *and* Y_n *are isomorphic for* n *large enough.*

Indeed, we can verify that the Cauchy sequence (Y_n, d_n) satisfies the hypothesis of Lemma 6.10.2, therefore (Y_n, d_n) converges to a metric space (Y, d) and it remains to verify that (Y, d) is in $\mathcal{M}(\delta, H)$. In order to verify the hypotheses of Lemma 6.10.2, we first observe that since (Y_n, d_n) is locally CAT(0) and satisfies the property of extension of geodesics, then (Y_n, d_n) has a universal cover (X, d) which is CAT(0), thus Busemann, cf. [5], Lemma 6.41. On the other hand, since the sequence (Y_n, d_n) is a Cauchy sequence and the diameter is 1-Lipschitz with respect to the Gromov–Hausdorff distance, there exists D such that $\text{diam}(Y_n, d_n) \leq D$. Hence, $(Y_n, d_n) \in \mathcal{M}(\delta, H, D)$ and by Lecture 8, Proposition 6.7.5, there exists $s_0 := s_0(\delta, H, D) > 0$ such that $\text{sys}_{\Gamma_n}(X_n) \geq s_0$. Lemma 6.10.2 then applies and (Y_n, d_n) converges to a compact metric space (Y, d). In order to end the proof of Theorem 6.10.1, we need to show that the limit space (Y, d) is in $\mathcal{M}(\delta, H)$, i.e. is locally CAT(0), verifies the property of extension of local geodesics, and has a universal cover which is δ-hyperbolic. This essentially follows from the continuity of the distance under Gromov Hausdorff convergence and from the fact that all these properties are defined by inequalities on expression involving the distance, we refer to [4], Theorem 4.5 for details. We finally observe that $\mathcal{M}(\delta, H, D)/ \sim$ is a closed subset of $\mathcal{M}(\delta, H)$ hence complete.

We now state a compactness Theorem.

Theorem 6.10.3 *The set $\mathcal{M}(\delta, H, D)/ \sim$ is compact for the Gromov–Hausdorff topology.*

Proof It suffices to show that $\mathcal{M}(\delta, H, D)/ \sim$ is precompact since it is already complete. This will follow from the precompactness criterium that we describe now.

For (Y, d) a compact metric space, and $\epsilon > 0$, let us denote by $N_\epsilon(Y, d)$ be the minimal cardinality of an ϵ-net of (Y, d). A subset $\mathcal{M}' \subset \mathcal{M}$ is precompact for the Gromov–Hausdorff distance if and only if for every ϵ, there exists N_ϵ such that for every $(Y, d) \in \mathcal{M}'$, then $\mathbf{N}_\epsilon(Y, d) \leq N_\epsilon$.

We now apply this criterium to $\mathcal{M}' = \mathcal{M}(\delta, H, D)$. Let us consider $(Y, d) \in \mathcal{M}(\delta, H, D)$ and $\pi : (X, d) \to (Y, d)$ the projection from the universal cover (X, d) of (Y, d). For every $\epsilon > 0$, we have $N_\epsilon(Y, d) \leq N_\epsilon(B(x, D))$, where $y = \pi(x)$ is any point in Y and $B(x, D) \subset X$ is the ball of radius D centered at x in X. We recall, from Lecture 4 that, for every $\lambda > 1$,

$$N_\epsilon\left(B(x, D)\right) = \mathrm{Cover}(x, \epsilon, D)$$
$$\leq \mathrm{Pack}\left(x, \frac{\epsilon}{2}, D + \frac{\epsilon}{2}\right)$$
$$\leq \mathrm{Pack}\left(x, \lambda\frac{\epsilon}{2}, \lambda\left(D + \frac{\epsilon}{2}\right)\right).$$

We deduce that for every λ such that $\frac{\lambda\epsilon}{2} > D$,

$$N_\epsilon\left(B(x, D)\right) \leq \mathrm{Pack}_\Gamma\left(x, \lambda\frac{\epsilon}{2} - D, \lambda\left(D + \frac{\epsilon}{2}\right)\right)$$
$$\leq \frac{\mu_x^\Gamma\left(B(x, \lambda(D + \frac{\epsilon}{2}))\right)}{\mu_x^\Gamma\left(B(x, \lambda\frac{\epsilon}{2} - D))\right)}$$
$$\leq \mathrm{Pack}\left(x, \lambda\frac{\epsilon}{2}, \lambda\left(D + \frac{\epsilon}{2}\right)\right).$$

Choosing $\lambda = \frac{4D}{\epsilon}$, we get

$$N_\epsilon\left(B(x, D)\right) \leq \frac{\mu_x^\Gamma\left(B(x, \frac{4D}{\epsilon}(D + \frac{\epsilon}{2}))\right)}{\mu_x^\Gamma\left(B(x, D)\right)}$$
$$\leq C(\delta, H, D, \epsilon),$$

where the last inequality comes from either Theorem 6.5.3 or 6.5.10 according to whether $\frac{4D}{\epsilon}(D + \frac{\epsilon}{2}) < 12(D + 2\delta)$ or $\frac{4D}{\epsilon}(D + \frac{\epsilon}{2}) \geq 12(D + 2\delta)$ with

$C(\delta, H, D, \epsilon)$ the corresponding constant. We therefore deduce that for every $(Y, d) \in \mathcal{M}(\delta, H, D)$ and $\epsilon > 0$,

$$N_\epsilon(Y, d) \leq N_\epsilon := C(\delta, H, D, \epsilon),$$

which concludes the prove of Theorem 6.10.3.   □

6.10.2 Topological Finiteness

When we restrict the set of metric spaces under consideration to be also topological manifolds, we show that this set contains only finitely many topological types of manifolds. The next definitions are due to S. Ferry and play an important role in finiteness Theorems, [14].

Definition 6.10.4 A continuous function $\rho : [0, R[\to \mathbf{R}_+$ such that $\rho(0) = 0$ and $\rho(t) \geq t$ is a contractibility function for a metric space (Y, d) if for every $y \in Y$ and every t, R with $0 < t \leq R$, the ball $B(y, t)$ is contractible in $B(y, \rho(t))$.

We also define

Definition 6.10.5 Let $\rho : [0, R[\to \mathbf{R}_+$ be a continuous function such that $\rho(0) = 0$. Let us define $\mathcal{M}^{man}(n, \rho)$ as the set of isometry classes of metric spaces (Y, d) such that Y is a topological manifold of dimension n with ρ as a contractibility function.

Theorem 6.10.6 ([14], Theorem 1) *Let $\mathcal{M}' \subset \mathcal{M}^{man}(n, \rho)$ be a subset such that $\overline{\mathcal{M}'}$ is compact in $\mathcal{M}/ \sim$, then $\mathcal{M}'$ contains finitely many topological types of manifolds.*

Let us define $\mathcal{M}^{man}(n, \delta, H, D)$ the set of isometry classes of metric spaces in $\mathcal{M}(\delta, H, D)$ such that Y is a topological manifold of dimension n. Theorem 6.10.6 has the following Corollary.

Corollary 6.10.7 *The set $\mathcal{M}^{man}(n, \delta, H, D)$ contains finitely many topological types of manifolds.*

Proof Let $\pi : (X, d) \to (Y, d) \in \mathcal{M}^{man}(n, \delta, H, D)$ be the universal cover of $(Y, d) \in \mathcal{M}^{man}(n, \delta, H, D)$ and Γ the fundamental group of Y. We have shown that the global systole of Γ is bounded below

$$\text{sys}_\Gamma(X) \geq s_0(\delta, H, D) > 0.$$

Therefore, since X is a CAT(0)-space, for every $x \in X$, $B(x, t) \subset X$ is contractible. Moreover, for all $0 < t \leq \frac{s_0}{2}$ the ball $B(x, t)$ is isometric to $B(y, t) \in Y$ where $y = \pi(x)$, therefore $\rho(t) : [0, \frac{s_0}{2}[\to \mathbf{R}_+$, defined by $\rho(t) = t$, can be chosen as a contractibility function of (Y, d).   □

6.10.3   Growth and Entropy of Groups

Let (Γ, S) be a finitely generated group and $(\mathcal{G}(\Gamma, S), d_S)$ its Cayley graph. Notice that left multiplication induces an isometric action of Γ on $(\mathcal{G}(\Gamma, S), d_S)$. We define the entropy of Γ with respect to S by

Definition 6.10.8 $\mathrm{Ent}(\Gamma, S) = \lim_{R\to\infty} \frac{1}{R} \log \#\{\gamma \in \Gamma \mid d_S(e, \gamma) \leq R\}$.

Notice that for a finitely generated group Γ, if $\mathrm{Ent}(\Gamma, S) > 0$ for one generating set S, then $\mathrm{Ent}(\Gamma, S) > 0$ for every S, and that $\mathrm{Ent}(\Gamma, S) > 0$ if and only if Γ has exponential growth. On the other hand, if Γ has polynomial growth, then $\mathrm{Ent}(\Gamma, S) = 0$ for every generating set S.

Examples 6.10.9

1. When Γ is the non abelian free group on k generators $\mathbf{F}_k$ and $\mathbf{S}_k$ the standard generating set, $\mathrm{Ent}(\mathbf{F}_k, \mathbf{S}_k) = \log(2k - 1)$.
2. When $\Gamma = \mathbf{Z}^k$, we have $\mathrm{Ent}(\mathbf{Z}^k, S) = 0$ for every generating set S since $\mathbf{Z}^k$ has polynomial growth.
3. Let us consider the matrix $A = \begin{pmatrix} 2 & 1 \\ 1 & 1 \end{pmatrix} \in SL(2, \mathbf{Z})$ and define the group
   $\Gamma_A = \mathbf{Z} \ltimes \mathbf{Z}^2$ where the product law is given by $(n, (p, q)) \cdot \left(n', (p', q')\right) = \left(n + n', (p, q) + (p', q')A^n\right)$. It can easily be shown that Γ_A is a solvable non nilpotent group of exponential growth. We leave it as an exercise.

However, the entropy depends on the finite generating set S and is in general impossible to compute. Taking the infimum over all finite generating sets of a group gives the "algebraic entropy" $\mathrm{Ent}\,\Gamma$ of Γ defined as follows.

Definition 6.10.10 $\mathrm{Ent}\,\Gamma := \inf_S \mathrm{Ent}(\Gamma, S)$, where S is a finite generating sets of Γ.

Answering a question of M. Gromov, Wilson constructed an example of a finitely generated group Γ of exponential *i.e.* $\mathrm{Ent}(\Gamma, S) > 0$ for every finite generating set S, but with vanishing algebraic entropy, $\mathrm{Ent}\,\Gamma = 0$. On the other hand, D. V. Osin has shown that non nilpotent solvable groups have positive algebraic entropy, cf. [25]. M. Koubi also proved that when Γ is a hyperbolic group, then $\mathrm{Ent}\,\Gamma > 0$, cf. [22].

The following observation is fundamental for getting lower bounds on the entropy.

Lemma 6.10.11 *Let (Γ, S) be a finitely generated group. Assume that there exist two elements $a, b \in \Gamma$ generating a non abelian free subgroup of Γ. Then,*

$$\mathrm{Ent}(\Gamma, S) \geq \frac{\log 3}{\sup\left(l_S(a), l_S(b)\right)},$$

where $l_S(a) = d_S(e, a)$ and $l_S(b) = d_S(e, b)$. Similarly, if the elements $a, b \in \Gamma$ generate a non abelian free semi-group, we have

$$\text{Ent}(\Gamma, S) \geq \frac{\log 2}{\sup (l_S(a), l_S(b))}.$$

Proof We prove the Lemma in the case where the elements a, b generate a non abelian free group; the second case is identical. We let $L_S(a, b) := \sup (l_S(a), l_S(b))$. We also write $\mathbf{F}_{a,b}$ for the non abelian free group on a and b and $S_{a,b}$ the canonical generating set of $\mathbf{F}_{a,b}$, namely, $S_{a,b} := \{a^{\pm 1}, b^{\pm 1}\}$. For $k \in \mathbf{N}$, we let $\mathbf{B}_{S_{a,b}}(k) := \{\gamma \in \mathbf{F}_{a,b} \mid l_{S_{a,b}}(\gamma) \leq k\}$ and $B_S(k) := \{\gamma \in \Gamma \mid l_S(\gamma) \leq k\}$, we have

$$B_{S_{a,b}}(k) \subset B_S (L_S(a, b)\, k),$$

therefore,

$$\log 3 = \lim_{k \to \infty} \frac{1}{k} \log \left(\# B_{S_{a,b}}(k)\right) \leq \lim_{k \to \infty} \frac{1}{k} \log \left(\# B_S (L_S(a, b)\, k)\right),$$

hence

$$\log 3 \leq L_S(a, b) \, \text{Ent}(\Gamma, S).$$

$\square$

This Lemma shows that getting a lower bound on the algebraic entropy of a group Γ amounts to finding, for *every* finite generating set S of Γ, two elements $a, b \in \Gamma$ generating a non abelian free group and with $L_S(a, b)$ not too large. For linear groups over a field $\mathbf{K}$ that is, $\Gamma \subset GL(n, \mathbf{K})$, the Tits alternative proves the existence of free subgroup as soon as Γ is not virtually solvable.

Theorem 6.10.12 ([27]) *Let Γ be a finitely generated group in $GL(n, \mathbf{K})$. Then, either Γ is virtually solvable or Γ contains a non abelian free subgroup.*

In particular, non virtually solvable linear groups have exponential growth. In the following statement, it is shown that for non virtually solvable linear groups, one can find pairs of elements generating non abelian free semi-groups at uniform bounded distance from the Identity with respect to all generating sets.

Theorem 6.10.13 ([13]) *Let Γ be a finitely generated linear group $GL(n, \mathbf{K})$ over a field of characteristic 0. If Γ is not virtually solvable, there exists $N \in \mathbf{N}^*$ such that for every generating set S, there exists $a, b \in B_S(N)$ such that a, b generates a non abelian free semi-group. In particular, the algebraic entropy of Γ is positive, $\text{Ent}\,\Gamma > 0$.*

For solvable groups, D. V. Osin proved the following alternative,

Theorem 6.10.14 ([25]) *Let Γ be a finitely generated (non necessarily linear) group. Then either Γ is virtually nilpotent or Γ has positive algebraic entropy.*

From this discussion, we get the following corollary for linear groups $\Gamma \in GL(n, \mathbf{K})$ over a field of characteristic 0: either Γ is virtually nilpotent or Ent $\Gamma > 0$.

A natural question is whether, given a class $\mathcal{C}$ of groups with exponential growth, there exists a universal positive constant $C > 0$ such that

$$\text{Ent } \Gamma \geq C,$$

for every group $\Gamma \in \mathcal{C}$.

6.11 Lecture 10

In this lecture we consider the class $\mathcal{C}$ of groups Γ which have an action on some δ-hyperbolic metric space (X, d) with $\text{Ent}(X, d) \leq H$ and such that diam $X/\Gamma \leq D$. We show that there exists a constant $C(\delta, H, D) > 0$ such that the algebraic entropy of every $\Gamma \in \mathcal{C}$ satisfies Ent $\Gamma \geq C(\delta, H, D)$.

6.11.1 Algebraic Entropy of Groups Acting on Hyperbolic Metric Spaces

For Γ a finitely generated group, S a subset of Γ and $N \in \mathbf{N}$, we denote by S^N the set of elements $\gamma \in \Gamma$ that can be written as a product of at most N elements in S. Given two elements a, b in Γ, we write $\langle a, b \rangle_+$ the semi-group generated by a and b.

Theorem 6.11.1 *There exists $N_0 := N_0(\delta, H, D) > 0$ such that for every group Γ acting properly discontinuously on a δ-hyperbolic non elementary metric space (X, d) with $\text{Ent}(X, d) \leq H$ and such that $\text{diam}(X, d) \leq D$, we have:*

1. *For every subset S of Γ such that the group $\langle S \rangle$ generated by S is not virtually cyclic, there exists $\gamma \in S^{3N_0}$ and $\sigma \in S$ such that one of the semi-groups $\langle \gamma^{14}, \sigma\gamma^{14}\sigma^{-1} \rangle_+$ or $\langle \gamma^{14}, \sigma\gamma^{-14}\sigma^{-1} \rangle_+$ is free.*
2. *The algebraic entropy of any finitely generated non virtually cyclic subgroup $\Gamma' \leq \Gamma$*

$$\text{Ent } \Gamma' \geq \frac{\log 2}{42N_0 + 2}.$$

In particular Ent $\Gamma \geq \frac{\log 2}{42N_0 + 2}$.

Proof We sketch the proof; the details can be found in [4], Proposition 5.18. The proof relies on the following theorem by Breuillard–Green–Tao that we recall for convenience.   □

Theorem 6.11.2 ([8]) *For every $p \in \mathbf{N} \setminus \{0\}$, there exists $N(p) \in \mathbf{N} \setminus \{0\}$ such that the following holds: for every group G and every generating set S of G, if there exists a finite subset $A \subset G$ such that*

1. $S^{N(p)} \subset A$
2. $\frac{\#A \cdot A}{\#A} \leq p$,

then G is virtually nilpotent.

We apply Theorem 6.11.2 to $G = \langle S \rangle \leq \Gamma$. Let $R_0 := 20(D + 2\delta)$ and $x \in X$. We define

$$A := \{g \in G \mid d(x, gx) \leq R_0\}.$$

By the triangle inequality, we have

$$A \cdot A \subset \{g \in G \mid d(x, gx) \geq 2R_0\},$$

We then have

$$\frac{\#A \cdot A}{\#A} \leq \frac{\mu_x^G(B(x, 2R_0))}{\mu_x^G(B(x, R_0))}$$

$$\leq \frac{\mu_x^\Gamma(B(x, 2R_0 + \frac{R_0}{2}))}{\mu_x^\Gamma(B(x, \frac{R_0}{2}))}$$

$$\leq p := p(\delta, H, D) = 3 \cdot 5^{\frac{25}{4}} e^{300H(D+2\delta)},$$

where the second inequality comes from the Bishop–Gromov Theorem for subgroups, Corollary 6.5.9 and the third one from the Bishop–Gromov Theorem 6.2.8 with (r, R) replaced by $(\frac{R_0}{2}, \frac{5}{2}R_0)$. Since the group G is not virtually cyclic, we have by Theorem 6.11.2

$$S^{N(p))} \not\subset A.$$

Therefore, setting $N_0 := N_0(\delta, H, D) = N(p(\delta, H, D))$, there exists $\gamma_0 \notin A$ such that

$$\gamma_0 \in S^{N_0} \quad \text{and} \quad d(x, \gamma_0 x) > R_0 \geq \frac{31\delta}{2}. \tag{6.13}$$

Now, given a finite subset Σ of G, we define

Definition 6.11.3 $L(\Sigma) := \inf_{x \in X} \max_{\gamma \in \Sigma} d(x, \gamma x)$, and $L(\gamma) := L(\{\gamma\})$.

For a non trivial isometry γ of (X, d), we define its "asymptotic displacement" $\ell(\gamma)$ by

Definition 6.11.4 $\ell(\gamma) := \lim_{k \to \infty} \frac{d(x, \gamma^k x)}{k}$.

The reader can easily check, using the triangle inequality, that the limit exists and does not depend on $x \in X$. Notice that

$$\ell(\gamma) \leq L(\gamma),$$

and that for $k \in \mathbf{Z}^*$,

$$\ell(\gamma^k) = |k|\ell(\gamma). \tag{6.14}$$

By (6.13), we have

$$L(S^{N_0}) \geq \frac{31\delta}{2}. \tag{6.15}$$

The following Lemma says that if a subset Σ of G has a large joint displacement, then there exists an element $\gamma \in \Sigma^3$ with large asymptotic displacement.

Lemma 6.11.5 *If $L(S^{N_0}) \geq \frac{31\delta}{2}$, then there exits $\gamma_1 \in S^{3N_0}$ such that $\ell(\gamma_1) \geq \delta$.*

Proof We skip the proof of this Lemma and refer to [5], Theorem 4.17. □

Now, since G is not virtually cyclic, there exists $\sigma \in S$ such that the group $\langle \gamma_1, \sigma \gamma_1 \sigma^{-1} \rangle$ is not virtually cyclic.

Claim 6.11.6 *One of the semi-groups $\langle \gamma_1^{14}, \sigma \gamma_1^{14} \sigma^{-1} \rangle_+$ or $\langle \gamma_1^{14}, \sigma \gamma_1^{-14} \sigma^{-1} \rangle_+$ is free.*

The Claim corresponds to the first part of Theorem 6.11.1. The second part is a consequence of Lemma 6.10.11. Let us assume for example that the semi-group $\langle \gamma_1^{14}, \sigma \gamma_1^{14} \sigma^{-1} \rangle_+$ is free and write $\alpha := \gamma_1^{14}$, $\beta := \sigma \gamma_1^{14} \sigma^{-1}$. Since $\gamma_1 \in S^{3N_0}$, we have $l_S(\alpha) \leq 14(3N_0)$ and $l_S(\beta) \leq 14(3N_0) + 2$. By the second part of Lemma 6.10.11, we get

$$\mathrm{Ent}(\Gamma, S) \geq \frac{\log 2}{42N_0 + 2}.$$

It remains to prove Claim 6.11.6. This Claim will be basically a consequence of the following "ping-pong Lemma" that we now formulate in our particular setting.

Lemma 6.11.7 *Let α, β be two hyperbolic isometries of (X, d). We assume that there exist subsets $U_\alpha^\pm, U_\beta^\pm$ of X and $x \in X \setminus (U_\alpha^+ \cup U_\beta^+)$ such that*

$$U_\alpha^+ \subset X \setminus (U_\beta^- \cup U_\beta^+),$$

$$U_\beta^+ \subset X \setminus (U_\alpha^- \cup U_\alpha^+),$$

Fig. 6.27 Ping-pong

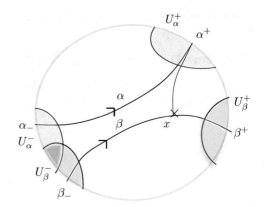

and such that $\alpha^k(\{x\} \cup U_\beta^+) \subset U_\alpha^+$ and $\beta^k(\{x\} \cup U_\alpha^+) \subset U_\beta^+$ for every $k \in \mathbf{N}^*$, then the semi-group $\langle \alpha, \beta \rangle_+$ generated by α and β is free.

Proof We want to prove that, given a non trivial word $w(\alpha, \beta)$ in α and β, then $w(\alpha, \beta) \neq Id$. Let us consider a non trivial word $w(\alpha, \beta)$. We can assume without loss of generality that the word $w(\alpha, \beta)$ starts with α. There are two cases (Fig. 6.27).

1. $w(\alpha, \beta) = \alpha^{n_1} \beta^{p_1} \cdots \alpha^{n_k} \beta^{p_k}$;
2. $w(\alpha, \beta) = \alpha^{n_1} \beta^{p_1} \cdots \alpha^{n_k}$

where $n_i, p_i \in \mathbf{N}^*$.

It is straightforward to check from the assumptions that $w(\alpha, \beta) \cdot x \in U_\alpha^+$. Therefore, since $x \notin U_\alpha^+$, we deduce that $w(\alpha, \beta) \neq Id$. □

We now briefly sketch how to deduce the claim from the ping-pong lemma. The details can be found in [5], Proposition 4.6 and 4.9. Writing $\alpha := \gamma_1^{14}$ and $\beta := \sigma \gamma_1^{14} \sigma^{-1}$, we observe that $\ell(\alpha) \geq 14\delta$ and $\ell(\beta) \geq 14\delta$ by Lemma 6.11.5 and (6.14). Now, the idea is the following: since the asymptotic displacement of α and β is large with respect to δ and since the fixed point sets $\{\alpha^\pm\}$, and $\{\beta^\pm\}$ of α, β are disjoint, one can construct neighbourhoods of their fixed points such as in the ping-pong Lemma for $\{\alpha, \beta\}$ or $\{\alpha, \beta^{-1}\}$. Given γ an isometry of (X, d) and $x \in X$, let us define the Dirichlet domain of γ at x as

$$D_\gamma(x) := \{y \in X \mid \min_{k \in \mathbf{Z}} d(y, \gamma^k x) = d(y, x)\},$$

and

$$U_\gamma^\pm(x) := \cup_{k \in \mathbf{N}^*} D_{\gamma^{\pm k}}(x). \tag{6.16}$$

We choose an axis c_β of β. We can show that, up to a change of orientation of α, there is a point x on c_β which is roughly a projection of α^+ onto c_β such that for every $p, q \in (\mathbf{Z}^* \times \mathbf{Z}^*) \setminus (\mathbf{Z}^- \times \mathbf{Z}^-)$,

$$d(\alpha^p x, \beta^q x) > \max\{d(x, \alpha^p x), d(x, \beta^q x)\} + 2\delta. \tag{6.17}$$

We skip the proof of (6.17) and refer to [5], Proposition 4.9, for the details. Now, we set $U_\alpha^\pm := U_\alpha^\pm(x)$ and $U_\beta^\pm := U_\beta^\pm(x)$ and prove that the hypotheses of the Ping-Pong lemma 6.11.7 are satisfied. By definition we have $x \notin U_\alpha^+ \cup U_\beta^+$ and $\alpha^k x \in U_\alpha^+$, $\beta^k x \in U_\beta^+$ for every $k \in \mathbf{N}^*$. Let $y \in U_\beta^+ \cup U_\beta^-$. We first observe that, by definition, there exists $k \in \mathbf{Z}^*$ such that $y \in D_\beta(\beta^k x)$ hence

$$d(y, \beta^k x) \leq d(y, x). \tag{6.18}$$

Let us recall the quadrangle property of δ-hyperbolic metric spaces (see Sect. 6.3.1) for the points y, $\alpha^p x$, x and $\beta^k x$ with $k, p \in \mathbf{Z}^*$,

$$d(y, x) + d(\alpha^p x, \beta^k x) \leq \max\{d(y, \beta^k x) + d(x, \alpha^p x) \; ; \; d(y, \alpha^p x) + d(x, \beta^k x)\} + 2\delta. \tag{6.19}$$

By (6.17) and (6.19), we deduce

$$d(y, x) < \max\{d(y, \beta^k x) \; ; \; d(y, \alpha^p x)\}, \tag{6.20}$$

and since $d(y, x) \geq d(y, \beta^k x)$, we get that $d(y, x) < d(x, \alpha^p x)$, for every $p \in \mathbf{Z}^*$, which means that

$$y \in D_\alpha(x) = X \setminus (U_\alpha^+ \cup U_\alpha^-).$$

Thus, $U_\beta^+ \cup U_\beta^- \subset X \setminus (U_\alpha^+ \cup U_\alpha^-)$. We similarly verify the same hypothesis with the roles of α and β reversed. This proves that the hypotheses of the Ping-Pong lemma are satisfied and concludes the proof of Theorem 6.11.1.   □

6.11.2   Entropy of a δ-Hyperbolic Space with a Group Action

So far, we have investigated properties of groups Γ acting by isometries on a δ-hyperbolic metric space (X, d) such that $\mathrm{Ent}(X, d) \leq H$ and $\mathrm{diam}(X/\Gamma)$. In the other way round, we have the

Theorem 6.11.8 *Let (X, d) be a non elementary δ-hyperbolic metric space and Γ a group acting properly by isometries on (X, d) such that* $\mathrm{diam}\, X/\Gamma \leq D$. *Then*

$$\mathrm{Ent}(X, d) \geq \frac{\log 2}{27\delta + 10D},$$

We skip the proof of this Theorem and refer to [5], Proposition 5.10.

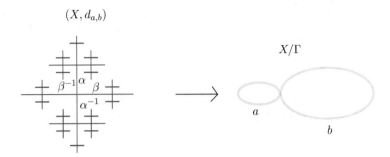

Fig. 6.28 Example

Coming back to the groups, let us consider a finitely generated group (Γ, S) which is δ-hyperbolic with respect to the distance d_S. As a direct Corollary of Theorem 6.11.8, we get

$$\mathrm{Ent}(\Gamma, S) \geq \frac{\log 2}{27\delta + 10}.$$

Example 6.11.9 Let us consider the regular tree of valence 4 which we see as the Cayley graph $X := \mathcal{G}(\mathbf{F}(\alpha, \beta), \mathbf{S}(\alpha, \beta))$ of the non abelian free group on two generators $\Gamma := \mathbf{F}(\alpha, \beta)$ with respect to the generating set $\{\alpha^{\pm 1}, \beta^{\pm 1}\}$. Given $a > 0$ and $b > 0$, we define the distance $d_{a,b}$ on X by setting the length of each edge $(\gamma, \alpha^{\pm 1}\gamma)$ equal to a and each edge $(\gamma, \beta^{\pm 1}\gamma)$ equal to b. For every $a > 0, b > 0$, the metric space $(X, d_{a,b})$ is δ-hyperbolic with $\delta = 0$ and the group $\Gamma = \mathbf{F}(\alpha, \beta)$ acts isometrically on $(X, d_{a,b})$. Notice that (Fig. 6.28)

$$\mathrm{diam}\left((X, d_{a,b})/\Gamma\right) = \frac{a + b}{2}.$$

Theorem 6.11.8 implies

$$\mathrm{Ent}(X, d_{a,b}) \geq \frac{\log 2}{5(a + b)}.$$

We leave to the reader to verify that, when $b = 1$ is fixed, then

$$\lim_{a \to 0} \mathrm{Ent}(X, d_{a,b}) = +\infty,$$

and

$$\lim_{a \to \infty} \mathrm{Ent}(X, d_{a,b}) = 0.$$

References

1. A.D. Aleksandrov, V.N. Berestovskij, I.G. Nikolaev, Generalized Riemannian spaces.Russ. Math. Surv. **41**(3), 1–54 (1986)
2. H. Bass, The degree of polynomial growth of finitely generated nilpotent groups. Proc. Lond. Math. Soc. (3) **25**, 603–614 (1972)
3. D. Burago, Yu. Burago, S. Ivanov, *A Course in Metric Geometry*, vol. 33 (American Mathematical Society (AMS), Providence, 2001)
4. G. Besson, G. Courtois, S. Gallot, A. Sambusetti, Compactness and finiteness theorems for Gromov-hyperbolic spaces (2020). Preprint
5. G. Besson, G. Courtois, S. Gallot, A. Sambusetti, Curvature-free Margulis lemma for Gromov-hyperbolic spaces (2020). arXiv: 1712.08386v2
6. D. Bakry, M. Emery, Diffusions hypercontractives. Sémin. de probabilités XIX, Univ. Strasbourg 1983/1984. Proc., Lect. Notes Math. **1123**, 177–206 (1985)
7. W. Ballmann, M. Gromov, V. Schroeder, Manifolds of nonpositive curvature. Progress in Mathematics, vol. 61 (Birkhäuser, Boston-Basel-Stuttgart, 1985), iv, p. 263
8. E. Breuillard, B. Green, T. Tao, The structure of approximate groups. Publ. Math. Inst. Hautes Étud. Sci. **116**, 115–221 (2012)
9. M.R. Bridson, A. Haefliger, *Metric Spaces of Non-Positive Curvature* (Springer, Berlin, 1999)
10. Yu. D. Burago, V.A. Zalgaller, *Geometric Inequalities. Transl. from the Russian by A. B. Sossinsky* (Springer, Berlin, 1988)
11. M. Coornaert, Th. Delzant, A. Papadopoulos, *Géométrie et théorie des groupes. Les groupes hyperboliques de Gromov. (Geometry and group theory. The hyperbolic groups of Gromov)* (Springer, Berlin, 1990)
12. G. Courtois, Lemme de Margulis à courbure de Ricci minorée (d'après Vitali Kapovitch et Burkhard Wilking), in *Séminaire Bourbaki. Volume 2013/2014. Exposés 1074–1088* (Société Mathématique de France (SMF), Paris, 2015), pp. 25–56
13. A. Eskin, S. Mozes, H. Oh, On uniform exponential growth for linear groups. Invent. Math. **160**(1), 1–30 (2005)
14. S. Ferry, Topological finiteness theorems for manifolds in Gromov-Hausdorff space. Duke Math. J. **74**, 95–106 (1994)
15. S. Gallot, *A Sobolev Inequality and Some Applications, Proceedings du Séminaire Franco-Japonais de Kyoto 1981* (Kaigai Publications, Tokyo, 1983), pp. 45–55
16. S. Gallot, Inégalités isopérimétriques, courbure de Ricci et invariants géométriques. II. C. R. Acad. Sci. Paris Sér. I Math. **296**(8), 365–368 (1983)
17. E. Ghys, P. de la Harpe eds., *Sur les groupes hyperboliques d'après Mikhael Gromov. (On the hyperbolic groups à la M. Gromov)* (Birkhäuser, Boston, 1990)
18. M. Gromov, Groups of polynomial growth and expanding maps (with an appendix by Jacques Tits). Publ. Math., Inst. Hautes Étud. Sci. **53**, 53–78 (1981)
19. M. Gromov, Hyperbolic groups. Essays in group theory. Publ. Math. Sci. Res. Inst. **8**, 75–263 (1987)
20. M. Gromov, *Metric structures for Riemannian and non-Riemannian spaces. Transl. from the French by S. M. Bates. With appendices by M. Katz, P. Pansu, and S. Semmes. Edited by J. Lafontaine and P. Pansu* (Birkhäuser, Basel, 2007)
21. Y. Guivarc'h, Groupes de Lie à croissance polynomiale. C. R. Acad. Sci. Paris Sér. A-B **272**, A1695–A1696 (1971)
22. M. Koubi, Croissance uniforme dans les groupes hyperboliques. Ann. Inst. Fourier (Grenoble) **48**(5), 1441–1453 (1998)
23. V. Kapovitch, B. Wilking, Structure of fundamental groups of manifolds with Ricci curvature bounded below (2011). arxiv 1105.5955
24. G.A. Margulis, *Discrete Groups of Motions of Manifolds of Non-Positive Curvature*. Proc. Int. Congr. Math., Vancouver 1974, vol. 2 (1975), pp. 21–34
25. D.V. Osin, The entropy of solvable groups. Ergod. Theory Dyn. Syst. **23**(3), 907–918 (2003)

26. G. Reviron, Topological rigidity under the hypothesis of bounded entropy and applications. Comment. Math. Helv. **83**(4), 815–846 (2008)
27. J. Tits, Free subgroups in linear groups. J. Algebra **20**, 250–270 (1972)
28. C. Villani, *Optimal Transport. Old and New*, vol. 338 (Springer, Berlin, 2009)

Chapter 7
All 4-Dimensional Smooth Schoenflies Balls Are Geometrically Simply-Connected

A Fast Survey of the Proof

Valentin Poénaru

Abstract In a famous paper, Barry Mazur showed (among other things) that a smooth n-dimensional Schoenflies ball, with one boundary point removed, is diffeomorphic to the n-ball, with one boundary point removed. Via the work of Smale and Milnor, that missing point was taken care of, except in dimension four, still mysterious to this day. In dimensions other then four, smooth Schoenflies balls are diffeomorphic to the standard ball. And then, the same dimension four is the only one where (in the compact case) simple connectivity does not imply geometric simple connectivity (GSC). We sketch here the proof that four-dimensional Schoenflies balls are GSC. Strangely enough, the proof requires infinite processes. We explain here, with only hints of proofs, the main ideas contained in a much longer paper where complete proofs are provided, available online, at the site arXiv, carrying the title *All smooth four-dimensional Schoenflies balls are geometrically simply connected*. See [10].

Keywords 4-Manifold · 4-Dimensional Schoenflies ball · Geometric simple connectivity · 3-Dimensional Poincaré conjecture

AMS Codes 57M05, 57M10 57N35

7.1 Introduction

We denote by Δ^2 the 2-skeleton of the 4^{d} smooth Schoenflies ball $\Delta^4_{\text{Schoenflies}}$ and what will be sketched here is the proof of the following result, which clearly implies that Δ^4 is GSC, since $\Delta^4 = N^4(\Delta^2) + \{3\text{-handles}\}$.

V. Poénaru (✉)
Université de Paris-Sud, Mathématiques, Orsay, France

© The Author(s), under exclusive license to Springer Nature Switzerland AG 2022
A. Papadopoulos (ed.), *Surveys in Geometry I*,
https://doi.org/10.1007/978-3-030-86695-2_7

Theorem 7.1.1 $N^4(\Delta^2)$ *is GSC. (And hence the smooth* 4^d *Schoenflies ball* Δ^4 *too.)*

The proof, of which the present survey only outlines the main ideas and also gives some hints, hoping that it may help reading the longer paper, which is available on arXiv at the site above, will rely heavily on Barry Mazur's celebrated result (see [4]) that

$$\Delta^n_{\text{Schoenflies}} - \{\text{a boundary point}\} \underset{\text{DIFF}}{=} B^n - \{\text{a boundary point}\}$$

which, in dimension four, the one of interest here (and the only one where $\Delta^n_{\text{Schoenflies}} \underset{\text{DIFF}}{=} B^n$ is not yet proved), is equivalent to the statement that int $\Delta^4_{\text{Schoenflies}} = R^4$.

The equivalence between the two statements

$$\Delta^4_{\text{Schoenflies}} - \{\text{boundary point}\} \underset{\text{DIFF}}{=} B^n - \{\text{boundary point}\},$$

and

$$\text{int } \Delta^4_{\text{Schoenflies}} \underset{\text{DIFF}}{=} R^4,$$

follows from the fact that there is a unique way of glueing, smoothly, a boundary R^3 to R^4. There are no wild Artin-Fox arcs in dimensions > 3.

There is a forerunner of the present survey, namely the old arXiv.org/abs/math. GT/0612554, from 2006; see here reference [7]. In point of fact, I actually had the 4^d smooth Schoenflies always behind the back of my mind, since the early seventies. And then, during a work session with Dave Gabai, at the IHES, in the 1990s, Mike Freedman made us the crucial suggestion, to connect the 3^d Poincaré Conjecture project on which we were working, with the 4^d DIFF Schoenflies problem. Although many of the basic ideas are already in that arXiv paper [7], there is a lot of progress here and now. Also, the arXiv paper [7] which treats simultaneously the GSC Schoenflies issue and the 3^d Poincaré Conjecture should certainly be much harder to read than the present work. Moreover, the long paper of which the present survey only gives a fast glimpse, is a vastly reworked and re-revised version of an older manuscript, summarized in the arXiv paper. And the old manuscript was hard to re-read, even for myself, in 2015, when I started working on the theorem above, with new ideas, after a long break.

It should be stressed that in this new version, the issue of showing that the Schoenflies DIFF 4^d-ball is GSC, is completely disentangled from any 3^d Poincaré Conjecture connection. In the arXiv paper [7] the proofs that $\Delta^4_{\text{Schoenflies}} \in$ GSC and that $\Delta^3 \times I \in$ GSC, were presented in parallel. [For those readers who might not know that, my approach to the 3^d Poincaré Conjecture was completely four-dimensional, so the connection with the 4^d Schoenflies was not that unnatural, in the 2006 paper. And one of the very important steps in my Poincaré Conjecture approach was to prove that for a homotopy 3-dimensional disc Δ^3 we have $\Delta^3 \times I \in$ GSC (this was the key to what I call the COHERENCE THEOREM, living at the core of my approach).]

In view of the context from which the present work stems, it should not be very surprising that the basic fact that

$$\partial \Delta^4_{\text{Schoenflies}} = S^3 \tag{7.0}$$

is nowhere explicitly used here. Actually it is not absent, but only hidden from view. I will explain this.

The only feature of the $\Delta^4 = \Delta^4_{\text{Schoenflies}}$ which the proof of our theorem explicitly uses, is the basic fact that

$$\text{int } \Delta^4 \underset{\text{DIFF}}{=} R^4, \tag{$*$}$$

a consequence of Barry's breakthrough result, as already said.

But then, here are two more or less immediate consequences of $(*)$.

(A) Let Δ^4 be any smooth compact bounded four-manifold satisfying the $(*)$. Then, since

$$\text{int } \Delta^4 \underset{\text{DIFF}}{=} \Delta^4 \cup (\partial \Delta^4 \times [0, 1)),$$

our Δ^4 itself already embeds smoothly in R^4, the initial Schoenflies assumption.

(B) Consider now a smooth embedding $B^4 \subset \text{int}\{\Delta^4 \text{ from A}\}$. Deleting the B^4 shows that

$$\partial \Delta^4 \times R \underset{\text{DIFF}}{=} S^3 \times R. \tag{$**$}$$

At this point, one can appeal to Grisha Perelman's famous result [1, 2, 5] and get that $\partial \Delta^4 = S^3$.

Question Is there a shorter road for going directly from $(**)$ to $\partial \Delta^4 = S^3$, rather than appealing to Perelman?

With all this, our hypothetical Δ^4 from the beginning of (A) *is* the smooth 4^d Schoenflies ball, anyway.

And, finally, the 3-handles of the $\Delta^4_{\text{Schoenflies}}$, which do not occur explicitly in the statement of our theorem, are actually present in our proof too. There is here a so-called RED 3^d-collapse which will play an essential role for achieving the desired GSC property. And 3^d collapsing flow means 3^d handles, which are then also present.

Since this is a survey paper it is only appropriate to explain to the reader the general context into which it fits, its history and its mathematical surroundings.

Towards 1900, when topology was in its infancy, the German mathematician A. M. Schoenflies raised the following question: If one embeds a copy of S^{n-1} into S^n, does it divide S^n into two copies of B^n or, put differently, what does that pair (S^n, S^{n-1}) look like? In those early days these distinctions which are so familiar to us today, a topological embedding, a locally flat topological embedding, a smooth, C^∞ embedding and, why not, a piecewise linear, PL-embedding, were inexistant.

Anyway, in the very early 1920s, Louis Antoine (who had lost his eyes in the Great War), showed that any topological embedding of S^1 into S^2 is standard, i.e., it divides S^2 into two discs. Remember, also, the Antoine's necklace, from the same vintage.

But, few years later J. W. Alexander showed that things get much more complicated in one dimension higher. For the embeddings $S^2 \subset S^3$ there is a big, nagging regularity issue. A simply topological embedding, like the so-called Alexander horned sphere, can divide S^3 into a copy of B^3 AND a wild set. That might have been the birth cry of wild topology, à la R. H. Bing. Then, a few years later, the same Alexander showed that for a C^∞ embedding $S^2 \subset S^3$ things are standard indeed. In those early days, that was already a hard theorem and the higher dimensions looked quite hopeless. And then, of course, people thought that if dimension $n = 3$ was hard, then $n = 4$ was harder and when n grows, they became even harder and harder. Smale's discovery that dimensions ≥ 5 were easier, was far in the future.

Anyway, after Alexander there was no progress for our problem, during more than 30 years. Then, in the late 1950s, there was a bolt from the blue sky, Barry Mazur (who was barely 18 years old at the time) solved, topologically at least, the Schoenflies problem for all n's, at one stroke (see [4]). At the time, his short proof seemed so strange and unbelievable, that it took the mathematical community 6 months to accept it. I have tried to explain its magic in my paper [8]. I will not state here Barry's full result. Suffices to say here that, among other things, Barry proved that if $S^{n-1} \subset S^n$ is a C^∞ embedding, and if X^n is any of the two compact bounded C^∞ manifolds into which S^{n-1} divides S^n, then there is a homeomorphism $X^n \xrightarrow{h} B^n$ which is C^∞ outside, *possibly*, of a boundary point. (I have already mentioned this result a few paragraphs above).

The X^n above is called the *smooth n-dimensional Schoenflies ball*, usually denoted Δ^n. Barry's breakthrough certainly opened the door for the glorious revolution of high-dimensional topology, by Milnor, Smale, Kervaire and others. Barry's result showed people that they did not have to be afraid of those high dimensions, since his proof worked, simultaneously for all. But then, what happens with that boundary point where h may not be smooth and hence the DIFF n-dimensional Schoenflies might fail to be true? Very soon after Barry's breakthrough, using very heavy artillery, Smale and Milnor showed that when $n \geq 5$, the n-dimensional Δ^n is diffeomorphic to B^n. Smale used here his h-cobordism theorem and Milnor used his surgery theory, recently developed then, by him and Kervaire. But for $n = 4$ the issue is still open today. Our theorem with which this survey starts is here a first step in the good direction.

And then, as I have explained in [11], the 4^d DIFF Schoenflies problem is itself a first step for understanding that big and mysterious abyss between DIFF and TOP in dimension four, and exactly four. To my mind that might be the biggest mystery in topology, today.

Let us turn now to geometric simple connectivity, a concept which occurred in connection with the Poincaré Conjecture, certainly in Smale's work. In [11] I have presented this GSC notion in its mathematical surroundings and historical context, and then, there are the papers [6] and [9] which show, respectively, the impact of this notion on both low-dimensional topology and on geometric group theory. For other insights concerning this crucial notion of GSC, see also [12–14].

So, for the time being, the 4^d Schoenflies problem $\Delta^4 \overset{?}{\underset{\text{DIFF}}{=}} B^4$ is still a big open problem and proving that $\Delta^4 \in$ GSC should be a first step towards its proof, like this was in Smale's proof of the high-dimensional Poincaré Conjecture. And, finally, the 4^d Smooth Schoenflies problem is a piece of the bigger and still very mysterious 4^d Smooth Poincaré Conjecture (see here [11] too).

Many thanks are due to Dave Gabai for the very many intense and useful conversations concerning the matters of the present paper, which we had during the years. It is likely that this paper would not have existed without his help. I would also like to thank Frank Quinn and Louis Funar for their comments and advices. Some of the crucial ideas in the present paper occurred to me in the middle of very intense conversations with Dave Gabai and Frank Quinn, a discussion turning around what later on became [7], in Princeton during the Spring of 2003. I also wish to thank Cécile Gourgues for the typing and Marie-Claude Vergne for the drawings, and the IHES for its friendly help.

This survey is an expanded version of a lecture I gave at the Encounter between Mathematicians and Theoretical Physicists, June 2015, Strasbourg.

7.2 The Two Collapsing Flows

We have here two "infinite collapsing flows" namely:

(i) The BLUE collapse $\Delta_1^4 \equiv \Delta^4 \cup \partial\Delta^4 \times [0, 1) \searrow$ pt and
(ii) The RED collapse $\Delta_1^4 \searrow \Delta^4$. [The compact collapse $\Delta^4 \cup \partial\Delta^4 \times [0, 1] \searrow \Delta^4$ will be of no use to us here.]

The (i) is a consequence of Barry's result, while the (ii) should be obvious.

There is one cell-decomposition of $\Delta_1^4 = R^4$ with a number of local features very useful for our present proof, but not to be described here, so that Δ^4 is a subcomplex and both collapses above make sense for *this* cell-decomposition. This is, of course, a variation on the DIFF Hauptvermutung theme.

But we will actually mostly work with 2^d versions of these flows. So, let X^4 be a cell-division of the open manifold $\Delta_1^4 = R^4$ with X^2 its 2-skeleton, and $\Delta^2 \subset X^2$ a subcomplex. [In the actual, long paper, the cell-division X^4 is chosen very carefully and that turns out to be essential for keeping the proof manageable in length.]

We will denote by $\Gamma(\infty)$ the 1-skeleton of X^2 and by $\Gamma(1) \subset \Gamma(\infty)$ the 1-skeleton of Δ^2. To $\Gamma(\infty)$, 2^d-cells are attached along a $\{link\} \subset \Gamma(\infty)$, so as to get $X^2 \supset \Delta^2$. This link comes with two distinct independent partitions

$$\{link\} = \sum_1^{\bar{n}} \Gamma_j + \sum_1^\infty C_i + \sum_1^\infty \gamma_k^0 \text{ (the RED partition)}$$

$$= \sum_1^\infty \eta_i + \sum_1^\infty \gamma_\ell^1 \text{ (BLUE partition).} \tag{7.1}$$

So we have

$$\Delta^2 = \Gamma(1) + \sum_1^{\bar{n}} D^2(\Gamma_j) \subset X^2 = \Gamma(\infty) + \sum_1^{\bar{n}} D^2(\Gamma_j) + \sum_1^\infty D^2(C_i)$$

$$+ \sum_1^\infty D^2(\gamma_k^0)$$

$$= \Gamma(\infty) + \sum_1^\infty D^2(\eta_i) + \sum_1^\infty D^2(\gamma_\ell^1).$$

Here D^2(curve) denotes the 2-cell cobounding the corresponding curve, the $D^2(\Gamma_j)$ are the 2-cells of Δ^2, $D^2(\gamma_k^0)$ are killed by the 3^d RED collapse, and $D^2(C_i)$ by the 2^d RED collapse. Similarly, $D^2(\gamma_\ell^1)$ are killed by the BLUE 3^d collapse and $D^2(\eta_i)$ by the 2^d BLUE collapse. Actually the RED 3^d collapse will be an important player too.

Inside $\Gamma(\infty)$ there are two sets of "spots" ($= $ 1-handles)

$$\text{RED 1-handles}: \quad R = (\underbrace{R_1, R_2, \ldots, R_n;}_{\text{in } \Gamma(1)} \underbrace{h_1, h_2, \ldots\ldots\ldots}_{\text{in } \Gamma(\infty)-\Gamma(1)}) \tag{7.2}$$

$$\text{BLUE 1-handles}: \quad B = (\underbrace{b_1, b_2, \ldots, b_{M'};}_{\text{in } \Gamma(1)} \underbrace{b_{M'+1}, b_{M'+2}, \ldots\ldots}_{\text{in } \Gamma(\infty)-\Gamma(1)})$$

and here $\Gamma(\infty) - R$, $\Gamma(\infty) - B$ and $\Gamma(1) - \sum_1^n R_i$ are trees; but $\Gamma(1) - \sum_1^{M'} b_i$ is, generally speaking not connected; we have $\bar{n} \geq n$, $M' \gg n$. If $X_0^2 \equiv X^2 - \sum_1^\infty \text{int } D^2(\gamma_k^0)$, then the RED 2^d collapse $X_0^2 \searrow \Delta^2$ is expressed by

$$C_i \cdot h_j = \text{easy id} + \text{nilpotent} \tag{7.3}$$

and, similarly, the BLUE 2^d collapse $X^2 - \sum_1^\infty \mathrm{int}\, D^2(\gamma^1) \searrow$ pt is expressed by

$$\eta_\alpha \cdot b_\beta = \text{easy id} + \text{nilpotent}. \tag{7.4}$$

Reminder. "Easy" id + nilpotent, in the context of the geometric intersection matrix $C_i \cdot h_j$ (3), for instance, means the following.

We have $C_i \cdot h_j = \delta_{ij} + a_{ij}$, where $a_{ij} \in Z_+$ and where $a_{ij} > 0$ implies that $i > j$. With the last inequality reversed, we get something which I call the "difficult" id + nilpotent (and there is no difference, of course, in the finite case). Of course, in the finite case the two notions are equivalent. The point is that easy id + nilpotent for the geometric intersection matrix of a handlebody decomposition (see [11, 13]) is the good way to express geometric simple connectivity (GSC), i.e., the fact that 1-handles are in cancelling position with the 2-handles (see here [6, 9, 11, 13, 14]).

The classical Whitehead manifold Wh^3 (see [15]), for instance, admits handlebody decompositions which have difficult id + nilpotent geometric intersection matrix, but none which are easy id + nilpotent. And it cannot be GSC because it comes with $\pi_1^\infty \mathrm{Wh}^3 \neq 0$. End of Reminder.

Concerning the RED 3^d collapse, the one which kills the 2-cells $D^2(\gamma_k^0)$, it is described as follows. For each $D^2(\gamma_k^0)$ there is a collapsibly triangulated 3-cell $B^3(\gamma_k^0)$ coming with $D^2(\gamma_k^0) \subset \partial B^3(\gamma_k^0)$ and also with a non-degenerate map

$$B^3(\gamma_k^0) \xrightarrow[g_k]{} X^3 (\equiv \text{3-skeleton of } X^4 = \Delta_1^4), \tag{7.4.1}$$

which has the following features.

- For $d_k^2 \equiv \partial B^3(\gamma_k^0) - \mathrm{int}\, D^2(\gamma_k^0)$, the $\gamma_k^0 \subset X_0^2 \equiv X^2 - \sum_k \mathrm{int}\, D^2(\gamma_k^0)$ extends to a non-degenerate simplicial map

$$d_k^2 \xrightarrow[g_k]{} X_0^2, \quad \partial d_k^2 = \gamma_k^0. \tag{7.4.2}$$

- The only contacts which the $g_k\, B^3(\gamma_k^0)$'s have with each other, are either telescopic embeddings, or boundary contacts.

Except for the RED 3^d collapse, the other collapses of dimension > 2, RED or BLUE, will be of no interest or use, for us.

Now, soon we will have to change the viewpoint, and then all the little theory above will be referred to as the *old* context. So, the X_0^2, X^2 will become just $X_0^2(\text{old}) \subset X^2(\text{old})$. In fact the old context is here only to help our exposition.

But before explaining our two successive changes of viewpoint which will be needed, here is a simple criterion for geometric simple-connectivity, the proof of which is elementary differential topology.

Lemma 7.2.1 *Let N_0^4 be a bounded smooth 4-manifold. Assume that there is a smooth embedding of a system of **discs***

$$\left(\sum_1^M \delta_i^2, \sum_1^M \partial\,\delta_i^2\right) \xrightarrow{\;\mathcal{J}\;} \left(\partial N_0^4 \times [0,1], \partial N_0^4 \times \{0\}\right), \tag{7.5}$$

*such that, inside $N_0^4 \cup \partial N_0^4 \times [0,1]$, the $\sum_1^M \delta_i^2$ are in **cancelling position** with the 1-handles of N_0^4.*

Then N_0^4 itself is GSC.

Remarks

(A) In (7.5) we have

$$\partial N_0^4 \times \{0\} = \partial N_0^4, \text{ inside } N_0^4 \cup \partial N_0^4 \times [0,1] \text{ of boundary } \partial N_0^4 \times \{1\}.$$

(B) Our cancelling condition can also be expressed as follows: $N^4\left(N_0^4 \cup \sum_1^M \delta_i^2\right)$ is GSC.]

7.3 The New Context and the Doubling

The present section will be purely 2-dimensional. For parametrizations, in addition to the coordinates (x, y, z, t) of R^4, we introduce two other axes $0 \geq \xi_0 \geq -1$ and $-\infty < \zeta < +\infty$. We will present two successive changes of viewpoint, namely

$$\text{old (section II)} \Longrightarrow \text{NEW} \Longrightarrow \text{DOUBLE}.$$

The $\Delta^2 \subset X^2(\text{old})$ is denoted $\Delta^2 \times (\xi_0 = 0)$ and to $\Gamma(1) = \Gamma(1) \times (\xi_0 = 0) \subset X^2(\text{old})$ we add $\Gamma(1) \times [0 \geq \xi_0 \geq -1]$. With this, we introduce the ABSTRACT 2-complex, not a priori embedded inside any ambient 4^d-space

$$X_0^2[\text{NEW}] \equiv \left\{\left[X^2(\text{old})(\text{section II}) - \sum_1^\infty \text{int } D^2(\gamma^0) - \sum_1^{\bar{n}} \text{int } D^2(\Gamma_i)(\xi_0 = 0)\right] \cup \right. \tag{7.6}$$

$$\underbrace{\cup}_{\Gamma(1)\times(\xi_0=0)} \Gamma(1) \times [0 \geq \xi_0 \geq -1]\Bigg\} \underbrace{\cup}_{\Gamma(1)\times(\xi_0=-1)} \Delta^2 \times (\xi_0 = -1) \subsetneq X^2[\text{NEW}] \equiv$$

$$\equiv X^2(\text{old}) \cup (\Gamma(1) \times [0 \geq \xi_0 \geq -1]) \cup \Delta^2 \times (\xi_0 = -1).$$

From now on, throughout this paper, it is $\Delta^2 \times (\xi_0 = -1)$ which will be *our* "Δ^2, 2-skeleton of $\Delta^4_{\text{Schoenflies}}$", and $\Gamma(\infty)$ is the 1-skeleton of $X^2_0[\text{NEW}]$. Forget the old meanings unless explicitly mentioned.

When, for defining $X^2_0[\text{NEW}]$ we delete the int $D^2(\gamma^0)$, int $D^2(\Gamma_j) \times (\xi_0 = 0)$, we leave a thin boundary collar $\gamma^0 \times [0, \varepsilon]$ in place, bringing the $\gamma^0 = \gamma^0 \times \{\varepsilon\}$ to the boundary. So $\partial X^2_0[\text{new}] = \sum_1^\infty \gamma^0_k + \sum_1^{\bar{n}} \Gamma_j \times (\xi_0 = 0)$.

The 2^d RED and BLUE collapsing flows extend to the context (7.6). For the RED flow notice the 2^d collapse, which replaces $X^2_0 \searrow \Delta^2$,

$$X^2_0[\text{NEW}] \searrow \Delta^2 \times (\xi_0 = -1), \tag{7.0}$$

which proceeds as follows: we start with the obvious collapse (see (7.3))

$$X^2(\text{old}) - \sum \text{int } D^2(\gamma^0) - \sum \text{int } D^2(\Gamma_i) \times (\xi_0 = 0) \searrow \Gamma(1) \times (\xi_0 = 0), \tag{7.1}$$

and then we continue with

$$\Gamma(1) \times [0 \geq \xi_0 \geq -1] \cup \Delta^2 \times (\xi_0 = -1) \searrow \Delta^2 \times (\xi_0 = -1). \tag{7.2}$$

We extend the R from (7.2) to a system of RED 1-handles R_0 of $X^2_0[\text{new}]$ defined by

$$R_0 \equiv \sum_1^\infty h_i(\text{from 2}) + \{\text{one NEW } h_i \text{ for each edge } e \subset \Gamma(1) \times (\xi_0 = 0)\}$$

$$+ \{\text{the } R_i \times [\xi_0 = -1], \text{ for } i = 1, 2, \ldots, n\} \text{ (These are the NEW } R_i\text{'s.)}.$$

If $e \subset \Gamma(1)$ is an edge, the $e \times [0 \geq \xi_0 \geq -1]$ become $D^2(C)$'s duals to the $h_i \in e \subset \Gamma(1) \times (\xi_0 = 0)$. Also, now the $D^2(\Gamma_j) \times (\xi_0 = -1)$ are the 2-handles of the $\Delta^2(\text{Schoenflies})$. There are no R_0's on the $P \times [0 \geq \xi_0 \geq -1]$'s.

There is also a 2^d BLUE collapsing flow, made precise in (7.8.2) below,

$$X^2[\text{NEW}] \searrow \text{pt}, \text{ NOT everywhere well defined}, \tag{7.8}$$

with a system of BLUE 1-handles

$$B_0 \equiv \left\{ \text{The } \sum_1^\infty b_i \text{ from (2), for } X^2 \text{ (old)} \right\}$$

$$+ \left\{ \text{one } b_i \text{ for each edge } e_i \subset \Gamma(1) \times (\xi_0 = -1), \text{ call this } \sum_1^M b_i \right\} \tag{7.8.0}$$

$$= B + \sum_1^M b_i .$$

There are no B_0's on $P \times [0 \geq \xi_0 \geq -1]$. Also, inside B_0 (7.8.0), the $\sum_1^\infty b_i$ and the $\sum_1^M b_i$ are independent of each other.

Here $M \geq M'$ (see (7.2)), and we will forget about M'. The labels η, γ^1 extend to X^2[NEW]. On the X^2(old) part these are just as before, in II) and then, by decree For the edges $e_i \subset \Gamma(1) \times (\xi_0 = -1)$, the $e_i \times [0 \geq \xi_0 \geq -1]$ are $D^2(\eta_i)$'s, duals to the $b_i \subset e_i$ $(1 \leq i \leq M)$. Next

$$\boxed{\text{The } D^2(\Gamma_j) \times (\xi_0 = -1) \text{ are, BLUE-wise, } D^2(\gamma^1)\text{'s.}} \qquad (8.1)$$

It will turn out that, this last prescription is the main achievement of the transformation old $\Rightarrow$ NEW. It will be very essential when we will want to implement the criterion of Lemma 7.2.1 to the Schoenflies context. Here is how, very explicitly, the (7.8) proceeds, well defined now as

$$X^2[\text{NEW}] - \text{int}\left(\underbrace{\sum_1^{\bar{n}} D^2(\Gamma_j \times (\xi_0 = -1)) + \sum_1^\infty D^2(\gamma^1)(\text{old})}_{\text{the } D^2(\gamma^1) \text{ (NEW)}} \right) \searrow \text{pt.} \qquad (7.8.2)$$

For this (7.8.2), we start with the collapse $\Gamma(1) \times [0 \geq \xi_0 \geq -1] \searrow \Gamma(1) \times (\xi_0 = 0)$, the inverse of (7.2), and then we continue normally, in X^2(old).

At this point, notice the contrast between the BLUE and the RED 2^d collapsing flows, when in the NEW context, in the region $[0 \geq \xi_0 \geq -1]$, schematized as

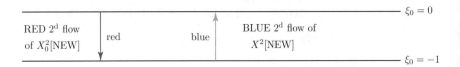

Here is finally, the double aim of our transformation

$$\text{NEW} \Longrightarrow \text{DOUBLE}$$

which will be very soon described. Notice, to begin with, that both in the old and in the NEW context, there is a priori a very complicated interaction of the two 2^d collapsing flows,

$$(\text{RED } 2^d \text{ flow}) \pitchfork (\text{BLUE } 2^d \text{ flow}) \neq \emptyset,$$

with dangerous loops like the one suggested below

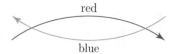

red

blue

Going to the DOUBLED context, eliminates on the one hand this potential trouble, while on the other hand keeping the framed condition from (8.1) with us.

We move now to the DOUBLED context, and as a preliminary to the DOU-BLING in question, we will introduce, on the ζ-axis $-\infty < \zeta < +\infty$, the quantities below

$$-\infty < r \text{ (for RED)} < b \text{ (for BLUE)} < +\infty$$

and also a $\beta > 0$ between r and b, coming with $0 < \beta \ll |b - r|$. The $|b - r|$ is assumed small with respect to the diameter of the 2-cells of $X^2[\text{NEW}]$.

We consider, next, the space $\Gamma(\infty) \times [r, b]$, which is made out of thin rectangles $e \times [r, b]$, where e is the generic edge of $\Gamma(\infty) \equiv \{1\text{-skeleton of } X^2[\text{NEW}]\}$.

There are four types of e's, namely

(I) $e \ni r \in R_0 - B_0$, (II) $e \ni r' \in R_0 \cap B_0$, (III) $e \ni b \in B_0 - R_0$, IV) $e \cap (R_0 \cup B_0) = \emptyset$. (7.9)

We denote by $c(r)$ the $\partial\{$rectangle of type I or IV$\}$ and $c(b)$ the $\partial\{$rectangle of type II or III$\}$. See here also Figs. 7.1 and 7.2, for the four kinds of rectangles $e \times [r, b]$, just mentioned.

We can introduce now the next ABSTRACT 2^d spaces

$$2X_0^2 \equiv (X_0^2[\text{NEW}] \times r) \underbrace{\cup}_{\Gamma(\infty) \times r} \{\Gamma(\infty) \times [r, b] \text{ with the interiors of the rectangles}$$

of type II, III deleted, leaving only a thin collar $c(b) \times [0, \varepsilon]$ in place, with $c(b) \times \{\varepsilon\}$

on the boundary$\} \underbrace{\cup}_{\Gamma(\infty) \times b} \left\{ X_b^2 \equiv \bigcup_1^\infty D^2(\eta_i) \times b \right\} \subsetneqq 2X^2 \equiv (X^2[\text{NEW}] \times r) \cup$

$(\Gamma(\infty) \times [r, b]) \cup X_b^2$. Here $\beta < \varepsilon$, and the $\eta_i = \eta_i$ [NEW] are $\{$the η_i's from (7.1) and the ones in (8.1)$\}$. (10)

Notice that

$$\partial(2X_0^2) = \sum_k \gamma_k^0 + \sum_j \Gamma_j \times (\xi_0 = 0) + \sum_i c(b_i) \quad (b_i \in B_0),$$

which are the free curves left behind by the surviving collars.

The general idea now is to keep the RED 2^d flow of $X_0^2[\text{new}]$ on the r-side of (10) and to transfer the BLUE 2^d flow of $X^2[\text{NEW}]$ on the b-side. The slightly more complicated detailed definitions are explained below.

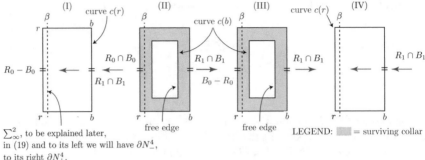

Σ_∞^2, to be explained later,
in (19) and to its left we will have ∂N_-^4,
to its right ∂N_+^4.

Fig. 7.1 The RED flow of $2X_0^2$. All the arrows here are RED. To the left of the edge $e \times r$ we have the flow $C \cdot h$, not represented in our drawing. To the right of the edges $e \times b$, we have the flow $\eta \cdot B$, which proceeds normally from the sites $R_1 \cap B_1$ in (II, III) and then, after having dealt with X_b^2, comes to hit the sites $R_1 \cap B_1$ in (I, IV). Next it continues through the (I, IV). Finally, we proceed normally on X_0^2 [NEW] $\times r$, leaving only $\Delta^2 \times (\xi_0 = -1)$ alive. So the RED flow on $2X_0^2$ proceeds, in order, like this: the flow $\eta \cdot B$ on X_b^2, collapse of the 2-cells (I), (IV) and finally, the RED flow on X_0^2[NEW] $\times r$. The families R_1, B_1, and do not mix the family R_1 with the element R_1 in (7.2), are defined by starting with the requirements $R_0 \subset R_1$, $B_0 \subset B_1$ and then proceeding like in our present Figs. 7.1 and 7.2

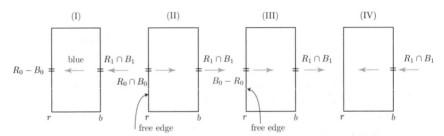

Fig. 7.2 The BLUE 2^d flow of $2X^2$. On the r-side there is NO 2^d-flow, on the b-side we have $\eta \cdot B$ and on $\Gamma(\infty) \times [r, b]$ the flow is given by the present blue arrows. This leads to (13) below. All the arrows in this figure are BLUE. We start at the free edges in II, III, next we handle the X_b^2 and we end with the rectangles I, IV. It is the (12) below which makes that there is no BLUE 2-flow in X^2[NEW] $\times r$

We have a RED collapsing flow

$$2X_0^2 \searrow \Delta^2 \times (\xi_0 = -1) \subset X_0^2[\text{new}] \times r \tag{7.11}$$

proceeding as follows: On X_0^2[new] $\times r$ we keep the RED 2-flow of X_0^2[NEW], the $C \cdot h$. On X_b^2 we have the flow $\eta \cdot b$ which is now, *by definition* the RED flow on X_b^2. To be precise, R_0, B_0 extend to larger families R_1, B_1, where *each* edge $e \times b \subset \Gamma(\infty) \times b$ carries now an element in $R_1 \cap B_1$. The text which comes with Fig. 7.1 gives more details concerning (7.11).

Figure 7.1 shows the contribution of $\Gamma(\infty) \cap [r, b]$ to the RED 2^d flow of $2X_0^2$.

For the γ_k^0's we have the obvious extensions of the (7.4.2), namely

$$d_k^2 \xrightarrow{\ g_k\ } X_0^2[\text{new}] \times r \subset 2X_0^2, \tag{7.11.1}$$

where every occurrence of $D^2(\Gamma_j) \times (\xi_0 = 0) \subset \{\text{the } g_k \, d_k^2 \text{ from (7.4.2)}\}$ gets replaced by the corresponding $(\Gamma_j \times [0 \geq \xi_0 \geq -1]) \cup D^2(\Gamma_j) \times (\xi_0 = -1)$. This way we get a 3^{d} RED flow on $2X^2$.

For the BLUE 2^{d} flow at the DOUBLED level, we have to move to $2X^2$ (10). We have already defined the families of 1-handles R_1, B_1, and now
We extend the framed formula in (8.1) to the following prescription: *All* the 2-cells of $X^2[\text{NEW}] \times r \subset 2X^2$ are $D^2(\gamma^1)$'s, BLUE-wise. [The 2^{d} BLUE flow of $2X^2$ stops at $\Gamma(\infty) \times r$.] \hfill (7.12)

Notice that the BLUE flow on $X^2[\text{NEW}]$ is NOT the restriction of the BLUE flow on $2X^2$. There is NO 3^{d} flow on $X^2[\text{NEW}] \times r$ either, our (12) is just a decree, for the space in (13) below. And, of course there are no $D^2(\gamma^1)$'s in X_b^2.

On the piece $X_b^2 \subset 2X^2$, the BLUE flow is given by $\eta \cdot B$ moved to the b-side. The BLUE flow on $\Gamma(\infty) \times [r, b]$ is displayed in Fig. 7.2.

For the BLUE flow defined this way, we have

$$2X^2 - \sum \overset{\circ}{D^2} \text{ (concerning, exactly, } \boldsymbol{ALL} \text{ the interiors of the 2-cells in } X^2[\text{NEW}] \times$$
r, which now are all $D^2(\gamma^1)$'s) $\searrow$ pt. \hfill (7.13)
In detail, this collapse proceeds as follows:

(i)   Starting at the free edges $(e \ni B_0) \times r$, the II, III in Fig. 7.2, one uses the (7.4) and collapses away the X_b^2.
(ii) Now one can collapse the 2-cells I, IV too (see Fig. 7.2).
(iii) Since on $X^2[\text{NEW}] \times r$ all the 2-cells are $D^2(\gamma^1)$'s (the extension of (8.1) by (12)), and since all the $B_0 \times r \in \Gamma(\infty) \times r$ are gone by (i) above, we are left exactly with

$$\Gamma(\infty) - B_0 = \Gamma(\infty) \times r - B_0 \times r \searrow \text{pt.}$$

Notice that now, at DOUBLED level, whenever a 2-cell carries both a BLUE and a RED collapsing arrow we always have the

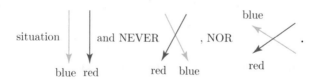

For any point $x \notin \operatorname{int}(D^2(\Gamma_j) \times (\xi_0 = -1))$, the RED 2^{d} flow of $2X_0^2$ incoming into x, defines an

$$\{\text{extended cocore of } x\} \subset 2X_0^2. \tag{7.14}$$

So, the {extended cocore of x} is the union of all the RED infinite collapsing flow lines which end at x. [The situations when this is not infinite, turn out to be trivial in our context.]

This is what I call a **full tree** based at x. By this I mean a tree which is PROPERLY embedded into $2X_0^2$ and which, outside of its root $x \in$ {extended cocore x}, SPLITS locally exactly in two, the $2X_0^2$. [By PROPER written with capital letters we mean (inverse image of a compact) = compact.]

The 1-skeleton of $2X^2$ is denoted by $2\Gamma(\infty)$, or sometimes $\Gamma(2\infty)$.

7.4   Four Dimensional Thickenings and Compactifications

The 2^{d} $2X_0^2$ as defined above is an ABSTRACT object not coming, naturally embedded nor even mapped into any ambient 4^{d}-space. We construct a 4^{d} regular neighbourhood $N^4(2X_0^2)$, containing a $N^4(2\Gamma(\infty))$ via a process of glueing together local well-defined pieces. For each vertex $P \in 2\Gamma(\infty)$ we will consider a $B^4(P)$ coming with a graph, embedded in $\partial B^4(P)$

$$\Gamma(P) \equiv \{\text{germ } 2X_0^2 \mid P\} \pitchfork S^3(P) (= \partial B^4(P)) \subset S^3(P). \tag{7.15}$$

This is, also, the link of the vertex $P \in 2X_0^2$. The edges of $\Gamma(P)$ are traces of the curves $\in$ {link} (analogue of (7.1), for $2X_0^2$) and next, joining appropriately vertices of adjacent $B^3(P_1)$, $B^3(P_2)$ with 1-handles, one builds up an $N^4(2\Gamma(\infty))$ which comes with a

$$\{\text{framed link}\} \subset \partial N^4(2\Gamma(\infty)), \tag{7.16}$$

and finally by adding 2^{d}-handles to (7.16) one gets the $N^4(2X_0^2)$. This construction is subjected to TWO CONDITIONS, listed below, the only restrictions we impose on it.
We want that

$$N^4(2X_0^2) \mid (\Delta^2 \times (\xi_0 = -1)) \underset{\mathrm{DIFF}}{=} \{N^4(\Delta^2), \text{ from our THEOREM}\}, \tag{7.17.I}$$

so as to keep our connection with Schoenflies alive.
For the curves $C = C(2X_0^2)$ and 1-handles $h = h(2X_0^2)$ we want to continue to have $C \cdot h = $ easy id $+$ nilpotent, like for $2X_0^2$. This will keep our geometry safe. (7.17.II)

But, *outside of* $\Delta^2 \times (\xi_0 = -1)$ we are allowed to cross curves inside the $\partial N^4(P)$'s (a step which does not conflict with (17.I) nor (17.II) either)

WE ARE NOT ALLOWED SUCH CROSSINGS for $\Gamma_j \times (\xi_0 = -1)$'s. (Here $C =$ "curve".)

This will mean allowable *local changes of topology* which will be very useful. Because of (17.II) the GLOBAL TOPOLOGY of $N^4(2X_0^2)$ will stay unchanged under the allowed crossings.

Our final aim in this paper is to get a smooth compactification $N^4(2X_0^2)^\wedge$ such that:

(A) We should have $N^4(2X_0^2)^\wedge \underset{\text{DIFF}}{=} N^4(\Delta^2)$. At this point, the fact that with the DOUBLING process we achieve that

$$(\text{Red } 2^{\text{d}} \text{ flow of } 2X_0^2) \pitchfork (\text{BLUE } 2^{\text{d}} \text{ flow of } 2X^2) = \emptyset$$

is important. Out of the 2^{d} flow of $X^2[\text{NEW}]$, RED and BLUE we can put now together the big 2^{d} RED collapsing flow on $2X^2$, which is behind our diffeomorphism above. And this also helps big for the next item, too, and so does also the framed condition in (8.1).

(B) To $N^4(2X_0^2)^\wedge$ we can apply the criterion from Lemma 7.2.1, and then, making use of (A) conclude that $N^4(\Delta^2)^\wedge \in$ GSC. The DOUBLING provides us with the spare parts for constructing the embedded exterior discs, but we need the BLUE flow of $X^2[\text{NEW}]$ (which is not the same as the one of $2X^2$), in order to put them together. $\square$

Now, at the level of $N^4(2X_0^2)$ the object in (7.14) becomes, with the same notations as in (7.14), used again, the following items:

We consider first the case when $x \in \Gamma(2\infty) \subset 2X_0^2$ and then lift it to $x \in \partial N^4(2X_0^2)$. This will be the case for the $R_1, R_2, \ldots, R_n \in \Gamma(1) \subset \Gamma(2\infty)$, in the compactification lemma 7.4.3 below. In this case, the {extended cocore x} $\subset N^4(2X_0^2)$, is a 3^{d} manifold which is embedded inside $N^4(2X_0^2)$, containing x on its boundary. It is, in fact the 3^{d} regular neighbourhood of (7.14). This 3^{d} regular neighbourhood is the following non compact 3-manifold for which we keep the same notation (7.14)

$$x \in \{\text{extended cocore } x\} \underset{\text{DIFF}}{=} B^3 - \{\text{a } \textit{tame} \text{ Cantor set} \subset \partial B^3\}, \qquad (*)$$

with an embedding {extended cocore x} $\subset N^4(2X_0^2)$ which is proper, and the full {extended cocore x} $\subset N^4(2X_0^2)$ is PROPER. We have, in this case, $x \in \partial$ {extended cocore}, $x \in \partial B^3$.

[Here proper means

$$\partial(\{\text{extended cocore } x\}) \subset \{\text{extended cocore } x\} \cap \partial N^4(2X_0^2),$$

and, as already said, PROPER means (inverse image of compact) = compact.]

In a second case, we have $x \in \text{int } D^2(C)$, when $C = C_i$ or γ_k^0. In this case, when we go to $N^4(2X_0^2)$, there is a properly embedded disc $D^2(x) \subset N^4(2X_0^2)$, centered at x and such that $D^2(x)$ and $D^2(C)$ meet transversally, $D^2(x) \pitchfork D^2(C)$.

Now, $D^2(x) \subset \partial$ {extended cocore}, with x like in ($*$) above, and now

$$\{\text{extended cocore } (x)\} - \text{int } D^2(x) \subset N^4(2X_0^2) \qquad (**)$$

is proper and {extended cocore (x)} $\subset N^4(2X_0^2)$ is PROPER. So, now

$$\partial\{\text{extended cocore } (x)\} - \text{int } D^2(x) = \partial\{\text{extended cocore}\} \cap \partial N^4(2X_0^2).$$

End of (18).

Our construction of $N^4(2X_0^2) \supset N^4(2\Gamma(\infty))$ is such that it comes with a PROPERLY embedded surface

$$\Sigma_\infty^2 \subset \partial N^4(2\Gamma(\infty)) \qquad (7.19)$$

inducing a SPLITTING

$$\partial N^4(2\Gamma(\infty)) = \partial N_-^4(2\Gamma(\infty)) \bigcup_{\Sigma_\infty^2} \partial N_+^4(2\Gamma(\infty)) \qquad (7.19.1)$$

with $\partial N_-^4, \partial N_+^4$ corresponding, roughly, to the r-side and b-sides in Figs. 7.1 and 7.2. So, "$-$" is RED and "$+$" is BLUE. The Σ_∞^2 is suggested, very schematically, by the dotted line in Fig. 7.1. Very importantly for us, there will be some CONFINEMENT conditions putting some curves of the link in ∂N_-^4 and others in ∂N_+^4. We will come back to that later on.

The little theory which we will develop next, in the present section, is developed with much more detail in our recent paper [13] (and see here also the reference [12]). So, here comes this little story.

When one considers the geometric intersection matrix of $2X_0^2$
$C_i \cdot h_j = \delta_{ij} + \text{nilpotent}$ (of the easy type), meaning $C_i \cdot h_j = \delta_{ij} + a_{ij}$, with $\{i\} = \{j\} = Z_+$ and with $a_{ij} > 0 \Longrightarrow i > j$, $\qquad (7.20)$
then $h_i \cup D^2(C_i) = \{\text{the } 4^d \text{ 1-handle}\} \cup \{\text{the } 4^d \text{ dual 2-handle}\} \underset{\text{DIFF}}{=} B^4$.

[We denote by h_i a 4^d 1-handle and by $D^2(C_i)$ the corresponding 4^d 2-handle. So, h_i and $D^2(C_i)$ are each a 4-ball and $h_i \cup D^2(C_i) = h_i \# D^2(C_i) \underset{\text{DIFF}}{=} B^4$ too.] We will define an oriented graph $\mathcal{M}$ having as vertices the indices i in (20) and if $a_{ij} \in Z_+$ is the off-diagonal term in (20) the #{arrows (= oriented edges) $i \to j$ in $\mathcal{M}$} $= a_{ij}$. For each vertex i of $\mathcal{M}$ we define the ABSTRACT 3-cell

$$\text{Box}\,(i) \equiv I_i \times D_i^2, \tag{7.21}$$

where

$$I_i = \underbrace{\partial\,(\text{core of the 2-handle } D^2(C_i))}_{\text{curve } C_i} - \{\text{core of the 1-handle } h_i\},$$

and

$$D_i^2 = \partial(\text{cocore of the 1-handle } h_i) - \{\text{the cocore of the 2-handle } D^2(C_i)\}.$$

When we have arrows $i \to j$ in $\mathcal{M}$, then this comes with natural maps which are smooth embeddings

$$I_j \xrightarrow{\lambda(i \to j)} I_i\,, \quad D_i^2 \xrightarrow{\mu(i \to j)} D_j^2\,, \text{ suggested in Fig. 7.3,}$$

out of which we build up the not everywhere well-defined smooth embedding, associated to $i \to j$

$$\text{Box}\,(i) \supset \lambda(i \to j)\, I_j \times D_i^2 \xrightarrow{\nu(i \to j) \equiv \text{id} \times \mu(i \to j)} I_j \times D_j^2 = \text{Box}\,(j). \tag{7.22}$$

Here, Box (i) **goes cleanly through** Box (j) in a Markov-partition like manner, suggested in Fig. 7.3.

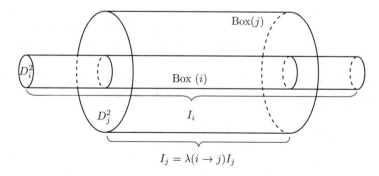

Fig. 7.3 The not everywhere well-defined map $\nu(i \to j)$ (7.22), from Box(i) to Box(j)

On the disjoined union

$$\mathcal{X}^3(C \cdot h) = \sum_i \text{Box}\,(i)$$

we consider the **equivalence relation** $\mathcal{R}(C \cdot h)$ generated by the pairs of points (p, q) where $p \in \lambda(i \to j)\, I_j \times D_i^2\ (= \{\text{domain of definition of the map } v(i \to j)\})$ $\subset \text{Box}(i)$, and $q = v(i \to j)(p) \in \text{Box}(j)$. We will introduce the following quotient space, endowed with the quotient-space topology

$$\chi^3(C \cdot h)/\mathcal{R}(C \cdot h)\,. \tag{7.22.1}$$

Lemma 7.4.1 *If $C \cdot h = easy\ id + nilpotent$ (as we know it is in our situation) then the quotient space $X^3(C \cdot h) = \mathcal{X}^3(C \cdot h)/\mathcal{R}(C \cdot h)$ has a natural structure of smooth not necessarily connected 3^{d} manifold with boundary, naturally equipped with a **lamination** $\mathcal{L}(C \cdot h) \subset X^3(C \cdot h)$ by lines.*

The transverse structure of $\mathcal{L}$ is

$$(R^2, \text{tame Cantor set}).$$

Comment Remember that, in the appropriate context, easy id + nil is another way of expressing GSC. Now the construction $X^3 = \mathcal{X}^3/\mathcal{R}$ makes sense in more general contexts which are far from easy id + nil. And then, the X^3 is a horrible space, not obviously directly connected to the manifold the geometric intersection matrix

$$\partial(\text{2-handles}) \cdot (\text{1-handles})$$

of which is used to produce the X^3 (or X^{n-1} in more general setting); this horrible space is, in some cases, non-commutative, in the sense of Alain Connes [3]. In a philosophical-speculative vein, what is suggested here is a possible connection, valid in the context of (22.1)

$$\text{GSC} \Longrightarrow \text{usual manifolds}$$

BRUTAL violation of GSC (as defined in [13]) $\Longrightarrow$ non-commutative spaces.

The BRUTAL violations of geometric simple connectivity (GSC) are explicitly defined in [13], where the connections above are more thoroughly investigated. End of Comment.

We go back now to $N^4(2X_0^2)$ which can be expressed as follows

$$
N^4(2X_0^2) = \left[\left(N^4(2\Gamma(\infty)) - \sum_1^\infty h_i \right) \cup \underbrace{\bigcup_{i \in \text{vertices of } \mathcal{M}} (h_i \cup D^2(C_i))}_{} \right]
$$
$$
\cup \sum_1^{\overline{n}} \{4^d \text{ 2-handles } D^2(\Gamma_j)\}. \tag{7.23}
$$

Here, the $D^2(\Gamma_j)$'s are attached to $N^4(\Gamma(1)) \subset N^4(2\Gamma(\infty)) - \sum_1^\infty h_i$, without touching to the non compact 4^d manifold with boundary below

$$
\text{LAVA} \equiv \bigcup_i h_i \cup D^2(C_i) \subset N^4(2X_0^2). \tag{7.24}
$$

Remember that the $D^2(\Gamma_j)$'s in the formula (7.23) are the Schoenflies 2-handles. We will denote, also

$$
\delta\,\text{LAVA} = \partial(\text{LAVA}) \cap \partial \left(N^4(2\Gamma(\infty)) - \sum_1^\infty h_i \right). \tag{7.25}
$$

Lemma 7.4.2 (Product Property of LAVA)

(1) *We have* $\delta\,\text{LAVA} \underset{\text{DIFF}}{=} X^3(C \cdot h)$, *from* Lemma 7.4.1.
(2) *There is a diffeomorphism of pairs (our PRODUCT PROPERTY)*

$$
(\text{LAVA}, \delta\,\text{LAVA}) = (X^3(C \cdot h) \times [0,1] - \mathcal{L}(C \cdot h) \times \{1\},\ X^3(C \cdot h) \times \{0\}), \tag{7.26}
$$

where the $\mathcal{L}(C \cdot h) \times \{1\} \subset X^3(C \cdot h) \times \{1\}$ *is the closed set of the lamination from* Lemma 7.4.1. *This* $\mathcal{L}(C \cdot h) \times \{1\}$ *lives at the infinity of* LAVA, *and also in* $N^4(2X_0^2)^\wedge$ *below.*

There is here also a set of end-points, à la Hopf-Freudenthal

$$
\varepsilon \left(N^4(2\Gamma(\infty)) - \sum_1^\infty h_i \right) \subset \partial N^4(\Gamma(1)),
$$

such that

$$N^4 \left(2\Gamma(\infty) - \sum_1^\infty h_i \right) \underset{\mathrm{DIFF}}{=} N^4(\Gamma(1)) - \varepsilon \left(N^4(2\Gamma(\infty)) - \sum_1^\infty h_i \right),$$

where $\varepsilon \left(N^4(2\Gamma(\infty)) - \sum_1^\infty h_i \right) \subset \partial N^4(\Gamma(1))$ is a tame embedding.

But there is also another set of end-points

$$\varepsilon \, (\mathrm{LAVA}) = \varepsilon \, (\delta \, \mathrm{LAVA}),$$

which comes with a natural continuous map

$$\varepsilon \, (\mathrm{LAVA}) \xrightarrow{\ \varphi\ } \varepsilon \left(N^4(2\Gamma(\infty)) - \sum_1^\infty h_i \right), \tag{7.27}$$

and each leaf of the lamination $\mathcal{L}$ is a copy of the real line R, having its two distinct end-points in two distinct points of $\varepsilon \, (\delta \, \mathrm{LAVA})$, going by φ to two distinct endpoints in $\varepsilon \left(N^4(2\Gamma(\infty)) - \sum_i^\infty h_i \right)$.

Lemma 7.4.3 (Compactification)

(1) *The two non-compact spaces in the RHS of (7.23) can be compactified as smooth compact 4-manifolds with boundary:*

$$N^4 \left(2\Gamma(\infty) - \sum_1^\infty h_i \right)^{\wedge} \underset{\mathrm{DIFF}}{=} N^4(\Gamma(1)), \tag{7.28.1}$$

and

$$(\mathrm{LAVA}, \delta \, \mathrm{LAVA})^{\wedge} = (X^3(C \cdot h) \times [0, 1] \cup \varepsilon \, (\delta \, \mathrm{LAVA}), (\delta \, \mathrm{LAVA})$$

$$\cup \varepsilon \, (\delta \, \mathrm{LAVA}) = X^3(C \cdot h) \cup \varepsilon \, (\delta \, \mathrm{LAVA})). \tag{7.28.2}$$

The compactification of the {extended cocore x} from B^3 inside $\mathrm{LAVA}^{\wedge}$ is

$$\mathrm{LAVA}^{\wedge} \supset \{extended \ cocore \ x\}^{\wedge} \underset{\mathrm{DIFF}}{=} B^3 \underset{\mathrm{proper}}{\subset} \mathrm{LAVA}^{\wedge}.$$

(2) *There is a diffeomorphism of pairs*

$$\left(\left(N^4\left(2\Gamma(\infty) - \sum_1^\infty h_i\right) \cup \text{LAVA}\right)^\wedge, \sum_1^n \{\text{extended cocore } R_i\}^\wedge\right)$$

$$\underset{\text{DIFF}}{=} n \# (S^1 \times B^3, (*) \times B^3). \tag{7.29}$$

[Here $\left(\left(N^4(2\Gamma(\infty)) - \sum_1^\infty h_i\right) \cup \text{LAVA}\right)^\wedge = \left\{\left(N^4(2\Gamma(\infty)) - \sum_1^\infty h_i\right)\right\}^\wedge \cup$
(LAVA)$^\wedge$ glued together along $(\delta \text{ LAVA})^\wedge$, and with each $x \in \varepsilon$ (LAVA) glued
to the corresponding $\varphi(x) \in \varepsilon \left(N^4(2\Gamma(\infty)) - \sum_1^\infty h_i\right)$.
Also, remember here that $R_1, R_2, \ldots, R_n \in \Gamma(1)$ are the RED 1-handles of
$N^4(\Delta^2)$.]

(3) *There is also a diffeomorphism which extends the (7.29)*

$$N^4(2X_0^2)^\wedge = \left(\left(N^4(2\Gamma(\infty)) - \sum_1^\infty h_i\right) \cup \text{LAVA}\right)^\wedge + \sum_1^{\bar{n}} D^2(\Gamma_j) \underset{\text{DIFF}}{=} N^4(\Delta^2). \tag{7.30}$$

This is the model of $N^4(\Delta^2)$ to which we will hang on to, until further notice.

7.5 Exterior Discs

We will be interested now in the family $\sum_j^M b_i \subset \Gamma(1) \times (\xi_0 = -1)$ from (7.8.0), which clearly comes with $M \gg n$, and with the corresponding curves $c(b_i)$, $i = 1, \ldots, M$, defined like in Fig. 7.1 in the situations (II, III). We have $c(b_i) \subset \partial N^4(2X_0^2)$.

The construction which comes next will use the BLUE flow of $X^2[\text{NEW}]$ **and** the geometry of $2X_0^2$. [The BLUE flow of $2X^2$ which has its active part localized outside $X^2[\text{new}] \times r$ is NOT the same as the BLUE flow of $X^2[\text{new}]$, remember.]

Let $b \in B_0$ be any of the b's occurring in one of the finite BLUE 2^d collapsing trajectories of $\sum_1^M b_i$, inside $X^2[\text{NEW}]$. So, the b_i ($i \leq M$) themselves are among these b's. To b, we associate a disc with holes $B^2(b) \subset 2X_0^2$, defined as follows:

We start with the 2-sphere

$$S^2(b) = (D^2(\eta) \times r) \cup (\eta \times [r, b]) \cup (D^2(\eta) \times b) \subset 2X^2 . \qquad (7.31)$$

Then, if RED-wise $\eta = \gamma_k^0$ we leave the collar $\gamma_k^0 \times [0, \varepsilon] \subset D^2(\eta) \times r$ alive, with $\gamma_k^0 = \gamma_k^0 \times \varepsilon$ on the boundary and, similarly, if $e \subset \eta$ is an edge, and if $e \in$ (II, III, Fig. 7.1) we leave only the collar $c(b) \times [0, \varepsilon]$ alive. Here, in this situation we have some $b \in e$. With this, by definition, $B^2(b)$ is what is left from $S^2(b)$ after these deletions. Now, if for all b_i, $i \leq M$, and the b's in their BLUE X^2[NEW]-collapsing trajectories, we put together the various $B^2(b)$'s, making use of the $\eta \cdot B$ of X^2[NEW] in the obvious way, then for each b_i-trajectory we get a big disc with holes, denoted $\mathcal{B}_i$, and which is such that

$$\partial \mathcal{B}_i = c(b_i) + \sum (\gamma_k^0 \text{'s}).$$

End of (31). Figure 7.4 should help understand our $S^2(b) \supset B^2(b)$.

Notice that in order to have all the spare parts for our $\mathcal{B}_i$ we needed the $2X_0^2$, i.e., we need the doubling process.

If we fill in the γ_k^0's with the d_k^2's (11.1) we get, finally, an abstract 2-disc, put together out of the spare parts $\mathcal{B}_i$ (= union of $B^2(b)$'s) and d_k^2's from (11.1), call it

$$\delta_i^2 = \mathcal{B}_i \cup \sum d_k^2 \text{'s, with } \partial \delta_i^2 = c(b_i). \qquad (7.32)$$

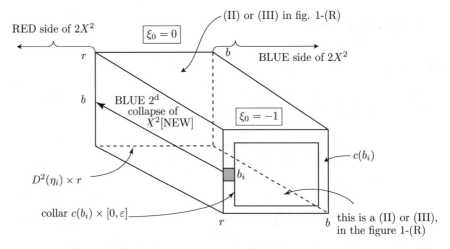

Fig. 7.4 We see here the $S^2(b_i)$, with b_i being any $b_i \in \sum_1^M b_i \subset B_0$. Notice the trivial case when the collapse from b_i stops at $\xi_0 = 0$. Also, for any $b \in B_0$, the $S^2(b)$ is like this $S^2(b_i)$. We are here in $2X_0^2$

So, here the 3^d RED collapsing flow, as felt by $X^2[\text{new}] \times r$ is also needed, hence the 3-handles of Δ^4 do play a role. This comes with a non-degenerate map

$$\sum_{j=1}^{M} \delta_j^2 \xrightarrow{\;\;G\;\;} 2X_0^2 \tag{7.33}$$

extending $\sum_{1}^{M} c(b_i) \subset \partial(2X_0^2) \subset 2X_0^2$.

The map G in (7.33) can be lifted, rel its boundary, to a generic immersion which winds tightly around $2X_0^2 \subset N^4(2X_0^2)$

$$\sum_{i=1}^{M} \delta_i^2 \xrightarrow{\;\;\mathcal{J}\;\;} N^4(2X_0^2). \tag{7.34}$$

This immersion $\mathcal{J}$ comes with ACCIDENTS, which are of the following two types:
Transversal contacts $z \in \text{Im}\,\mathcal{J} \pitchfork 2X_0^2 \subset N^4(2X_0^2)$, (like $R^2 \pitchfork R^2 \subset R^4$). (7.34.1)
Double points $x \in \mathcal{J}M^2(\mathcal{J})$. (There are NO triple points.) (7.34.2)

Very importantly, as a consequence of the framed part of (8.1), we have
The transversal contacts $z \in \mathcal{J}\delta_i^2 \pitchfork (D^2(\Gamma_j) \times (\xi_0 = -1))$ can only come from the $\sum d_k^2$ part of δ_i^2 and NOT from the $\mathcal{B}_i^2 \subset \delta_i^2$. (7.35)

The point here is that BLUE-wise $D^2(\Gamma_j) \times (\xi_0 = -1)$, which only contributes to the $\sum_{j=1}^{M} \delta_j^2$ with its b_i's, is a $D^2(\gamma^1)$ and hence it is NOT concerned by the BLUE 2^d flow of $X^2[\text{NEW}]$. Only the mythical BLUE 3^d flow concerns it, but that is not an active player in our story.

We have a decomposition $\delta_i^2 = \text{body}\,\delta_i^2 + (\delta_i^2 - \text{body}\,\delta_i^2)$ where body δ_i^2 is the part of δ_i^2 which, via G, stays close to the 1-skeleton of $2X_0^2$, $2\Gamma(\infty)$ while $\delta_i^2 - \text{body}\,\delta_i^2$ follows the (lateral surface of the) 2-handles, and with this, the passage from G (7.33) to $\mathcal{J}$ (7.34) consists, to begin with, of two successive steps:
A lift of $\sum \text{body}\,\delta_i^2$ to a *not* everywhere well-defined map

$$\sum_{1}^{M} \text{body}\,\delta_i^2 \xrightarrow{\;\;F\;\;} \partial N^4(2\Gamma(\infty)) \tag{7.36.1}$$

and

A lift of the various connected components of $\sum_i (\delta_2^i - \text{body } \delta_i^2)$ to parallel copies

of the respective {cure $D^2(\text{curve})$} $\subset$ {lateral surface of the 4^{d} 2-handle $D^2(\text{curve})$}. (36.2)

All the ACCIDENTS are produced by the lift part F (36.1) of the $\mathcal{J}$ (7.34). Typically, inside $\partial N^4(2\Gamma(\infty))$, the F is not well-defined at the **punctures**

$$p \in \{\text{Curve "}C\text{" (which can be } C_\ell \text{ or } \Gamma_j \times (\xi_0 = -1)),$$

$$\text{attaching zone of a 2-handle}\} \pitchfork F \text{ body } \delta^2.$$

$$\text{Here we are like } R \pitchfork R^2 \subset R^3 .$$

The map F is an immersion without triple points, resting on the curves C_i, $\Gamma_j \times (\xi_0 = -1) \subset \partial N^4(2\Gamma(\infty))$ and having CLASPS and RIBBONS like in Fig. 7.5. When one lifts the CLASP to $N^4(2X_0^2)$, and then one goes from F to $\mathcal{J}$, what one gets is something like in Fig. 7.6. There one sees a section in 4^{d}, corresponding to our Fig. 7.5, namely

$$\left(\sum_i \mathcal{J}\delta_i^2\right) \cap \{\text{a plane } (x, t) \text{ containing the CLASP,}$$

$$\text{and transversal to the } \partial N^4(2\Gamma(\infty))\}.$$

As a consequence of (8.1) the CLASPS are, at worst, like in Fig. 7.5, where only one of the curves is a $\Gamma_j \times (\xi_0 = -1)$ and NOT both. There are also RIBBONS, which I have not drawn but for which the analogue of Fig. 7.6 is Fig. 7.7. The point

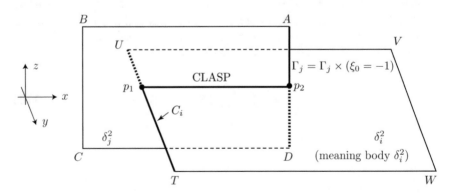

Fig. 7.5 A typical CLASP $F\delta_i^2 \pitchfork F\delta_j^2$ of (36.1) inside $\partial N^4(2\Gamma(\infty))$. The p_1, p_2 are **punctures** where F (36.1) is **not** well-defined. There is a similar figure for RIBBONS, which I have not drawn. The F body δ^2 rests on C_i, Γ_j and continues beyond $[A, B, C, D]$, $[U, V, W, T]$, appearing all in thin lines (plain or dotted)

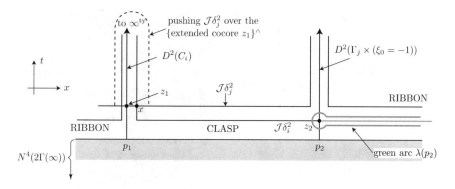

Fig. 7.6 We see here $z_1 \in \mathcal{J}\delta_j^2$ ⫛ $D^2(C_i)$ and $z_2 \in \mathcal{J}\delta_i^2$ ⫛ $D^2(\Gamma_j \times (\xi_0 = -1))$. At x, in $R^4(x, y, z, t)$, the $\mathcal{J}\delta_j^2$ is in $R^2(x, z)$ and $\mathcal{J}\delta_i^2$ in $R^2(t, y)$. The CLASP from Fig. 7.5 generates in 4^d, for the map $\mathcal{J}$ from (7.34), two transversal contacts z_1, z_2 and one double point x. The z_2 has no extended cocore. The z_1 has one, while the z_2 hasn't because $z_2 \in \text{int } D^2(\Gamma_j) \times (\xi_0 = -1)$

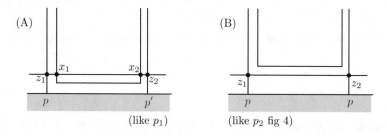

Fig. 7.7 The two possibilities for lifting a RIBBON to 4^d. There is no local geometrical obstruction for choosing one option or the other. It is the GLOBAL POLICY of getting rid of ACCIDENTS which decides, for each individual RIBBON, the appropriate policy, (A) or (B)

is that the (CLASPS) ∪ (RIBBONS) form several highly connected and non-simply connected system. Figure 7.6 suggests two RIBBONS adjacent to the CLASP, corresponding, respectively, to (a) and (b) in Fig. 7.7.

There are **two mechanisms** for destroying the ACCIDENTS of $\mathcal{J}$:
Pushing over the {extended cocore z}$^\wedge$ $\subset$ $N^4(2X_0^2)^\wedge$, when z has an extended cocore, like in the LHS of Fig. 7.6. [But notice that, even if both z_1 and z_2 would have such an extended cocore, which is not the case in Fig. 7.6, since $z_2 \in D^2(\Gamma_i \times (\xi_0 = -1)) \subset \Delta^2 \times (\xi_0 = -1)$, which is our Δ^2, the process still could NOT be applied to both z_1 and z_2. But the important fact, consequence of (8.1), is that at least one of the z_1, z_2 **has** an extended cocore.] In Fig. 7.6, this kills both z_1 and x. (7.37.1)

Also, the process described does not change $\partial\delta_i^2$, but the next one does so.
We join the puncture (let us say p_2 in Fig. 7.5) to $\partial\delta_i^2$, via an arc $\lambda(p_2) \subset \mathcal{J}\delta_i^2$, which I call the **green arc**, like in Fig. 7.8a and then we modify the $\mathcal{J}\delta_i^2$ (Fig. 7.6), like it is suggested in Fig. 7.8b, where the shaded area is the projection of the modified $\mathcal{J}\delta_i^2$. (7.37.2)

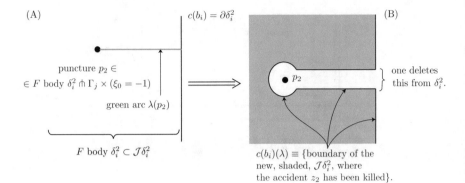

Fig. 7.8 One sees here the process $c(b_i) \Longrightarrow c(b_i)(\lambda)$. Destroying transversal contacts $z \in \mathcal{J}\delta_i^2 \pitchfork$ $2X_0^2$, via the green arc process

Lemma 7.5.1 *We can destroy all the ACCIDENTS of* (7.34) *by an appropriate combination of the mechanisms* (37.1), (37.2).

We can actually do better than that, and this will be explained in the next section.

7.6   Confinement

We go back to the SPLITTING (19.1). There are two kinds of 1-handles h_i (for $2\Gamma(\infty) - (\Gamma(1) \times (\xi_0 = -1)))$, the $h_i(r) \subset X_0^2[\text{new}] \times r$ and the $h_k(b) \subset X_b^2$. Each h means a pair

$$(B^3 \times [-\varepsilon, \varepsilon], S^2 \times [-\varepsilon, \varepsilon]) \subset (N^4(2\Gamma(\infty)), \partial N^4(2\Gamma(\infty))). \qquad (7.38)$$

Here, the $S^2(b) \times [-\varepsilon, \varepsilon] \subset \partial N_+^4(2\Gamma(\infty))$, while $\sum_\infty^2$ splits each $S^2(r)$ into

$$S^2(r) = \frac{1}{2} S^2(r)_+ \cup \frac{1}{2} S^2(r)_-, \text{ with } \frac{1}{2} S^2(r)_\pm \subset \partial N_\pm^4(2\Gamma(\infty)). \qquad (7.38.1)$$

With this comes a partition of LAVA into two pieces

$$(\text{LAVA}_r, \delta\,\text{LAVA}_r) \equiv \left(\sum_i (h_i(r) \cup D^2(C_i(r))), \delta\,\text{LAVA} \cap \partial\,\text{LAVA}_r \right),$$

$$(\text{LAVA}_b, \delta\,\text{LAVA}_b)$$

$$\equiv \left(\sum_j (h_j(b) \cup D^2(C_j(b))), \partial\,\text{LAVA}_b \cap \partial \left(N^4(2\Gamma(\infty)) - \sum_\ell h_\ell(b) \right) \right),$$

$$(7.38.2)$$

coming with a

$$\text{GLUEING TERM} \equiv \delta \text{LAVA}_b \cap \partial \text{LAVA}_r. \tag{7.38.3}$$

Individually, each of the two LAVA's has the PRODUCT PROPERTY from (7.26), Lemma 7.4.2, and the global property of LAVA comes from the following two successive collapses $\text{LAVA}_b \searrow \delta \text{LAVA}_b$, leaving the GLUEING TERM as a free boundary of LAVA_r, and then $\text{LAVA}_r \searrow \delta \text{LAVA}_r$. Also $\text{LAVA}_b \subset \{\text{The} + \text{side of}$ the SPLITTING of $\partial N^4(2\Gamma(\infty))$ by $\sum^2_\infty\}$.

Most of LAVA_r lives on $-$ side of the SPLITTING and, also, it is there that the non-trivial LAVA moves which one needs, will take place. But then, because of (7.38.1), LAVA_r $\textit{has}$ to trespass on the $+$ side too.

We consider now $\textbf{all}$ the curves $\mathcal{C} = \{C \text{ or } \Gamma\}$ such that $D^2(\mathcal{C})$ contains a transversal contact $z \in D^2(\mathcal{C}) \pitchfork \mathcal{J} \sum_i \delta_i^2$ and which, when Lemma 7.5.1 is being applied, require the mechanism (37.2) of the green arcs. This will certainly be the case for $D^2(\Gamma_j \times (\xi_0 = -1))$, for instance.

Lemma 7.6.1

(1) *There is an $\textbf{admissible}$ $\textbf{subdivision}$ of $2X_0^2$, respecting the all-important (8.1) and which breaks each $D^2(\mathcal{C})$, for the generic curve $\mathcal{C} = \{C_i \text{ or } \Gamma_j \times (\xi_0 = -1)\}$, into*

$$D^2(\mathcal{C}) = D^2(\textit{little } \mathcal{C}) \cup D^2(\textit{remaining } \mathcal{C}),$$

so that $z \in D^2(\textit{little } \mathcal{C})$ and $D^2(\textit{remaining } \mathcal{C})$ is free of ACCIDENTS.

(2) *With this, without contradicting anything said so far, we can enforce the following CONFINEMENT CONDITIONS, referring to the $\{link\}(1)$ and its later avatars until $2X_0^2(10)$,*

$$\sum \{\textit{remaining } \Gamma_j \times (\xi_0 = -1)\} + \sum \{\textit{remaining } C_i\} + \sum \gamma_k^0 \subset \partial N_-^4(2\Gamma(\infty)),$$
$$\sum \{\textit{little } \Gamma_j \times (\xi_0 = -1)\} + \sum \{\textit{little } C_i\} + \sum \eta_i \times b + \sum c(r) \subset \partial N_+^4(2\Gamma(\infty)) \supset$$
$$\sum_1^M c(b_i)(\lambda) \text{ (defined like in Fig. 7.8b, boundary of the external discs } \delta_i^2). \tag{7.39}$$

So, now the interesting things are on the BLUE side.

The subdivisions used in this lemma cannot be of the general unrestricted type. For those, here is the effect on a $D^2(\gamma^1)$, for instance,

$$D^2(\gamma^1) \Longrightarrow \{\text{one smaller } D^2(\gamma^1)\} + \{\text{many } D^2(\eta)\text{'s}\}.$$

This, as such, would certainly violate the framed formula in (8.1), which is sacrosancted. So, special, "admissible" subdivisions have to be used. I will not explain them here.

In (39) all the curves on the LHS of the "$\subset$"'s are attaching zones of internal 2-handles of $2X_0^2$, while the $\sum_1^M c(b_i)(\lambda)$, with $c(b_i)$ like in Fig. 7.1 and with $c(b_i)(\lambda)$ like in Fig. 7.8b, are boundaries of external discs $\sum_1^M \delta_i^2$. The $c(r)$'s in (39) are also like in Fig. 7.1, now (I) and (IV) rather than (II) and (III).

It is in the last section VIII, that we will see why both SPLITTING and CONFINEMENT are such important items for our whole approach.

At the end of Lemma 7.5.1 we find ourselves with a system of external discs

$$\left(\sum_1^M \delta_i^2, \sum_1^M (\partial \delta_i^2 = c(b_i)(\lambda)) \right) \xrightarrow{\mathcal{J}} \left(\partial N^4 (2X_0^2)^\wedge \times [0, 1], \partial N^4 (2X_0^2) \right),$$

(7.40)

which is free of ACCIDENTS, i.e., smoothly embedded.

The next lemma makes use of the BLUE collapse of $2X^2$.

Lemma 7.6.2 (The Little Blue Diagonalization) *There is a transformation confined inside $\partial N_+^4 (2\Gamma(\infty))$ and dragging along the $\sum_1^M \delta_i^2$, via covering isotopy,*

$$\sum_1^M c(b_i)(\lambda) \Longrightarrow \sum_1^M \eta_i(\text{green}) \subset \partial N_+^4(2\Gamma(\infty)),$$

(7.40.1)

where now $\eta_i(\text{green}) = \partial \delta_i^2$, with the following features
For $1 \leq i, j \leq M$ we have $\eta_i(\text{green}) \cdot b_j = \delta_{ij}$ and also

$$\eta_i(\text{green}) \cdot \left(B_1 - \sum_1^M b_i \right) = 0.$$

(7.41)

It is this (41) which is our "little BLUE diagonalization", to be soon superseded by a BIG BLUE DIAGONALIZATION, where in $\eta_i(\text{green}) \cdot b_j = \delta_{ij}$ we substitute the $\{$extended cocore $b_j\}^\wedge$'s, to the b_j's. Now, superficially, it would look as if (41) could be plugged into the criterion from Lemma 7.2.1, **EXCEPT THAT**, on the one hand, the $\sum_1^M \delta_i^2$ are certainly not attached to $N^4(\Delta^2)$, but to the $N^4(2X_0^2)$ and then, on the other hand, even with the (7.30) in mind, it is not clear how to make sense of the $\sum_1^M b_i$ as the system of 1-handles for $N^4(2X_0^2)^\wedge \underset{\text{DIFF}}{=} N^4(\Delta^2)$.

But before facing these problems, here are some more general *"PHILO-SOPHICAL COMMENTS"*. In this paper there are a number of sacro-sancted principles never to be transgressed and then other things which we can violate or transgress, when necessary. Here are the sacro-sancted ones:

(I) The CONFINEMENT conditions and the PRODUCT PROPERTY OF LAVA are both sacro-sancted. The steps following next, including the LAVA MOVES are supposed to abide strictly to this.

(II) But then, the condition $C \cdot h = $ id $+$ nilpotent, will have to be violated. Of course, $C \cdot h = $ id $+$ nil was essential for putting LAVA, with its PRODUCT PROPERTY, into place, and the LAVA MOVES, which we have not yet met, will never violate the PRODUCT PROPERTY. And once this property, i.e., the collapse LAVA $\searrow \delta$ LAVA is well with us, we can do without the $C \cdot h = $ id $+$ nil. The product property, itself, will define our necessary {extended cocores}. But retain that the violation of $C \cdot h = $ id $+$ nil, will be compensated for, by appropriate LAVA MOVES.

7.7 Balancing Blue and Red

At this point, we are by definition, at time $t = 0$ in our game and various global moves at times $t > 0$ will follow. Our $\Gamma(1) = \{1\text{-skeleton of } \Delta^2\}$ has n RED 1-handles and $M \gg n$ BLUE 1-handles; the $\Gamma(1) - R$ is a tree, while $\Gamma(1) - B$ is a *disconnected* collection of trees, and we need to correct this *disbalance* $M - n > 0$. We start now with

$$N^4(2X_0^2) = \left[\left(N^4(2\Gamma(\infty)) - \sum_1^\infty h_i\right) \cup \text{LAVA}\right] + \sum_1^{\bar{n}} D^2(\Gamma_j), \qquad (7.42)$$

see here (7.30) too, and we will say that this corresponds to time $t = 0$ of our process in this paper. At the level of the (7.42) there is a transformation $\left(\text{time } (t = 0) \Longrightarrow \text{time } \left(t = \frac{1}{2}\right)\right)$ consisting of a finite number of handle slidings and also of promotions of some $M - n$ 1-handles h_i as RED 1-handles of $N^4(\Delta^2)$ and of their dual $D^2(C)$'s as 2-handles of $N^4(\Delta^2)$, i.e., as honorific "$D^2(\Gamma_i) \times (\xi_0 = -1)$"'s. This process changes (7.42) into

$$N^4(2X_0^2)\left(t = \frac{1}{2}\right) = \left[\left(N^4(2\Gamma(\infty)) - \sum_1^\infty h_i\right) \cup \text{LAVA}\right]\left(t = \frac{1}{2}\right) + \sum_1^{\bar{\bar{n}}} D^2(\Gamma_j),$$

$$\qquad (7.43)$$

with $\bar{\bar{n}} = \bar{n} + (M - n)$, satisfying Lemma 7.7.1 below. [But, to begin with, here are some EXPLANATIONS. The promotion of the $M - n$ 1-handles h_i, will change

$N^4(\Gamma(1))$ into an $N^4(\Gamma(1)) \left(t = \frac{1}{2} \right)$ where

$$ \#\,\text{RED 1-handles} = \#\,\text{BLUE 1-handles} = M \,. $$

This is our balancing.

BUT the former $D^2(C)$'s which become now additional $\sum\limits_{\bar{n}+1}^{\bar{\bar{n}}} D^2(\Gamma_j \times (\xi_0 = -1))$'s, are NOT directly attached to $N^4(\Gamma(1)) \left(t = \frac{1}{2} \right)$ so this promoting is really only honorific, at least as long as we stay in the naïve context. Actually, there is NO naïve $N^4(\Delta^2) \left(t = \frac{1}{2} \right)$ having as 1-skeleton the $N^4(\Gamma(1)) \left(t = \frac{1}{2} \right)$, which we can make sense of, and this means, of course, that the newly promoted 2-handles do not live at $\xi_0 = -1$ either. To really make geometric sense of this promotion, we **have** to make fully use of the infinitistic context of $\left(N^4(2X_0^2) \left(t = \frac{1}{2} \right) \right)^{\wedge}$, like in the Lemma 7.7.1 below.]

Lemma 7.7.1

(1) *We have a diffeomorphism*

$$ \left(\left(\left[\left(N^4(2\Gamma(\infty)) - \sum_{1}^{\infty} h_i \right) \cup \text{LAVA} \right] \left(t = \frac{1}{2} \right) \right)^{\wedge}, \right. $$

$$ \left. \sum_{1}^{M} \{\text{extended cocore } b_i\}^{\wedge} \right) \underset{\text{DIFF}}{=} \tag{7.44} $$

$$ = \underset{i=1}{\overset{M}{\#}} (S_i^3 \times B_i^3, (*) \times B_i^3). $$

(2) *There is a second diffeomorphism, extending (7.44),*

$$ \left(N^4(2X_0^2) \left(t = \frac{1}{2} \right) \right)^{\wedge} = \left(\left[\left(N^4(2\Gamma(\infty)) - \sum_{1}^{\infty} h_i \right) \cup \text{LAVA} \right] \left(t = \frac{1}{2} \right) \right)^{\wedge} \tag{7.45} $$

$$ + \sum_{1}^{\bar{\bar{n}}} D^2(\Gamma_j) \underset{\text{DIFF}}{=} N^4(\Delta^2). $$

Comments We have now a diffeomorphic model of $N^4(\Delta^2)$ superseding the one in (7.30) where, on the one hand the $\sum\limits_{1}^{M} \{\text{extended cocore } b_i\}^{\wedge}$ **are** now the 1-handles and we will forget about the RED 1-handles of $N^4(\Delta^2)$ from now on;

and the $\sum_1^M \delta_i^2$ are a system of embedded exterior discs like in (40.1), satisfying the little diagonalization conditions (41). BUT it is the {extended cocore b_i}$^\wedge$, for $1 \leq i \leq M$, which are now our system of 1-handles for "$N^4(\Delta^2)$" and *no longer* the $\sum_1^M b_i$. With this, as a consequence of (41), what we find for the relevant geometric intersection matrix, is now the following item

If $1 \leq i, j \leq M$, the η_i (green) $\cdot$ {extended cocore b_j}$^\wedge = \delta_{ij}$ + {some *parasitical* off-diagonal terms $h_k \in R_0 - B_0$, namely the $h_k \subset$ {extended cocore b_j}, which are touched by the η_i (green)}.   (7.46)

So, our parasitical contacts mean,

$$\eta_i(\text{green}) \cdot (h_k \in R_0 - B_0 (= R_1 - B_1)) \neq \emptyset\,,$$

and now diagonality really should mean absence of these unwanted terms. We do not have that now, so what we have gotten, so far, is still not good enough for being plugged into Lemma 7.2.1. More work will be necessary.

An Idea of How One Achieves the BALANCING The idea is very simple, indeed. Consider $\Gamma(1) = \Gamma(1) \times (\xi_0 = -1) \subset 2\Gamma(\infty)$ and the disconnected $\Gamma(1) - B_1 \subset 2\Gamma(\infty) - B_1$. Take two distinct connected components $X_1, X_2 \subset \Gamma(1) - B_1$ and join them by a geodesic arc $g_1 \subset 2\Gamma(\infty) - B_1$, like in Fig. 7.9a. It is easy to see that $g_1 \cap R_1 \neq \emptyset$ and that

$$g_1 \cap R_1 \subset \{h_1, h_2, \ldots\} \subset R_0\,.$$

We denote $g_1 \cap R_1 = \{y; x_1, \ldots, x_k\}$, where in the RED order of $C \cdot h$ we have $y > $ {all the x_i's}, meaning that none of the RED collapsing trajectories go like $x_i \xrightarrow{C \cdot h} y$, they can only go like $y \xrightarrow{C \cdot h} x_i$. This means that we can slide y over the $x_1, \ldots, x_k$'s and join X_1, X_2, like in Fig. 7.9b, without violating $C \cdot h = \text{id} + \text{nil}$.

Actually, the b-part of $2X_0^2$ plays hardly any active role in our balancing process, and we can happily work here with a $\Gamma \equiv 2\Gamma(\infty) \mid r = \{\Gamma(\infty), \text{1-skeleton of } X_0^2[\text{NEW}]\}$. But then, do not forget that the b-part $X_b^2 \subset 2X_0^2$ plays a big role when putting up the $\sum_1^M \delta_i^2$, in getting the $N^4(2X_0^2)^\wedge \underset{\text{DIFF}}{=} \Delta^4_{\text{Schoenflies}}$, and in the proof of the Lemma 7.6.2 too.

The y gets promoted as a 1-handles of $N^4(\Delta^2)$ and then, purely honorifically, $D^2(C)$ also gets promoted as 2-handle of $N^4(\Delta^2)$. This operation does not violate the $C \cdot h = \text{id} + \text{nil}$ and LAVA gets redefined by excluding $y \cup D^2(C(y))$, and it keeps the PRODUCT PROPERTY. This is how the definition of the LHS of (7.45) should be read.

If $X_1 \cup g_1 \cup Y_1$ is not connected, we restart, a.s.o.

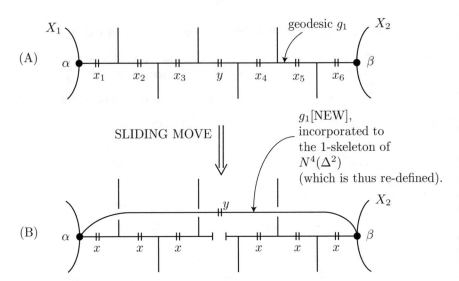

Fig. 7.9 The main step in the BALANCING of RED/BLUE. When all the finitely many steps are completed, then

$$\#\{\text{1-skeleton of } N^4(\Delta^2)\} \cap R_0 = \#\{\text{1-skeleton of } N^4(\Delta^2)\} \cap B_0 = M, \qquad (*)$$

and the two graphs $\{\text{1-skeleton of } N^4(\Delta^2)\} - R_0$, $\{\text{1-skeleton of } N^4(\Delta^2)\} - B_0$, are trees. In the LHS of $(*)$ we find $\{R_1, R_2, \ldots, R_n\} + \{\text{the promoted } h\text{'s}\}$ and in the RHS we find the $\sum_1^M b_i$. Also, in (A) we are in $\Gamma(2\infty)(t = 0)$ and in (B) in $\Gamma(2\infty)(t = \frac{1}{2})$

Here are some explanations for this figure. Since Γ only refers to the r-side of $2X_0^2$, the β refers to one of the infinitely many passages to the rest of $\Gamma(2\infty) \left(t = \frac{1}{2} \right)$. If we denote by $\Gamma(\infty)_0$ the 0-skeleton of $\Gamma(\infty)$ and we consider

$$\Gamma(\infty)_0 \times [r, b] \subset \Gamma(\infty) \times [r, b] \quad (\text{see (10)}),$$

then β is a spot on some $P \times [r, b]$, $P \in \Gamma(\infty)_0$, close to $P \times r \in \Gamma(\infty) \times r$. Only one such spot has been drawn, but there are infinitely many. These sites β are swept by the move in Ministep II below making that the move

$$B^3(r) \Longrightarrow B^3(b)$$

suggested in Fig. 7.11 below is certainly NOT an isotopy, but a rather non-trivial business. We substitute, by decree, a piece of LAVA with another piece, which was not LAVA, before. But this happens so that all our desired compatibilities should be respected, as well as the sacro-san<ted principles. The fat points ($= \bullet$) in this figure live at infinity.

The green G's ($\neq$) are finitely many 1-handles, distinct from R and B, called GREEN GATES. They isolate a sufficiently large, **compact** piece of X_1 from infinity. And all our action will be concentrated in this piece, shaded in the drawing.

The little red branches

are all that is left from our R's, which all have been cut at the level of $\Gamma - R$ (not the $r(1)$, of course). So, their ends have to be joined in pairs, when we move from $(\Gamma - R) \cup r(1)$ to the full $(2\Gamma(\infty)) \left(t = \frac{1}{2} \right)$.

There are **finitely** many such red branches touching that part of X_1 which lives between $r(1), b(1)$ and the gates G. We call them the "**active red 1-handles**".

7.8   Change of Colour

In this section I will explain how one gets rid of the parasitical terms in (46). And the full (41) will be necessary.

At the time $t = \frac{1}{2}$ of Lemma 7.7.1, the $\Gamma(2\infty)$ has changed into $\Gamma(2\infty) \left(t = \frac{1}{2} \right)$ and the $\Gamma(1)$ (1-skeleton of $N^4(\Delta^2)$) has become $\Gamma(1) \left(t = \frac{1}{2} \right)$.

The $\Gamma(1) \left(t = \frac{1}{2} \right) \supset \sum_1^M b_i$ and now $\Gamma(1) \left(t = \frac{1}{2} \right) - \sum_1^M b_i$ is a tree. We decide also that $\sum_1^M b_i \subset R_0 \cap B_0$, **by decree**, and we concentrate now on the rest of $(R_1$ and $B_1) \cap \Gamma(2\infty) \left(t = \frac{1}{2} \right)$. Here, all the $R_1 - R_0 = B_1 - B_0$ and we have

$$R_1 - R_0 = B_1 - B_0 \subset R_1 \cap B_1 .$$

In order to simplify things, I will use the notation

$$\left(\Gamma(2\infty) \left(t = \frac{1}{2} \right); R_1, B_1 \right) = (\Gamma; R, B). \tag{7.47}$$

In the context of (7.47) we put up the following INDUCTIVE PROCEDURE.

(1) We pick up a $b(1) \in B - R$ and we notice that $b(1) \subset \Gamma - R$ splits $\Gamma - R$ as follows

$$\Gamma - R = X_1 \cup b(1) \cup Y_1, \text{ with } X_1, Y_1 \text{ disjoined trees, glued together by } b(1). \tag{7.48}$$

Notice that $\Gamma(1)(t = \frac{1}{2}) \supset \sum_1^M b_i \subset R_0 \cap B_0 \subset R \cap B$ and that, once $b(1) \in B - R$, this excludes that $b(1) \subset \Gamma(1)(t = \frac{1}{2})$, so our $\Gamma(1)(t = \frac{1}{2})$ cannot see both X_1 and Y_1.

(2) There has to be an $r(1) \in R - B$ joining X_1 to Y_1 and if we define $R(1) \equiv R - r(1) + b(1)$, then it is easy to see that $\Gamma - R(1)$ is again a tree, just like $\Gamma - B$.

(3) So we replace now $(\Gamma; R, B)$ by $(\Gamma; R(1), B)$ and pick up an element $b(2) \in \Gamma - R(1)$, going from (7.48) to the higher level, similar decomposition

$$\Gamma - R(1) = X_2 \cup b(2) \cup Y_2.$$

(4) There has to be an $r(2) \in R(1) - B$ joining X_2 to Y_2. We define $R(2) = R(1) - r(2) + b(2)$ a.s.o. This process can clearly continue indefinitely, from here on.

Combinatorial Lemma 1

(1) *When one continues this inductive procedure indefinitely, then one finds that the end-product*

$$R(\infty) \equiv R - r(1) - r(2)\ldots + b(1) + b(2) + \ldots$$

comes with $R(\infty) = B$. So, our infinite process turns RED into BLUE.

(2) *In other words, our inductive procedure ends up with a bijection (which for the time being is purely ABSTRACT)*

$$R \xrightarrow[\approx]{\Phi} B \qquad (7.49)$$

such that $\Phi \mid R \cap B = identity$, $\Phi(r(n)) = b(n)$.

Our inductive process has the feature of changing into BLUE all the RED guys, keeping fixed the $R \cap B$.

One should notice now that in the context of (46) there is only a FINITE number of parasitical terms $h_k \in R - B$. This follows immediately from the fact that

$$\sum_1^M \text{length } \eta_i (\text{green}) < \infty.$$

So, on the road from R to $R(\infty)$ let us pick up a sufficiently high $R(N) = R - \sum_{n=1}^N r(n) + \sum_1^N b(n)$ such that all the parasitical h_k's should be among the $r(1), r(2), \ldots, r(N)$. In other words, the isomorphic transformation

$$R \xrightarrow{\Phi(N)} R(N) \qquad (7.50)$$

turns BLUE all the RED parasitical terms $h_k \in R - B$.

Theorem 7.8.1 *Starting from the time* $t = \frac{1}{2}$ *of Lemma 7.7.1, there is a transformation*

$$\text{time}\left(t = \frac{1}{2}\right) \Longrightarrow \text{time } (t = 1)$$

which realizes geometrically, in 4^d, *the isomorphism*

$$R \xrightarrow[\approx]{\Phi(N)} R(N).$$

This transformation proceeds by ADMISSIBLE LAVA MOVES (which consists of so-called LAVA bridges and LAVA dilatations), of slidings and isotopies of 1-handles, dragging everything along by covering isotopy. It also consists of geometric transformations, concerning the B^3's *cocores of* $r(i)$ *and* $b(i)$ *respectively, for* $i \leq N$,

$$B^3(r) \Longrightarrow B^3(b) \quad \text{(see Fig. 7.11)}$$

with $B^3(r)$ *leaving LAVA (and disappearing from our scene altogether), while* $B^3(b)$ *enters LAVA, by decree. At the local level of Fig. 7.11 these may look like isotopies of 1-handles cocore, which they are certainly not. They just give flesh and bone to the abstract Combinatorial Lemma 1.*

   The ADMISSIBLE LAVA MOVES introduce things like LAVA BRIDGES, which are no longer gotten by putting together spare parts h_i, $D^2(C_i)$, *while the 1-handle slidings will violate, generally speaking, the condition* $C \cdot h = \text{id} + \text{nilpotent}$. *But the admissible LAVA moves preserve the PRODUCT PROPERTY. Very importantly too, in order to achieve everything we need in this theorem the transformation (7.50) has to be always compatible with the SPLITTING by* $\sum^2_\infty$, $\partial N^4(2\Gamma(\infty)) = \partial N^4_+(2\Gamma(\infty)) \underset{\sum^2_\infty}{\cup} \partial N^4_-(2\Gamma(\infty))$ *and with the corresponding CONFINEMENT conditions. Actually all the active part of the LAVA now concerns the* $-$ *side of* $N^4(2\Gamma(\infty))$, *meaning in the RED side.*

   [These kind of compatibilities were respected by the change $(t = 0) \Rightarrow \left(t = \frac{1}{2}\right)$ *too, although we did not mention it explicitly at the time.]*

(1) *Like in Lemma 7.7.1, we find now, again, that*

$$\left(\left[\left(N^4(2\Gamma(\infty)) - \sum_1^\infty h_i\right) \cup LAVA\right](t = 1)\right)^\wedge, \sum_1^M \{\text{extended cocore } b_i\}^\wedge\right)$$

$$\underset{\text{DIFF}}{=} \overset{M}{\underset{i=1}{\#}} (S_i^1 \times B_i^3, (*) \times B_i^3), \tag{7.51}$$

and that

$$(N^4(2X_0^2)(t=1))^\wedge = \left(\left[\left(N^4(2\Gamma(\infty))\right) - \sum_1^\infty h_i\right) \cup LAVA\right](t=1)\right)^\wedge$$

$$+ \sum_1^{\bar{\bar{n}}} D^2(\Gamma_j) \underset{\text{DIFF}}{=} N^4(\Delta^2). \qquad (7.52)$$

These formulae will replace from now on (7.44), (7.45) to which they are superior, since by now we also have the next item.

(2) *(THE BIG BLUE DIAGONALIZATION) The system of exterior embedded discs*

$$\left(\sum_1^M \delta_i^2, \sum_1^M (\partial \delta_i^2 = \eta_i(green))\right) \xrightarrow{\partial} (\partial(N^4(2X_0^2)(t=1))^\wedge$$

$$\times [0,1], \partial(N^4(2X_0^2)(t=1))^\wedge \times \{0\})$$

has now the property that

$$\eta_i(green) \cdot \{extended\ cocore\ b_j\}^\wedge = \delta_{ij}. \qquad (7.53)$$

(3) *Combining the (1), (2) above with Lemma 7.2.1, it follows that $N^4(\Delta^2)$ is GSC, i.e., our main THEOREM, from the beginning, is proved.*

Very Sketchy Idea of Proof Let us go back to the decomposition $\Gamma - R = X_1 \cup b(1) \cup Y_1$ which, with a lot of additional embellishments is presented in Fig. 7.10. Actually, since most of the present action does not concern the X_b^2 piece of $2X_0^2$, (see (10)), our Γ is now $\left(\Gamma(2\infty)\left(t=\frac{1}{2}\right)\right) \mid (r\text{-side of } 2X_0^2)$.

In Fig. 7.10, the $\Gamma(1)\left(t=\frac{1}{2}\right) - R$ lives deep inside the Y_1-side. This is actually how we distinguish the active part X_1 from the passive Y_1. Our $\Gamma(1)(t=\frac{1}{2}) - R$ is a tree.

There are finitely many steps in the process

$$R \xRightarrow{\Phi(N)} R(N),$$

typically the substitution $r(1) \Rightarrow b(1)$ concerning Fig. 7.10. Here is how such a step will be realized geometrically; but this will be just a schematic view of the real thing, which is too long for this survey.

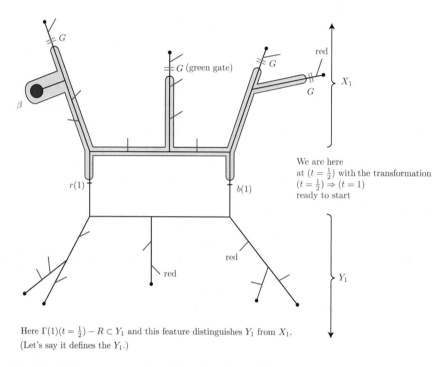

Here $\Gamma(1)(t = \frac{1}{2}) - R \subset Y_1$ and this feature distinguishes Y_1 from X_1.
(Let's say it defines the Y_1.)

Fig. 7.10 We have represented here $(\Gamma - R) \cup r(1)$, with $\Gamma - R = X_1 \cup b(1) \cup Y_1$

Ministep I Think of Fig. 7.10, thickened in dimension four, as being part of the 1-skeleton $N^4(2\Gamma(\infty)) \left(t = \frac{1}{2} \right)$ of (7.43). We start by sliding all the finitely many active red 1-handles and gates G, over $r(1)$, pushing them on the other side with respect to $r(1)$. This movement, where the 2-handles follow and which may contradict the $C \cdot h = \text{id} + \text{nil}$, requires admissible LAVA MOVES, concerning LAVA$_r$ (see (38.2)) and taking place on the $-$ side of the basic SPLITTING by $\sum_\infty^2$. Thereby the PRODUCT PROPERTY is preserved.

This move creates a clean COLOUR CHANGING CYLINDER $B^3 \times [r, b]$ corresponding to the part of Fig. 7.10 which is shaded and surrounded by a dotted contour ▇▇▇. The cylinder joins the $B^3(r)$, cocore $B^3(r)$ of $r(1)$, to the $B^3(b)$, cocore of $b(1)$. Also the cylinder comes naturally split into $\pm$ parts, like Fig. 7.11 may suggest, with one dimension less. At the end of our STEP I, it is separated from the rest of the world only by the $B^3(r)$ and $B^3(b)$.

Fig. 7.11 The
colour-changing cylinders

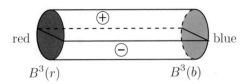

Ministep II Here comes the "isotopy" of 1-handle cocores, which in terms of Fig. 7.11 moves $B^3(r)$ to the position $B^3(b)$. Since this has to sweep through the sites β from Fig. 7.10, the step is anything but trivial. But we give no details here.

The SPLITTING + CONFINEMENT allow us to treat things very differently on the $+$ side and $-$ side. The curves confined on the $\pm$ side are NOT ENTANGLED with the ones on the $\mp$ side. Essentially, on the $+$ side, curves and LAVA$_b$ stay all put, and only the GLUEING TERMS (38.3) are being moved around. For instance, the $+$ side of $B^3(r)$, which is LAVA$_r$, like the whole of $B^3(r)$, may be touched by LAVA$_b$, which can be then harmlessly unglued and then glued somewhere else (on LAVA$_r$).

On the $-$ side, everything follows solidarily the move, thinking now of the move in question as a translation if $B^3(r)$ in Fig. 7.11 to the right, until it gets superposed to $B^3(b)$.

Before our MINISTEP II, $B^3(r)$ is LAVA and $B^3(b)$ NOT. After the step, $B^3(b)$ enters LAVA, *by decree*, while $B^3(r)$ becomes a new "site", without any special role or meaning. It is certainly not part of LAVA, any longer.

Ministep III By now $r(1)$ has been replaced by $b(1)$ and so it is no longer a 1-handle, just a "site". We essentially proceed now with Ministep I *in reverse* and, since $r(1)$ is just a "site", devoid of any kind of special geometrical meaning, this is now a mere isotopy, not a sliding of handles over other handles. We just restore Γ as it was in the beginning, before our STEP I has been undertaken.

This 3-step move has realized the change

$$R \Longrightarrow R(1) = R - r(1) + b(1).$$

Then we continue in the same vein, until the $\Phi(N)$ has been fully realized, geometrically.

We end here our sketchy idea of the proof of Theorem 7.8.1 which itself is the geometric 4-dimensional realization of the change of RED into BLUE.

Our sketch of proof for the theorem with which this paper starts is also ended.

References

1. L. Bessières, G. Besson, M. Boileau, S. Maillot, J. Porti, *Geometrization of 3-manifolds*. EMS Tract. Math., vol. 13 (European Mathematical Society, Zürich, 2010)
2. G. Besson, Une nouvelle approche de la topologie de dimension 3, d'après R. Hamilton et G. Perelman, *Séminaire Bourbaki*, 57eme année, N^0 947 (2005)
3. A. Connes, *Non-commutative Geometry* (Academic Press, New York, 1994)
4. B. Mazur, On embeddings of spheres. BAMS **65**, 59–65 (1959)
5. J. Morgan, G. Tian, The Ricci flow and the Poincaré Conjecture. AMS Clay Math. Inst. **3**, 1–521 (2007)
6. V. Poénaru, Geometric simple connectivity and low-dimensional topology. Proc. Steklov Inst. Math. **247**, 195–208 (2004)
7. V. Poénaru, On the 3-dimensional Poincaré Conjecture and the 4-dimensional Schoenflies problem, Prépublication d'Orsay 2006–25 (2006), pp. 1–38
8. V. Poénaru, What is ... an infinite swindle. Notices AMS **54**(5), 619–622 (2007)
9. V. Poénaru, Geometric simple connectivity and finitely-presented groups, arXiv:1404.4283[Math-GT] (2014). Soon, a revised and better version of this paper should be available, hopefully much more reader-friendly than the old one
10. V. Poénaru, All smooth four-dimensional Schoenflies balls are geometrically simply connected, arXiv:1609.05094 (2016)
11. V. Poénaru, *A Glimpse into the Problems of the Fourth Dimension* in the volume Geometry in History, ed. by S. G. Dani, A. Papadopoulos (Springer Nature Switzerland AG, Cham, 2019), pp. 687–704
12. V. Poénaru, On geometric group theory, in *Topology and Geometry – A Collection of Essays Dedicated to Vladimir G. Turaev*, ed. by A. Papadopoulos (EMS Press, 2021), pp. 399–432
13. V. Poénaru, Classical differential topology and non-commutative geometry, in *Surveys in Geometry I*, ed. by A. Papadopoulos (Springer, Cham, 2022), pp. 309–341. https://doi.org/10.1007/978-3-030-86695-2_8
14. V. Poénaru, C. Tanasi, Some remarks in geometric simple connectivity. Acta Math. Hungarica **81**, 1–12 (1998)
15. J.H.C. Whitehead, A certain open manifold whose group is unity. Q. J. Math. Oxf. Ser. **6**, 268–279 (1935)

Chapter 8
Classical Differential Topology and Non-commutative Geometry

Valentin Poénaru

Abstract The message of this chapter is that very innocent looking, and actually quite useful, constructions in standard differential topology can, in the very precise situation of, what I call, BRUTAL VIOLATION of geometric simple connectivity, lead to non-commutative spaces. Geometric simple connectivity is a venerable, fundamental concept in geometric topology and its occurrence here has been a surprise to me. Geometric simple connectivity is also an important concept for geometric group theory (see [7, 11]). Also, my paper [9] contains a rather complete account of the whereabouts of this concept of "geometrically simply connected" (GSC). In slightly more detail, here is what this chapter does. In a purely general set-up, for a handlebody decomposition of a non-compact smooth four-manifold M, a certain auxiliary space is constructed, for the aim of investigating the asymptotic structure of M. In the GSC situation this turns out to be a smooth 3-manifold, endowed with a lamination. This is very useful for our Schoenflies paper, in this same volume. But, when GSC is BRUTALLY violated, then the auxiliary space is non-commutative, à la Alain Connes. We believe that this is a link to a conjectural new category, encompassing all finitely presented groups, which should occur there as the rationals amidst the irrationals.

Not more is proved in this paper than the strict minimum necessary to show that the strange connection mentioned above is really there.

Keywords Geometric simply connected · Handlebody decomposition · 4-Manifold

AMS Codes 57M05, 57M10 57N35

V. Poénaru (✉)
Université de Paris-Sud, Mathématiques, Orsay, France

© The Author(s), under exclusive license to Springer Nature Switzerland AG 2022   309
A. Papadopoulos (ed.), *Surveys in Geometry I*,
https://doi.org/10.1007/978-3-030-86695-2_8

8.1   Introduction

We will present here our little story in the context of C^∞ non-compact four-manifolds with non-empty boundary. This is how these ideas first occurred in my work, and this is the way I am familiar with them too. But they make sense in any dimension, and even in the compact case. But I did not yet have the leisure to look carefully into that.

On the other hand, I do believe that the kind of ideas presented here are likely to have an impact in geometric group theory where they should nicely fit with my conjecture of the existence of a much larger category engulfing the category of finitely presented groups, which should fit inside it like the rational numbers among the real numbers or like the periodic functions inside the almost periodic ones. The Penrose aperiodic tilings and also the aperiodic crystals should reflect objects in this category, but I will not dwell more on this issue here. See here [11] for slightly more details and for the context of the conjecture. And, anyway, much more work remains to be done here.

To be now more concrete, in this paper we start with some, otherwise arbitrary smooth 4-manifold, noncompact and with non-empty boundary, call it M^4. For this M^4 there is also a given handlebody decomposition, which we will not change. My initial, original aim had been, some years ago to try to understand the asymptotic structure of this M^4. For this purpose, I developed a certain CONSTRUCTION, which will be presented explicitly in Sect. 8.2. This construction produces a topological space, which I call $\chi(\partial h^2 \cdot \delta h^1)$, for reasons which will become clear later. This space $\chi(\partial h^2 \cdot \delta h^1)$ is ABSTRACT, in the sense that, a priori, it looks totally unrelated to M^4, except that it is that given handlebody decomposition of M^4 which dictated the construction of $\chi(\partial h^2 \cdot \delta h^1)$.

And interestingly strange things happen with our abstract space, right now I will tell them informally, but later on they will be told much more formally, and these will be the theorems which the present paper proves.

When we are in the GEOMETRICALLY SIMPLY-CONNECTED case, then $\chi(\partial h^2 \cdot \delta h^1)$ is a smooth non-compact 3-manifold, canonically endowed with a lamination by lines, and this χ has a lot to do with the actual asymptotic structure of M^4. When we are in the situation where geometric simple-connectivity is violated, various things may happen, some of them wild, BUT when we are in a case of BRUTAL VIOLATION OF GEOMETRIC SIMPLE CONNECTIVITY, a notion which will be explicitly defined, then $\chi(\partial h^2 \cdot \delta h^1)$ is a NON-COMMUTATIVE SPACE in the sense of A. Connes [1].

So, what we seem to see here is a rather weird connection, at least when we are in the context of our CONSTRUCTION, namely, geometric simple connectivity (GSC) $\Longrightarrow$ usual manifolds and spaces, a connection which continues as long as GSC is NOT brutally violated, but then

BRUTAL violation of geometric simple connectivity

$\Longrightarrow$ NON-COMMUTATIVE spaces.

Now, in the rest of this first introductory section, I will present some material, most of which is more or less standard, but which will be absolutely necessary, at least in the form in which it is presented here, for the more technical rest of the paper.

At least so as to fix the notation, I will remind the reader the basics of handlebody theory. A handle H^λ of dimension n and of *index* λ, is a copy of B^n factorized as follows

$$B^n = B^\lambda \times B^{n-\lambda}, \tag{8.0}$$

where we distinguish the *attaching zone* $= \partial B^\lambda \times B^{n-\lambda}$ and the *lateral surface* $= B^\lambda \times \partial B^{n-\lambda}$. We will use the notation

$$\partial H^\lambda = \text{attaching zone}, \qquad \delta H^\lambda = \text{lateral surface}.$$

And for our present purposes, a bit more detail will be necessary. Let $*_\lambda \in B^\lambda$ and $*_{n-\lambda} \in B^{n-\lambda}$ be the centers of the respective balls. Then, we define $h^\lambda \equiv B^\lambda \times (*_{n-\lambda})$, the *core* of the handle in (0), and also the *cocore* (of the λ-handle (0)) $\equiv (*_\lambda) \times B^{n-\lambda}$. (8.1)
For $n = 3$ and $\lambda = 1$ these things are illustrated in Fig. 8.1.

Now a side comment. As everybody knows, handlebody theory is essentially equivalent to the Morse theory of real C^∞ functions. But I do not well see how I could tell the present story in Morse-theoretical terms. It would require some contorsions which I am not able to do, or at least not comfortably so.

We will be interested here in the following smooth 4-manifold which is non-compact with non-empty boundary, for which we write down a *handlebody decomposition* of the form below:

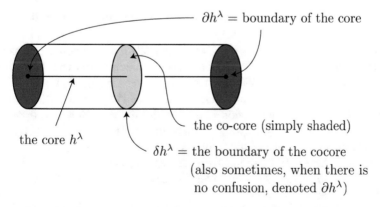

$\partial h^\lambda = $ boundary of the core

the core h^λ

the co-core (simply shaded)

$\delta h^\lambda = $ the boundary of the cocore (also sometimes, when there is no confusion, denoted ∂h^λ)

Fig. 8.1 On this figure, the attaching zone (doubly shaded) and the lateral surface (= the rest of the outer boundary) should be clear

$M^4 = N^4$ (4^d regular neighbourhood of an infinite tree T with all its end points at infinity) $+ \sum_{i=1}^{\infty} H_i^1 + \sum_{j=1}^{\infty} H_j^2 +$ {other handles of index $\lambda = 2$ and 3}, where one is given a CANONICAL ISOMORPHISM $\{i\} = \{j\}$. And the elements of this set are called **states**. $\hspace{4cm}$ (8.2)

It is also assumed in the context of (2), that every individual contact of the attaching zone ∂H_j^2 with the lateral surface of H_i^1, i.e., each connected component of {attaching zone of H_j^2} $\cap$ {lateral surface of H_i^1}:

$$(S^1 \times B^2) \cap (I \times S^2) \hspace{4cm} (*)$$

(and here S^1, boundary of the core h_j^2 of H_j^2 is also denoted ∂h_j^2, while I is the core h_i^1 of H_i^1), is a complete copy of $I \times B^2$, going from one end of h_i^1 to the other. The geometric meaning of the canonical isomorphism above, i.e., for $\{i\} = \{j\}$, is that for any $i = j$, in the context of (*) there is, singled out from possibly several similar contacts a canonical contact (also called **privileged** contact)

$$I \times B^2 \subset (\partial h_j^2 \times D^2) \cap (h_j^1 \times S^2) \hspace{3cm} (8.2.1)$$

(same index j ($= i$) twice). End of (2.1).

Once this contact is singled out, in the handlebody decomposition (2) we see, for each i, a piece

$$B_i^4 \equiv \underbrace{H_i^1 \cup H_i^2}_{\text{privileged } I \times B^2} , \hspace{3cm} (8.2.2)$$

the boundary of which may touch itself more than once in M^4, along more non-canonical $I \times B^2$'s. This leaves us, for each i, with a **submersive** map $B_i^4 \xrightarrow{\psi_i} M^4$ s.t. $\psi_i \mid \text{int } B_i^4$ injects. End of (8.2.2)

The (2) + (2.1) + (8.2.2) are the given initial data for the action in this paper.

With all these things, the handlebody decomposition comes with the following GEOMETRIC INTERSECTION MATRIX, introduced in (3) below.

We start by observing, that in the conditions above, $\partial h_j^2 \cap \delta h_i^2$ consists of a (finite) number of transversal intersection points. With this, we define our **geometric intersection matrix** by $\partial h_j^2 \cdot \delta h_i^1 \equiv$ {the number of points in $\partial h_j^2 \cap \delta h_i^1$, counted without orientations, hence without any $\pm$'s} $= \delta_{ji} + a_{ji}, a_{ji} \in Z_+$. The Kroenecker delta accounts here for the canonical (= privileged) contact $I \times B^2 \subset \partial H_i^2 \cap \partial H_i^1$ while, of course, an $a_{ii} \geq 0$ accounts (if really > 0) for the diagonal but not privileged contacts $\partial H_i^2 \cdot \partial H_i^1$. $\hspace{3cm}$ (8.3)

The matrix $\delta_{ji} + a_{ji}$ above is called "geometric" because when we look into the contacts $\partial h_j^2 \cdot \delta h_i^1$ we ignore orientations, i.e., the $\pm$ signs coming with them. My cousin, Sir Christopher Zeeman used to say that with $\pm$ signs we have algebraic topology and without, we have geometric topology.

To the geometric intersection matrix (to which, for the time being we do not impose any other restrictions), we associate an oriented *graph* *G*, accounting for all the non-canonical part of the geometric intersection matrix. And, from now on we will also assume that the a_{ii} are always $a_{ii} = 0$, so G accounts for the non-diagonal part of the matrix. Here is how G is defined. The vertices of G are the elements of the set of indices $\{i\} \approx \{j\}$ and remember that we call them *states*. And, there are exactly a_{ji} oriented arrows (edges) going from state j to state i, i.e.

$$\text{For } \forall i, j, \; i \neq j, \; \#\{j \to i\} \equiv a_{ji}. \tag{8.3.1}$$

In (3) we assumed that $i, j \in Z$ but very importantly, in the next definition we restrict things to $i, j \in Z_+$.

Definition 8.1.1 The manifold M^4 is said to be *geometrically simply connected* (and we will also use the label GSC), if there exists a handlebody decomposition like in (2) satisfying the additional condition that the matrix $\delta_{ji} + a_{ji}$ is of the following form (which we call easy id + nilpotent), $\delta_{ji} + a_{ji}$ with

$$a_{ji} > 0 \Longrightarrow j > i, \tag{$*_1$}$$

and in this definition of GSC we definitely assumed that the set of indices is $\{i\} = \{j\} = Z_+$ and NOT Z. End of Definition 8.1.1.

This definition is very important and it deserves some comments.

Comments

(I) GSC really should mean that the 1-handles are in cancelling position with the 2-handles. And, for infinite handlebody decompositions the easy id + nil is the good way to express it, AS LONG AS IN $(*_1)$ THE INDEXING IS Z_+. For instance, in the conditions of Fig. 8.2, even with an indexing satisfying $(*_1)$ there is NO good cancellation, hence we cannot be GSC.

Fig. 8.2 We see here a *G*-trajectory modelled in R. Actually this oriented graph, just by itself, can occur in the geometric context $M^4(2)$ as the G associated to a geometric intersection matrix. Here there is no initial element (no incoming arrows) and no final element (no outgoing arrows). Depending on how we orient the R and accordingly to this choice index the states, we may have been either $(*_1)$ or the dual condition

$$a_{ji} > 0 \Longrightarrow i > j. \tag{$*_2$}$$

But we are NOT GSC here, since the indexing cannot be Z_+

(II) There is also **difficult id + nil** where $(*_1)$ is replaced by $(*_2)$ above, namely,
      by

$$a_{ji} > 0 \Longrightarrow j < i \quad \text{(which also implies } a_{ii} = 0, \forall i\text{)},$$

and here the indexing is Z or Z_+. Whatever that is, we do not have handle-
cancellation and we are NOT GSC with difficult id + nil. The Whitehead
manifold Wh^3 (see [14] and [9]), for instance, has handlebody-positions which
have the **difficult id + nil property** and it cannot have the cancelling
feature, because $\pi_1^{\infty} Wh^3 \neq 0$. In Fig. 8.3b we have suggested for this Wh^3

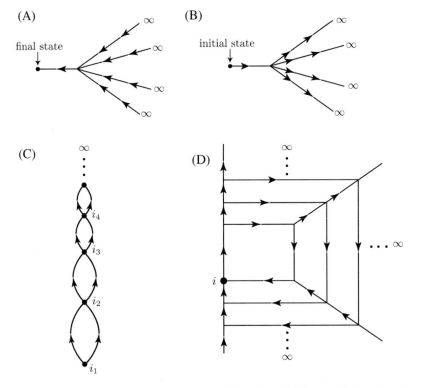

Fig. 8.3 The oriented graph (A) corresponds to easy id + nil, and indices are Z_+, i.e., it concerns
an M^4 which is GSC. All the other graphs violate the easy id + nil. The (B) is the simplest version
of difficult id + nil. The (C) which is the G corresponding to a certain handlebody decomposition
of the Whitehead manifold Wh^3 (see [9]) is also of the difficult id + nilpotent type. In both (B)
and (C) the indexing is again $\{i\} = \{j\} = Z_+$. The (D), where the indexing of states has to be
Z, contains an infinite spiral going infinitely many times through the same state (= site) i. It is
not id + nilpotent of any kind and it illustrates a BRUTAL violation of GSC, leading, actually, as
we shall see, to non-commutative spaces. More will be said in the main text concerning the spaces
which come with these graphs

the G of a handlebody decomposition with $(*_2)$ and s.t. $\{i\} = \{j\} = Z_+$. We will come back to it.

(III) The GSC is a very general notion, applicable to any handlebody decomposition, finite or not, in any dimension. Of course, in the finite case we have an equivalence

$$\text{easy id} + \text{nil} \Longleftrightarrow \text{difficult id} + \text{nil}.$$

All this being said, we restrict ourselves in this paper to handlebodies of type (2), for convenience reasons.

Here are some reminders concerning GSC, and for more details see [9].

For smooth compact manifolds of dimensions $n \geq 5$ we have the basic implication $\pi_1 = 0 \Longrightarrow$ GSC. This is crucial for Smale's h-cobordism theorem [13] and his proof of the high-dimensional Poincaré Conjecture. This result extends also to the open case, if one adds the condition $\pi_1^\infty = 0$ (see [12] for a proof). It also follows from G. Perelman's celebrated work (see [3]) that, in the compact case, when $n = 3$, then again $\pi_1 = 0 \Longrightarrow$ GSC. But the implication $\pi_1 = 0 \Longrightarrow$ GSC fails in the following two cases.

- If $n = 4$ OR
- If the manifold is non-compact, with $\partial \neq \emptyset$. (8.4)

Again, see [9] for these things. Anyway when, sometimes in those situations • or •• one still wants to establish GSC, then hard work is necessary. This was the case in my own work in low-dimensional topology (specifically $n = 4$), see here [5, 8], or in geometric group theory. See here [7].

The GSC concept seems to speak a lot to me and more often than not, when toying with mathematical objects, in low-dimensional topology, or in group theory, I seem to bump into GSC.

(IV) The GSC concept makes good sense not only for handlebody decompositions, but for cell-complexes too. And it so that GSC which occurs in geometric group theory (see [4, 7, 10]). But all that is beyond our present scope here. End of Comments.

After all this introductory material, the next Sect. 8.2 will present our construction and also will also state our two main results, the first concerning the GSC case and the second the case when GSC is brutally violated.

Figure 8.3 presents some oriented graphs associated to various geometric intersection matrices. We constantly assume here, and from now on, that $a_{ii} = 0$, $\forall i$.

8.2   The Construction

We will develop now a CONSTRUCTION, the prospective aim of which is to understand the asymptotic structure of the smooth, non-compact 4-manifold with non-empty boundary M^4, from (2) above. Actually, our construction makes sense in any dimension and in the compact case too. But we restrict here ourselves to the 4-dimensional, non-compact case. The $B_i^4 = H_i^2 \cup H_i^1$'s for all $i \in \{i\} \approx \{j\}$, illustrated in Fig. 8.4, concerned with the diagonal terms of the geometric intersection matrix. Now we are concerned with the off-diagonal part, i.e., with the oriented edges of the oriented graph G. But we need some notation, to begin with.

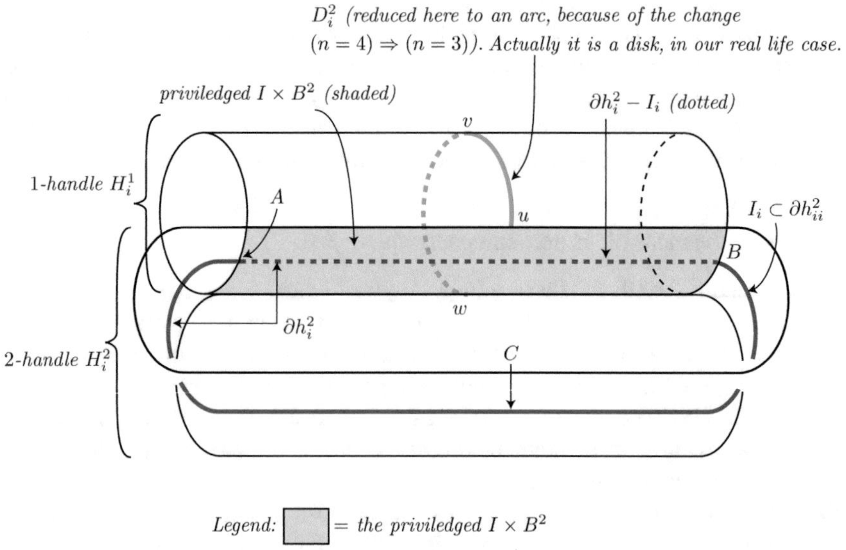

Fig. 8.4 We see here the $H_i^2 \cup H_i^1 = B_i^4$. For typographical reasons, the $H_i^1 = I \times B^3$ appears here as a $I \times D^2$ and the $H_i^2 = B^2 \times B^2$ appears as a $D^2 \times I$. We have shaded the privileged $I \times B^2$ in (2.1), appearing here as a mere shaded rectangle. We also have

$$I_i \equiv [A, C, B] \,(\text{in red}) = \partial((\text{core } H_i^2) = h_i^2) - I \times B^2$$

and (with one dimension less than the real one),

$$D_i^2 \equiv [u, v, w] \,(\text{in green}) = \partial(\text{cocore } H_i^1) - I \times B^2.$$

Other, more symmetrical definitions for I_i, D_i^2 will be provided in the main text

Lemma 8.2.1

(1) To each state $i \in \{i\} \approx \{j\}$ and the corresponding

$$B_i^4 \equiv \underbrace{H_i^2 \cup H_i^1}_{privileged \ I \times B^2} \ , \tag{8.5}$$

already mentioned in the introduction, in (8.2.2), we will attach a 3^d cell Box(i)
via the process developed below.
  To begin with, we have an arc

$$I_i \equiv \partial h_i^2 - \{the \ privileged \ I \times B^2\}, \tag{8.5.0}$$

and noticing the obvious isomorphism

$$\partial h_i^2 \cap (I \times B^2) \approx h_i^1 \ , \tag{8.5.1}$$

we can write (8.5.0) in the more symmetrical form

$$I_i = \partial h_i^2 - h_i^1 = \partial(core \ H_i^2) - (core \ of \ H_i^1) \ . \tag{8.5.2}$$

*In a 3^d situation (and I cannot make 4^d drawings), the I_i is correctly represented
in Fig. 8.4. Then, we define the disc*

$$D_i^2 = \underbrace{\partial(cocore \ H_i^1)}_{\delta h_i^1} - \{the \ privileged \ I \times B^2\}, \tag{8.6}$$

appearing like an arc in Fig. 8.4, and noticing the obvious isomorphism

$$\partial(cocore \ H_i^1) \cap (I \times B^2) = \delta h_i^1 \cap (I \times B^2) \approx cocore \ H_i^2 \ , \tag{8.6.0}$$

we can write (8.6) in the more symmetrical form

$$D_i^2 = \partial(cocore \ H_i^1) - (cocore \ H_i^2) \ . \tag{8.6.1}$$

*With one dimension less than the realistic four, D_i^2 appears as the green arc
$[u, v, w]$ in Fig. 8.4.*
  To these I_i, D_i^2 we attach the ABSTRACT 3^d box

$$\mathrm{Box} \, (i) \equiv D_i^2 \times I_i \ . \tag{8.7}$$

*We call this "ABSTRACT" because, contrary to the I_i, D_i^2, this disc is **not** part
of our M^4 (in (2)). It will be the basic building brick of our construction.*

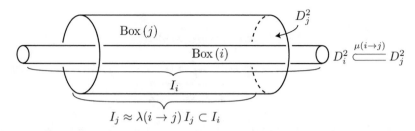

Fig. 8.5 We see here the passage $v(i \rightarrow j)$ of Box (i) through Box (j). Distinct passages are disjoint and parallel (when the target is the same Box (i). Im $v(i \rightarrow j)$ is the intersection of the two Boxes

(2) *Every time we have an off-diagonal arrow* $i \rightarrow j$ *in the geometric intersection matrix, and actually ALL our arrows are off-diagonal, this comes with maps, which are* **inclusions**

$$I_j \xrightarrow{\lambda(i \rightarrow j)} I_i \tag{8.8}$$

$$D_i^2 \xrightarrow{\mu(i \rightarrow j)} D_j^2 \, .$$

Together, these two maps induce a third one, at the level of the corresponding Boxes, namely, another inclusion $v(i \rightarrow j)$

$$\text{Box}\,(i) \supset (\lambda(i \rightarrow j)I_j) \times D_i^2 \xrightarrow{v(i \rightarrow j) \equiv \mathrm{id} \times \mu(i \rightarrow j)} I_j \times D_j^2 = \text{Box}\,(j)\,. \tag{8.8.1}$$

This like the boxes themselves, is purely **ABSTRACT** *too not corresponding to anything physical in* $M^4(2)$. *But it means that* Box (i) *goes through* Box (j), *like in Fig. 8.5.*

(3) *Via the* $v(i \rightarrow j)$ *above, for each arrow* $i \rightarrow j$ *there is a* **clean passage** *of* Box (i) *through* Box (j), *like in a Markov partition. Figure 8.5 illustrates the passage in question. End of Lemma*

Proof We only need to make explicit the embeddings λ, μ. Figure 8.6, which is in the same style as Fig. 8.4, illustrates the geometry accompanying an arrow $i \rightarrow j$. The map $\lambda(i \rightarrow j)$ should be readable from Fig. 8.6, via the following successive isomorphisms

$$I_j = \partial h_j^2 - h_j^1 = [a, \overline{c}, b] \underset{\lambda}{\approx} [A, \overline{C}, B] \subseteq I_i \,.$$

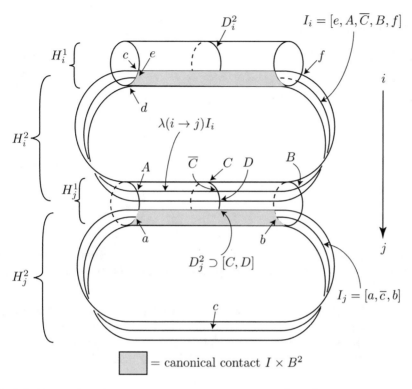

Fig. 8.6 Here $[AB] = \lambda(i \to j)I_j \subset I_i$ and $[CD] = \mu(i \to j)D_i^2 \subset D_j^2$

Similarly $\mu(i \to j)$ corresponds to

$$D_i^2 = \partial(\text{cocore } H_i^1) - I \times B^2(\text{at } i) \approx [c, d] \approx [DC] \underset{\mu}{\subset} D_j^2.$$

(Try to do these things with Morse theory!) End of Proof.

So far, for the maps λ, μ in (8.8) there was a lot of margin of freedom as far as their precise definition was concerned. There is a lot of isotopic margin of freedom. But soon metric conditions will be imposed too, making the situation more rigid.

Definition 8.2.2 Under the above conditions, we will say that we have a **brutal violation of GSC** when the oriented graph G contains closed (oriented) cycles, like in Fig. 8.7 below.

End of Definition 8.2.2.

Fig. 8.7 A closed oriented
cycle which the graph G may
contain. And then, we have a
BRUTAL VIOLATION OF
GSC, by definition

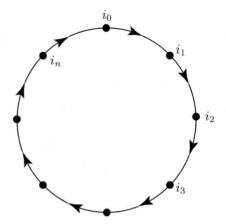

In our present infinite case (2), the various possible geometric intersection
matrices (3) and/or their associated oriented graphes G break down into three 2-
by-2 disjoint classes:

(1) id + nilpotent (including the *easy* id + nilpotent which is our GSC case when
    the indexing is Z_+).
(2) The intermediary mixed case, which I will not try to unwrap further.
(3) The case of BRUTAL violation of GSC (actually brutal violation of id + nil,
    also).

Notice that between GSC (which is part of 1) and its brutal violation we have a
whole intermediary world of violation of GSC which is not brutal. And it will not
be very much discussed here.

In all the rest of paper, when precise things will proved, we will completely
concentrate on the following two situations, completely ignoring all the others:

(I) The GSC case (i.e., metric easy id + nil with ordering Z_+) and
(II) A precise given cycle, like in Fig. 8.7.

All our product structures, metric conditions and other refinements to follow, will
exclusively concern these two cases, unless the contrary is explicitly stated, like in
Fig. 8.11.

Lemma 8.2.3

*(1) When, in the presence of a cycle like above, we look at the succession of Markov
partition type passages*

$$\text{Box}(i_0) \text{ through } \text{Box}(i_1), \quad \text{Box}(i_1) \text{ through } \text{Box}(i_2), \ldots, \text{Box}(i_n)$$

$$\text{through } \text{Box}(i_0), \tag{8.9}$$

then their composition is a self-passage of Box(i_0) *through itself*

$$\text{Box}(i_0) \text{ through } \text{Box}(i_0) , \tag{8.10}$$

which we represent like in Fig. 8.5 denoting now by Box(i) *the LHS of (8.10) and* Box(j) *the RHS. This composite self-passage corresponds to inclusion maps which we denote* $\overline{\lambda}$ *and* $\overline{\mu}$ *(or sometimes with* $\lambda, \mu,$ *again):*

$$\text{Box}(i) \supset \overline{\lambda}(i \to j)I_j \times D_i^2 \xrightarrow{\overline{\nu}(i \to j) \equiv \text{id} \times \overline{\mu}(i \to j)} I_j \times D_j^2 = \text{Box}(j) .$$
$$\tag{8.11}$$

But here, of course $i = i_0 = j$ *and hence there is here also a diffeomorphism* Box(j) $\xrightarrow{\varphi}$ Box(i), *sending* I_j *to* I_i *and* D_j^2 *to* D_i^2. *It will be assumed, from now on, that* φ *is always of the product form*

$$\varphi = \varphi \mid I_j \times \varphi \mid D_j^2 ,$$

hence we have diffeomorphisms

$$I_j \xrightarrow{\varphi|I_j} I_i \quad and \quad D_j^2 \xrightarrow{\varphi|D_j^2} D_i^2 .$$

Our φ *(which is there only in the presence of cycles) and the* (λ, μ) *(which are always there) have nothing to do with each other, a priori.*

(2) *From now on, on our particular* Box(i_0) *(and actually on all the* Box(k)*'s for the cases I, II) we will choose metrics, which we take always to be of the product form*

$$\text{metric on } \text{Box}(i) = \text{metric on } I_i \times \text{metric on } D_i^2 .$$

In the context of λ *and* μ *this will also come with the following requirements, which are very natural and can be implemented without any special effort.*

The two embeddings

$$I_j \xrightarrow{\lambda(i \to j)} I_i \quad and \quad D_i^2 \xrightarrow{\mu(i \to j)} D_j^2 \tag{8.12}$$

are isometries of source and image. We are now in the case I.

When it comes to the diffeomorphism φ, *in the case II, we will ask for the following natural conditions too, which will be incorporated in our definition of* χ, *as well as the metric conditions above. (But of course* φ *is there only when there* *are cycles.) The condition is that φ respects the product structure of* Box(j) *and* Box(i), *and moreover that*

φ *decreases the distances on D^2's, i.e., for $D_j^2 \xrightarrow[\approx]{\varphi} D_i^2$, if $x, y \in D_j^2$ then, with* $\varphi(x), \varphi(y) \in D_i^2$, *we have*

$$\text{dist}(x, y) > \text{dist}(\varphi(x), \varphi(y)). \quad \text{(Here } i \rightarrow j.) \qquad (8.13)$$

Conversely φ increases the distances on I, i.e., if $x, y \in I_j$, then with $\varphi(x), \varphi(y) \in I_i$, we have

$$\text{dist}(x, y) < \text{dist}(\varphi(x), \varphi(y)).$$

Notice that this means that, on D^2's, the φ^{-1} INCREASES the distances. Notice, also, that when one contemplates the Markov type passage of Box(i) *through* Box(j), *as well as the various identification occurring (let us say at the level of Fig. 8.6) when this passage is defined, then these conditions are completely natural.* End of Lemma 8.2.3.

The G in Fig. 8.3d obviously displays a brutal violation of GSC. Not only does it have oriented trajectories closing at the site i, the flow even goes infinitely many times through it.

Finally we can present the CONSTRUCTION.

It should be understood that all the various metric fixing above have to be incorporated too, in the definition coming next, at least when we are in the situations I, II.

Definition 8.2.4 (Our CONSTRUCTION) We introduce now the following *quotient-space*

$$\chi(\partial h^2 \cdot \delta h^1) \equiv \sum_i \text{Box}(i)/\mathcal{R}$$

where $\sum_i$ is the infinite *disjoint union* of all Boxes and $\mathcal{R}$ the *equivalence relation* generated by the pairs

$$(p, q) \in \sum_i \text{Box}(i)$$

s.t. $p \in (\lambda(i \rightarrow j)I_j) \times D_i^2 = \{$domain of definition of the map $\nu(i \rightarrow j)\} \subset$ Box(i), and $q \in \nu(i \rightarrow j)p \in$ Box(j).

The pair (p, q) above is such that $(p, q) \in \mathcal{R}$. Then we complete this by adding more pairs to $\mathcal{R}$, according to the rule

$$\{(p, q), (q, r)\} \in \mathcal{R} \implies (p, r) \in \mathcal{R},$$

a.s.o. We endow $\chi(\partial h^2 \cdot \delta h^1)$ with the standard quotient space topology. End of Definition.

Lemma 8.2.5

*(1) Assume that we are in the simplest possible case of a brutal violation of the GSC condition, when there is a site $i_0 \in \{i\} \subset G$, with a unique cycle passing through it like in Fig. 8.7. We will also ask that all states in our cycle should not have any other cyclic passing through them. So now $\mathrm{Box}(i_0)$ passes Markov-partition like through itself. For our purposes it will be convenient to say that there are two copies of $\mathrm{Box}(i_0)$, call them $\mathrm{Box}(i)$ and $\mathrm{Box}(j)$, with $\mathrm{Box}(i)$ passing through $\mathrm{Box}(j)$, and the two being connected by a diffeomorphism $\mathrm{Box}(j) \xrightarrow[\approx]{\varphi} \mathrm{Box}(i)$, with all the features from Lemma 8.2.3. In this case, $\mathrm{Box}(i_0)$ contributes to $\chi(\partial h^2 \cdot \delta h^1)$ with the quotient space below, which is a **subset** of $\chi(\partial h^2 \cdot \delta h^1)$*

$$\chi(\partial h^2 \cdot \delta h^1) \mid \mathrm{Box}(i_0) = (\mathrm{Box}(i) + \mathrm{Box}(j))/\{\text{the equivalence relation } \mathcal{R}_1\},$$
(8.14)

where $\mathcal{R}_1$ is the equivalence relation generated by (14.1) (generators of the equivalence relation $\mathcal{R}_1$)

- *The pairs $(x, \varphi(x))$, when $x \in \mathrm{Box}(j)$, $\varphi(x) \in \mathrm{Box}(i)$ and then also $(y, \varphi^{-1}(y))$, when $y \in \mathrm{Box}(i)$, $\varphi^{-1}(y) \in \mathrm{Box}(j)$, and*
- • *The pairs (p, q) with $p \in \mathrm{Box}(i)$, $q \in \mathrm{Box}(j)$ which are such that $q = \nu(i \to j)(p)$, like in Definition 8.2.4.* (8.14.1)

So, if $(a, b) \in \mathcal{R}$ and we are in the appropriate context, we also have $(a, b) \in \mathcal{R}_1$.

With all these things, when we are in the context of a G with a unique closed cycle, then for the respective $(\chi(\partial h^2 \cdot \delta h^1), \mathcal{R})$ and for $\mathrm{Box}(i_0) \in \text{cycle}$, we find that

$$\mathcal{R} \mid \mathrm{Box}(i_0) = \mathcal{R}_1.$$

(2) When we have more than one cycle going through a given state i, then the situation is more complicated, and it will not be examined here in detail. Anyway, in the general case, $\mathcal{R} \mid \mathrm{Box}(i_0)$ is just a QUOTIENT of $\mathcal{R}_1$, and $(\mathrm{Box}(i) + \mathrm{Box}(j))/\mathcal{R}_1$ no longer a subspace of $\chi(\partial \chi^2 \cdot \delta h^1)$, just a quotient space of a subset of the more complicated $\chi(\partial h^2 \cdot \delta h^1)$.

We can finally state the results of this paper.

Theorem 8.2.6

(1) When the geometric intersection matrix $\partial h^2 \cdot \delta h^1$ is of the type easy id + nilpotent with indices Z_+, i.e., when we are in the GSC situation for M^4, then $\chi(\partial h^2 \cdot \delta h^1)$ has a natural structure of C^∞, not necessarily connected non-compact 3-manifold N^3, with $\partial \neq \emptyset$, which comes equipped with a lamination $\mathcal{L}$ by lines (with transversal structure $(R^2, \text{Cantor set}))$.

(2) In this same GSC situation, our $\chi(\partial h^2 \cdot \delta h^1)$ lives naturally at the infinity of M^4, so that:

- *$\chi(\partial h^2 \cdot \delta h^1) - \mathcal{L}$ can be added to the boundary of M^4 and*
- *each leaf of $\mathcal{L}$ joins (at infinity) two endpoints of the tree T in (2).*

(3) Moreover, this is exactly the contribution of

$$N^4(T) + \sum_{i=1}^{\infty} H_i^1 + \sum_{j=1}^{\infty} H_j^2 \subset M^4$$

to the infinity of M^4

Something on the lines of Theorem 8.2.6 occurred in my paper [8], when it was a very minor ingredient in the proof that all smooth four-dimensional Schoenflies balls (see [2] and [6]) are GSC.

The proof of Theorem 8.2.6 will be given in the next section. But let us go now back to Fig. 8.3. Figure 8.3a corresponds to the GSC case and Theorem 8.2.6 concerns this case. In (b), (c), (d) GSC is violated. Now in (b) and (c) we are in the case when the geometric intersection matrix is of the type difficult id + nil. One can show that when we are in the case (b) then the corresponding space $\chi(\partial h^2 \cdot \delta h^1)$ is, quite naturally, a very simple-minded smooth non-compact 3-manifold coming with $\pi_1 = 0$ and nicely embeddable in R^3.

The situation of the G in Fig. 8.3c is the following. This corresponds to a handlebody decomposition of the Whitehead manifold Wh^3 which is of the type difficult id + nil. This decomposition comes from [9].

The corresponding $\chi(\partial h^2 \cdot \delta h^1)$ is now again a non-compact 3-manifold (hence a C^∞ manifold) with infinitely generated π_1. But, as it will be a bit discussed later, there is here something tricky.

The smooth structure can, by no means, be compatible with the natural decomposition into Boxes which $\chi(\partial h^2 \cdot \delta h^1)$ has.

So, in a certain sense, this case is wild.

Finally, as already said, in Fig. 8.3d we see a brutal violation of GSC, albeit a very serious one, since we have an infinite orbit passing infinitely many times through a same state i.

The next theorem tells us what happens with $\chi(\partial h^2 \cdot \delta h^1)$ when we have a brutal violation of GSC.

Theorem 8.2.7

(1) In the situation when there is a BRUTAL VIOLATION OF GSC, in the sense of Definition 8.2.2, the topological space $\chi(\partial h^2 \cdot \delta h^1)$ is such that the C^0 functions on it do not (always) manage to separate points.

(2) Hence in this case the $\chi(\partial h^2 \cdot \delta h^1)$ is a non-commutative space and the algebra of functions on $\chi(\partial h^2 \cdot \delta h^1)$ is now the non-commutative algebra of 2×2 matrices of 2-valued C^0 functions

$$\begin{pmatrix} f(p,p) & f(p,q) \\ f(q,p) & f(q,q) \end{pmatrix}$$

with the multiplication law given by the convolution product

$$(f \times g)(p,q) = \sum_{(p,r),(r,q)\in\mathcal{R}} f(p,r)g(r,q)$$

(see [1]).

The proof is given in the next section and I will end the present one with some questions, which I am certainly not able to answer, and which I leave to the reader.

Question 1 Can one give a reasonably nice description of the algebras occurring as function algebras of the spaces $\chi(\partial h^2 \cdot \delta h^1)$, when they are non-commutative?

My knowledge of function algebras is insufficient even for formulating more precisely this question.

Question 2 In the case when the GSC is violated, and in particular when it is brutally violated, is there some earthly connection between $\chi(\partial h^2 \cdot \delta h^1)$ and M^4, in particular the asymptotic structure of M^4?

The next question should not be very hard to answer, but I must confess I did not have the leisure to think more about it.

What happens with this little theory if one goes from the indices $\lambda = 1$ and $\lambda = 2$, to the general case λ and $\lambda + 1$?

8.3 The Proofs

We start with the

Proof of Theorem 8.2.6 We are now in the context of $M^4(2)$ when we are in the GSC situation (which also implies, remember, that all $a_{ii} = 0$). And now we look at the oriented graph G. We consider now, for each state i a Bloc(i) defined as follows $G \supset \text{Bloc}(i) \equiv \{\text{the union of all the incoming and outcoming arrows from the state } i, \text{ like in Fig. 8.8b}\}.$ (8.15)

And these arrows can all come with multiplicities, of course.

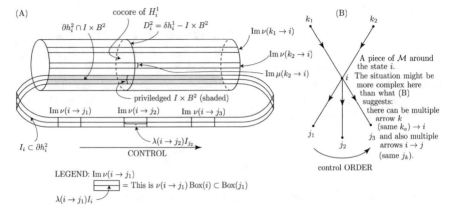

Fig. 8.8 In (B) we see a detail of G, around the state i and in (A) we suggest the geometry coming with it, in particular with the outgoing part from i. Compare (A) with Fig. 8.4

Notice that, because of the dimensions involved ($\lambda = 2$ and $\lambda = 1$);

$$(2\text{-handle}) \longrightarrow (1\text{-handle}),$$

there is a natural linear order on the outgoing arrows, which we call *control*, displayed in Fig. 8.8a. (To have it well defined, we will assume that all the 1-handle cores h_i^1 and hence all the I_i's are *oriented*. Then via the canonical (or privileged $I \times B^2 \subset$ lateral surface of H_i^1, this orientation is extended to ∂h_i^2.

Once control and orientation are noth with us, we can be a bit more precise about what it is going on. Let us say we have an arrow $i \rightarrow j$, with both states i and j coming with their orientations for $\{h^1$ and $\partial h^2\}$ and when i comes with its control, then two pieces of information can and have to be added now:
At our given contact $i \rightarrow j$ are the two orientations compatible OR incompatible? (8.16.1)
In the control order of the outgoing arrows from i, where does our given contact sit? (8.16.2)

When we have our various arrows $i \rightarrow j_1, j_2, \ldots$ and the orientations, it is very easy to adjust the passages of $\text{Box}(i)$ through the successive boxes $\text{Box}(j_1), \text{Box}(j_2) \ldots$ so that both (16.1), (16.2) should be OK, meaning that they will not come with head-aches for us, at least in the GSC or the unique cycle situation.

The lemma which comes next (together with its proof) will provide the proof for (1) in our Theorem 8.2.6 and also develop much more that point.

Lemma 8.3.1 *We consider now a noncompact smooth manifold $M^4(2)$ with a handlebody decomposition GSC (like in Definition 8.1.1) i.e., we have a geometric interaction matrix like in $(*_1)$ from Definition 8.1.1, indexed by $\{i\} = \{j\} = \mathbb{Z}_+$.*

(1) We can choose now compatible Riemannian metrics on the various Box(*i*)'s, *which are of product-type, i.e.*

$$metric\ on\ B(i) = (metric\ on\ I_i) \times (metric\ on\ D_i^2).$$

Various conditions listed below will be satisfied for the metrics.
When we have an arrow $i \rightarrow j$, *then the following universal conditions are satisfied*

$$\text{length } I_i > \text{ length } I_j \quad and \quad diam\ D_i^2 < diam\ D_j^2, \qquad (8.17)$$

and without them the passage of Box(*i*) *through* Box(*j*) *would not be possible, reason for calling them universal. Whenever we have an infinite trajectory of G of the type* $i_1 \leftarrow i_2 \leftarrow i_3 \leftarrow i_4 \leftarrow \cdots$, *then besides (17) we also have*

$$\lim_{n=\infty} \text{length } I_n = \infty, \quad \lim_{n=\infty} diam\ D_n^2 = 0. \qquad (8.17.1)$$

(2) The $\chi(\partial h^2 \cdot \delta h^1)$ *is a not necessarily connected, noncompact* C^∞ *3-manifold with* $\partial \chi(\partial h^2 \cdot \delta h^1) \neq 0$ *and inside it each* Box(*i*) *keeps its topological identity intact, meaning that we cannot find distinct points* $p, q \in$ Box(*i*), $p \neq q$ *which are such that* p *and* q *are equivalent via* $\mathcal{R}$ *(i.e., they are the same point in* $\chi(\partial h^2 \cdot \delta h^1)$).
This 3-manifold has a Riemannian metric satisfying (1).
(3) More precisely, in our present GSC situation, the $\chi(\partial h^2 \cdot \delta h^1)$ *has a description as a union of Boxes, passing through each other, each Box being smooth and endowed with a product metric and, very importantly, this description is* **compatible** *with the global* C^∞ *and Riemannian structures of the 3-manifold* $\chi(\partial h^2 \cdot \delta h^1)$.

[As we shall see we loose this compatibility when we move to the $\chi(\partial h^2 \cdot \delta h^1)$ for the Whitehead manifold, which does not admit a description as union of Boxes, compatible with its C^∞ structure.]

Proof In our present GSC situation, we can give a general description of how the oriented graph G looks like.
We have a set of initial states

$$i_{11}, i_{12}, i_{13}, \ldots \qquad (8.18)$$

which, since they are all initial, have to be independent of each other (i.e., no arrows in between them). Next we have a set of states

$$i_{21}, i_{22}, i_{23}, \ldots$$

328   V. Poénaru

which only send arrows to the i_{1n}'s and also among themselves

....

We have, after the i_{1m_1}'s, i_{2,m_2}'s, ..., $i_{n-1\,m_{n-1}}$'s, a next set of states

$$i_{n1}, i_{n2}, i_{n3}, \ldots \qquad (*)$$

which only send arrows to i_{pq}, when $p < n$ or among themselves, a.s.o., indefinitely. End of (18)

Notice that this situation can produce *overlaps* of intermediary arrows, like in Fig. 8.9. Such things may be a source of difficulties with the Feynman diagrams occurring in QFT, but not here, as it will turn out.

Let us assume now, inductively, that the intermediary

$$\chi(\partial h^2 \cdot \delta h^1) \Big| \sum_{p<n} i_{pq} \qquad (8.18.1)$$

with given n, is already constructed with both the smooth manifold and the product metrics. We want to go to the next step

$$\chi(\partial h^2 \cdot \delta h^1) \Big| \sum_{p\leq n} i_{pq} . \qquad (8.18.2)$$

We have to take care now of the various $(*)$ in (18) above. Each $i_{n\alpha}$ comes with two kinds of outgoing arrows:
Outgoing internal arrows $i_{n\alpha} \to i_{n\beta}$, $\alpha \neq \beta$; and outgoing external arrows $i_{n\alpha} \to i_{p\gamma}$, $p < n$. $\qquad (8.18.3)$

Fig. 8.9 An overlap

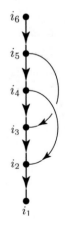

The internal arrows allow us to give a canonical meaning to the indexing of the i_{nm}'s, a bit like (18), but NOT quite,
We have $i_{n1}(1), i_{n1}(2), \ldots$ without having incoming internal arrows. Next, we have

$$i_{n2}(1), i_{n2}(2), \ldots$$

which send internal arrows to the $i_{n1}(\ldots)$'s, but NOT internal ones among themselves. Next, still, we have

$$i_{n3}(1), i_{n3}(2), \ldots$$

which send internal arrows to the $i_{n1}(\ldots)$ and $i_{n2}(\ldots)$'s, but NOT among themselves. And this continues then in an obvious way.   (8.18.4)
Next, we can construct the following first extension of (18.1):

$$\chi(\partial h^2 \cdot \delta h^1)\Big|\left(\sum_{p<n} i_{pq} + \sum_m i_{n1}(m)\right). \qquad (8.18.5)$$

For that, we construct, abstractly first, the metric $\mathrm{Box}(i_{n1}(m))$'s, sufficiently long and sufficiently thin, satisfying of course (17) (and aspiring at the very end when all the $i_{np}, i_{n+1,p} \ldots$ are taken care of to (17.1)), and let them pass through the needed Boxes of (18.1).
Next we move to the $i_{n2}(\ldots)$'s, $\ldots$ a.s.o. Figure 8.10 may help vizualizing what we are doing. End of Proof of Lemma 8.3.1.

We have not really finished with (1) in Theorem 8.2.6, we still have to construct the lamination $\mathcal{L}$.
Our oriented graph G (easy id + nil and indexed by Z_+) has, generally speaking, infinitely many trajectories going into a given state i. Such a trajectory looks like below:

$$\underbrace{i \equiv i_0}_{\substack{\text{initial element}\\\text{of the trajectory}}} \longleftarrow i_1 \longleftarrow i_2 \longleftarrow i_3 \longleftarrow \cdots \qquad (8.19)$$

Given such a trajectory, the composite arrow of i_n into i will be denoted by $\lambda(i_n \to i)$, avoiding a much heavier notation (Fig. 8.11).

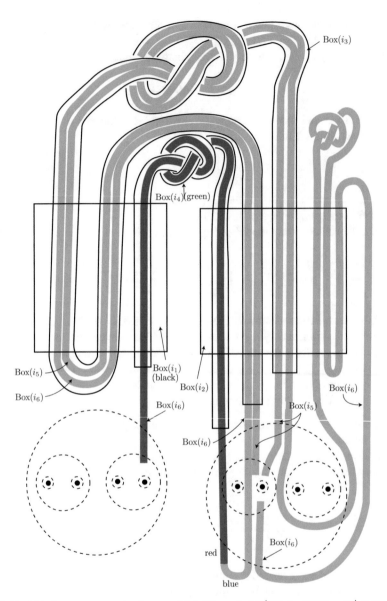

Fig. 8.10 This figure accompanies Lemma 8.3.1. We are in $\partial N^4(T)$ for $N(T) \subset X^4$ (8.20). We see the image $J\chi(\partial \chi^2 \cdot \delta h^1)$ (see (8.27)) and the fat points suggest the endpoints of T, living at the infinity of $\partial N^4(T)$. The figure is borrowed from [8]. Notice that, inside a given Box, the Boxes going through it have passages all parallel with each other. All the complicated knotting occurs outside these passages. We are in the GSC situation

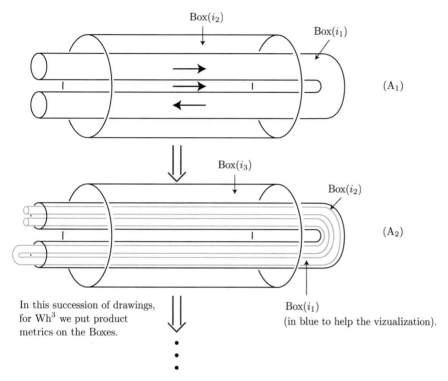

In this succession of drawings, for Wh^3 we put product metrics on the Boxes.

⇓

Box(i_1)
(in blue to help the vizualization).

•
•
•

Fig. 8.11 We consider the Wh^3 (Whitehead manifold) with the handlebody decomposition having the geometric intersection matrix suggested in Fig. 8.3c. It turns out that when we consider the two passages $i_n \to i_{n+1}$ (Fig. 8.3c) for one of them the orientations are compatible and for the other incompatible. This condition (suggested in (A$_1$)) is satisfied by our drawings

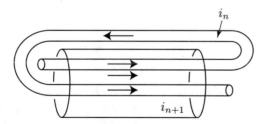

The opposite situation would be like in the little drawing below, when both passages come with compatible orientations. What we see in (A$_1$) is metrically correct and this metric will stay. In Box(i_3), we submit (A$_1$) to the contortions represented in (A$_2$). To make this metrically possible we have to shrink lengths in the horizontal direction by a factor of 2, but we did not try to draw that. Anyway, what the topology is (A$_1$) $\Rightarrow$ (A$_2$) $\Rightarrow \cdots$ imposes are the metric conditions

$$\text{length } I_1 > 2 \text{ length } I_2 > 4 \text{ length } I_3 > 8 \text{ length } I_4 > \cdots$$

and, similarly

$$\text{diam } D_1^2 < 2 \text{ diam } D_2^2 < 4 \text{ diam } D_3 < 8 \text{ diam } D_4 < \cdots$$

which also means that

$$\lim_{n=\infty} \text{ length } I_n = 0, \qquad \lim_{n=\infty} \text{ diam } D_n^2 = \infty.$$

We are here, with the Whitehead manifold, in the difficult id + nilpotent case, but the violation of GSC is NOT of the brutal kind. The main text will say more about what is going on here.

Lemma 8.3.2 *Our 3-manifold* $\chi(\partial h^2 \cdot \delta h^1)$ *from (17) comes naturally endowed with a lamination by lines* $\mathcal{L}$, *which is defined as follows, and in the next formula i is not necessarily initial, but generic*

$$\mathcal{L} \mid \text{Box}(i) = \sum_{i=i_0 \leftarrow i_1 \leftarrow i_2 \leftarrow i_3 \leftarrow \cdots} \left(\bigcap_{i_n > 0} \nu(i_n \to i)(\lambda(i_n \to i) \, I_i \times D_{i_n}^2) \right),$$

$$(8.19.1)$$

a formula which should make the transversal structure $(R^2, \text{Cantor set})$ *of* $\mathcal{L}$ *transparent. Next, of course, our lamination* $\mathcal{L}$ *is defined by*

$$\mathcal{L} \equiv \bigcup_i (\mathcal{L} \mid \text{Box}(i)).$$

$$(8.19.2)$$

End of Lemma 8.3.2.

So, by now the pair $(\chi(\partial h^2 \cdot \delta h^1), \mathcal{L})$ is with us, and hence the (1) in Theorem 8.2.6 has been proved, and we move to (2).

So, next, we look into the connection with the asymptotic structure of M^4. We will concentrate on a piece of M^4, already mentioned in (3) in Theorem 8.2.6, namely, the following submanifold

$$X^4 \equiv N^4(T) \cup \sum_{i=1}^{\infty} H_i^1 \cup \sum_{j=1}^{\infty} H_j^2 \subset M^4.$$

$$(8.20)$$

The embedding above is PROPER, and so, in informal language, we find that infinity of $X^4 \subset$ infinity of M^4, and it is the infinity of X^4 which interests us now.

For each B_i^4 (see (8.5) and Fig. 8.4) we will distinguish the following two subsets of ∂B_i^4, namely,

$\mathcal{B}_i^1 = \{\delta H_i^2 \text{ (lateral surface of } H_i^1)\} - \{\text{priviledged } I \times B^1\} \approx h_i^1 \times D^2$, and
$\mathcal{B}_i^2 = \{\text{attaching zone of } H_i^1\} \cup (\{\text{attaching zone of } H_i^2\} - \{\text{ priviledged } I \times B^2\})$,
coming with
$$(8.21)$$

$$\partial B_i^4 = \mathcal{B}_i^1 \cup \mathcal{B}_i^2 \cup \{\text{lateral surface of } H_i^2\}.$$

$$(8.22)$$

This formula can be vizualized (but in the unrealistic dim = 3) in Fig. 8.4. Each of the two $\mathcal{B}_i^1 \cup \mathcal{B}_i^2$ and {lateral surface of H_i^2} are here a copy of $S^1 \times D^2$, making up for $\partial B_i^4 = S^3$, in a Hopf manner.

Lemma 8.3.3 *There is a flattening retraction*

$$B_i^4 \xrightarrow{\quad \Phi(i) \quad} B_i^2 \,, \tag{8.23}$$

suggested in Fig. 8.12, and which induces a diffeomorphism $\Phi(i) \mid B_i^1$

$$
\begin{array}{ccc}
B_i^1 \subset B_i^4 & \xrightarrow{\ \Phi(i)\ } & B_i^2 \,. \\
\Big| & & \Big\uparrow \\
& \xrightarrow{\ \Phi(i)|B_i^1\ } &
\end{array}
\tag{8.24}
$$

Notice the three diffeomorphisms

$$\text{Box}(i) \underset{\text{DIFF}}{=} I_i \times D_i^2 \xrightarrow{(I_i \approx h_i^1) \times \text{id } D_i^2} h_i^1 \times D_i^2 \underset{\text{DIFF}}{=} B_i^1 \xrightarrow[\text{DIFF}]{\Phi(i)|B_i^1} B_i^2 \,, \tag{8.25}$$

giving us three distinct diffeomorphic models for the same object, namely, Box(i), B_i^1, B_i^2, and in the proof of our Theorem 8.2.6 (in the sequel of that proof, actually) one should be prepared to move freely and happily between these models, and also a fourth one too, namely, {attaching zone of H_i^2}−{priviledged $I \times B^2$}. The important point here is that, while Box(i) $= I \times D^2$ is ABSTRACT, and in this abstract form

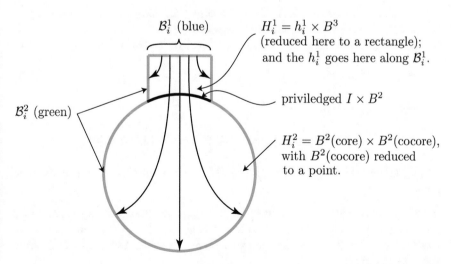

B_i^1 (blue)

$H_i^1 = h_i^1 \times B^3$
(reduced here to a rectangle);
and the h_i^1 goes here along B_i^1.

priviledged $I \times B^2$

B_i^2 (green)

$H_i^2 = B^2(\text{core}) \times B^2(\text{cocore})$,
with $B^2(\text{cocore})$ reduced
to a point.

Fig. 8.12 One sees here the flattening retraction $\Phi(i)$, suggested by the arrows. This is a profile view

it serves us for defining $\chi(\partial x^2 \cdot \delta h^1)$, the other three models

$$\mathcal{B}_i^1, \mathcal{B}_i^2 \text{ and } \{\text{attaching zone of } H_i^2\} - I \times B^2$$

are very concrete, and help us prove geometric facts about X^4.

With these things, one should think now of the flattening restriction, in the form

$$B_i^4 \xrightarrow{\ \ \Phi(i)\ \ } \text{Box}(i).$$
$$(8.26)$$

Moreover, we can perceive an embedding, when there is an arrow $i \to j$, namely, $\text{Box}(i) \subset \partial B_j^4$, by using the identification

$$\text{Box}(i) = I_i \times D_i^2 \approx (\lambda(i \to j)I_j \times \mu(i \to j)D_i^2) =$$

$= \{$ contact of ∂B_i^4 with ∂B_j^4, corresponding to the (precise) arrow $i \to j\} \subset \partial B_j^4 \cap \partial B_i^4 \subset \partial B_j^4$ (another concrete model!).   (8.26.1)

Our Theorem 8.2.6 is completely proved once we have shown the following lemma, which actually supersedes 2 in Theorem 8.2.6:

Lemma 8.3.4

(1) There is a PROPER smooth embedding

$$\chi(\partial h^2 \cdot \delta h^1) \xrightarrow{\ \ \mathcal{J}\ \ } \partial N^4(T)$$
$$(8.27)$$

and this $\mathcal{J}$ is such that in $\mathcal{J}(\mathcal{L})$ each leaf of $\mathcal{J}(\mathcal{L})$ joins two distinct ends of T. [PROPER means $(\text{compact})^{-1} = \text{compact}.]$

(2) Up to diffeomorphism, the 4-manifold X^4 is described by the following formula

$$X^4 = N^4(T) \cup \{\chi(\partial h^2 \cdot \delta h^1) \times [0, 1] - \mathcal{L} \times \{1\}\},$$
$$(8.27.1)$$

with the union being made along

$$\partial N^4(T) \supset \mathcal{J}(\chi(\partial h^2 \cdot \delta h^1) = \chi(\partial h^2 \cdot \delta h^1)) \times \{0\}.$$

This lemma certainly proves (2) (and (3)) in Theorem 8.2.6, and actually quite more.

Proof Let us consider some arbitrary state $i \in \{i\}$ and also an initial state i_α, i.e., a state without outgoing arrows. In our present situation, GSC $\Leftrightarrow$ (easy id + nilpotent, indexed by Z_+) such states *have* to exist. We also consider a trajectory of G, from some given i to i_α,

$$i \equiv i_0 \longrightarrow i_1 \longrightarrow i_2 \longrightarrow \cdots \longrightarrow i_\alpha.$$
$$(8.28)$$

Making use of (8.24), (8.25), more specifically, using now all the various different CONCRETE models of Box(i), described a bit earlier, we can consider the successive retractions and inclusion map

$$B^4(i) = B^4(i_0) \xrightarrow{\Phi(i_0)} \text{Box}(i_0) \subset B^4_{i_1} \xrightarrow{\Phi(i_1)} \text{Box}(i_1) \subset B^4_{i_2} \xrightarrow{\Phi(i_2)} \qquad (8.29)$$

$$\longrightarrow \text{Box}(i_2) \longrightarrow \cdots \xrightarrow{\Phi(i_{\alpha-1})} \text{Box}(i_{\alpha-1}) \subset B^4(i_\alpha) \xrightarrow{\Phi(i_\alpha)} \text{Box}(i_\alpha)$$

$$\approx \{\text{attaching zone of } H^2_\alpha\} \cap \partial N^4(T).$$

The composition of these maps can be read as a canonical diffeomorphism which *is $\mathcal{J}$ | Box(i)*:

$$
\begin{array}{c}
\text{Box}(i) \xrightarrow[\approx]{\mathcal{J}_i} \partial B^4(i_\alpha) \cap \partial N^4(T) \subset \partial N^4(T). \\[2pt]
\Big| \qquad\qquad\qquad\qquad\qquad \uparrow \\
\underline{\qquad\qquad\qquad\qquad\qquad\qquad\qquad} \\
\mathcal{J}|\text{Box}(i)
\end{array}
\qquad (8.30)
$$

The various $\mathcal{J}_i$'s are compatible with each other, and this defines the $\mathcal{J}$. Next, by a second big composition of maps one can perceive a retraction, call it again $\Phi(i)$, since it *is $\Phi(i)$*, in disguise,

$$B^4_i \xrightarrow{\Phi(i)} \mathcal{J}\text{Box}(i) \subset \partial N^4(T). \qquad (8.31)$$

We have $X^4 = N^4(T) \cup \bigcup\limits_{i=1}^{\infty} B^4_i$, where for the time being the union between the two pieces in the RHS of the formula is the obvious thing induced by the handlebody decomposition of X^4, for various i's the $\Phi(i)$'s are all compatible with each other and, together, they define a BIG PROPER FLATTENING RETRACTION

$$\bigcup_{i=1}^{\infty} B^4_i \xrightarrow{\Phi(\infty)} \mathcal{J}\chi(\partial h^2 \cdot \delta h^1) \subset \partial N^4(T). \qquad (8.32)$$

It is not hard to see now, that in the formula $X^4 = N^4(T) \cup \bigcup\limits_{i=1}^{\infty} B^4_i$, we have

$$N^4(T) \cap \bigcup_{i=1}^{\infty} B^4_i = \partial N^4(T) \cap \partial \left(\bigcup_{i=1}^{\infty} B^4_i \right) = \mathcal{J}\chi(\partial h^2 \cdot \delta h^1),$$

giving a more interesting meaning of "$\cup$" in $N^4(T) \cup \bigcup_i B_i^4$. By a process of successive blow-ups, out of the pair $(N^4(T), \mathcal{J}\chi(\partial h^2 \cdot \delta h^1) \equiv \chi(\partial h^2 \cdot \delta h^1) \times \{0\})$, we can reconstruct formula (27.1). We start by renumbering our states with a total Z_+-order compatible with the natural one, coming from Definition 8.1.1. We first let $B^4(i_1)$ (first state in the new order) grow out of $N^4(T)$, performing a first growth process

$$N^4(T) \Longrightarrow N^4(T) \cup B^4(i_1),$$

where on the right hand side things are naturally glued along $\Phi(\infty) \mid B^4(i_1)$. Next, we perform, by the same kind of recipe (but the notation would start getting heavy, so I omit them) the next growth process

$$N^4(T_i) \cup B_{i_1}^4 \Rightarrow N^4(T_i) \cup B^4(i_1) \cup B^4(i_2).$$

This process continues indefinitely, providing in with the formula (27.1). $\square$

So, by now we have proved Theorem 8.2.6 and actually more, since in (27.1) we gave a complete description of the manifold $X^4 \subset M^4$ and of its contribution to the infinity of M^4. So, in agreement with our declared policy of not proving more than necessary for our main message, we will not try to pursue a more thorough investigation of the space $\chi(\partial h^2 \cdot \delta h^1)$, in the completely general situation when GSC is violated but not brutally.

Nevertheless, I feel I owe to the reader a couple of words concerning the Whitehead manifold Wh^3. But, before that, let us look a bit more closely at what we have achieved for $\chi(\partial h^2 \cdot \delta h^1)$ in the GSC case.

(i) To begin with, the $\chi(\partial h^2 \cdot \delta h^1)$ is here a smooth 3-manifold, non-compact, with $\partial \chi(\partial h^2 \cdot \delta h^1) \neq \emptyset$ and not necessarily connected.

(ii) And this manifold has a Riemannian structure, satisfying the product condition for its Boxes (and (17), (17.1)). Actually the whole space χ is union of Boxes, glued together Markov partition-like. Incidentally too, since $N^4(T) \underset{\mathrm{DIFF}}{=} B^4 -$ closed tame set (Cantor?), the $\chi(\partial h^2 \cdot \delta h^1)$ happily embeds is R^3, with all these structures preserved.

Let us contemplate now Fig. 8.11 which shows us how to construct $\chi(\partial h^2 \cdot \delta h^1)(\mathrm{Wh}^3)$. What we see is a recipe for constructing an infinite sequence of objects $Y_n^3 = \mathrm{Box}(i_n) \cup \{\text{all the passages of the Boxes coming with } \mathrm{Box}(i_p), p < n\}$. For this infinite sequence of smooth objects condition (17.1) should be satisfied too, as the legend of the figure says. Notice also, that each of the Y_n^3's is a copy of B^3 with 1-handles attached, the number of 1-handles going to infinity, when $n \to \infty$.

So, for Wh^3, our $\chi(\partial h^2 \cdot \delta h^1)$ is highly non simply-connected.

Forget now about the metrics and just contemplate the infinite sequence of 3-manifolds, successively embedded into each other

$$Y_1^3 \subset Y_2^3 \subset Y_3^3 \subset \cdots, \qquad (*)$$

which Fig. 8.11 suggests. We forget here about the Boxes going through each other and about the product metric and only look out the compact 3-manifolds Y_n^3. Without overstretching our minds we can see that what we get TOPOLOGICALLY, is a 3-manifold. And since every 3-manifold has, canonically, a unique DIFF structure, we do get a C^∞ 3-manifold like in (i) above. With just a bit of more mind-stretching, this DIFF structure can be perceived directly at level $(*)$. Anyway, there is no problem with the (i) above, for Wh^3. Now let us look at (ii) at the Boxes with their product metric structure. So, let us say we have now, in a compatible way, a fixed product metric structure for each $\mathrm{Box}(i_n) \equiv I_{i_n} \times D_{i_n}$. This functions well at the individual level of each Y_n^3 in the sequence $(*)$. The problem occurs when we want to corral together the whole $(*)$.

With our metric structure in place, each I_{i_n} has a given finite length. When we go through the whole infinite sequence, the I_{i_n} undergoes infinitely many contorsions and it cannot stay smooth (C^∞) at the fully assembled level.

So, for Wh^3 we cannot have both a smooth 3-manifold $\chi(\partial h^2 \cdot \delta h^1)$ AND the product metric structure in individual Boxes. In this sense, the case of $\chi(\partial h^2 \cdot \delta h^1)$ for Wh^3 is wild.

Proof of Theorem 8.2.7 We will restrict ourselves to the case of a unique cycle from Lemma 8.2.5, and show that, then, continuous C^0 function on $\chi(\partial h^2 \cdot \delta h^1) \mid$ $\mathrm{Box}(i_0)$ cannot separate points. Hence, in the unique cycle case at least, continuous functions cannot separate points. When more than one cycle goes through $\mathrm{Box}(i_0)$ (like for instance in Fig. 8.3d), then $\chi(\partial h^2 \cdot \delta h^1) \mid \mathrm{Box}(i_0)$ is a more complicated quotient space of the previous ones, where it should be even harder for continuous functions to separate points from each other. So, things like Theorem 8.2.7 should work then too. But that simplest case of the unique cycle is enough for the main message of this paper. Anyway, we have now the passage from (8.13), B_{i_0} through itself. In Fig. 8.13 we have represented, albeit schematically, the passage of $\mathrm{Box}(i)$ through $\mathrm{Box}(j)$. But keep in mind that, here, we have $\mathrm{Box}(j) \approx \mathrm{Box}(i_0) \approx \mathrm{Box}(i)$, and so there is here also the diffeomorphism $\mathrm{Box}(j) \xrightarrow{\varphi} \mathrm{Box}(i)$, from Lemma 8.2.3 and, keep in mind too that we are in th simple context of Lemma 8.2.5.

We consider now the following composite self-map of D_j^2 into itself, and here D_i^2, D_j^2 are like in Fig. 8.13:

$$D_j^2 \xrightarrow{\varphi} D_i^2 \xrightarrow{\mu(i \to j)} D_j^2 \ .$$

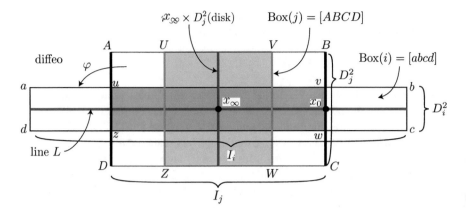

Fig. 8.13 The passage of Box(i) through Box(j). Here $[u, v, w, z]$ (shaded blue) is the intersection Box$(j) \cap$ Box$(i) \subset$ Box(i) created by the passage. The $[U, V, W, Z]$ (shaded green) is

$$\varphi^{-1}[u, v, w, z] \subset \text{Box}(j).$$

Here, in Box$(i_0) = $ Box$(j) = $ Box(i), the equivalence relation $\mathcal{R} \mid$ Box(i_0) performs, exactly, the following things. WHen $p \in [U, V, W, Z] \subset$ Box(j) and $q \in [u, v, w, z]$, where φ sends $[U, V, W, Z]$ into $[u, v, w, z]$, if $q = \varphi(p)$, then $(p, q) \in \mathcal{R}$. The vertical line through x_∞ is red

Then, via Brouwer, this composite map which we call φ again, has a fixed point $x_0 \in D_j^2$, so $\varphi(x_0) = x_0$. We will also denote $L \equiv x_0 \times I_i \subset$ Box(i). This occurs as a horizontal red line in Fig. 8.13. We can consider the shorter line $L \cap [u, v, w, z]$,

$$\text{Box}(i) \supset L \cap [u, v, w, z] \subset \text{Box}(j). \tag{8.33}$$

The diffeomorphism φ^{-1} is product, since φ is so, and hence we have a self-map of $L \cap [u, v, w, z]$ into itself, actually a contraction,

$$\text{Box}(i) \supset L \supset L \cap [u, v, w, z] \xrightarrow{\ \varphi^{-1}\ } L \cap \text{Box}(j) = L \cap [u, v, w, z]. \tag{8.34}$$

Keep constantly in mind now that, according to (13), the map $D_j^2 \xrightarrow{\ \varphi\ } D_i^2$, which respects the product structure, *decreases* the distances on the D^2's and *increases* them on the I's. So φ^{-1} decreases distances in I's as stated.

Let us explain this formula (8.34). Since for $x_0 \in$ Box(j), $\varphi(x_0) \in$ Box(i) we find that $\varphi(x_0) = x_0$, we also find that $\varphi^{-1}(x_0) = x_0$. Here, of course,

$\text{Box}(i) \xrightarrow[\text{DIFF}]{\ \varphi^{-1}\ } \text{Box}(j)$. This implies that

$$L \cap \text{Box}(i) \xrightarrow{\ \varphi^{-1}\ } L \cap \text{Box}(j) = L \cap [u, v, w, z].$$

When we restrict the LHS to $L \cap [u, v, w, z]$ we find our

$$L \cap [u, v, w, z] \xrightarrow{\ \varphi^{-1}\ } L \cap [u, v, w, z] .$$

And, actually, what *this* does on distances will not matter.

Again, by Brouwer, the (8.34) has a fixed point, call it x_∞, for reasons to become soon clear, i.e., we have now

$$\varphi^{-1}(x_\infty) = x_\infty \in \mathrm{Box}(j) \cap \mathrm{Box}(i) . \qquad (8.35)$$

When we consider $x_\infty \times D_i^2$ we find that

$$x_\infty \times D_i^2 \xrightarrow{\ \varphi^{-1}\ } x_\infty \times D_j^2, \quad \text{with} \quad x_\infty \times D_i^2 = (x_\infty \times D_j^2) \cap [u, v, w, z] .$$

$$(8.36)$$

Since (8.36) is important, we will look into the seams of its proof. Saying that φ is product means that we have diffeomorphisms

$$I_j \xrightarrow{\ \varphi_I\ } I_i , \quad D_j^2 \xrightarrow{\ \varphi_{D^2}\ } D_i^2 , \quad \text{(and do not mix this up with } (\lambda, \mu) \text{)},$$

such that, for $(x, y) \in \mathrm{Box}(j)$, with $x \in I_j$ and $y \in D_j^2$, we have

$$\varphi(x, y) = (\varphi_I(x), \varphi_{D^2}(y)) .$$

With this, if $(x_\infty, y) \in (x_\infty \times D_i^2) \subset \mathrm{Box}(i)$, then (and we think here in terms of $x_\infty \times D_i^2 \subset x_\infty \times D_j^2$), we also have

$$\varphi^{-1}(x_\infty, y) = (\varphi_I^{-1}(x_\infty) = x_\infty, \ \varphi_{D^2}^{-1}(y) \in D_j^2) ,$$

and this is exactly what our (8.36) says. Notice, also, that

$$x_\infty \times D_i^2 = (x_\infty \times D_j^2) \cap [u, v, w, z] .$$

The disc $x_\infty \times D_i^2$ is represented as a vertical red line in Fig. 8.13, which continues with the longer $x_\infty \times D_j^2$.

Since x_∞ is a fixed point for φ^{-1} (see (8.35)), it is also a fixed point for φ and we are interested now in inverting (8.36), which becomes $\varphi \mid x_\infty \times D_j^2$, i.e., the map

$$x_\infty \times D_j^2 \xrightarrow{\ \varphi\ } x_\infty \times D_i^2 \subset x_\infty \times D_j^2 , \qquad (8.37)$$

a map which has x_∞ as a fixed point ($x_\infty \in D_j^2$) and which DECREASES distances (so that Box(j) can pass through Box(i), with all our product-structure conditions respected). Moreover $\varphi(x_\infty) = x_\infty$ and, if we consider the context of Lemma 8.2.5, x_∞ is NOT related to any other point by the equivalence relation $\mathcal{R}_1$, outside of itself.

In the discussion which follows, D_i^2, D_j^2 mean $x_\infty \times D_i^2 \subset x_\infty \times D_j^2$. Consider now a sequence of points $x_1, x_2, x_3, \ldots$ in D_j^2, such that

$$\lim_{n=\infty} x_n = x_\infty, \text{ (which means almost all are in } D_i^2 \subset D_j^2 \text{), and} \qquad (8.38.1)$$

there are no relations $(x_n, x_m) \in \mathcal{R}$, for $n \neq m$. $\qquad (8.38.2)$

With x_∞ being the unique fixed point of φ, if the $x_1, x_2 \ldots$ are unrelated by φ, and once we have (38.2) too, there are no relations $(x_n, x_m) \in \mathcal{R}$, $n \neq m$ either. So, finally the relations $(x_n, x_m) \in \mathcal{R}_1$ are excluded too (and also $(x_n, x_\infty) \in \mathcal{R}_1$).

For each x_n we consider now its infinite φ-orbit,

$$x_n, \varphi(x_n), \varphi^2(x_n), \varphi^3(x_n), \ldots \qquad (8.39)$$

and, since φ decreases distances, with n being given we also find

$$\lim_{m=\infty} \varphi^m(x_n) = x_\infty .$$

One should notice two things here. First, by (38.1) all the points x_n, except for finitely many, are in $x_\infty \times D_i^2 \subset x_\infty \times D_j^2$. Next, once

$$\varphi(x_n) \in x_\infty \times (D_i^2 \cap D_j^2) \subset x_\infty \times D_j^2 ,$$

for the map (8.37) all the iterates $\varphi^m(x_n)$ fall in $x_\infty \times D_i^2 = x_\infty \times (D_i^2 \cap D_j^2) \subset x_\infty \times D_j^2$ (when φ is defined), too.

For given n, the whole orbit (8.39) is a unique point of the space $\chi(\partial h^2 \cdot \delta h^1)$, distinct from the distinguished unique point x_∞, infinitely closed to it, but still **not** equal to it.

This is true for all the various x_n-orbits, for $n = 1, 2, \ldots$ and no continuous function on $\chi(\partial h^2 \cdot \delta h^1)$ can separate them. We have infinitely many points, all distinct, and all infinitely close to x_∞.

Much more could and should be said of course. But we said enough for the message to be clear.

Acknowledgments I would like to thank the IHES for its friendly help, Cécile Gourgues for the typing and Marie-Claude Vergne for the drawings.

References

1. A. Connes, *Noncommutative Geometry* (Academic Press, New York, 1994)
2. B. Mazur, On embeddings of spheres. BAMS **65**, 59–65 (1959)
3. J. Morgan, G. Tian, *The Ricci Flow and the Poincaré Conjecture*. AMS Clay Math. Inst. (American Mathematical Society, Providence, 2007)
4. D.E. Otera, V. Poénaru, Topics in geometric group theory I, in *Handbook of Group Actions* (vol. V), ed. by L. Ji, A. Papadopoulos, S.-T. Yau. ALM 48 (Higher Education Press/International Press, Beijing/Boston, 2020), pp. 301–346
5. V. Poénaru, Geometric simple connectivity and low-dimensional topology. Proc. Steklov Inst. Math. **247**, 195–208 (2004)
6. V. Poénaru, What is . . . and infinite swindle. Notices AMS **54**(5), 619–622 (2007)
7. V. Poénaru, Geometric simple connectivity and finitely presented groups, ArXiv/1404.4283 [Math.GT] (2014)
8. V. Poénaru, All smooth four-dimensional Schoenflies balls are geometrically simply connected, ArXiv [1609.05094] (2016)
9. V. Poénaru, A glimpse into the problems of the fourth dimension, in *Geometry in History*, ed. by S.G. Dani, A. Papadopoulos (Springer Nature Switzerland, Cham, 2019), pp. 687–704
10. V. Poénaru, Topics in geometric group theory II, in *Handbook of Group Actions* (vol. V), ed. by L. Ji, A. Papadopoulos, S.-T. Yau, ALM 48 (Higher Education Press/International Press, Beijing/Boston, 2020), pp. 347–398
11. V. Poénaru, On geometric group theory, in *Topology and Geometry – A Collection of Essays Dedicated to Vladimir G. Turaev*, ed. by A. Papadopoulos (EMS Press, 2021), pp. 399–432
12. V. Poénaru, C. Tanasi, Some remarks on geometric simple connectivity. Acta Math. Hung. **81**, 1–12 (1998)
13. S. Smale, On the structure of manifolds. Am. J. Math. **84**, 387–399 (1962)
14. J.H.C. Whitehead, A certain open manifold whose group is unity. Q. J. Math. Oxf. Ser. **6**, 268–279 (1935)

Chapter 9
A Short Introduction to Translation Surfaces, Veech Surfaces, and Teichmüller Dynamics

Daniel Massart

Abstract We review the different notions of translation surfaces that are necessary to understand McMullen's classification of $GL_2^+(\mathbb{R})$-orbit closures in genus two. We start by recalling the different definitions of a translation surface, in increasing order of abstraction, starting with cutting and pasting plane polygons, ending with Abelian differentials. We then define the moduli space of translation surfaces and explain its stratification by the type of zeroes of the Abelian differential, the local coordinates given by the relative periods, its relationship with the moduli space of complex structures and the Teichmüller geodesic flow. We introduce the $GL_2^+(\mathbb{R})$-action, and define the related notions of Veech group, Teichmüller disk, and Veech surface. We explain how McMullen classifies $GL_2^+(\mathbb{R})$-orbit closures in genus 2: we have orbit closures of dimension 1 (Veech surfaces, of which a complete list is given), 2 (Hilbert modular surfaces, of which again a complete list is given), and 3 (the whole moduli space of complex structures). In the last section, we review some recent progress in higher genus.

Keywords Translation surface · Veech surface · Abelian differential · Moduli space

AMS Classification 37D40, 30F60, 32G15

9.1 Introduction

Translation surfaces are a great pedagogical tool at almost every level of education. At the elementary, or secondary, level, you are impeded by the lack of embedding into 3-space, but billiards provide plenty of access points into geometry. At the undergraduate level, they help in topology, to explain quotient spaces, or in

D. Massart (✉)
Institut Montpelliérain Alexander Grothendieck, CNRS, Université de Montpellier, Montpellier, France
e-mail: daniel.massart@umontpellier.fr

© The Author(s), under exclusive license to Springer Nature Switzerland AG 2022   343
A. Papadopoulos (ed.), *Surveys in Geometry I*,
https://doi.org/10.1007/978-3-030-86695-2_9

differential geometry, to explain atlases and transition maps, or in complex analysis, with meromorphic functions. At the graduate level, the doors of abstraction burst open, and in a few light steps you get to the enchanted gardens of moduli spaces, Teichmüller theory, and elementary algebraic geometry. The beauty of the subject lies in the interplay between the basic (cutting and pasting little pieces of paper) and the abstract (moduli spaces of Riemann surfaces).

There are already many wonderful introductions to translation surfaces (for instance, [13, 18, 35, 55], or [53]), and this one is in no way meant as a substitute to any of them, but it could be used as a stepping stone. It is specifically aimed at beginning graduate students. Mostly, it consists of what, given 2 (or 5) hours and a blackboard, I would tell a student looking for a thesis subject and eager to know what translation surfaces are about.

I have tried to include (sketches of) proofs of two types of results: first, those that are too elementary to be included in the aforementioned introductory papers, but can still cause a good deal of head-scratching; and some, not uncommon in the subject, that are of the "takes genius to see it, but not that hard once you know it" variety.

I thank Erwan Lanneau for many useful conversations, the editor of the present volume, and Pablo Montealegre for their careful reading of the manuscript, and Smail Cheboui for letting me use some of the beautiful drawings of [4] (the ugly ones are mine).

9.2 Three Different Definitions of a Translation Surface

9.2.1 The Most Hands-on Definition: Polygons with Identifications

Let P be a polygon in the Euclidean plane, not necessarily convex, maybe not even connected, but with its sides pairwise parallel and of equal length. Glue each side to a parallel side of equal length. There are several ways of doing this, so let's be a bit more careful.

First, let us make the convention that we orient the sides of a polygon so that the interior of the polygon lies to the left.

Now let us identify the sides in pairs, in such a way that the resulting surface is orientable (see Fig. 9.1 for a non-orientable example). One way to think of it is to imagine that the upside of the polygon is painted red, while the downside is painted black. The surface is orientable if it has one red side, and one black side. This means that each edge is glued to another edge in such a way that the colors match, so the arrows on the edges do not match.

Let us assume, furthermore, that we glue each edge to a parallel edge of equal length. Let n be the number of edges. Thus there exists some permutation σ of $\{1, \ldots, n\}$ such that $v_{\sigma(i)} = \pm v_i$, for all $i = 1, \ldots, n$. Since we must oppose the

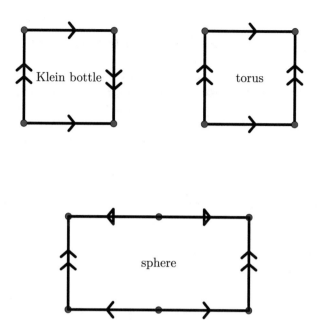

Fig. 9.1 Here the arrows indicate the gluing, not the orientation of the boundary, so arrows must match under the gluing. Note that in the case of the sphere, the edges are not glued by translation, but with a half-turn

arrows when gluing, there are two possible cases: either $v_{\sigma(i)} = -v_i$, in which case the two edges are glued by translation (see Fig. 9.1, top right), or $v_{\sigma(i)} = v_i$, in which case the edges are glued with a half-turn (see Fig. 9.1, bottom).

Formally, you define an equivalence relation $\sim$ on P, and you are considering the quotient space $X = P/\sim$. Although the polygon itself may not be connected, we require the resulting quotient space to be connected, just because if it is not, we study its connected components separately. What you get is a compact, orientable manifold of dimension two (a surface, for short).

Here is why the quotient space is a manifold.

What we need is to find, for each $x \in X$, a neighborhood U_x of x in X, and a chart $\phi_x : U_x \longrightarrow \mathbb{C}$, such that for any x, y in X, if $U_x \cap U_y \neq \emptyset$, then $\phi_x \circ \phi_y^{-1}$ preserves whatever structure it is that you want your manifold to come with (if you want a differentiable manifold, you require $\phi_x \circ \phi_y^{-1}$ to be differentiable, if you want a complex manifold, you require it to be holomorphic, and so on).

If x is the image in X of an interior point $\bar{x}$ of P, then x has a neighborhood U_x in X which is the homeomorphic image in X of an open neighborhood $\bar{U}_x$ of $\bar{x}$ in P (hence in $\mathbb{C}$). We take the inverse quotient map $U_x \to \bar{U}_x$ as the chart ϕ_x.

Shorter and somewhat sloppier version: the quotient map, restricted to the interior of P, is a homeomorphism. Identify x with its pre-image. Take a neighborhood U_x of x which is contained in the interior of P, and take the identity as a chart.

If x is the image in X of an interior point of an edge of P, then x has only two pre-images in P, and x has a neighborhood in X whose pre-image in P is the reunion of two half-disks of equal radius and parallel diameters. If the two edges are glued by translation, we define a chart by the identity on one of the half-disks, and a translation on the other half-disk. If the two edges are glued with a half-turn, we define a chart by the identity on one of the half-disks, and a translation composed with a half-turn on the other half-disk.

If x is the image in X of a vertex of P, then all pre-images in P of x are vertices $P_1, \ldots, P_k$ of P, because we glue edges of equal length, and x has a neighborhood U_x in X whose pre-image in P is a union of circular sectors S_i, $i = 1, \ldots, k$, with radius r and straight boundaries e_i and f_i, so that f_i is parallel to e_{i+1}, and f_k is parallel to e_1. The angle θ_i of the sector S_i is the angle between two adjacent (at P_i) edges of P. Note that since f_k is parallel to e_1, the angles θ_i sum to a multiple of π, say $p\pi$. Furthermore, if the edges f_k and e_1 are glued by a translation, the angles θ_i sum to a multiple of 2π. If the edges f_k and e_1 are glued with a half-turn, the angles θ_i sum to an odd multiple of π. Denote

$$\Theta_i = \sum_{j=1}^{i-1} \theta_j.$$

We define a chart which takes each S_i to a circular sector with vertex at 0, by

$$P_i + \rho \exp i(\theta + \Theta_i) \longmapsto \rho \exp i \frac{2}{p}(\theta + \Theta_i), \text{ for } 0 \le \rho \le r, 0 \le \theta \le \theta_i.$$

We say a point in X which is the image of an interior point of P is of type I, a point in X which is the image of an interior point of an edge of P is of type II, and a point in X which is the image of a vertex of P is of type III.

If x and y are both of type I, and $U_x \cap U_y \ne \emptyset$, then $\phi_x \circ \phi_y^{-1}$ is the identity, which preserves every structure imaginable.

If x and y are both of type I or II, and $U_x \cap U_y \ne \emptyset$, then $\phi_x \circ \phi_y^{-1}$ is either the identity, or a translation, or a "half-translation" $z \mapsto -z + c$, all of which preserve almost every structure imaginable.

If x is of type III and y is of type I, II, or III and $U_x \cap U_y \ne \emptyset$, then $\phi_x \circ \phi_y^{-1}$ is holomorphic (it is essentially a branch of the $p/2$-th root). If both x and y are of type III, we choose U_x and U_y so that $U_x \cap U_y = \emptyset$.

Thus we have defined a complex structure on X (an atlas with holomorphic transition maps), but this does not tell the whole story. Denote $\Sigma = \{x_1, \ldots, x_k\}$ the set of all images in X of the vertices of P, then if all edges are glued by translation, $X \setminus \Sigma$ has an atlas whose transition maps are translations, whence the name **translation surface**. Among the three surfaces in Fig. 9.1, only the torus is a translation surface.

If some pair of edges is glued with a half-turn, $X \setminus \Sigma$ has an atlas whose transition maps take the form $z \mapsto \pm z + c$, in which case X is called a **half-translation**

surface. The sphere in Fig. 9.1 is a half-translation surface. The Klein bottle is not orientable, so it is neither a translation nor a half-translation surface.

Points of type III in X with a total angle $> 2\pi$ will hereafter be called **singularities**. While being of type III depends on the polygon, being a singularity only depends on the quotient space X.

The fact that translations and the $z \mapsto -z$ map preserve almost every structure imaginable on the plane (complex, metric, you name it) entails that translation and half-translation surfaces come with a lot of structure, and part of the appeal of this theory is that you can view it from so many different angles.

9.2.2   Main Examples of Translation Surfaces

Before proceeding further, let us introduce our favorite examples. Probably the first thing that comes to mind after seeing the definition is an even-sided, regular polygon.

The surface obtained by identifying opposite sides of a square is a torus.

The surface obtained by identifying opposite sides of a regular hexagon is also a torus, as may be seen by computing the Euler characteristic: one two-cell (the hexagon itself), three edges (one for each pair of opposite edges of the hexagon), and two vertices (if the vertices of the hexagon are cyclically numbered, all even (resp. odd)-numbered vertices are identified into one point), so the Euler characteristic is zero.

For $n > 1$, the surface obtained by identifying opposite sides of a regular $4n$-gon has genus n, with all vertices identified into one point, with a total angle $(4n - 2)\pi$. The surface obtained by identifying opposite sides of a regular $4n+2$-gon has genus n, with two points of type III, both with angle $2n\pi$.

Another surface of interest is the double $(2n + 1)$-gon, which, from the combinatorial viewpoint, is the same as the regular $4n$-gon, so it has genus n, with all vertices identified into one point, with a total angle $(4n - 2)\pi$. See the double pentagon on Fig. 9.2.

Our second main example is a generalization of the square: imagine that instead of just one square, you have a finite collection of same-sized squares, each of which has its edges labeled "right", "left", "top", "bottom", and glue each right (resp. left) edge with a left (resp. right) edge, and each top (resp. bottom) edge with a bottom (resp. top) edge. Such surfaces are called **square-tiled**, or sometimes **origami**. We have already seen the one-square surface, which is a torus. Two-square surfaces are also tori (compute the Euler characteristic). The first examples of genus 2 are three-squared (see Fig. 9.3). They have only one point of type III, with total angle 6π. A four-squared, genus 2 surface, with two points of type III, each with total angle 4π, is shown in Fig. 9.4.

Note that according to common usage, square-tiled surfaces are translation surfaces, so the Klein bottle and the sphere in Fig. 9.1 are not square-tiled surfaces, even though they are actually tiled by squares.

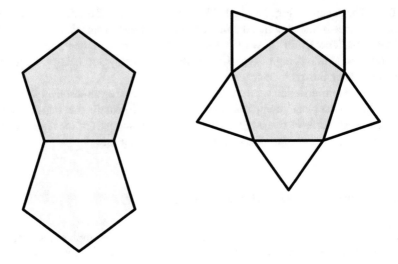

Fig. 9.2 The double pentagon: different polygons, same surface

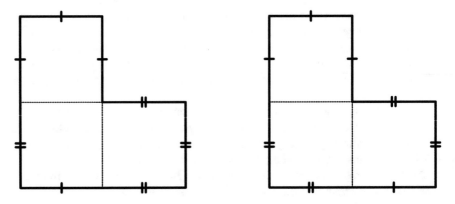

Fig. 9.3 Two different three-squared surfaces of genus two, with $St(3)$ on the left

Unlike regular polygons, which are somewhat few and far between, square-tiled surfaces are swarming all over the place. In fact, any translation surface may be approximated, in a sense that will be made precise later, by square-tiled surfaces, just because any line may be approximated uniformly by a stair-shaped broken line made of horizontal and vertical segments.

A useful feature of square-tiled surfaces is that they are covers of the square torus, ramified over one point. Of course there is nothing special about the square here, each regular polygon comes with its own family of ramified covers.

It is usually not a good idea to try to vizualize the surface X in space, if only because, having non-positive curvature, it does not embed isometrically in $\mathbb{R}^3$. Computing the Euler characteristic is the way to go. It is interesting, however,

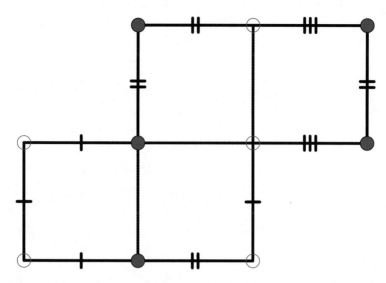

Fig. 9.4 $St(4)$, a four-square surface of genus 2, with two singularities

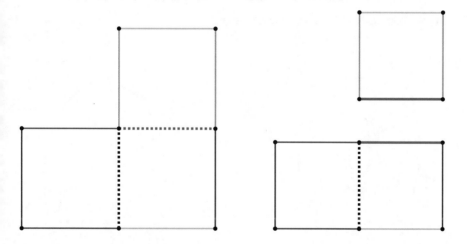

Fig. 9.5 Identifications for the surface $St(3)$, stage 1

when it lets you vizualize a decomposition of the surface into flat cylinders. Figures 9.5, 9.6, and 9.7 show a topological embedding into $\mathbb{R}^3$ of the surface on the left of Fig. 9.3, called $St(3)$ after [42]. Figures 9.8, 9.9, and 9.10 show a topological embedding into $\mathbb{R}^3$ of the surface of Fig. 9.4, called $St(4)$ after [42].

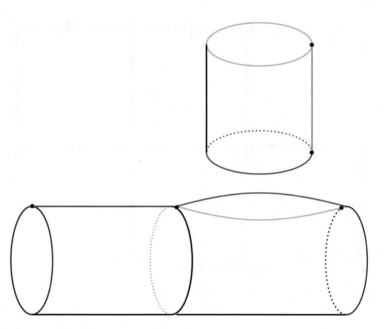

Fig. 9.6 Identifications for the surface $St(3)$, stage 2

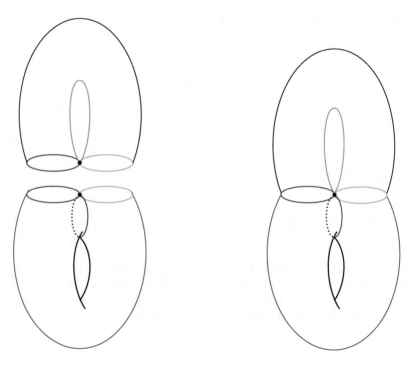

Fig. 9.7 Identifications for the surface $St(3)$, stage 3

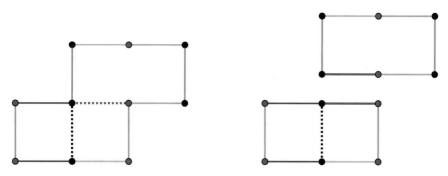

Fig. 9.8 Identifications for the surface $St(4)$, stage 1

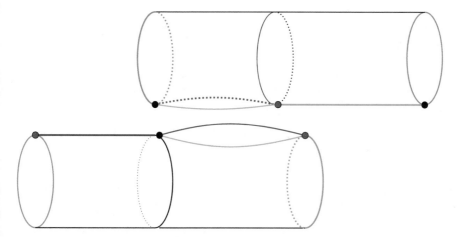

Fig. 9.9 Identifications for the surface $St(4)$, stage 2

9.2.3 Definition Through an Atlas

What if we took the property we have just proved as a definition? Let us say that
a translation surface is a compact manifold X of dimension two, such that there
exists a finite subset $\Sigma = \{x_1, \ldots, x_k\}$, called the singular set of X, and an atlas
of X, such that the transition map between any two charts whose domains do not
meet Σ is a translation, and the transition map between a chart whose domain meets
the singular set, and a chart whose domain does not, is $z \mapsto z^k$ for some $k \in \mathbb{N}$.
Is that an equivalent definition? Meaning, from this data, can we extract a polygon
with identifications, so that when we perform the identifications, we get back the
translation atlas?

Well, among the many structures translations preserve, there is the Euclidean
metric. Therefore we can equip our surface $X \setminus \Sigma$ with a Riemannian metric which is
locally Euclidean, so its geodesics are locally straight lines. The Riemannian metric

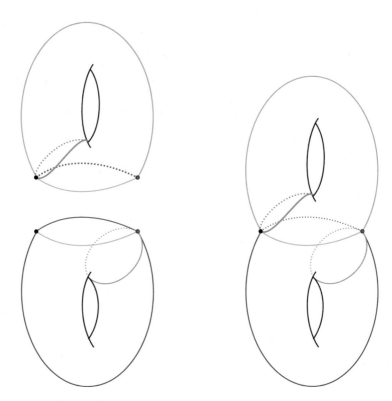

Fig. 9.10 Identifications for the surface $St(4)$, stage 3

may not extend to the whole of X, but the distance function does. In particular we can draw geodesics between any two singularities (elements of Σ).

Now, draw geodesics between singularities, as many (but finitely many) of them as you like, as long as the connected components of the complement in X of those geodesics are simply connected. Being simply connected, they may be developed to the plane, to polygons. Then, apply the construction of Sect. 9.2.1 to this polygon (remember, I never said the polygon from which X is glued should be connected).

The atlas we obtain from gluing back may not be the same atlas we started with, but they share a common maximal atlas. So, with the usual polite fiction of a manifold as a maximal atlas, the new definition is equivalent to the first one.

Although we shall try to ignore it as much as possible, we have to mention the following annoying

Fact 9.2.1 *Given a homeomorphism f of the surface X, and an atlas $(U_i, \phi_i)_i$ on X, the charts ϕ_i may be pre-composed with f, thus giving a new atlas $(U_i, \phi_i \circ f)_i$, which in general is not compatible with the first one.*

This fact is annoying because it only makes a difference if you tag each point in X with a label and are interested in tracking each individual label. The usual way around it is through another polite fiction: decide two maximal atlases are equivalent if they may be deduced from each other by pre-composition with a homeomorphism (sometimes it is convenient to restrict to homeomorphisms which are isotopic to the identity map). Then define your structure as this most unfathomable object: an equivalence class of maximal atlases.

9.2.4 Definition of a Translation Surface as a Holomorphic Differential

We have seen that a translation surface (as in our first definition) has a complex structure, which is essentially a protractor (a way to measure angles between tangent vectors at any point). But it actually has much more: a graduated ruler (since we can measure distances), and a compass, whose needle points North, wherever you are, except at the singular set. This is because our polygon is a subset of the Euclidean plane, so we can choose any direction we want as the North, and since all identifications are translations (except at the singular set), this goes down to the quotient space X. At a vertex of P, the needle of the compass gets a little dizzy, but not in the same way as an actual compass at the magnetic pole of the Earth would: while the latter has infinitely many directions to choose from, our compass only has finitely many. If the angle around the singular point x_i is $2k\pi$, then there are, so to say, k different norths at x_i.

Formally, the package (protractor, ruler, compass) is called a **holomorphic, or Abelian, differential**, and usually denoted ω. If you have a complex manifold X, a holomorphic differential is just, for every $x \in X$, the choice of a complex linear map $l(x)$ from the tangent space to X at x, to $\mathbb{C}$, with the condition that $l(x)$ depends homomorphically on x. Our X's have complex dimension one, and the only complex linear maps from $\mathbb{C}$ to itself are the maps $z \mapsto \lambda z$, with $\lambda \in \mathbb{C}$. When we think of the identity as a holomorphic differential on $\mathbb{C}$, we denote it dz, so we will usually think of a holomorphic differential ω on X as $f(z)dz$, with f holomorphic, in local charts, with the usual compatibility requirement that if two chart domains intersect, and ω reads $f(z)dz$ in one chart, and $g(z)dz$ in the other, and $T(z)$ is the transition map, then $f(T(z))T'(z) = g(z)$ (f and g should be thought of as derivatives, since $f(z)dz$ is a differential, so we just apply the chain rule for derivatives).

If you would like to see it put another way: if $x \in X$, and v is a tangent vector to X at x, and we have two different charts ϕ and ψ at x, such that ω reads $f(z)dz$ in the chart ϕ, and $g(z)dz$ in the chart ψ, then $\omega(x).v$ reads $f(\phi(x))\phi'(x).v$ in the chart ϕ, and $g(\psi(x))\psi'(x).v$ in the chart ψ. Since the evaluation of ω does not depend on the chart, we have $f(\phi(x))\phi'(x) = g(\psi(x))\psi'(x)$. Now let T be the transition map $\phi \circ \psi^{-1}$, then, setting $z = \psi(x)$, we have $f(T(z))\phi'(\psi^{-1}(z)) = g(z)\psi'(\psi^{-1}(z))$, which is $f(T(z))T'(z) = g(z)$.

Not every compact manifold has a non-zero holomorphic differential, for instance, if you try and extend dz to the sphere $\mathbb{C}P^1$, you get a double pole at

infinity, so the needle of your compass will act tipsy at infinity, like at the North
Pole. Translation surfaces are different from $\mathbb{C}P^1$: the holomorphic differential dz
in the plane goes down to a holomorphic differential on the quotient space X, albeit
with zeroes. The order of the zeroes is given by the angle at the singularities, as
follows from the chain rule: if g is a chart around a singular point, and f is a regular
chart, then the transition map is $z \mapsto z^k$, so $kf(z^k)z^{k-1} = g(z)$, in particular g has
a zero of order $k - 1$ at the singular point.

Assume ω reads $g(z)dz$ in some chart (U, ϕ) at $x_0 \in X$, with $\phi(x_0) = 0$ and
$g(0) \neq 0$, so we say that x_0 is a regular point of ω. Then we may take a primitive of
ω as a new chart at x_0:

$$\psi : V \longrightarrow \mathbb{C}$$
$$x \longmapsto \int_{x_0}^x \omega$$

where V is some neighborhood of x_0, with $V \subset U$. The map ψ is a local
biholomorphism because ω does not vanish at x_0. Let G be the local primitive of g,
defined in V, such that $G(0) = 0$. Note that $\psi(x) = G(\phi(x))$, so the transition map
$\psi \circ \phi^{-1}$ between ϕ and ψ is G. Thus, by the chain rule, assuming ω reads $f(z)dz$
in the chart ψ, we have

$$f\left(\psi \circ \phi^{-1}(z)\right) G'(z) = g(z)$$

so f is constant $= 1$, that is, ω reads dz in the chart ψ.

Now assume we have two different charts, at different, regular, points x and y,
whose domains intersect, and assume that in both charts ω reads dz. Let ϕ be the
transition map between the two charts. Then, by the chain rule, we have $\phi' = 1$,
that is, ϕ is a translation. Therefore, from a holomorphic differential, we recover a
translation atlas, thus proving the equivalence of our three definitions.

9.2.5 Definition of a Half-Translation Surface as a Quadratic Differential

Now assume that instead of a translation surface, we have a half-translation surface.
What could play the role of a holomorphic differential in that case? Well, the $z \mapsto$
$-z$ map does not preserve the linear form dz, but it does preserve the quadratic form
dz^2.

If you have a complex manifold X, a **holomorphic quadratic differential** is to
a holomorphic differential what a quadratic form is to a linear form, so it is, for
every $x \in X$, the choice of a complex-valued quadratic form $l(x)$ from the tangent
space to X at x, to $\mathbb{C}$, with the condition that $l(x)$ depends holomorphically on x.
Our X's have complex dimension one, and the only complex-valued quadratic forms
from $\mathbb{C}$ to itself are the maps $z \mapsto \lambda z^2$, with $\lambda \in \mathbb{C}$. When we think of the square

map as a quadratic differential on $\mathbb{C}$, we denote it dz^2, so we will usually think of a holomorphic quadratic differential ω on X as $f(z)dz^2$, with f holomorphic, in local charts, with the usual compatibility requirement that if two chart domains intersect, and ω reads $f(z)dz^2$ in one chart, and $g(z)dz^2$ in the other, and $\phi(z)$ is the transition map, then $f(\phi(z))(\phi'(z))^2 = g(z)$ (f and g should be thought of as squares of derivatives). Actually we should be allowing simple poles for f as well, for instance the half-translation sphere in Fig. 9.1 has four simple poles, but we shall not dwell on that.

Note that given a holomorphic differential ω, written $f(z)dz$ in local coordinates, the formula $f(z)^2dz^2$ yields a quadratic holomorphic differential, so the set of holomorphic differentials (modulo identification of a differential with its polar opposite) identifies with a subset of the set of quadratic holomorphic differentials. It is a proper subset, however, because not every holomorphic function is a square.

While not every quadratic differential is a square, every quadratic differential becomes a square in a suitable double cover of X, by the classical Riemann surface construction of the square root function. For instance, the half-translation sphere in Fig. 9.1 is covered, twofold, by a 4-square torus, in fact the involution of the covering is the hyperelliptic involution of the torus. It is often convenient to deal with translation surfaces only, taking double covers when need be.

9.2.6 Dynamical System Point of View

So far we have seen a translation surface as a geometric, or complex-analytic, object, but it is interesting to view it as a dynamical system.

We have seen that the North is well-defined on a translation surface X, outside the singularities; so we have a local flow ϕ_t on X, usually called the **vertical flow**, defined by "walk North for t miles". The flow is complete (i.e well-defined for any $t \in \mathbb{R}$) if we remove the finite, hence negligible, union of orbits which end in a singularity (which we call **singular orbits**).

Of course we can multiply our Abelian differential by $e^{i\theta}$ for any θ, thus turning the vertical by θ, so in fact we have a family of flows indexed by $\theta \in [0, 2\pi]$.

We ask the usual dynamical questions: are there periodic orbits? if so, can we enumerate them? are there invariant measures not supported on periodic orbits? If so, how many of them?

In the case of the square torus, the questions are readily answered: by taking the first return map on a closed transversal, we see that the dynamical properties of the flow in a given direction θ are those of a rotation on the circle. That is, when $\theta/\pi \in \mathbb{Q}$, every orbit is periodic, with the same period.

When $\theta/\pi \in \mathbb{R} \setminus \mathbb{Q}$, every orbit is equidistributed, that is, the proportion of its time it spends in a given interval is the proportion of the circle this interval occupies. In particular the flow in the direction θ is uniquely ergodic, that is, it supports a unique invariant measure, up to a scaling factor.

In the case of square-tiled surfaces, the answer is equally satisfying. Recall that any square-tiled surface X is a ramified cover of the square torus, so the flow in the direction θ on X projects to the flow in the direction θ on the square torus. Therefore, if $\theta/\pi \in \mathbb{Q}$, every orbit of the flow in the direction θ on X projects to a closed orbit in the square torus, so it must be periodic itself. The period need not be the same for all orbits, though, for instance the vertical flow on $X = St(3)$ has orbits of period 1 and 2 (see Fig. 9.3). In fact $X = St(3)$, minus the set of singular orbits, is the reunion of two cylinders, one made with orbits of length 2, and the other made of orbits of length 1.

The answer when $\theta/\pi \in \mathbb{R} \setminus \mathbb{Q}$ is a bit trickier, because measures cannot be projected, they may only be lifted, and it is unclear why every invariant measure in X should be the lift of an invariant measure in the torus, but as in the torus case, orbits are equidistributed.

This dichotomy between directions which are **completely periodic**, meaning that the surface decomposes into cylinders of periodic orbits, and directions which are uniquely ergodic, is called the **Veech dichotomy**. In the next sections we are going to see a powerful criterion for unique ergodicity, and apply it to find a larger class of surfaces which satisfy the Veech dichotomy.

In general it is hard to prove that a given surface (say, the double pentagon) satisfies the Veech dichotomy. On the other hand, it is easy to find a surface which does not satisfy the Veech dichotomy (see Fig. 9.11).

In [26] an example is given of a translation surface with a direction θ such that the flow in the direction θ is minimal (every orbit is dense) but not uniquely ergodic, that is, it supports several distinct invariant measures.

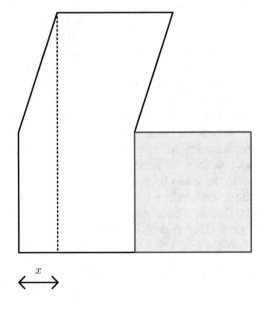

Fig. 9.11 Take the surface $St(3)$ and shear the top edge to the right by x, keeping the same gluing pattern. The shaded square projects to the surface as a cylinder of vertical periodic geodesics of length 1, while the rest projects to a cylinder where every vertical geodesic is dense, if x is irrational

9.3   Moduli Space

Now that we know what a translation structure on a surface is, we would like to know what it means for two translation structures to be the same, or almost the same. Sameness is easy to define, if hard to visualize, through atlases: we say two translation structures (as atlases) are equivalent if they share a common (equivalence class of) maximal translation atlas. Observe that a necessary condition for sameness is that the number of singular points, and the angle around each singular point, are the same.

It is a bit harder to see sameness from polygons, because two polygons which look very different may yield the same surface after identifications, see Fig. 9.2. The correct definition of sameness is that two polygons yield the same translation surface if one may be cut and pasted into the other, provided the pieces are only re-arranged by translation (no rotation or flip allowed).

It is not always easy to tell when two polygons do *not* yield the same surface, either. The two polygons of Fig. 9.3 are really different surfaces, because the one on the right is foliated, in the vertical direction, by closed geodesics of length 3, while the one on the left is not.

9.3.1   Strata

The previous discussion suggests lumping together all translation surfaces with the same number of singularities, and equal angles around each singularities. The set of all such translation surfaces is called a **stratum** (term coined by Veech [50], to be explained later). It is usually denoted $\mathcal{H}(k_1, \ldots, k_n)$, where $2\pi(k_i + 1)$ is the angle around the i-th singular point. For instance $\mathcal{H}(2)$ is the stratum of surfaces with only one singular point, of angle 6π, while $\mathcal{H}(1, 1)$ is the stratum of surfaces with two singular points, each with angle 4π.

Note that if two translation surfaces lie in the same stratum, they must have the same genus. To see this, let us assume, to begin with, that X is obtained by identifying the sides of a connected $2N$-gon, with the vertices identifying into the n singularities of an Abelian differential ω.

Then the Euler characteristic $\chi(X)$ of X is $1 - N + n$.

Let us say the angle around each singularity x_i is $2(k_i + 1)\pi$, then the sum of the interior angles of the polygon equals the sum of the angles around the singularities, that is,

$$\sum_{i=1}^{n} 2(k_i + 1)\pi = (2N - 2)\pi,$$

whence

$$\sum_{i=1}^{n} k_i = N - 1 - n = -\chi(X) = 2\text{genus}(X) - 2.$$

Now, assume that X is obtained by identifying the sides of a $2N$-gon with p connected components, each with N_i vertices, for $i = 1, \ldots, p$. Then

$$\chi(X) = p - \frac{1}{2} \sum_{i=1}^{p} N_i + n,$$

and the sum of the interior angles of the polygons is

$$\sum_{i=1}^{p} (N_i - 2)\pi = \sum_{i=1}^{n} 2(k_i + 1)\pi$$

whence

$$\sum_{i=1}^{n} k_i = \frac{1}{2} \sum_{i=1}^{p} N_i - p - n = -\chi(X).$$

For instance, the stratum $\mathcal{H}(0)$ consists of all flat tori. In genus 2, we have the strata $\mathcal{H}(2)$ or $\mathcal{H}(1, 1)$, and those are the only strata of genus 2, because the only way the number 2 can be partitioned is as $1 + 1$ or $2 + 0$. The surfaces in Figs. 9.2 and 9.3 lie in $\mathcal{H}(2)$, while the surface in Fig. 9.4 lies in $\mathcal{H}(1, 1)$.

The regular $4n$-gon, and the double $(2n + 1)$-gon, lie in the stratum $\mathcal{H}(2n - 2)$. The regular $(4n + 2)$-gon lies in the stratum $\mathcal{H}(n - 1, n - 1)$. On the other hand, square-tiled surfaces are dense in every stratum.

Now we want to define what it means for translation surfaces to be close, that is, we are looking for a topology on the set of translation surfaces. In fact we can do even better: we may find local coordinates on a given stratum, thus making the stratum (almost) a manifold.

9.3.2 Period Coordinates

Fix a basis for the $\mathbb{Z}$-module $H_1(X, \Sigma, \mathbb{Z})$. Such a basis may be chosen as (the relative homology classes of) $\alpha_1, \ldots, \alpha_{2g}, c_1, \ldots, c_{n-1}$, where g is the genus of X, $\alpha_1, \ldots, \alpha_{2g}$ are simple closed curves (based at x_1) which generate the absolute homology $H_1(X, \mathbb{Z})$, and for each $i = 1, \ldots, n - 1$, c_i is a simple arc joining x_1 to x_{i+1}.

Definition 9.3.1 The **period coordinates** of the holomorphic differential ω, with respect to the basis $\alpha_1, \ldots, \alpha_{2g}, c_1, \ldots, c_{n-1}$, are the $2g + n - 1$ complex numbers

$$\int_{\alpha_1} \omega, \ldots, \int_{\alpha_{2g}} \omega, \int_{c_1} \omega, \ldots, \int_{c_{n-1}} \omega.$$

The complex numbers $\int_{\alpha_1} \omega, \ldots, \int_{\alpha_{2g}} \omega$ are called the **absolute periods** of ω, because the homology classes $[\alpha_1], \ldots, [\alpha_{2g}]$ live in the absolute homology $H_1(X, \mathbb{Z})$, while the complex numbers $\int_{c_{n-1}} \omega, \ldots, \int_{c_{n-1}} \omega$ are called the **relative periods** of ω, because the homology classes $[c_1], \ldots, [c_{n-1}]$ live in the relative homology $H_1(X, \Sigma, \mathbb{Z})$.

Why this is actually a set of local coordinates on the stratum of X is a theorem of [48] (see also [13, section 2.3], and also [54], [50]). Let us briefly explain the idea. Assume two holomorphic differential ω_1 and ω_2 lie in the same stratum and have the same period coordinates. Assume that ω_1 and ω_2 are close enough, in some sense, that they can be considered as smooth (not necessarily holomorphic) differential forms on the same surface X (seen as a differentiable manifold), with the same homology basis $\alpha_1, \ldots, \alpha_{2g}, c_1, \ldots, c_{n-1}$. Each holomorphic differential ω_1 and ω_2 gives you a Euclidean metric on X, for which geodesic representatives of $\alpha_1, \ldots, \alpha_{2g}, c_1, \ldots, c_{n-1}$ can be found. Cut X open along these geodesics, in each of the two Euclidean metrics, you get a pair of $2(2g + 2n - 1)$-gons whose sides are represented by the period coordinates of the holomorphic differentials. Since ω_1 and ω_2 have the same period coordinates, the two polygons are isometric, by a translation, so ω_1 and ω_2 give the same translation structure.

Example Assume (X, ω) is a square-tiled surface. Then the homology basis may be chosen so that the curves $\alpha_1, \ldots, \alpha_{2g}, c_1, \ldots, c_{n-1}$ lie in the sides of the squares. Thus the period coordinates lie in $\mathbb{Z}[i]$. Conversely, if all period coordinates lie in $\mathbb{Z}[i]$, then X may be cut into a plane polygon, all of whose vertices lie in $\mathbb{Z}[i]$, so X is actually a square-tiled surface.

9.3.3 Dimension of the Strata

Thus we know the (complex) dimension of the stratum $\mathcal{H}(k_1, \ldots, k_n)$ is $2g + n - 1$. The largest stratum is the one with $2g - 2$ singularities of index 1, its dimension is $4g - 3$. The smallest stratum is the one with only one singularity of index $2g - 2$, its dimension is $2g$. The strata are not disconnected from each other, each stratum but the largest one lies in the closure of one or several of the larger ones: for instance, if we have a sequence (X, ω_m) of translation surfaces in the stratum $\mathcal{H}(k_1, \ldots, k_n)$, and $\int_{c_1} \omega_m \longrightarrow 0$ when $m \longrightarrow \infty$, then the limit surface lies in $\mathcal{H}(k_1 + k_2, k_3, \ldots, k_n)$, because the singularities x_1 and x_2 merge in the limit. This is the reason they are called strata, because they are nested in each other's closure, like matriochka. The union of all strata, with the topology given by the period coordinates, is called the **moduli space of translation surfaces of genus** g, denoted $\mathcal{H}_g$. It contains the largest (or principal) stratum, denoted $\mathcal{H}(1^{2g-2})$, as a dense open subset, so its dimension is $4g - 3$.

While the global topology of $\mathcal{H}_g$ is mysterious, we have a simple compactness criterion (**Mumford's compactness theorem**, [37]) for subsets of $\mathcal{H}_g$: a sequence X_n of elements of $\mathcal{H}_g$, of uniformly bounded area, goes to infinity if and only if there exist closed geodesics in X_n whose lengths go to zero.

9.3.4 Quadratic Differentials

The theory of translation surfaces becomes much more interesting when you think about it in connection with Teichmüller theory.

The set of complex structures on surfaces of genus g is a well-studied object, going back to Riemann (see [38], reprinted in [39], pp.88–144, or, for the faint of heart, [15]). It is worth noting that Riemann actually got there when studying Abelian differentials.

It is called the **moduli space of complex structures of genus** g, and denoted $\mathcal{M}_g$. It has a topology, but actually it has much more structure than that, in fact it is, except at some special points which correspond to very symmetrical surfaces, a complex manifold of dimension $3g - 3$, provided $g > 1$. The genus one case is special and will be discussed in more detail in Sect. 9.7.1. The cotangent space to moduli space at a given X is the vector space of holomorphic quadratic differentials (see [15, Theorem 6.6.1]).

Recall that we have just computed the dimension of $\mathcal{H}_g$, which we found to be $4g - 3$, while the cotangent bundle of $\mathcal{M}_g$ has dimension $6g - 6$, so the squares of Abelian differentials have codimension $2g - 3$ in the space of quadratic differentials.

9.3.4.1 Dimension of Strata of Quadratic Differentials

Following [21], we denote $\mathcal{Q}_g$ the space of all quadratic differentials of genus g, and $\mathcal{Q}(k_1, \ldots, k_n)$ the set of quadratic differentials with n zeroes, of respective orders $k_1, \ldots, k_n$, which are not squares.

Computations identical to those of Sect. 9.3.1 show that $\mathcal{Q}_g$ stratifies as the union of $\mathcal{Q}(k_1, \ldots, k_n)$, with $k_1 + \ldots + k_n = 4g - 4$.

It is interesting to compute the dimension of the strata of non-square quadratic differentials, because there is one notable difference with the Abelian case. In the case of the Abelian stratum $\mathcal{H}(k_1, \ldots, k_n) \subset \mathcal{H}_g$, we cut our surface into a $2(2g + n - 1)$-gon, and basically said that sides being pairwise equal, you need $2g + n - 1$ complex parameters to determine an element of $\mathcal{H}(k_1, \ldots, k_n)$. Let us try to apply the same argument to quadratic differentials.

Let us take an element of $Q(k_1, \ldots, k_n)$, and cut it into a polygon, whose vertices correspond to the singularities of the quadratic differential. Let us assume for simplicity that the polygon is connected. Then the sum of the interior angles of the polygon equals the sum of the angles around the singularities, which is $\sum_{i=1}^{n}(k_i + 2)\pi$. Thus the polygon has $\sum_{i=1}^{n}(k_i + 2) + 2 = 4g - 4 + 2n + 2$ edges. Since the edges are pairwise equal, this gives us $2g + n - 1$ complex parameters. But then, the largest stratum, with $n = 4g - 4$ and $k_i = 1, i = 1, \ldots, 4g - 4$, would have dimension $6g - 5$, instead of $6g - 6$ as befits the cotangent bundle of $\mathcal{M}_g$.

To understand this, we must take a closer look at how we identify the sides of a polygon. Recall that we made the convention that we orient the sides of a polygon so that the interior of the polygon lies to the left. With this convention, a necessary and sufficient condition for plane vectors $v_1, \ldots, v_n$ to be the oriented edges of a polygon is that $v_1 + \ldots + v_n = 0$.

Next, recall that we identify the sides in pairs, in such a way that the resulting surface is orientable, so each edge is glued to another edge in such a way that the arrows on the edges do not match.

Recall, furthermore, that we glue each edge to a parallel edge of equal length, so there exists some permutation σ of $\{1, \ldots, n\}$ such that $v_{\sigma(i)} = \pm v_i$, for all $i = 1, \ldots, n$. Since we must oppose the arrows when gluing, there are two possible cases: either $v_{\sigma(i)} = -v_i$, in which case the two edges are glued by translation, or $v_{\sigma(i)} = v_i$, in which case the edges are glued with a half-turn.

In the former case, we may choose v_i freely without being constrained by the relation $v_1 + \ldots + v_n = 0$, while in the latter case, it imposes a condition on v_i:

$$-2v_i = v_1 + \ldots + \hat{v}_i + \ldots + \hat{v}_{\sigma(i)} + \ldots + v_n$$

where the hats mean that $v_{\sigma(i)}$ and v_i are omitted in the sum.

So, when some pair of sides is glued with a half-turn, this pair of sides is determined by the others, so we lose one complex parameter. Now, saying that the quadratic differential is not a square is precisely saying that some pair of sides is glued with a half-turn, since we have seen that when every side is glued by translation, we get an Abelian differential. Therefore, the dimension of the stratum $Q(k_1, \ldots, k_n)$ is $2g + n - 2$.

9.3.5 Another Look at the Dimension of $\mathcal{H}_g$

Here is another way to understand the fact that $\dim \mathcal{H}_g = 4g - 3$. A translation structure, or holomorphic differential, consists of the following data: a complex structure on a surface X, and a complex-valued differential form, holomorphic for the given complex structure. As we have seen, a complex structure is determined by $3g - 3$ complex parameters.

Besides, the vector space of holomorphic (for a given complex structure) differentials has complex dimension g. This is because, given a complex structure X, by the Hodge theorem, the complex vector space $H^1(X, \mathbb{C})$, which has dimension $2g$, is the direct sum of two isomorphic summands, the subspace of holomorphic differentials, and the subspace of anti-holomorphic differentials. Thus a translation structure is determined by $3g - 3 + g = 4g - 3$ complex parameters, and $\mathcal{H}_g$ may be viewed as a real vector bundle over $\mathcal{M}_g$, with fiber $H^1(X, \mathbb{R})$.

9.4   The Teichmüller Geodesic Flow

9.4.1   Teichmüller's Theorem

Take two complex structures X_1 and X_2 of genus g. Then Teichmüller's theorem says there exists a holomorphic quadratic differential q on X_1, and a number $t \in \mathbb{R}$, such that, assuming for simplicity that q is the square of a holomorphic differential ω, the 1-differential form with real part $e^t \text{Re}\, \omega$ and imaginary part $e^{-t} \text{Im}\, \omega$ is holomorphic with respect to the complex structure X_2.

Then one may define, at least locally, a distance function by $d(X_1, X_2) := |t|$, and this distance comes with geodesics (shortest paths): if $s \in [0, t]$, and X_s is the complex structure which makes the complex differential with real part $e^s \text{Re}\, \omega$ and imaginary part $e^{-s} \text{Im}\, \omega$ holomorphic, then $d(X_1, X_s) = |s|$, so the path $s \mapsto X_s$ is a geodesic.

Teichmüller's distance has an infinitesimal expression, like a Riemannian metric, and more precisely, it is a Finsler metric (a Finsler metric is to a Riemannian metric what a norm is to a Euclidean norm). This was discovered by Teichmüller, see [45, p. 26], translated in [46], or just [15, Theorem 6.6.5].

A quadratic differential ω induces a Riemannian (flat except at the singularities) metric on X, in particular it comes with a volume form, so it makes sense to evaluate the total volume of X with respect to q, denoted $\text{Vol}(X, q)$. The map $(X, q) \mapsto \text{Vol}(X, q)$, restricted to the tangent space to $\mathcal{M}_g$ at X, is a norm: it is 1-homogeneous, positive except at $\omega = 0$, and satisfies the triangle inequality. The first two properties are immediate, the last one may warrant a short proof.

First, observe that if the quadratic differential q is $f(z)dz^2$ in some chart, then the volume form of q, seen as a flat metric, is $\frac{i}{2} |f(z)|\, dz \wedge d\bar{z}$, because $dz \wedge d\bar{z} = -2i\, dx \wedge dy$. This expression is coordinate-invariant, because if q is $g(z)dz^2$ in some other chart, and T is the transition map, then $f(z) = g(T(z))T'(z)^2$, so

$$|f(z)|\, dz \wedge d\bar{z} = |g(T(z))|\, T'(z)\overline{T'(z)}dz \wedge d\bar{z} = |g(T(z))|\, (T'(z)dz) \wedge (\overline{T'(z)d\bar{z}})$$

so the total volume of q may be calculated in local coordinates. Now, assume we have two quadratic differentials $q_1 = f_1(z)dz^2$ and $q_2 = f_2(z)dz^2$, so $q_1 + q_2 = (f_1 + f_2)dz^2$, the triangle inequality in $\mathbb{C}$ yields $|f_1 + f_2| \leq |f_1| + |f_2|$, and, integrating over X, we get $\text{Vol}(X, q_1 + q_2) \leq \text{Vol}(X, q_1) + \text{Vol}(X, q_2)$.

This norm endows $\mathcal{M}_g$ with a metric: if we have a C^1 path $X(t)$ in $\mathcal{M}_g$, its derivative $\dot{X}(t)$ is a quadratic differential, holomorphic at $X(t)$, and we just say that the velocity of X at time t is $\mathrm{Vol}(X, \dot{X}(t))$. This is not a Riemannian metric, because the norm is not Euclidean, the specific name is Finsler metric. In any case it induces a geodesic flow on the cotangent bundle of $\mathcal{M}_g$: start at a complex structure X, in the direction given by a quadratic differential ω, and go to X_t such that the complex quadratic differential with real part $e^t \mathrm{Re}\,\omega$ and imaginary part $e^{-t} \mathrm{Im}\,\omega$ is holomorphic with respect to the complex structure X_t.

Most interesting to us here is the fact that $\mathcal{H}_g$, which is a submanifold of the cotangent bundle of $\mathcal{M}_g$, is invariant by the geodesic flow. In fact, we shall see, in Part 9.5, that $\mathcal{H}_g$ is invariant by a much more interesting action.

9.4.2   Masur's Criterion

In dynamical systems, we have a saying, which goes "sow in the parameter space, reap in the phase space". This is particularly relevant to translation surfaces, because the parameter space $\mathcal{H}_g$ comes with so much structure: a geodesic flow, invariant measures, affine coordinates. . . Masur's criterion (possibly the single most useful result in the theory) is a striking example.

Theorem 9.4.1 (Masur's Criterion, [25]) *Given an Abelian differential ω, the orbit of ω under the Teichmüller geodesic flow is recurrent if and only if the vertical flow of ω is uniquely ergodic.*

The idea is very clearly explained in [36]. We'll see several applications of Masur's criterion, here is one:

Theorem 9.4.2 ([20]) *Given an Abelian differential ω, for almost every θ, the vertical flow of $e^{i\theta}\omega$ is uniquely ergodic.*

Once we have Masur's criterion, it is easy to get a weak version of Theorem 9.4.2: *for almost every* Abelian differential ω, for almost every θ, the vertical flow of $e^{i\theta}\omega$ is uniquely ergodic. The proof is basically just Poincaré's Recurrence Theorem. Of course, to apply it we need an invariant measure of full support and finite total volume, so we apply it, not on $\mathcal{H}_g$, but on the subset $\mathcal{H}_g^1 \subset \mathcal{H}_g$ of Abelian differentials of area 1. It is a theorem by Masur and Veech [24, 47] that the Lebesgue measure induced on $\mathcal{H}_g^1$ by the period coordinates has finite total volume.

9.5   Orbits of the $GL_2^+(\mathbb{R})$-Action

Every translation surface may be viewed as a polygon in the Euclidean plane with parallel sides of equal length pairwise identified. The group $\mathrm{GL}_2^+(\mathbb{R})$ acts linearly on polygons, mapping pairs of parallel sides of equal length to pairs of parallel sides of equal length, so it acts on translation surfaces.

Let us take a basis $\mathcal{B} = (\alpha_1, \ldots, \alpha_{2g}, c_1, \ldots, c_{n-1})$ of $H_1(X, \Sigma, \mathbb{Z})$. If we cut X into a polygon along the curves $\alpha_1, \ldots, \alpha_{2g}, c_1, \ldots, c_{n-1}$, the period coordinates of (X, ω) in the basis $\mathcal{B}$ are the sides of the polygon (identifying a complex number and its affix). So, if $A \in GL_2^+(\mathbb{R})$, the period coordinates, in the basis $\mathcal{B}$, of the surface $A.(X, \omega)$, are the complex numbers

$$A.\int_{\alpha_1} \omega, \ldots, A.\int_{\alpha_{2g}} \omega, A.\int_{c_1} \omega, \ldots, A.\int_{c_{n-1}} \omega,$$

where A acting on a complex numbers means that it acts on its affix. The matrix A induces a homeomorphism (which is actually a diffeomorphism outside the singularities), from the translation surface (X, ω), to the surface $A.(X, \omega)$.

First, let us observe that strata are invariant under $GL_2^+(\mathbb{R})$. This is because acting by an element of $GL_2^+(\mathbb{R})$ on a polygon does not change the order in which sides are identified, which determines the singularities, and the angles around the singularities.

The $SL_2(\mathbb{R})$-action contains the geodesic flow, in the following sense: the geodesic, with respect to the Teichmüller metric, which has initial position X, and initial velocity ω, is the orbit of (X, ω) under the diagonal subgroup of $SL_2(\mathbb{R})$,

$$\left\{ \begin{pmatrix} e^t & 0 \\ 0 & e^{-t} \end{pmatrix} : t \in \mathbb{R} \right\}.$$

It is usually preferred to deal with the $GL_2^+(\mathbb{R})$-action rather than with the geodesic flow, for the following reason, which is a theorem by Eskin and Mirzakhani [9], henceforth referred to as the Magic Wand, after [56]. While the invariant sets for the geodesic flow may be wild (for instance, Teichmüller disks contain geodesic laminations, which are locally the product of a Cantor set with an interval), the closed invariant sets of the $GL_2^+(\mathbb{R})$-action are always nice submanifolds (again, ignoring singularities which occur at very symmetrical surfaces), locally defined by affine equations in the period coordinates. Any attempt at informally describing the proof, assuming the author could do it, would be as long as this paper. A recurring theme is the analogy with Ratner's theorems.

Now, by a result by Masur and (independently) Veech [24, 47], the geodesic flow is ergodic, with respect to a measure of full support in $\mathcal{H}_g$, so almost every orbit is dense; consequently, almost every $GL_2^+(\mathbb{R})$-orbit is dense. In fact, any stratum supports a full-support, ergodic measure. This means finding interesting (meaning: other than strata closures and $\mathcal{H}_g$ itself), closed invariant subsets won't be easy.

9.6 Veech Groups

It may happen that for some polygon P and some element A of $GL_2^+(\mathbb{R})$, both P and $A.P$, after identifications, are just the same translation surface X. This happens when $A.P$ can be cut into pieces, and the pieces re-arranged, by translations, into P.

The **Veech group** of (X, ω) is the subgroup of $\mathrm{GL}_2^+(\mathbb{R})$ which preserves X. Since the Veech group must preserve volumes, it is a subgroup of $\mathrm{SL}_2(\mathbb{R})$, and it turns out to be a Fuchsian group (see [18], Lemma 2). We sometimes denote it $\mathrm{SL}(X, \omega)$ after McMullen.

Take a matrix A in $\mathrm{SL}(X, \omega)$, then it defines a homeomorphism from X to itself, which we again denote A for simplicity. Let A_* be the linear automorphism of $H_1(X, \Sigma, \mathbb{Z})$ induced by A. Then the period coordinates of (X, ω), in a basis $\mathcal{B}$ of $H_1(X, \Sigma, \mathbb{Z})$, are exactly the period coordinates of $A.(X, \omega)$ in the basis $(A_*)^{-1}(\mathcal{B})$. Thus, modulo a change of basis, the set of periods is invariant under the Veech group.

Note that $\mathrm{SL}(X, \omega)$ is never cocompact, by the following argument: a translation surface always has a closed geodesic, for topological reasons; up to a rotation, we may assume this geodesic is horizontal. Let l be its length. Applying the Teichmüller geodesic flow for a time t, the length of the closed geodesic becomes $e^{-t}l$, so the Teichmüller geodesic goes to infinity in the moduli space, therefore $\mathrm{SL}_2(\mathbb{R})/\mathrm{SL}(X, \omega)$ cannot be compact.

The next best thing to being co-compact is to have finite co-volume, Veech groups of finite co-volume are going to play an important part. Note that for a Fuchsian group Γ to have finite co-volume, its ends at infinity must be finitely many cusps (see Section 4.2 of [19]).

Now let us see some examples.

9.6.1 The Veech Group of the Square Torus is $\mathrm{SL}_2(\mathbb{Z})$

Figure 9.12 shows that the matrices

$$T = \begin{pmatrix} 1 & 1 \\ 0 & 1 \end{pmatrix} \text{ and } S = \begin{pmatrix} 0 & -1 \\ 1 & 0 \end{pmatrix},$$

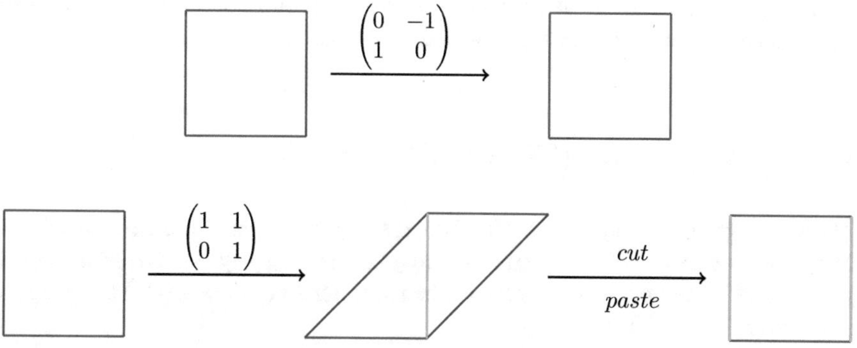

Fig. 9.12 Action of T and S on the torus $\mathbb{T}^2$

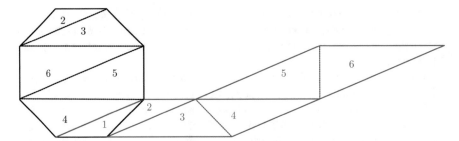

Fig. 9.13 The octagon is cut and pasted into a slanted stair shape

which generate $SL_2(\mathbb{Z})$, lie in the Veech group of the square torus. Conversely, every element of the Veech group preserve the set of periods, so, in the case of the torus, it must preserve $\mathbb{Z}[i]$, hence it must lie in $SL_2(\mathbb{Z})$.

9.6.2 Veech Groups of the Regular Polygons

It is obvious that the rotation by π/n is in the Veech group of the regular $2n$-gon. If you view the double $(2n+1)$-gon as a star, as in Fig. 9.2, you see that the rotation by $2\pi/(2n+1)$ is in the Veech group of the double $(2n+1)$-gon.

Figure 9.13 explains how to find a parabolic element in the Veech group of the regular n-gon: first, the regular n-gon may be cut and pasted into a slanted stair-shape. Then the matrix

$$\begin{pmatrix} 1 & -2\cot\frac{\pi}{n} \\ 0 & 1 \end{pmatrix}$$

brings the slanted stair-shape to its mirror image with respect to the vertical axis, and the mirror image may be cut and pasted back to a regular n-gon.

In fact, Veech proved in [49] that the Veech groups of the aforementioned surfaces are generated by the rotation, and by the parabolic element.

9.6.3 Veech Groups of Square-Tiled Surfaces

The Veech group of a square-tiled surface, if we assume that the squares are copies of the unit square in $\mathbb{R}^2$, is a subgroup of $SL_2(\mathbb{Z})$, because any element of the Veech group must preserve the set of periods, hence it must preserve $\mathbb{Z}[i]$. In fact it has finite index in $SL_2(\mathbb{Z})$ (see [14]).

Figure 9.14 shows why the parabolic element T^2 lies in the Veech group of the surface $St(3)$. Figure 9.15 shows why the parabolic element T does not lie in the

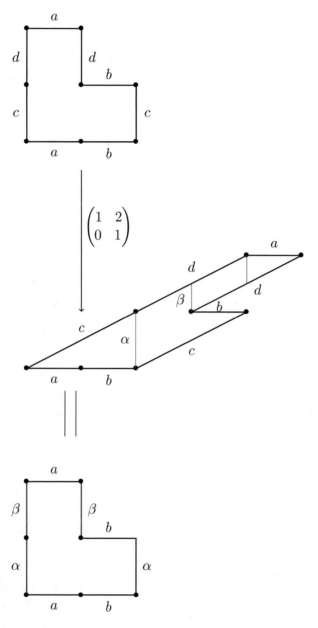

Fig. 9.14 Action of T^2 on $St(3)$

Fig. 9.15 Action of T on
$St(3)$

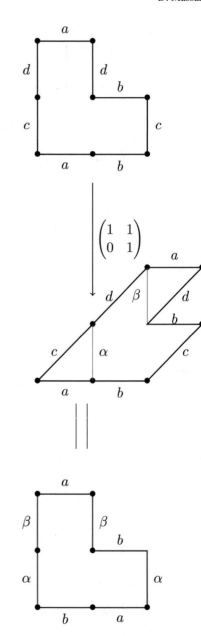

Veech group of the surface $St(3)$, in fact it takes $St(3)$ to the surface on the right in
Fig. 9.3. The Veech group of $St(3)$ is generated by T^2 and S, the rotation of order
4. It has index 3 in $SL_2(\mathbb{Z})$, the right cosets are those of id, T and its transpose.

In [41] an algorithm is given to compute the Veech group of any square-tiled
surface. It is implemented in the Sagemath package [8].

9.7 Teichmüller Disks

The orbit of (X, ω) under $\mathrm{GL}_2^+(\mathbb{R})$, or, more properly, its projection to the Teichmüller space $\mathcal{T}_g$, is called the **Teichmüller disk** of (X, ω). The concept (under the name "complex geodesic", which is still used sometimes) originates in [45], §121. Here is why it is called a disk.

Observe that if ω is a holomorphic differential on a Riemann surface X, and $A \in \mathrm{GL}_2^+(\mathbb{R})$ is a similitude, then $A.\omega$ is still holomorphic with respect to the complex structure of X, so if we are looking at the projection to the Teichmüller space $\mathcal{T}_g$, the action of $\mathrm{GL}_2^+(\mathbb{R})$ factors through the quotient space of $\mathrm{GL}_2^+(\mathbb{R})$ by similitudes. Recall that said quotient space is the hyperbolic disk $\mathbb{H}^2$.

The **Teichmüller curve** of (X, ω) is the quotient of the hyperbolic disk $\mathbb{H}^2$ by $\mathrm{SL}(X, \omega)$, which is a Fuchsian group. The fact that $\mathrm{SL}(X, \omega)$ is a Fuchsian group tells us that this quotient is a hyperbolic surface, possibly with some conical singularities at symmetrical surfaces (any symmetry of X is an element of $\mathrm{SL}(X, \omega)$ which fixes X).

Since the $\mathrm{GL}_2^+(\mathbb{R})$-action contains the Teichmüller geodesic flow as a subgroup action, Teichmüller curves, which are $\mathrm{GL}_2^+(\mathbb{R})$-orbits, are invariant under the geodesic flow. Furthermore, Teichmüller curves, as hyperbolic manifolds, are isometrically embedded in $\mathcal{M}_g$ (see [40]).

Figure 9.15 shows that the two surfaces of Fig. 9.3 lie in the same Teichmüller curve.

9.7.1 Example: The Teichmüller Disk of the Torus

The Teichmüller disk of the torus is special because it is also the Teichmüller *space* of the torus.

A torus, as a complex curve, is $\mathbb{C}/\Lambda$, where $\Lambda = \mathbb{Z}u \oplus \mathbb{Z}v$ is a lattice in $\mathbb{R}^2$, u and v being two non-colinear vectors in $\mathbb{R}^2$. Since $\mathbb{Z}u \oplus \mathbb{Z}v = \mathbb{Z}v \oplus \mathbb{Z}u$, we may assume that (u, v) is a positive basis of $\mathbb{R}^2$. Thus the matrix M whose columns are the coordinates of u and v, in that order, lies in $\mathrm{GL}_2^+(\mathbb{R})$. The Abelian differential we consider on $\mathbb{C}/\Lambda$ is $dz = dx + i\,dy$. A matrix $A = \begin{pmatrix} a & b \\ c & d \end{pmatrix} \in \mathrm{GL}_2^+(\mathbb{R})$ acts on dz by pull-back:

$$\forall \begin{pmatrix} x \\ y \end{pmatrix} \in \mathbb{R}^2,\ A^* dz \begin{pmatrix} x \\ y \end{pmatrix} = dz \left(A.\begin{pmatrix} x \\ y \end{pmatrix} \right) = dz \begin{pmatrix} ax + by \\ cx + dy \end{pmatrix} = ax + by + i(cx + dy)$$

so $A^* dz = a.dx + b.dy + i(c.dx + d.dy)$. The definition of the pull-back means that the complex 1-form $A^* dz$ on $\mathbb{C}/\Lambda$, which is non-holomorphic unless A happens to be $\mathbb{C}$-linear, becomes the Abelian differential dz on the torus $\mathbb{C}/A.\Lambda$, where $A.\Lambda = \mathbb{Z}A.u \oplus \mathbb{Z}A.v$.

Let (e_1, e_2) be the canonical basis of $\mathbb{R}^2$. The straight segments $\{te_i : 0 \le t \le 1\}$, for $i = 1, 2$, become closed curves in $\mathbb{C}/\mathbb{Z}^2 = \mathbb{C}/\mathbb{Z}e_1 \oplus \mathbb{Z}e_2$. We again denote e_1 and e_2, respectively, the homology classes of those closed curves. For any torus $\mathbb{C}/\Lambda$, with $\Lambda = \mathbb{Z}u \oplus \mathbb{Z}v$, $M \in GL_2^+(\mathbb{R})$ being the matrix whose columns are the coordinates of u and v, in that order, M may be viewed as a diffeomorphism from $\mathbb{C}/\mathbb{Z}^2$ to $\mathbb{C}/\Lambda$. This allows us to consider (e_1, e_2) as a basis of $H_1(\mathbb{C}/\Lambda, \mathbb{R})$: e_1 (resp. e_2) is the homology class of the closed curve $\{tu, \text{resp.} tv : 0 \le t \le 1\}$ in $\mathbb{C}/\Lambda$.

We compute period coordinates in the basis (e_1, e_2), and since $\omega = dz$, the period coordinates of $\mathbb{C}/\Lambda$ are $z(u)$ and $z(v)$, where $z(u)$ is the complex number whose affix is u. So, considering periods as vectors in $\mathbb{R}^2$, $GL_2^+(\mathbb{R})$ acts on periods linearly on the left.

If $\phi : \mathbb{R}^2 \longrightarrow \mathbb{R}^2$ is a similarity (a non-zero $\mathbb{C}$-linear map), then $\mathbb{C}/\Lambda$ and $\mathbb{C}/\phi(\Lambda)$ are bi-holomorphic. So, if we are interested in the projection of the Teichmüller disk to the moduli space of complex structures, we might as well consider the quotient of $GL_2^+(\mathbb{R})$ by the subgroup H of similarities, acting on the left. Beware that $GL_2^+(\mathbb{R})$ must then act on the quotient from the right.

Say two matrices $M_1, M_2 \in GL_2^+(\mathbb{R})$ are equivalent under H if there exists $P = \begin{pmatrix} a & c \\ -c & a \end{pmatrix} \in H$, such that $PM_1 = M_2$. We use the following representative of an equivalence class:

$$\text{if } u = \begin{pmatrix} a \\ c \end{pmatrix}, v = \begin{pmatrix} b \\ d \end{pmatrix}, M = \begin{pmatrix} a & b \\ c & d \end{pmatrix} \in GL_2^+(\mathbb{R}), \text{ then}$$

$$\frac{1}{a^2 + c^2} \begin{pmatrix} a & c \\ -c & a \end{pmatrix} \begin{pmatrix} a & b \\ c & d \end{pmatrix} = \frac{1}{a^2 + c^2} \begin{pmatrix} a^2 + c^2 & ab + cd \\ 0 & ad - bc \end{pmatrix}.$$

Observe that the matrix $\frac{1}{a^2+c^2} \begin{pmatrix} a & c \\ -c & a \end{pmatrix}$, viewed as a map from $\mathbb{C}$ to $\mathbb{C}$, is $z \mapsto \frac{1}{z(u)} z$. Geometrically, this means that the complex number which represents (the biholomorphism class of) $\mathbb{C}/\Lambda$, where $\Lambda = \mathbb{Z}u \oplus \mathbb{Z}v$, is just $[z(u) : z(v)] \in \mathbb{C}P^1$, so the canonical representative is $\left[1 : \frac{z(v)}{z(u)}\right] \in \mathbb{C}P^1$.

Since (u, v) is a positive basis of $\mathbb{R}^2$, $\frac{z(v)}{z(u)}$ lies in the upper half-plane $\mathbb{H}^2$, so we have a bijection

$$\Psi : H \backslash GL_2^+(\mathbb{R}) \longrightarrow \mathbb{H}^2.$$

Take an element of $H \backslash \mathrm{GL}_2^+(\mathbb{R})$, represented by a matrix $\begin{pmatrix} 1 & x \\ 0 & y \end{pmatrix}$, and an element $\begin{pmatrix} a & b \\ c & d \end{pmatrix}$ of $\mathrm{GL}_2^+(\mathbb{R})$. Then, recalling that $\mathrm{GL}_2^+(\mathbb{R})$ acts on the quotient from the right,

$$M := \begin{pmatrix} 1 & x \\ 0 & y \end{pmatrix} \begin{pmatrix} a & b \\ c & d \end{pmatrix} = \begin{pmatrix} a + cx & b + dx \\ cy & dy \end{pmatrix}.$$

Set $z = x + iy$, so the affixes of the columns of the matrix M are $a + cz$ and $b + dz$, then the map Ψ takes the equivalence class of the matrix M to the point $\frac{dz+b}{cz+a}$ in $\mathbb{H}^2$. This means that the map Ψ is equivariant, with respect to the right action of $\mathrm{GL}_2^+(\mathbb{R})$ on right cosets, and the right action of $\mathrm{GL}_2^+(\mathbb{R})$ on $\mathbb{H}^2$ defined by

$$\mathbb{H}^2 \times \mathrm{GL}_2^+(\mathbb{R}) \longrightarrow \mathbb{H}^2 \tag{9.1}$$

$$\left(z, \begin{pmatrix} a & b \\ c & d \end{pmatrix} \right) \longmapsto \frac{dz + b}{cz + a}. \tag{9.2}$$

9.7.1.1 Hyperbolic Metric vs. Teichmüller Metric

So far we have identified the Teichmüller disk of the torus with the hyperbolic plane. We already know that the Veech group of the square torus is $\mathrm{SL}_2(\mathbb{Z})$, so the Teichmüller curve of the torus is the *modular curve* $\mathbb{H}^2 / \mathrm{SL}_2(\mathbb{Z})$ (more on which can be found in [1]).

The hyperbolic plane, as a set, is not particularly interesting; what is interesting about it is the hyperbolic metric. We shall now see that Teichmüller's metric, on the Teichmüller space of the torus viewed as the hyperbolic plane, is precisely the hyperbolic metric, up to a multiplicative factor, which we shall ignore.

We shall use the fact that the hyperbolic metric, up to a multiplicative factor, is the only Finsler metric on $\mathbb{H}^2$ invariant under the action of $\mathrm{SL}_2(\mathbb{R})$. This is because $\mathrm{SL}_2(\mathbb{R})$ acts transitively on the unit tangent bundle of $\mathbb{H}^2$, so once we know the length of one tangent vector at some point, we know the length of every tangent vector, at every point.

By the equivariance of the map Ψ, and since we are not interested in the multiplicative factor, all we have to check is the invariance of the Teichmüller metric under the action of $\mathrm{SL}_2(\mathbb{R})$.

Now recall the Teichmüller norm, on the fiber, over X, of the holomorphic quadratic differential bundle, is just the total area. Since $\mathrm{SL}_2(\mathbb{R})$ preserves area, it preserves the Teichmüller metric. Thus we have proved that the Teichmüller metric is a multiple of the hyperbolic metric on $\mathbb{H}^2$. This was originally observed in §5 of [45].

D. Massart

9.7.1.2 Hyperbolic Geodesics

Let us investigate a little bit the relationship between hyperbolic geodesics and flat tori. The subgroup

$$\mathcal{G} := \left\{ g_t = \begin{pmatrix} e^t & 0 \\ 0 & e^{-t} \end{pmatrix} : t \in \mathbb{R} \right\}$$

of $SL_2(\mathbb{R})$ is mapped by Ψ to the geodesic

$$\mathbb{R} \longrightarrow \mathbb{H}^2$$

$$t \longmapsto e^{-2t} i.$$

For any $A = \begin{pmatrix} a & b \\ c & d \end{pmatrix} \in SL_2(\mathbb{R})$,

$$\mathcal{G}A = \left\{ \begin{pmatrix} e^t a & e^t b \\ e^{-t} c & e^{-t} d \end{pmatrix} : t \in \mathbb{R} \right\}$$

is mapped by Ψ to the geodesic

$$\mathbb{R} \longrightarrow \mathbb{H}^2$$

$$t \longmapsto \frac{e^{2t} ab + e^{-2t} cd + i(ad - bc)}{e^{2t} a^2 + e^{-2t} c^2}$$

whose endpoints at infinity are $\frac{b}{a}$ and $\frac{d}{c}$.

Here is a geometric interpretation of the endpoints. Consider the complex differential form $\eta := A^* dz = a.dx + b.dy + i(c.dx + d.dy)$. Then, for any $t \in \mathbb{R}$,

$$(g_t A)^* dz = e^t \Re \eta + i e^{-t} \Im \eta,$$

that is, $e^t \Re \eta + i e^{-t} \Im \eta$ is holomorphic on the torus $\mathbb{C}/\Lambda$, where $\Lambda = g_t.A.\mathbb{Z}^2$. So η^2 is the quadratic differential given by Teichmüller's theorem (see Sect. 9.4.1), associated with the geodesic $\Psi g_t.$, whose real part is contracted along the geodesic, while its imaginary part is expanded. The endpoints at infinity of the geodesic are the (reciprocals of the opposites of the) slopes of the respective kernels of the real and imaginary parts of η.

9.7.1.3   Horocycles

The subgroup

$$\mathcal{N} := \left\{ n_t = \begin{pmatrix} 1 & t \\ 0 & 1 \end{pmatrix} : t \in \mathbb{R} \right\}$$

of $SL_2(\mathbb{R})$ is mapped by Ψ to the horocycle

$$\mathbb{R} \longrightarrow \mathbb{H}^2$$
$$t \longmapsto i + t.$$

For any $A = \begin{pmatrix} a & b \\ c & d \end{pmatrix} \in SL_2(\mathbb{R})$,

$$\mathcal{N}A = \left\{ \begin{pmatrix} a + tc & b + td \\ c & d \end{pmatrix} : t \in \mathbb{R} \right\}$$

is mapped by Ψ to the horocycle

$$\mathbb{R} \longrightarrow \mathbb{H}^2$$
$$t \longmapsto \frac{(a+tc)(b+td) + cd + i(ad - bc)}{(a+tc)^2 + c^2}$$

whose endpoint at infinity is $\frac{d}{c}$, and whose apogee is $\frac{d}{c} + i\frac{ad-bc}{c^2}$.
  The pull-back, by $n_t A$, of the real 1-form dy, is $c.dx + d.dy$, because

$$(n_t A)^* dy \begin{pmatrix} x \\ y \end{pmatrix} = dy \left(n_t A \begin{pmatrix} x \\ y \end{pmatrix} \right) = dy \begin{pmatrix} (a+tc)x + (b+td)y \\ cx + dy \end{pmatrix} = cx + dy,$$

in particular $(n_t A)^* dy$ is constant along the horocycle $\Psi(\mathcal{N}A)$. The point at infinity of the horocycle is (minus the reciprocal of) the slope of the kernel of the constant 1-form.
  The reason for the annoying recurrence of "minus the reciprocal of. . ." is that the matrix $\begin{pmatrix} d & b \\ c & a \end{pmatrix}$ is conjugate, by $\begin{pmatrix} 1 & 0 \\ 0 & -1 \end{pmatrix}$, which acts on the hyperbolic plane by $z \mapsto -\bar{z}$, to the inverse matrix $A^{-1} = \begin{pmatrix} d & -b \\ -c & a \end{pmatrix}$.
  If $\frac{d}{c} \in \mathbb{Q}$, then all the closed geodesics whose velocity vectors are tangent to the kernel of $c.dx + d.dy$ are closed, and of length $c^2 + d^2$. So the torus $n_t A$, for $t \in \mathbb{R}$, may be seen as a cylinder of girth $c^2 + d^2$, with the boundaries identified. When t varies in $\mathbb{R}$, the identification varies but the closed geodesics remain fixed.

9.7.1.4 Relationship Between the Asymptotic Behaviour of Hyperbolic Geodesics, and the Dynamic Behaviour of Euclidean Geodesics

The set of slopes of closed geodesics in $\mathbb{C}/\mathbb{Z}^2$ is $\mathbb{Q}$. The Veech group $SL_2(\mathbb{Z})$ acts transitively on $\mathbb{Q}$. Given a hyperbolic geodesic $t \mapsto \gamma(t)$ in the modular surface $\mathbb{H}^2/SL_2(\mathbb{Z})$, the following four points are equivalent:

- $\gamma(t)$ goes to infinity in $\mathbb{H}^2/SL_2(\mathbb{Z})$, when $t \to \infty$
- the lifts of γ to $\mathbb{H}^2$ have a rational endpoint when $t \to \infty$
- all orbits of the vertical flow of the Abelian differential associated with γ are closed
- γ is invariant under the parabolic subgroup of $SL_2(\mathbb{Z})$ generated by $\begin{pmatrix} 1 & 0 \\ 1 & 1 \end{pmatrix}$.

9.7.2 Examples of Higher Genus Teichmüller Disks

9.7.2.1 Three-Squared Surfaces

The Veech group of the surface $St(3)$ is an index 3 subgroup of $SL_2(\mathbb{Z})$. It has a fundamental domain which is an ideal triangle in $\mathbb{H}^2$, with ideal vertices $-1, 1, \infty$. The parabolic transformation T^2 identifies the two vertical boundaries, and the rotation S identifies the two halves of the semi-circular boundary. The point i, which corresponds to the identity matrix, that is, to the surface $St(3)$ itself, is a singular point: a conical point with angle π. This is because it is invariant by S, which acts as the involution $z \mapsto -\frac{1}{z}$ on the hyperbolic plane. The other two three-squared translation surfaces, which correspond to the matrices T and ST, are both (since they induce the same complex structure) mapped by Ψ, as defined in Sect. 9.7.1, to the point $1 + i$ (which is identified with $1 - i$ by T^2).

The Teichmüller curve of $St(3)$ has two cusps, one at ∞, and the other at ± 1 (which become the same point in the Teichmüller curve). Since $St(3)$ is square-tiled, the set of slopes of closed geodesics is $\mathbb{Q}$. The difference with the torus is that the Veech group does not act transitively on $\mathbb{Q}$, it has two orbits, one for each cusp. The orbit of ∞ consists of all fractions $\frac{p}{q}$, with $p \neq q \mod 2$. The orbit of ± 1 consists of all fractions $\frac{p}{q}$, with $p = q = 1 \mod 2$.

The first cusp is called a two-cylinder cusp, while the second cusp is called a one-cylinder cusp, for the following reason. If a geodesic, in the Teichmüller curve, escapes to infinity in the two-cylinder cusp, then any lift of this geodesic to $\mathbb{H}^2$ corresponds, by Teichmüller's theorem, to a quadratic differential, the trajectories of whose real part are closed, and decompose $St(3)$ into two cylinders of closed geodesics. For instance, the vertical direction admits two cylinders, one, obtained by identifying the green boundaries in Fig. 9.5, is made of vertical closed geodesics of length 2, and the other, obtained by identifying the blue boundaries in Fig. 9.5, is made of vertical closed geodesics of length 1. If a geodesic, in the Teichmüller curve, escapes to infinity in the one-cylinder cusp, then any lift of this geodesic to $\mathbb{H}^2$ corresponds, by Teichmüller's theorem, to a quadratic differential, the trajectories of whose real part are closed, pairwise homotopic, and of equal length. For instance, all geodesics in the 45° direction, except those that hit the singular point, are closed and of length $3\sqrt{2}$.

See [5] for more on this Teichmüller curve. Note that the three-square Teichmüller curve has genus zero, topologically it is a sphere minus two points, one for each cusp. The genus, and the number of cusps, of Teichmüller curves of square-tiled surfaces may be arbitrarily large. See [16] for the stratum $\mathcal{H}_2$; given a square-tiled surface in any stratum, the genus and number of cusps of its Teichmüller curve may be computed with [8].

9.7.2.2   Regular Polygons

We have seen that the Veech group of the regular polygons is generated by a parabolic element, and a rotation. The Veech group of the double $2n + 1$-gon has a fundamental domain which looks like Fig. 9.16, left. It has only one cusp, at infinity in Fig. 9.16. This is reflected in the fact that the Veech group acts transitively on the set of directions of closed geodesics. It is an n-cylinder cusp, meaning that in any direction of closed geodesic, the surface decomposes into n cylinders.

The point $(0, 1)$ in Fig. 9.16, left, is the image by Ψ of the identity matrix, that is, the double $2n + 1$-gon. In the Teichmüller curve it is singular, more precisely it

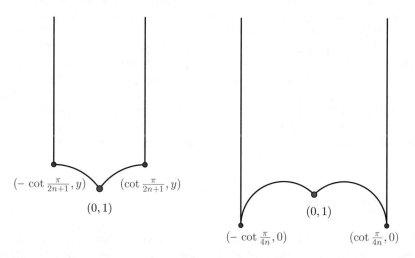

Fig. 9.16 Fundamental domains for the Veech groups of the double $2n+1$-gon, on the left, and the $4n$-gon, on the right. The number y is such that the circular boundary meets the vertical boundary perpendicularly. The angle at the point $(0, 1)$ is $\frac{2\pi}{2n+1}$ (left), and $\frac{\pi}{2n}$ (right)

is a conical point with angle $\frac{2\pi}{2n+1}$. This reflect the fact that the double $2n + 1$-gon is invariant by a rotation of order $2n + 1$.

The corner of coordinates $(\pm \cot \frac{\pi}{2n+1}, y)$, is the image by Ψ, when $n = 2$, of the golden L (see [6]), and when $n > 2$, of stair-shaped polygons (see [22]). Those polygons are invariant by S, which acts as the involution $z \mapsto -\frac{1}{z}$ on the hyperbolic plane, this is why the corner becomes a conical point with angle π in the Teichmüller curve.

The main difference between the Teichmüller curve of the double $2n + 1$-gon, and Teichmüller curve of the regular $4n$-gon, is that the latter has two cusps, at ∞ and $(\pm \cot \frac{\pi}{4n}, 0)$ on Fig. 9.16, right. See [43] for more on the Teichmüller curve of the regular octagon, where the fundamental domain is drawn in the hyperbolic disk rather than in the hyperbolic half-plane.

The Teichmüller curve of the octagon, however, is different from the three-square Teichmüller curve, in that both cusps are two-cylinders, meaning that if a geodesic, in the Teichmüller curve, escapes to infinity in any cusp, then any lift of this geodesic to $\mathbb{H}^2$ corresponds, by Teichmüller's theorem, to a quadratic differential, the trajectories of whose real part are closed, and decompose the octagon into two cylinders of closed geodesics.

Then, you may ask, how do we distinguish between the cusps ? First, look at the horizontal direction in the octagon (Fig. 9.13). We leave it to the reader to check that the cylinder made up with the triangles 5 and 6 has height 1 and length $1 + \sqrt{2}$, while the cylinder made up with the triangles 1, 2, 3 and 4 has height $\frac{1}{\sqrt{2}}$ and length $2 + \sqrt{2}$.

The **module** of a cylinder is the ratio of its height to its length. So we see that for the horizontal direction, the ratio of the modules of the cylinders is 2. What is important about this ratio of modules, is that it is invariant by $GL_2^+(\mathbb{R})$, simply because linear maps, while they may not preserve length, preserve ratios of length of collinear segments. Now we leave it to the reader to check that in the direction of the short red diagonal, one cylinder is made up of the triangles 1 and 2, while the other is made of triangles 3, 4, 5, 6, and the ratio of their modules is not 2. This means that no element of the Veech group takes the horizontal direction to the direction of the short red diagonal, so the Veech group, acting on the set of directions of closed geodesics, has at least two orbits, therefore there are at least two cusps.

9.8 Veech Surfaces

When facing a daunting dynamical system, and looking for closed invariant sets, the first thing to do is to look for *small* closed invariant sets: if your dynamical system is an $\mathbb{R}$-action (a flow), you are going to look for fixed points, or periodic orbits. In the case of an action by a larger group, you are going to look for closed (topologically speaking) orbits. If the $GL_2^+(\mathbb{R})$-orbit of a holomorphic differential (X, ω) is closed in $\mathcal{H}_g$, we say (X, ω) is a **Veech surface**.

9.8.1 The Smillie-Weiss Theorem

What Veech was after was Veech groups with finite co-volume. Later it was proved in [44] that having a closed Teichmüller disk is equivalent to having a finite-covolume Veech group.

What follows, while very far from a proof, is meant to convey a bit of the idea.

First, let us assume that for some Abelian differential (X, ω), $\Gamma = SL(X, \omega)$ has finite co-volume in $\mathbb{H}^2$. Then $SL_2(\mathbb{R})/\Gamma$ consists of a compact part K and finitely many cusps $C_1, \ldots, C_n$ (see [19]). Therefore $K.(X, \omega)$ is compact, so it is closed in $\mathcal{H}_g$, and $C_i.(X, \omega)$ is closed in $\mathcal{H}_g$ unless there exists a sequence A_n of matrices in C_i, which converges to infinity in $SL_2(\mathbb{R})$, and such that $A_n.(X, \omega)$ does not go to infinity in $\mathcal{H}_g$. But recall that every cusp of $\mathbb{H}^2/\Gamma$ is stabilized by a parabolic element of the Veech group, which in turns yields a decomposition of X into cylinders of closed geodesics. When A_n goes to infinity, since $A_n \in C_i$, the length of the closed geodesics must go to zero, which means that $A_n.(X, \omega)$ leaves every compact set of $\mathcal{H}_g$.

Conversely, let us assume that for some Abelian differential (X, ω), the orbit $GL_2^+(\mathbb{R}).(X, \omega)$ is closed in $\mathcal{H}_g$. First, let us observe that since $GL_2^+(\mathbb{R}).(X, \omega)$ is closed in $\mathcal{H}_g$, the embedding of $SL_2(\mathbb{R})/\Gamma$ into $\mathcal{H}_g$ is proper (the inverse image of any compact set is compact). This can be seen from the Magic Wand Theorem of [9]: since $GL_2^+(\mathbb{R}).(X, \omega)$ is closed, it must be an embedded submanifold, in particular, it is properly embedded. This means that the ends at infinity of $GL_2^+(\mathbb{R}).(X, \omega)$ in $\mathcal{H}_g$ are exactly the images in $\mathcal{H}_g$ of the ends at infinity of $SL_2(\mathbb{R})/\Gamma$.

Now let us ask ourselves, what kind of ends at infinity a closed Teichmüller disk may have?

The first two examples that come to mind are a hyperbolic funnel (an end at infinity of a quotient of the hyperbolic disk by a hyperbolic element), and a cusp (the thin end a quotient of the hyperbolic disk by a parabolic element). One difference between the two is that the former has infinite volume, while the latter has not. Another difference is that geodesics that go to infinity in a cusp are all asymptotic to each other, while there exists a point x in $\mathbb{H}^2$, and an open interval I of directions such that any geodesic going from x with direction in I goes to infinity in the funnel. Now we apply Theorem 9.4.2: among those directions, there must be a uniquely ergodic one, let us call it θ (meaning that the vertical flow of $e^{i\theta}\omega$ is uniquely ergodic). But then, by Masur's Criterion, geodesics corresponding to uniquely ergodic directions do not go to infinity. This contradiction shows that the ends at infinity of a Teichmüller disk are cusps.

Of course, we also have to prove that there are but finitely many cusps. Note that by Hubert and Schmidt [17], a Veech group may be infinitely generated, but in that case it is not the Veech group of a Veech surface. Here we draw the Magic Wand again: orbits closures are algebraic subvarieties of $\mathcal{H}_g$, so orbit closures of dimension one cannot have infinitely many cusps. Now, since cusps have finite volume, a closed Teichmüller disk is the union of a compact part, and finitely many cusps, so it has finite volume.

9.8.2 The Veech Alternative

The Veech alternative (see [49]) is another great example of sowing in the parameter space and reaping in the phase space. It says that every direction on a Veech surface is either uniquely ergodic, or completely periodic. This means that if (X, ω) is a Veech surface, for every $\theta \in \mathbb{R}$, the vertical flow of $(X, e^{i\theta}\omega)$ is either uniquely ergodic, or every orbit which does not hit a singularity is periodic. Beware this does not mean the flow itself is periodic, for the periods of the orbits may not be commensurable.

Let us give some flavour of the proof. Assume the vertical flow on a Veech surface (X, ω) is not uniquely ergodic, so the orbit of ω under the Teichmüller geodesic flow goes to infinity. Now, since (X, ω) is a Veech surface, the only ends at infinity of its Teichmüller disk are cusps; and a geodesic which goes to infinity in a cusp must be invariant by the parabolic element which generates the cusp.

9.8.2.1 Examples of Veech Surfaces

The first examples found, by Veech [49], are the surfaces obtained by identifying opposite sides of a regular $2n$-gon.

Once people got interested in Veech surfaces, it was immediately realized (see [14]) that square-tiled surfaces are Veech surfaces, just because their Veech groups are (conjugate to) finite index subgroups of $SL_2(\mathbb{Z})$, which has finite co-volume.

In general, finding the Veech surfaces in a given stratum is a hard problem. However, in genus two, it was solved by McMullen in [28–33], see Sect. 9.9.2.

9.9 Classification of Orbits in Genus Two

The Magic Wand, magic as it is, does not tell us everything about the orbit closures. For instance, in most strata, all we know is that

- by ergodicity, almost every orbit is dense
- the union of all orbits of square-tiled surfaces (each of which is closed) is dense.

When are there orbit closures of intermediate dimension? Are there closed orbits (i.e. Veech surfaces) besides those of square-tiled surfaces? Speaking of square-tiled surfaces, how do they distribute in Teichmüller disks? For instance, could it be that each square-tiled surface is the only square-tiled surface in its Teichmüller disk?

The answers are known in genus two: we have a complete list of non-square-tiled Veech surfaces [3, 29], a complete list of orbit closures of intermediate dimension [33], and, in the stratum $\mathcal{H}(2)$, a complete list of orbits of *primitive* (more on this later) square-tiled surfaces [16].

9.9.1 Square-Tiled Surfaces

Observe that a square may be subdivided into little rectangles, and then we may use some matrix in $GL_2(\mathbb{R})$ to square the rectangles, so a square-tiled surface may be square-tiled in many ways.

Following [16], we say a square-tiled surface is primitive if it is not obtained from another square-tiled surface by subdivising squares into rectangles and then applying a matrix to square the rectangles.

Then [16] says that for $n > 4$, there are exactly two Teichmüller disks of primitive n-square surfaces $\mathcal{H}(2)$. For $n = 3, 4$, there is only one Teichmüller disk of primitive n-square surfaces. Each Teichmüller disk contains finitely many primitive n-square surfaces, the exact number is given by the index of the Veech group in $SL_2(\mathbb{Z})$. In genus two it turns out that this index is never 1, however, there exist square-tiled surfaces of higher genus whose Veech group is $SL_2(\mathbb{Z})$ (see "Eierlegende Wollmilchsau" and "Ornythorynque" in [13]). Even in genus two the situation is not fully understood, for instance, we don't know how many Teichmüller disks of primitive n-square surfaces there are in $\mathcal{H}(1, 1)$.

9.9.2 A Homological Detour

9.9.2.1 The Tautological Subspace

Since the complex-valued 1-form ω is holomorphic, the real-valued 1-forms $\Re(\omega)$ and $\Im(\omega)$ are closed, by the Cauchy-Riemann relations. We denote S the 2-dimensional subspace of $H^1(X, \mathbb{R})$ generated by the cohomology classes of $\Re(\omega)$ and $\Im(\omega)$.

Recall that $H^1(X, \mathbb{R})$ has a symplectic structure: let $[X] \in H^2(X, \mathbb{R})$ be the fundamental class, then

$$\forall a, b \in H^1(X, \mathbb{R}), \exists c \in \mathbb{R}, a \wedge b = c\,[X].$$

The symplectic 2-form is then defined as $(a, b) \mapsto c$. It is Poincaré dual to the intersection form (the bilinear form Int on $H_1(X, \mathbb{Z})$ such that $\text{Int}(h, k)$ is the total intersection, counted with signs, of any two representatives of h and k in transverse position) on $H_1(X, \mathbb{R})$, that is, if $P : H_1(X, \mathbb{R}) \longrightarrow H^1(X, \mathbb{R})$ is the Poincaré map, then

$$\forall h, k \in H_1(X, \mathbb{R}), P(h) \wedge P(k) = \text{Int}(h, k)\,[X].$$

The subspace S is symplectic, meaning that the 2-form $. \wedge .$, restricted to $S \times S$, is non-degenerate. Indeed, recall that in an appropriate chart, $\omega = dz = dx + i\,dy$, so

$$\Re(\omega) \wedge \Im(\omega) = dx \wedge dy.$$

The symplectic orthogonal $S^\perp$ of S is the set of cohomology classes c such that $c \wedge \Re(\omega) = c \wedge \Im(\omega) = 0$. We leave it as an exercise for the reader that in a symplectic vector space, the symplectic orthogonal of a symplectic subspace is also symplectic.

Now let us take a look at how the Veech group Γ acts on S. Let γ be an element of Γ, and let $\begin{pmatrix} a & b \\ c & d \end{pmatrix} \in SL_2(\mathbb{R})$ be the matrix of the derivative of γ, seen as a diffeomorphism of X minus its singular set. Let (x, y) be a point in X, in local coordinates, and let (u, v) be a tangent vector to X at (x, y). Then

$$\gamma^*(dx)_{(x,y)}.(u, v) = dx(d\gamma_{(x,y)}.(u, v)) = dx(au + bv, cu + dv) = au + bv$$
$$\gamma^*(dy)_{(x,y)}.(u, v) = dy(d\gamma_{(x,y)}.(u, v)) = dy(au + bv, cu + dv) = cu + dv$$

so the action of Γ by pull-back on S, endowed with the basis $[dx], [dy]$, is given by the same matrix as the linear action of γ on $\mathbb{R}^2$. For this reason S is sometimes called the **tautological subspace** of $H^1(X, \mathbb{R})$.

9.9.2.2   The Trace Field

The **trace field** $K(\Gamma)$ of a subgroup Γ of $SL_2(\mathbb{R})$ is the subfield of $\mathbb{R}$ generated by the traces of the elements of Γ. We are going to see that

Lemma 9.9.1 *When Γ is the Veech group of some translation surface of genus two,* $K(\Gamma) = \mathbb{Q}\left[\sqrt{d}\right]$ *for some $d \in \mathbb{N}^*$.*

Of course, for most surfaces, the Veech group is trivial, so the trace field is $\mathbb{Q}$. The trace field is most interesting when the Veech group is large, especially for Veech surfaces.

It is proved in [14] that for a Veech surface of any genus, having $\mathbb{Q}$ as its trace field is equivalent to being in the Teichmüller disk of a square-tiled surface. In the genus two case, this means that if d is a square, then X is parallelogram-tiled (that is, X lies in the Teichmüller disk of a square-tiled surface).

Take $\gamma \in \Gamma$, and consider the induced automorphism γ^* of $H^1(X, \mathbb{R})$. Since γ^* is induced by a homeomorphism of X, it preserves the integer lattice $H^1(X, \mathbb{Z})$ of $H^1(X, \mathbb{R})$, and the same goes for $(\gamma^*)^{-1}$. Thus the endomorphism $\gamma^* + (\gamma^*)^{-1}$ of $H^1(X, \mathbb{R})$ may be represented by an integer matrix, in particular its eigenvalues are algebraic numbers.

But recall that γ^*, restricted to S, is just γ, so $\gamma^* + (\gamma^*)^{-1}$, restricted to S, is just $\gamma + \gamma^{-1} = Tr(\gamma).Id$. Therefore the trace of γ is a double eigenvalue of $\gamma^* + (\gamma^*)^{-1}$, so the square of the minimal polynomial χ of $Tr(\gamma)$ divides the characteristic polynomial of $\gamma^* + (\gamma^*)^{-1}$, hence χ has degree at most two.

We have just proved that for every $\gamma \in \Gamma$, there exists $d \in \mathbb{N}^*$ such that $Tr(\gamma) \in \mathbb{Q}\left[\sqrt{d}\right]$. The same argument applies to any linear combination of traces of elements

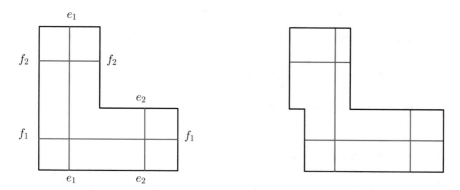

Fig. 9.17 On the left, the surface $L(a, 1)$. The sides e_1 and f_1 have length 1, while the sides e_2 and f_2 have length $a - 1$. The surface decomposes into two cylinders of horizontal closed geodesics, the lower one of width a and height 1, and the upper one of width 1 and height $a - 1$. On the right, the top rectangle in the surface $L(a, 1)$ has been translated to the left. Sides are still identified by vertical or horizontal translations. The resulting surface lies in the stratum $\mathcal{H}(1, 1)$, like $St(4)$, but it has the same absolute periods as $L(a, 1)$. The closed curves with respect to which the periods are computed are drawn in red

of Γ. We claim that d may be chosen independently of γ. If it were not the case, say, $Tr(\gamma) \in \mathbb{Q}\left[\sqrt{d}\right]$ and $Tr(\gamma') \in \mathbb{Q}\left[\sqrt{d'}\right]$ for some $\gamma' \neq \gamma$ and $d' \neq d$, then we could find a non-quadratic element in $K(\Gamma)$ (think, for instance, that $\sqrt{2} + \sqrt{3}$ is not quadratic). Therefore, $K(\Gamma) = \mathbb{Q}\left[\sqrt{d}\right]$ for some $d \in \mathbb{N}^*$.

Example Consider the surface $L(a, 1)$ on the left of Fig. 9.17, where $a = \frac{1+\sqrt{d}}{2}$. For $d = 5$, this surface lies in the Teichmüller curve of the double pentagon, it is usually called the Golden L (see [6, 7]). For $d = 2$, it can be shown to lie in the Teichmüller curve of the regular octagon (see [43]).

The matrix $A := \begin{pmatrix} 1 & 4a \\ 0 & 1 \end{pmatrix}$ acts on the lower cylinder, whose height is 1 and whose width is a, as the 4-th power of the horizontal Dehn twist, because

$$A. \begin{pmatrix} 0 \\ 1 \end{pmatrix} = \begin{pmatrix} 0 \\ 1 \end{pmatrix} + 4 \begin{pmatrix} a \\ 0 \end{pmatrix},$$

and it acts on the upper cylinder, whose height is $a - 1$ and whose width is 1, as the $(d - 1)$-th power of the horizontal Dehn twist, because

$$A. \begin{pmatrix} 0 \\ a - 1 \end{pmatrix} = \begin{pmatrix} 4a(a - 1) \\ a - 1 \end{pmatrix} = \begin{pmatrix} 0 \\ a - 1 \end{pmatrix} + (d - 1) \begin{pmatrix} 1 \\ 0 \end{pmatrix}.$$

Thus the matrix A lies in the Veech group of $L(a, 1)$. The same arguments apply to the vertical cylinders, so the matrix $\begin{pmatrix} 1 & 0 \\ 4a & 1 \end{pmatrix}$ also lies in the Veech group, and so does their product $\begin{pmatrix} 1 + 16a^2 & 4a \\ 4a & 1 \end{pmatrix}$, whose trace is $2 + 16a^2 = 2 + 4d + 8\sqrt{d}$, hence the trace field of $L(a, 1)$ is $\mathbb{Q}\left[\sqrt{d}\right]$.

9.9.2.3   The Jacobian Torus

For any manifold X, the **Jacobian torus** $\mathrm{Jac}(X)$ of X is the quotient of $H^1(X, \mathbb{R})$ by the integer lattice $H^1(X, \mathbb{Z})$. When X is a Riemann surface, $H^1(X, \mathbb{R})$ identifies with the space of Abelian differentials, by sending an Abelian differential to the cohomology class of its real part, and $H^1(X, \mathbb{Z})$ acts by translations, which are holomorphic, so $\mathrm{Jac}(X)$ is actually a complex manifold.

The Jacobian torus also has a symplectic structure, given by the wedge pairing, and the symplectic structure is compatible with the complex structure, in the following way: any complex subspace in $H^1(X, \mathbb{R})$ is also a symplectic subspace. If E is any complex subspace of $H^1(X, \mathbb{R})$, then the symplectic orthogonal $E^\perp$ is also a complex subspace. For instance, the canonical subspace S and its symplectic orthogonal are both complex subspaces of $H^1(X, \mathbb{R})$.

An **endomorphism** of $\mathrm{Jac}(X)$ is a $\mathbb{C}$-linear endomorphism of $H^1(X, \mathbb{R})$, which preserves the integer lattice, so it quotients to a self-map of $\mathrm{Jac}(X)$, and is self-adjoint with respect to the symplectic form. Such a thing actually exists, here is an example.

Assume X is a translation surface of genus two, and γ is an element of the Veech group of X, such that its trace is irrational (hence quadratic since the genus is two). Think of the example we have just seen on the surface $L(a, 1)$.

Then both γ^* and $(\gamma^*)^{-1}$ are $\mathbb{R}$-linear endomorphisms of $H^1(X, \mathbb{R})$ which preserve the integer lattice, so the same goes for $\gamma^* + (\gamma^*)^{-1}$. Recall that $T :=$ $\gamma^* + (\gamma^*)^{-1}$, restricted to S, is just $Tr(\gamma).Id$, so $t := Tr(\gamma)$ is a double eigenvalue of T. Thus the Galois conjugate $\bar{t}$ of t is also a double eigenvalue of T. But both γ^* and $(\gamma^*)^{-1}$ are symplectic, so, since they preserve S, they must preserve the symplectic orthogonal $S^\perp$. Therefore, since the dimension of $S^\perp$ is two (recall that the genus of X is two), we get that T, restricted to $S^\perp$, is $\bar{t}Id$. This proves that T is $\mathbb{C}$-linear, since it may be expressed by the matrix $\begin{pmatrix} t & 0 \\ 0 & \bar{t} \end{pmatrix}$ in some complex basis.

This is the part that breaks down in higher genus, and the reason why McMullen's results in genus two have been dubbed miraculous (see [55]). Let us see exactly how it breaks down. Assume X is a translation surface of genus $g > 2$, with a trace field of degree g (the maximal possible degree). Let γ be an element of the Veech group of X, whose trace is an algebraic integer of degree g. Then the $g - 1$ Galois conjugates of $Tr(\gamma)$ are double eigenvalues of $T = \gamma^* + (\gamma^*)^{-1}$, so $S^\perp$ decomposes as a direct sum of 2-dimensional real subspaces, on each of which

T acts by homothety. The trouble is, those 2-dimensional real subspaces have no reason to be complex subspaces, even though $S^\perp$ is a complex subspace, so we cannot conclude that T is $\mathbb{C}$-linear. In some cases (see [23]) T happens to be $\mathbb{C}$-linear, so McMullen's methods may be applied.

To prove that T is an endomorphism of $\mathrm{Jac}(X)$, we still have to prove that T is self-adjoint with respect to the symplectic form. Since γ is a homeomorphism of X, γ^* preserves the symplectic form, so the adjoint of γ^* is $(\gamma^*)^{-1}$. Similarly the adjoint of $(\gamma^*)^{-1}$ is γ^*, so $T = \gamma^* + (\gamma^*)^{-1}$ is self-adjoint.

9.9.2.4   Real Multiplication

Let X be a translation surface of genus two, with trace field $\mathbb{Q}\left[\sqrt{d}\right]$ for some non-square $d \in \mathbb{N}$. Let γ be an element of the Veech group of X, with irrational trace, and let $T = \gamma^* + (\gamma^*)^{-1}$.

The endomorphisms of $\mathrm{Jac}(X)$ form a ring, and the subring generated by Id and T is isomorphic to a finite index subring of the ring of integers of $\mathbb{Q}\left[\sqrt{d}\right]$, the isomorphism being the trace of the restriction to S. Of course there is another isomorphism, which is the trace of the restriction to $S^\perp$.

For any translation surface, when the ring of endomorphisms of $\mathrm{Jac}(X)$ has a subring isomorphic to (some finite index subring of) the integer ring of some number field K, we say $\mathrm{Jac}(X)$ has **real multiplication** by K. So, what we have proved so far is

Lemma 9.9.2 *The Jacobian of any non-square-tiled Veech surface of genus two has real multiplication by some $\mathbb{Q}\left[\sqrt{d}\right]$, with $d \in \mathbb{N}$ non-square.*

For instance, the double pentagon has real multiplication by $\mathbb{Q}\left[\sqrt{5}\right]$, while the octagon has real multiplication by $\mathbb{Q}\left[\sqrt{2}\right]$. More generally, for any non-square $d \in \mathbb{N}$, the surface $L(a, 1)$, with $a = \frac{1+\sqrt{d}}{2}$, has real multiplication by $\mathbb{Q}\left[\sqrt{d}\right]$.

Now, what is the point of all this, you may ask? The point is that there is a way to know which surfaces have Jacobians with real multiplication, and the stroke of genius was to think of looking at the problem this way.

Let us denote W_d the projection to the moduli space $\mathcal{M}_2$ of the set of translation surfaces of genus two whose Jacobian has real multiplication by $\mathbb{Q}\left[\sqrt{d}\right]$.

Theorem 9.9.3 *For $d \in \mathbb{N}$ non-square, W_d is a two-dimensional algebraic subvariety of $\mathcal{M}_2$.*

Proof We will focus on the dimension, and refer the reader to [28] for the algebraicity statement. Let us forget about Jacobians for a moment, and consider the set of all complex tori of dimension two (also known, poetically, as principally polarized Abelian varieties). This set may be viewed as the set of all complex structures on $\mathbb{R}^4/\mathbb{Z}^4$, which are compatible with the symplectic structure on $\mathbb{Z}^4$, that is, denoting $(.,.)$ the canonical scalar product and $. \wedge .$ the symplectic structure, for any $a, b, a \wedge b = (a, ib)$.

Among all tori, let us consider those whose endomorphism ring contains a copy of (a finite index subring of) the integer ring of $\mathbb{Q}\left[\sqrt{d}\right]$. Let ϕ be a generator of said subring, which then is the $\mathbb{Z}$-module of rank two $\mathbb{Z}[\phi] = \{a + b\phi : a, b \in \mathbb{Z}\}$. The embedding of $\mathbb{Z}[\phi]$ into the ring of endomorphisms of $\mathbb{Z}^4$ makes $\mathbb{Z}^4$ a $\mathbb{Z}[\phi]$-module, necessarily of rank two. Tensorizing by $\mathbb{Q}$, we may view $\mathbb{Q}^4$ as a two-dimensional vector space over $\mathbb{Q}\left[\sqrt{d}\right] = \mathbb{Q}[\phi]$.

Let $\bar{\phi}$ be the Galois conjugate of ϕ, so $P_\phi = (X - \phi)(X - \bar{\phi})$ is the minimal polynomial of ϕ, and let T_ϕ be the endomorphism of $\mathbb{Q}^4$ associated with ϕ. Then $P_\phi(T_\phi) = 0$. Since T preserves $\mathbb{Z}^4$, $T_\phi \neq \phi Id$, so T_ϕ has both ϕ and $\bar{\phi}$ as eigenvalues, which entails that, as an endomorphism of a two-dimensional $\mathbb{Q}[\phi]$-vector space, T_ϕ is diagonalizable. Both eigenspaces have dimension one over $\mathbb{Q}[\phi]$, which is dimension two over $\mathbb{Q}$. Since T is self-adjoint, the eigenspaces are mutually orthogonal, and they are symplectic, for if they were not, since their dimension is two, then the restriction of the symplectic form to either of them would vanish, and since they are orthogonal, the symplectic form on $\mathbb{Z}^4$ would be zero, a contradiction.

The complex structure on $\mathbb{R}^4/\mathbb{Z}^4$ is then completely determined by the requirement that the eigenspaces be complex lines, and by the choice of a complex structure for each eigenspace, compatible with the symplectic structure. Each choice is determined by one complex parameter (recall our discussion of the Teichmüller space, which is also the Teichmüller disk, of the torus), so the complex structure on $\mathbb{R}^4/\mathbb{Z}^4$ is determined by two complex parameters.

Next, observe that almost all complex tori are Jacobians: essentially, those who are not Jacobians of regular surfaces, are Jacobians of degenerate surfaces, for instance a bouquet of two 1-dimensional tori. So the set of Jacobians of regular surfaces has dimension two over $\mathbb{C}$. □

McMullen's result actually says much more: the image, in the Teichmüller space $\mathcal{T}_2$, of the set of all translation surfaces of genus two whose Jacobian has real multiplication by $\mathbb{Q}\left[\sqrt{d}\right]$, for $d \in \mathbb{N}$ non-square, is an holomorphically embedded copy of $\mathbb{H}^2 \times \mathbb{H}^2$. Its image W_d in the moduli space $\mathcal{M}_2$ is called a **Hilbert modular surface**.

Now we'd like to understand how Hilbert modular surfaces behave with respect to the $SL_2(\mathbb{R})$-action, and the strata.

Lemma 9.9.4 *For any non-square $d \in \mathbb{N}$, W_d is $SL_2(\mathbb{R})$-invariant.*

Proof Let (X, ω) be a translation surface of genus two whose Jacobian has real multiplication by $\mathbb{Q}\left[\sqrt{d}\right]$, for $d \in \mathbb{N}$ non-square, let T be an endomorphism of the Jacobian of X, and let γ be an element of $SL_2(\mathbb{R})$. Then $(\gamma^*)^{-1}$ conjugates T with an endomorphism of the Jacobian of $\gamma.(X, \omega)$, so $\gamma.(X, \omega)$ has real multiplication by $\mathbb{Q}\left[\sqrt{d}\right]$. $\qquad\square$

Recall that we denote by P the canonical projection $\mathcal{H}_2 \longrightarrow \mathcal{M}_2$.

Lemma 9.9.5 *The intersection $P(\mathcal{H}(2)) \cap W_d$ is not empty.*

Proof Consider the surface $L(a, 1)$, with $a = \frac{1+\sqrt{d}}{2}$. $\qquad\square$

Lemma 9.9.6 *The intersection $P(\mathcal{H}(1, 1)) \cap W_d$ is not empty.*

Proof Consider the surface on the right of Fig. 9.17, obtained from $L(a, 1)$ by shifting the upper cylinder to the left. This surface has the same absolute periods as $L(a, 1)$, hence it has the same Jacobian, since Jacobians ignore relative periods. Hence it lies in W_d. $\qquad\square$

By Lemmata 9.9.4 and 9.9.5, $P(\mathcal{H}(2)) \cap W_d$ is a reunion of $\mathrm{SL}_2(\mathbb{R})$-orbits, so its dimension is at least one. If its dimension were two, since two is the dimension of W_d, and the latter is connected as a quotient of $\mathbb{H}^2 \times \mathbb{H}^2$, then $\mathcal{H}(2) \cap W_d$ would be the whole of W_d. But this is impossible by Lemma 9.9.6. So the dimension of $P(\mathcal{H}(2)) \cap W_d$ is one, and since it is an algebraic subvariety of W_d, it must be a finite union of closed orbits. There, we have found Veech surfaces in $\mathcal{H}(2)$, with trace field $\mathbb{Q}\left[\sqrt{d}\right]$ for any non-square $d \in \mathbb{N}$, and proved that there are only finitely many of them, for each d. We have also proved that the surface $L(a, 1)$ is a Veech surface for $a = \frac{1+\sqrt{d}}{2}$, just by computing its trace field.

Again, McMullen's results actually say much more: the intersection $P(\mathcal{H}(2)) \cap W_d$ contains exactly one orbit, except when $d = 1 \mod 8$, $d \neq 9$, in which case there are exactly two (see [29]). The intersection $P(\mathcal{H}(1, 1)) \cap W_d$ contains no closed orbit, except when $d = 5$, in which case it contains exactly one, the Teichmüller curve of the regular decagon (see [30, 31]). An orbit closure is either the orbit itself, in which case it is a Teichmüller curve, of which we have a complete list, or W_d for some d, or a whole stratum (see [33]). Note that in genus two, both strata project surjectively to the moduli space $\mathcal{M}_2$ (see [12], III.7.5, Corollary 1), so an orbit which is dense in its stratum projects to a dense subset of $\mathcal{M}_2$.

9.10 What Is Known in Higher Genus?

First, a bit of vocabulary: a Veech surface is said to be **geometrically primitive** if it is not a ramified cover of a Veech surface of lower genus. For instance, square-tiled surfaces, other than tori, are not geometrically primitive since they cover the torus. Veech's family (the regular polygons), as well as Ward's (see [51]), are geometrically primitive.

A Veech surface is said to be **algebraically primitive** if its trace field has maximal possible degree, which is the genus of the surface. McMullen's family in genus 2 is both algebraically and geometrically primitive. In Veech's family (the regular polygons), as well as in Ward's (see [51]), infinitely many, but not all, are algebraically primitive. For instance, the regular $2n$-gon, or the double n-gon, are primitive if n is prime. On the other hand, the regular 20-gon, which has genus five, has a trace field of degree 4, because $\cot \frac{\pi}{20}$ is algebraic of degree 4. See [52], Theorem 1.7, for a precise statement.

The general philosophy of Veech surface hunters seems to be that algebraically and geometrically primitive Veech surfaces are scarce, and you should expect at most finitely many of them in a given stratum of genus > 2. In [27] this very statement is proved, for the minimal stratum of each genus > 2. In fact, in [10], it is proved that in any stratum, there are at most finitely many Veech surfaces with a trace field of degree > 2.

In [2] a family of Veech surfaces is found, which generalizes Veech's family. In [23, 32], the authors find a way around the fact that McMullen's techniques do not generalize in higher genus, and discover a family of Veech surfaces in genera 3 and 4, which generalizes McMullen's family, in the sense that their trace fields are quadratic (so they are not algebraically primitive). In [11, 34], another family of Veech surfaces in genus 4 is found, again with quadratic trace fields.

References

1. P. Arnoux, Le codage du flot géodésique sur la surface modulaire. Enseign. Math. (2) **40**(1–2), 29–48 (1994)
2. I. Bouw, M. Möller, Teichmüller curves, triangle groups, and Lyapunov exponents. Ann. Math. **172**, 139–185 (2010)
3. K. Calta, Veech surfaces and complete periodicity in genus two. J. Am. Math. Soc. **17**(4), 871–908 (2004)
4. S. Cheboui, Intersection algébrique sur les surfaces à petits carreaux, Ph.D. Thesis, Université de Montpellier, 2021
5. S. Cheboui, A. Kessi, D. Massart, Algebraic intersection in a family of Teichmüller disks, preprint. https://arxiv.org/abs/2007.1084
6. D. Davis, S. Lelièvre, Periodic paths on the pentagon, double pentagon and golden L. Preprint arXiv:1810.11310
7. D. Davis, D. Fuchs, S. Tabachnikov, Periodic trajectories in the regular pentagon. Moscow Math. J. **11**(3), 1–23 (2011)
8. V. Delecroix et al., Surface Dynamics - SageMath package, Version 0.4.1 (2019). https://doi.org/10.5281/zenodo.3237923
9. A. Eskin, M. Mirzakhani, Invariant and stationary measures for the $SL(2, \mathbb{R})$ action on moduli space. Publ. Math. Inst. Hautes Études Sci. **127**, 95–324 (2018)
10. A. Eskin, S. Filip, A. Wright, The algebraic hull of the Kontsevich-Zorich cocycle. Ann. Math. (2) **188**(1), 281–313 (2018)
11. A. Eskin, C. McMullen, R. Mukamel, A. Wright, Billiards, quadrilaterals and moduli spaces. J. Am. Math. Soc. **33**(4), 1039–1086 (2020)
12. H. Farkas, I. Kra, *Riemann Surfaces*, 2nd edn. Graduate Texts in Mathematics, vol. 71 (Springer, New York, 1992)

13. G. Forni, C. Matheus, Introduction to Teichmüller theory and its applications to dynamics of interval exchange transformations, flows on surfaces and billiards. J. Mod. Dyn. **8**(3–4), 271–436 (2014)
14. E. Gutkin, C. Judge, Affine mappings of translation surfaces: geometry and arithmetic. Duke Math. J. **103**(2), 191–213 (2000)
15. J. Hubbard, *Teichmüller Theory and Applications to Geometry, Topology, and Dynamics.* Teichmüller Theory, vol. 1 (Matrix Editions, Ithaca, NY, 2006)
16. P. Hubert, S. Lelièvre, Prime arithmetic Teichmüller discs in $\mathcal{H}(2)$. Israel J. Math. **151**, 281–321 (2006)
17. P. Hubert, T. Schmidt, Infinitely generated Veech groups. Duke Math. J. **123**(1), 49–69 (2004)
18. P. Hubert, T. Schmidt, *An Introduction to Veech Surfaces,* ed. by B. Hasselblatt, A. Katok. Handbook of Dynamical Systems, vol. 1B (Elsevier B. V., Amsterdam, 2006), pp. 501–526 (short version of the lecture notes from the summer school at Luminy, 2003, https://www. math.uchicago.edu/~masur/hs.pdf)
19. S. Katok, *Fuschian Groups* (University of Chicago Press, Chicago, 1992)
20. S. Kerckhoff, H. Masur, J. Smillie, Ergodicity of billiard flows and quadratic differentials. Ann. Math. (2) **124**(2), 293–311 (1986)
21. E. Lanneau, Promenade dynamique et combinatoire dans les espaces de Teichmüller, mémoire d'habilitation. https://www-fourier.ujf-grenoble.fr/~lanneau/articles/hdr-lanneau.pdf
22. E. Lanneau, D. Massart, Algebraic intersection in regular polygons. https://arxiv.org/abs/2110. 14235
23. E. Lanneau, D. Nguyen, Teichmüller curves generated by Weierstrass Prym eigenforms in genus 3 and genus 4. J. Topol. **7**(2), 475–522 (2014)
24. H. Masur, Interval exchange transformations and measured foliations. Ann. Math. (2) **115**(1), 169–200 (1982)
25. H. Masur, Hausdorff dimension of the set of nonergodic foliations of a quadratic differential. Duke Math. J. **66**(3), 387–442 (1992)
26. H. Masur, S. Tabachnikov, *Rational Billiards and Flat Structures,* ed. by B. Hasselblatt, A. Katok. Handbook of Dynamical Systems, vol. 1A (North-Holland, Amsterdam, 2002), pp. 1015–1089
27. C. Matheus, A. Wright, Hodge-Teichmüller planes and finiteness results for Teichmüller curves. Duke Math. J. **164**(6), 1041–1077 (2015)
28. C. McMullen, Billiards and Teichmüller curves on Hilbert modular surfaces. J. Am. Math. Soc. **16**(4), 857–885 (2003)
29. C. McMullen, Teichmüller curves in genus two: Discriminant and spin. Math. Ann. **333**, 87–130 (2005)
30. C. McMullen, Teichmüller curves in genus two: the decagon and beyond. J. Reine Angew. Math. **582**, 173–199 (2005)
31. C. McMullen, Teichmüller curves in genus two: torsion divisors and ratios of sines. Invent. Math. **165**(3), 651–672 (2006)
32. C. McMullen, Prym varieties and Teichmüller curves. Duke Math. J. **133**(3), 569–590 (2006)
33. C. McMullen, Dynamics of $SL_2(\mathbb{R})$ over moduli space in genus two. Ann. Math. (2) **165**(2), 397–456 (2007)
34. C. McMullen, R. Mukamel, A. Wright, Cubic curves and totally geodesic subvarieties of moduli space. Ann. Math. (2) **185**(3), 957–990 (2017)
35. M. Möller, *Affine Groups of Flat Surfaces,* ed. by A. Papadopoulos, in Handbook of Teichmüller Theory, vol. II (European Mathematical Society (EMS), Zürich, 2009), pp. 369–387
36. T. Monteil, Introduction to the theorem of Kerckhoff, Masur and Smillie Oberwolfach Report, No. 17, 989–994 (2010). https://lipn.univ-paris13.fr/~monteil/papiers/fichiers/kms.pdf
37. D. Mumford, A remark on Mahler's compactness theorem. Proc. Am. Math. Soc. **28**(1), 289–294 (1971)
38. B. Riemann, Theorie der Abel'schen Functionen. J. Reine Angew. Math. **54**, 115–155 (1857)

39. B. Riemann, *Gesammelte mathematische Werke, wissenschaftlicher Nachlass und Nachträge* (Springer-Verlag/BSB B. G. Teubner Verlagsgesellschaft, Berlin/Leipzig, 1990)
40. H. Royden, *Invariant Metrics on Teichmüller Space*. Contributions to Analysis (a collection of papers dedicated to Lipman Bers) (Academic Press, New York, 1974), pp. 393–399
41. G. Schmithüsen, An algorithm for finding the Veech group of an origami. Exp. Math. **13**(4), 459–472 (2004)
42. G. Schmithüsen, Examples for Veech groups of origamis, in *The Geometry of Riemann Surfaces and Abelian Varieties*. Contemp. Math., vol. 397 (American Mathematical Society, Providence, RI, 2006), pp. 193–206
43. J. Smillie, C. Ulcigrai, Geodesic flow on the Teichmüller disk of the regular octagon, cutting sequences and octagon continued fractions maps, in *Dynamical Numbers-interplay Between Dynamical Systems and Number Theory*. Contemp. Math., vol. 532 (American Mathematical Society, Providence, RI, 2010), pp. 29–65
44. J. Smillie, B. Weiss, Characterizations of lattice surfaces. Invent. Math. **180**(3), 535–557 (2010)
45. O. Teichmüller, Extremale quasikonforme Abbildungen und quadratische Differentiale. Abh. Preuss. Akad. Wiss. Math.-Nat. Kl. **1939**(22), 197 pp. (1940)
46. O. Teichmüller, *Extremal Quasiconformal Mappings and Quadratic Differentials*. Translated from the German by Guillaume Théret. IRMA Lect. Math. Theor. Phys., vol. 26, ed. by A. Papadopoulos. Handbook of Teichmüller Theory, vol. V (European Mathematical Society, Zürich, 2016), pp. 321–483
47. W. Veech, Gauss measures for transformations on the space of interval exchange maps. Ann. Math. (2) **115**(1), 201–242 (1982)
48. W. Veech, The Teichmüller geodesic flow. Ann. Math. (2) **124**(3), 441–530 (1986)
49. W. Veech, Teichmüller curves in moduli space, Eisenstein series and an application to triangular billiards. Invent. Math. **97**(3), 553–583 (1989)
50. W. Veech, Moduli spaces of quadratic differentials. J. Analyse Math. **55**, 117–171 (1990)
51. C. Ward, Calculation of Fuchsian groups associated to billiards in a rational triangle. Ergodic Theory Dynam. Syst. **18**(4), 1019–1042 (1998)
52. A. Wright, Schwarz triangle mappings and Teichmüller curves: the Veech-Ward-Bouw-Möller curves. Geom. Funct. Anal. **23**(2), 776–809 (2013)
53. A. Wright, Translation surfaces and their orbit closures: an introduction for a broad audience. EMS Surv. Math. Sci. **2**(1), 63–108 (2015)
54. A. Zemljakov, A. Katok, Topological transitivity of billiards in polygons. Mat. Zametki **18**(2), 291–300 (1975)
55. A. Zorich, Flat surfaces, in *Frontiers in Number Theory, Physics, and Geometry*, vol. I (Springer, Berlin, 2006), pp. 437–583
56. A. Zorich, Le théorème de la baguette magique de A. Eskin et M. Mirzakhani. Gaz. Math. **142**, 39–54 (2014)

Chapter 10
Teichmüller Spaces and the Rigidity of Mapping Class Group Actions

Ken'ichi Ohshika

Abstract In this chapter, we survey the rigidity of mapping class group actions on Teichmüller spaces and curve complexes. We start from a classical result of Royden on the rigidity of the mapping class group action on Teichmüller space, together with its infinitesimal version. We then present Ivanov's rigidity theorem on the mapping class group action on the curve complex, and show that this gives an alternative proof of Royden's theorem. In the final part, we touch upon a recent result by Huang–Ohshika–Papadopoulos on the infinitesimal rigidity of the mapping class group action on Teichmüller space with Thurston's asymmetric metric.

Keywords Teichmüller space · Teichmüller metric · Thurston's metric · Rigidity · Infinitesimal rigidity · Finsler structure · Mapping class group · Curve complex

1991 Mathematics Subject Classification 30F60, 57K20

The notion of Teichmüller space was first introduced by Teichmüller [29] as a space of quasi-conformal deformations of a Riemann surface S, which serves as a basepoint, although the same space had already been considered by Fricke–Klein [9] (see also [5]) from a different viewpoint regarding it as a space of representations into $\mathrm{PSL}_2\mathbb{R}$. Formally, a point in the Teichmüller space is a pair of a Riemann surface Σ and an orientation-preserving homeomorphism from the base topological surface $f : S \to \Sigma$. Two such pairs $(\Sigma_1, f_1), (\Sigma_2, f_2)$ are regarded as the same when there is a holomorphic bijection from Σ_1 to Σ_2 which is homotopic to $f_2 \circ f_1^{-1}$. Teichmüller showed that within a given homotopy class of homeomorphisms between Riemann surfaces, there exists a unique most efficient quasi-conformal homeomorphism. He defined the half of the logarithm

K. Ohshika (✉)
Department of Mathematics, Faculty of Science, Gakushuin University, Toshima-ku, Tokyo, Japan
e-mail: ohshika@math.gakushuin.ac.jp

© The Author(s), under exclusive license to Springer Nature Switzerland AG 2022
A. Papadopoulos (ed.), *Surveys in Geometry I*,
https://doi.org/10.1007/978-3-030-86695-2_10

of the maximal dilatation of this most efficient quasi-conformal homeomorphism homotopic to $f_2 \circ f_1^{-1}$ to be the distance between the two points (Σ_1, f_1) and (Σ_2, f_2). Thus, Teichmüller space was equipped with a metric coming from the dilatation constant. This metric, which is now usually called the Teichmüller metric, is known not to be a Riemannian metric, but a Finsler metric. In the first section of this article, we shall give a fairly detailed account on the theory of quasi-conformal mappings and how the Teichmüller distance is defined.

The mapping class group acts naturally on Teichmüller space, by taking (Σ, f) to $(\Sigma, f \circ g^{-1})$ for a mapping class $[g]$. The topic of the second section is the rigidity of this action, which is due to Royden [25]. He showed that every isometry of Teichmüller space is induced from a (unique unless the genus is 2) action of an extended mapping class. He obtained this by showing a stronger theorem, which should be called the infinitesimal rigidity, saying that every complex-linear isometry between cotangent spaces of Teichmüller space is a scalar multiple of the action of a mapping class. We shall explain this theorem and sketch how it was proved in Sect. 10.2.

There is a simplicial complex, called the curve complex and introduced by Harvey [12], which is a combinatorial analogue of Teichmüller space. The curve complex $\mathcal{CC}(S)$ of a closed surface S of genus greater than 1 is defined to be a simplicial complex whose vertices are essential simple closed curves on S and where $n + 1$ vertices span an n-simplex when they are represented by disjoint simple closed curves. The extended mapping class group of S acts on $\mathcal{CC}(S)$ by simplicial automorphisms. Ivanov [15] proved that these are the only simplicial automorphisms of $\mathcal{CC}(S)$. We shall explain how Ivanov proved this in Sect. 10.3.

Ivanov's theorem can be regarded as a combinatorial analogue of Royden's theorem. In the following section, we shall describe Ivanov's alternative approach to Royden's theorem making use of his theorem on curve complexes. We give a fairly detailed account of his argument which appeared in [16].

In the last section, we shall deal with another metric on Teichmüller space, which was introduced by Thurston [30] and which is called Thurston's asymmetric metric. This metric, which is not symmetric, is defined using hyperbolic metrics instead of complex structures on Riemann surfaces. The distance is defined using Lipschitz maps instead of quasi-conformal maps. The unit spheres in cotangent spaces of Teichmüller space in this case are identified with the projective lamination space, which was introduced by Thurston in his work on compactification of Teichmüller space. At the end of the paper, we shall touch upon recent work of Huang–Ohshika–Papadopoulos [13] on the infinitesimal rigidity of Teichmüller space with Thurston's asymmetric metric, which is an analogue of Royden's infinitesimal rigidity theorem in the setting of Thurston's asymmetric metric.

This article is based on a lecture course given by the author at the CIMPA conference on "Finsler geometry and its applications", which took place in Banaras Hindu University in India from 5 to 15 December 2019. The author would like to express his sincere gratitude to the organisers, Athanase Papadopoulos and Bankteshwar Tiwari, and the CIMPA for providing him with this marvellous occasion to give a series of lectures to promising young mathematicians, and to the BHU for its cordial hospitality.

10.1 Teichmüller Metric

In this section, we review the definition of the Teichmüller metric on Teichmüller space, and its properties. We first recall the formal definition of Teichmüller space, upon which we have already touched in the introduction, without giving a topology or a metric for the moment.

Definition 10.1.1 Let S be a closed oriented surface (of topologically finite type) of genus greater than 1. (Throughout this paper, we always assume S to be such a surface of genus $g \geq 2$ although we can consider Teichmüller spaces for more general surfaces.) We consider the set of pairs (Σ, f), where Σ is a Riemann surface and f a homeomorphism from S to Σ. We identify two such pairs (Σ_1, f_1), (Σ_2, f_2) when there is a holomorphic bijection from Σ_1 to Σ_2 which is homotopic to $f_2 \circ f_1^{-1}$. The quotient set is defined to be the *Teichmüller space* of S as a set, and is denoted by $\mathcal{T}(S)$.

10.1.1 Quasi-Conformal Homeomorphisms

To define a metric on $\mathcal{T}(S)$, we next introduce the notion of quasi-conformal map.

Definition 10.1.2 Let U and V be open regions in the complex plane $\mathbb{C}$, and K a real number greater than or equal to 1. An orientation-preserving homeomorphism $f : U \to V$ is called a *K-quasi-conformal map* when the following two conditions are satisfied.

(1) f is ACL (absolutely continuous on lines), that is, on almost every line parallel to either the real axis or the imaginary axis, f is absolutely continuous. This implies that f regarded as a two variable function setting $z = x + iy$ has partial derivatives $\dfrac{\partial f}{\partial x}$ and $\dfrac{\partial f}{\partial y}$ almost everywhere.

(2) Setting f_z to be $\dfrac{1}{2}(\dfrac{\partial f}{\partial x} - i\dfrac{\partial f}{\partial y})$, and $f_{\bar{z}}$ to be $\dfrac{1}{2}(\dfrac{\partial f}{\partial x} + i\dfrac{\partial f}{\partial y})$, we have

$$|f_{\bar{z}}(z)| \leq \frac{K-1}{K+1}|f_z(z)|$$

for almost every $z \in U$.

We note that $f_{\bar{z}} = 0$ is equivalent to the Cauchy–Riemann equation. Weyl's lemma implies that f is a 1-quasi-conformal map if and only it is a holomorphic bijection.

A direct calculation shows the following.

Lemma 10.1.3 *Suppose that $f : U \to V$ is differentiable at $z \in U$. Then $Jf(z) = |f_z(z)|^2 - |f_{\bar{z}}(z)|^2$, where J denotes the Jacobian.*

If f is orientation preserving and differentiable at z, then $Jf(z) > 0$. Therefore, we get the following.

Corollary 10.1.4 *Suppose that $f : U \to V$ is an ACL homeomorphism and orientation preserving. Then $|f_z(z)| > |f_{\bar{z}}|$ for every z for which these partial derivatives are defined.*

We are now going to look at the geometric meaning of K-quasi-conformality. Suppose that f is totally differentiable as a function from $\mathbb{R}^2$ to $\mathbb{R}^2$ at $z = x + iy$, and set $f(x + iy) = u(x, y) + iv(x, y)$. Then as a two-variable function, we have

$$
Df(x, y) = \begin{pmatrix} \frac{\partial u}{\partial x}(x, y) & \frac{\partial u}{\partial y}(x, y) \\ \frac{\partial v}{\partial x}(x, y) & \frac{\partial v}{\partial y}(x, y) \end{pmatrix}.
$$

Now on the unit circle of the tangent space at (x, y), the vector $\begin{pmatrix} \cos\theta \\ \sin\theta \end{pmatrix}$ is sent by $Df(x, y)$ to $\begin{pmatrix} \frac{\partial u}{\partial x}(x, y)\cos\theta + \frac{\partial u}{\partial y}(x, y)\sin\theta \\ \frac{\partial v}{\partial y}(x, y)\cos\theta + \frac{\partial v}{\partial y}(x, y)\sin\theta \end{pmatrix}.$

On the other hand, we have

$$
\frac{1}{2}(f_z(z)e^{i\theta} + f_{\bar{z}}(z)e^{-i\theta})
$$
$$
= \frac{\partial u}{\partial x}(x, y)\cos\theta + i\frac{\partial v}{\partial x}(x, y)\cos\theta + \frac{\partial u}{\partial y}(x, y)\sin\theta + i\frac{\partial v}{\partial y}(x, y)\sin\theta,
$$

and it turns out that the right-hand side is equal to the complex expression of the vector above. Now, express $f_z(z)$ and $f_{\bar{z}}(z)$ by polar coordinates as $f_z(z) = |f_z(z)|e^{i\alpha}$, $f_{\bar{z}}(z) = |f_{\bar{z}}(z)|e^{i\beta}$. Then we have $f_z(z)e^{i\theta} + f_{\bar{z}}(z)e^{-i\theta} = |f_z(z)|e^{i(\alpha+\theta)} + |f_{\bar{z}}(z)|e^{i(\beta-\theta)}$.

This expression implies that fixing z and moving θ, the absolute value $|f_z(z)e^{i\theta} + f_{\bar{z}}(z)e^{-i\theta}|$ takes the maximum $|f_z(z)| + |f_{\bar{z}}(z)|$ when $\alpha + \theta = \beta - \theta \mod 2\pi$, i.e. when $\theta = \dfrac{\alpha - \beta}{2}$ or $\theta = \dfrac{\alpha - \beta}{2} + \pi$, and the minimum $|f_z(z)| - |f_{\bar{z}}(z)|$ when $\alpha + \theta = \beta - \theta + \pi \mod 2\pi$, i.e. when $\theta = \dfrac{\alpha - \beta}{2} \pm \dfrac{\pi}{2}$. Therefore, the differential $Df(z)$ sends the unit tangent circle to an ellipse whose excentricity is equal to $\dfrac{|f_z(z)| + |f_{\bar{z}}(z)|}{|f_z(z)| - |f_{\bar{z}}(z)|}$, which is called the *dilatation* of f at z, and the eigenvectors are orthogonal. Thus the condition that f is K-quasi-conformal is equivalent to the condition that the inequality

$$\frac{|f_z(z)| + |f_{\bar{z}}(z)|}{|f_z(z)| - |f_{\bar{z}}(z)|} \leq K$$

holds almost everywhere.

Now we introduce a complex version of the dilatation.

Definition 10.1.5 For an ACL map $f : U \to V$, we define the *complex dilatation* or the *Beltrami differential* $\mu_f(z)$ of f by $\mu_f(z) := \dfrac{f_{\bar{z}}(z)}{f_z(z)}$. Then we have $\mu(z) < 1$ for almost every z by Corollary 10.1.4. We can see that the dilatation of f defined above is equal to $K_f(z) = \dfrac{1 + |\mu_f(z)|}{1 - |\mu_f(z)|}$. The map f is K-quasi-conformal if and only if $K_f(z) \leq K$ holds for almost every z.

Definition 10.1.6 Suppose that f as above is K-quasi-conformal. We define its *maximal dilatation* to be $K(f) := \text{ess. sup}_{z \in U} K_f(z)$.

A natural problem arising from this definition is the following: Given two Jordan domains U and V, and the same number of points $P_1, \ldots P_k \in \partial U$ and $Q_1, \ldots, Q_k \subset \partial V$ lying there in the clockwise order, find a quasi-conformal homeomorphism from U to V taking P_j to Q_j for every $j = 1, \ldots k$, which realises the smallest possible maximal dilatation.

Grötzsch gave a complete answer as follows to this problem in the case when U and V are rectangles and $P_1, \ldots, P_4; Q_1, \ldots, Q_4$ are their vertices. (See Example 1 of Grötzsch [10], its English translation [11], and the very informative comments by Alberge-Papadopoulos [2].) Consider two rectangles, R of width a and height b, and R' of width a' and height b', and put them on the complex plane so that the bases lie on the real axis and the lower left vertices are at the origin. Then there is an affine map A from R to R' defined by

$$A(z) = \frac{1}{2}\left(\frac{a'}{a} + \frac{b'}{b}\right) z + \frac{1}{2}\left(\frac{a'}{a} - \frac{b'}{b}\right) \bar{z}.$$

Grötzsch proved that this map $A(z)$ attains the smallest possible maximal dilatation, and its value can be easily calculated:

$$K_A = \max\left\{\frac{a'b}{ab'}, \frac{ab'}{a'b}\right\}.$$

His proof, which is rather short, relies on what is now called the "length-area method". This result of Grötzch can be regarded as the starting point of Teichmüller's work.

To close this subsection, we present a lemma which is often used in Teichmüller theory.

Lemma 10.1.7 *Let $f: U \to V$ and $g: V \to W$ be quasi-conformal maps. Then we have $K(f^{-1}) = K(f)$ and $K(g \circ f) \leq K(f)K(g)$. In particular, if g is conformal, then $K(g \circ f) = K(f)$, and if f is conformal, then $K(g \circ f) = K(g)$.*

An intuitive proof of this lemma can be given easily, although a formal one is not so straightforward. For generic points of $z \in U$ and $w \in V$, we have $K_f(z) \leq K(f)$ and $K_g(w) \leq K(g)$. As was explained before, $K_f(z)$ is equal to the distortion, that is, the ratio of the smaller eigenvalue to the larger eigenvalue, of df on the unit circle in the tangent space at z. Therefore, we should have $K_{f^{-1}}(f(z)) = K_f(z)$ and $K_{gf}(z) \leq K_g(f(z))K_f(z)$, and the first statement is obtained by taking the essential supremum. The second statement also follows immediately.

10.1.2 Teichmüller Distance

Now, we turn to Teichmüller's work and see how it gives rise to a distance function on Teichmüller space. Teichmüller's original work is contained in [26, 27], whose English translations, by Théret and A'Campo-Neuen respectively, can be found in [28, 29]. Although these papers are known to be very hard to read, the commentaries written by Alberge–Papadopoulos–Su and by A'Campo-Neuen–A'Campo–Alberge–Papadopoulos [1, 3] serve as very useful guides.

We first give the definition of quasi-conformal maps between Riemann surfaces.

Definition 10.1.8 Let R and R' be two homeomorphic Riemann surfaces. An orientation-preserving homeomorphism $f: R \to R'$ is said to be K-quasi-conformal if for every $z \in R$ and for all local coordinates ϕ around z for R and ψ around $f(z)$ for R', we have $K_{\psi \circ f \circ \phi^{-1}}(z) \leq K$. We note that by Lemma 10.1.7, this definition does not depend on the choice of local coordinates for R and R'. The infimum of K such that f is K-quasi-conformal is denoted by $K(f)$.

Teichmüller introduced the following distance on Teichmüller space.

Definition 10.1.9 For two points $x_1 = (\Sigma_1, f_1), x_2 = (\Sigma_2, f_2)$ in the Teichmüller space $\mathcal{T}(S)$, we define their (Teichmüller) distance by

$$d_T(x_1, x_2) := \inf_{g \simeq f_2 \circ f_1^{-1}} \frac{1}{2} \log K(g).$$

Lemma 10.1.10 *This function $d_T: \mathcal{T}(S) \times \mathcal{T}(S) \to \mathbb{R}$ defines a metric on $\mathcal{T}(S)$.*

Proof Before checking that the axioms of a metric hold, we note that d_T takes finite values. Indeed, if we choose a diffeomorphism from Σ_1 to Σ_2 homotopic to $f_2 \circ f_1^{-1}$, which is known to be always possible, its maximal dilatation is finite since we are assuming S to be compact and since the dilatation function is continuous.

Now we start to check the axioms. Suppose that $d_T(x_1, x_2) = 0$. Then there is a sequence of quasi-conformal homeomorphisms $g_n: \Sigma_1 \to \Sigma_2$ homotopic to

$f_2 \circ f_1^{-1}$ with $K(g_n) \longrightarrow 1$. By a standard but not so trivial argument, we can show that $\{g_n\}$ converges to a conformal homeomorphism.

The axiom of symmetry and the triangle inequality can be derived from Lemma 10.1.7. □

10.1.3 Teichmüller Maps

Teichmüller proved in [29] that between two points in the Teichmüller space, there is a unique quasi-conformal homeomorphism realising their distance. We now state this result formally.

Theorem 10.1.11 *For any two points* $(\Sigma_1, f_1), (\Sigma_2, f_2) \in \mathcal{T}(S)$*, there is a unique quasi-conformal homeomorphism* $g \colon \Sigma_1 \to \Sigma_2$ *homotopic to* $f_2 \circ f_1^{-1}$ *such that for any quasi-conformal homeomorphism* $h \colon \Sigma_1 \to \Sigma_2$ *homotopic to* $f_2 \circ f_1^{-1}$*, we have* $K(g) \leq K(h)$.

Such a quasi-conformal homeomorphism is called the *Teichmüller map* from (Σ_1, f_1) to (Σ_2, f_2).

Teichmüller constructed this map concretely. We shall review his construction relying on a more modern approach.

Recall that we have a universal covering $p \colon U \to \Sigma_1$ by the upper half complex plane. The fundamental group $\pi_1(S)$ acts on U by linear fractional transformations with real coefficients. A quasi-conformal homeomorphism $g \colon \Sigma_1 \to \Sigma_2$ can be lifted to $\tilde{g} \colon U \to \Sigma_2$ such that $g \circ p = \tilde{g}$. By considering a local coordinate on Σ_2, we can talk about $\tilde{g}_z$ and $\tilde{g}_{\bar{z}}$, which may depend on the choice of local coordinates on Σ_2. Still, for any (local) holomorphic isomorphism h, we can calculate

$$\mu_{h \circ \tilde{g}}(z) = \frac{(h \circ \tilde{g})_{\bar{z}}(z)}{(h \circ \tilde{g})_z(z)} = \frac{h'(\tilde{g}(z))\tilde{g}_{\bar{z}}(z)}{h'(\tilde{g}(z))\tilde{g}_z(z)} = \mu_{\tilde{g}}(z),$$

and hence $\mu_{\tilde{g}}(z)$ is independent of the choice of local coordinates.

Now let us see how $\mu_{\tilde{g}}(z)$ changes under covering translations. Let γ be an element of $\pi_1(S)$ and regard it as a covering translation $\gamma \colon U \to U$. Since $\tilde{g}$ is a lift of g, we have $\tilde{g} \circ \gamma = \tilde{g}$. Therefore, we have

$$\mu_{\tilde{g}}(z) = \frac{\tilde{g}_{\bar{z}}(z)}{\tilde{g}_z(z)} = \frac{(\tilde{g} \circ \gamma)_{\bar{z}}(z)}{(\tilde{g} \circ \gamma)_z(z)} = \frac{\tilde{g}_{\bar{z}}(\gamma z)\overline{\gamma'(z)}}{\tilde{g}_z(\gamma z)\gamma'(z)} = \mu_{\tilde{g}}(\gamma z)\frac{\overline{\gamma'(z)}}{\gamma'(z)}.$$

Thus, we have shown that the complex dilatation μ of the lift of a quasi-conformal map from Σ_1 to Σ_2 satisfies

$$\mu \circ \gamma = \mu \frac{\gamma'}{\overline{\gamma'}}$$

for any $\gamma \in \pi_1(S)$.

Now we define holomorphic quadratic differentials, which play an important role both in Teichmüller's construction and Royden's rigidity theorem, an example of the main topic of this article.

Definition 10.1.12 We consider the situation where Σ is a Riemann surface homeomorphic to S, and U is the upper half complex plane regarded as the universal cover of Σ on which $\pi_1(S)$ acts by linear fractional transformations. A holomorphic function $\phi\colon U \to \mathbb{C}$ is called a *holomorphic quadratic differential* on Σ when it satisfies

$$\phi(\gamma z)\gamma'(z)^2 = \phi(z) \tag{10.1.1}$$

for every $\gamma \in \pi_1(S)$.

The condition (10.1.1) above also means that the degree-2 holomorphic differential form $\phi(z)dz^2$ is well defined on Σ. Indeed, if we have such a form $\phi(z)dz^2$ on Σ, then by lifting it to U, we have the invariance property

$$\phi(\gamma z)d(\gamma z)^2 = \phi(z)dz^2. \tag{10.1.2}$$

Since the left hand side is equal to $\phi(\gamma z)\gamma'(z)^2 dz^2$, it follows that Eq. (10.1.2) is equivalent to (10.1.1).

Now, the existence of Teichmüller map is proved by using holomorphic quadratic differentials as follows.

Theorem 10.1.13 *Let Σ_1 be a Riemann surface homeomorphic to S, ϕ a holomorphic quadratic differential on Σ_1, and k a real number in $[0, 1)$. Then there is a quasi-conformal homeomorphism $f\colon \Sigma_1 \to \Sigma_2$ to some Riemann surface Σ_2 homeomorphic to S such that*

$$\mu_f(z) = k\frac{\bar{\phi}(z)}{|\phi(z)|}, \tag{10.1.3}$$

and f is the Teichmüller map from Σ_1 to Σ_2.

The proof, which needs a certain knowledge of differential equations, is rather involved. Instead of giving a formal proof, we shall see the geometric meaning of the theorem.

First, we consider a simple picture. Suppose that we have two complex planes, one being the z-plane and the other being the w-plane. We consider an affine map which expands the real axis by $K \geq 1$ and leaves the imaginary axis as it is. The map can be expressed as

$$w = \frac{K+1}{2}z + \frac{K-1}{2}\bar{z}.$$

Now the Euclidean structure on the w-plane is given by

$$|dw|^2 = dwd\bar{w} = \left(\frac{K+1}{2}\right)^2 |dz + kd\bar{z}|^2,$$

where $k = \dfrac{K-1}{K+1}$. This means that this map has complex dilatation (see Definition 10.1.5) constantly equal to k.

Next, we let $z = \psi(\zeta)$ be a holomorphic coordinate change on Σ_1, and we pull back the form $|dw|^2$ given above to the ζ-plane by ψ. Then we obtain

$$\left(\frac{K+1}{2}\right)^2 |dz + kd\bar{z}|^2 = \left(\frac{K+1}{2}\right) |\psi'(\zeta)^2| \left| d\zeta + k\frac{\bar{\psi}(\zeta)^2}{|\psi'(\zeta)|^2} d\bar{\zeta} \right|^2.$$

We note that if ψ is a $\pi_1(S)$-invariant homomorphic function defined on U, then $\psi'(\zeta)^2$ is a holomorphic quadratic differential. In fact, the invariance of ψ implies $\psi(\gamma\zeta) = \psi(\zeta)$, which in turn implies

$$\gamma'(\zeta)^2\psi'(\gamma(\zeta))^2 = (\psi \circ \gamma)'(\zeta)^2 = \psi'(\zeta)^2.$$

Therefore if we set ϕ to be $(\psi')^2$, then the Beltrami differential of the affine map defined above viewed through ζ is given by

$$\mu(\zeta) = k\frac{\bar{\psi}(\zeta)^2}{|\psi'(\zeta)|^2} = k\frac{\bar{\phi}(\zeta)}{|\phi(\zeta)|}.$$

Keeping this calculation in mind, we now turn to looking at the statement of Theorem 10.1.13. We shall show that except at zeroes of ϕ, Teichmüller maps are affine maps. To see this, we construct from ϕ a coordinate corresponding to the z-coordinate above. Suppose that we have a ζ-coordinate for the universal cover U of Σ_1, and the holomorphic quadratic differential given in Theorem 10.1.13 is expressed as $\phi(\zeta)$ with respect to this coordinate. We define a new coordinate (away from zeroes) by

$$z = \Phi(\zeta) = \int \sqrt{\phi(\zeta)}d\zeta,$$

which is well defined only up to constant. By taking the square of the exterior differential, we have $dz^2 = \phi(\zeta)d\zeta^2$. By considering the affine map given for the coordinates z and w before, and using the z coordinate we have now, we see that if a local chart of Σ_2 is given by w, then Eq. (10.1.3) is equivalent to the condition that w is an affine map expanding the real direction by $K = \dfrac{1+k}{1-k}$.

Theorem 10.1.13 only states that for each holomorphic quadratic differential (and k), there is a Teichmüller map satisfying Eq. (10.1.3). Using the facts that the space of holomorphic quadratic differentials is homeomorphic to $\mathbb{R}^{-3\chi(S)}$, and that the solution of (10.1.3) moves continuously with respect to ϕ, the following can be proved, which implies Theorem 10.1.11.

Theorem 10.1.14 *Let* (Σ_1, f_1) *and* (Σ_2, f_2) *be any two points in* $\mathcal{T}(S)$. *Then there is a holomorphic quadratic differential* ϕ *such that the map given by Theorem 10.1.13 is the Teichmüller map from* (Σ_1, f_1) *to* (Σ_2, f_2).

Moreover, we can show that by varying k in (10.1.3), we get a family of marked Riemann surfaces constituting a geodesic with respect to the Teichmüller distance.

Theorem 10.1.15 *Let* (R_0, g) *be a point in* $\mathcal{T}(S)$. *For a holomorphic quadratic differential* ϕ *on* Σ_0, *set* $\mu_t = \dfrac{e^{2t} - 1}{e^{2t} + 1} \dfrac{\bar{\phi}}{|\phi|}$ *for* $t \geq 0$, *and consider the Teichmüller map* $f_t \colon R_0 \to R_t$ *satisfying* $\dfrac{(f_t)_{\bar{z}}}{(f_t)_z} = \mu_t$ *given by Theorem 10.1.13. Then* $(R_t, f_t \circ g) \in \mathcal{T}(S)$ *gives a geodesic ray issued from* (R_0, g) *with respect to the Teichmüller distance. Furthermore, by defining* $\mu_{-t} = \dfrac{1 - e^{2t}}{e^{2t} + 1} \dfrac{\bar{\phi}}{|\phi|}$, *and considering* f_t *as above, we get a geodesic ray in the negative direction, and together with the ray above, we obtain a geodesic line passing through* (R_0, g).

Once we have Theorem 10.1.13, the fact that $\{(R_t, f_t \circ g)\}$ constitutes a geodesic ray can be proved by a simple computation. Such geodesic rays and geodesic lines are said to be *directed by* the holomorphic quadratic differential ϕ.

10.2 Mapping Class Group Actions

For a closed orientable surface S, its *mapping class group*, denoted by MCG(S), is defined to be the set of isotopy classes of orientation-preserving self-homeomorphisms of S, in which the composition of maps gives a group structure. We also need its index-2 extension, consisting of isotopy classes of homeomorphisms including orientation-reversing ones. We call this extension the *extended mapping class group* of S, and denote it by MCG$^e(S)$.

Given a point $(R, f) \in \mathcal{T}(S)$ and an element $[g] \in$ MCG$^e(S)$, the pair $(R, f \circ g^{-1})$ defines another point in $\mathcal{T}(S)$. (We take g^{-1} to make the action from the left.) It is obvious that this definition is independent of the choice of representative of $[g]$. In this way, MCG$^e(S)$ acts on $\mathcal{T}(S)$ from the left, and we denote the action of $[g]$ by $g_* \colon \mathcal{T}(S) \to \mathcal{T}(S)$. The following is the most fundamental property of this action.

Theorem 10.2.1 *The extended mapping class* MCG$^e(S)$ *acts on* $\mathcal{T}(S)$ *by isometries with respect to the Teichmüller metric.*

Proof Let $(R_1, f_1), (R_2, f_2)$ be two points in $\mathcal{T}(S)$, and $h \colon R_1 \to R_2$ the Teichmüller map between them. Then, since $(f_2 \circ g^{-1}) \circ (f_1 \circ g^{-1})^{-1} = f_2 \circ f_1^{-1} \simeq h$

for any $[g] \in \text{MCG}^e(S)$, the quasi-conformal map h is also the Teichmüller map between $(R_1, f_1 \circ g^{-1})$ and $(R_2, f_2 \circ g^{-1})$. Therefore, g_* acts on $\mathcal{T}(S)$ by an isometry.

□

In the late 1970s, Thurston [31] gave a classification of elements of the mapping class group into three types as follows. Bers [4] gave an interpretation of Thurston's classification in terms of actions on Teichmüller space with the Teichmüller metric.

Theorem 10.2.2 *Elements of* $\text{MCG}(S)$ *(except for the identity) are classified into three types as follows.*

(1) $[g] \in \text{MCG}(S)$ *is said to be* periodic *when there exists a positive integer n such that* $[g]^n = [\text{id}]$.

(2) $[g] \in \text{MCG}(S)$ *is said to be* reducible *when there exists a non-contractible simple closed curve C on S and a positive integer n such that $g^n C$ is homotopic to C.*

(3) $[g] \in \text{MCG}(S)$ *is said to be* pseudo-Anosov *when neither* (1) *nor* (2) *holds.*

This classification can be interpreted as follows if we regard the elements of $\text{MCG}(S)$ as acting on $\mathcal{T}(S)$. A periodic mapping class has a fixed point in $\mathcal{T}(S)$. A pseudo-Anosov mapping class stabilises a Teichmüller geodesic line, and acts on it as a translation. A reducible mapping class has neither a fixed point nor an invariant geodesic line.

Now we state the rigidity theorem for the mapping class group action for Teichmüller space with the Teichmüller metric due to Royden [25] and Earle-Kra [6]. (See also Earle-Markovic [7] for an alternative approach.)

Theorem 10.2.3 *If the genus of S is greater than 2, the isometry group* $\text{Isom}(\mathcal{T}(S))$ *is equal to the extended mapping class group* $\text{MCG}^e(S)$ *which is regarded as a group of isometries by Theorem 10.2.1. If the genus is equal to 2, then* $\text{Isom}(\mathcal{T}(S))$ *is isomorphic to the quotient of* $\text{MCG}^e(S)$ *by an order-2 subgroup generated by the class represented by a hyper-elliptic involution.*

Royden proved that any analytic automorphism of $\mathcal{T}(S)$ with respect to its natural complex structure induced from the space of Beltrami differentials is an isometry with respect to the Teichmüller metric. Then he showed that any holomorphic isometry can be represented as an action of a mapping class, making use of the "infinitesimal rigidity", which we shall explain below. Earle and Kra gave a generalisation of Royden's result to more general surfaces of finite type, and also showed that in fact any isometry without the assumption of analyticity is represented as an action of an extended mapping class, again relying on the infinitesimal rigidity.

To describe the infinitesimal rigidity, we now define a norm on the space of holomorphic quadratic differentials, and show that the space can be identified with a cotangent space. Let (Σ, f) be a point in $\mathcal{T}(S)$. Let $QH(\Sigma)$ be the space of holomorphic quadratic differentials on Σ, which is regarded as a complex linear space. The norm on $QH(\Sigma)$ is defined by $\|\phi\| = \int_\Sigma |\phi(z)||dz|^2$. Recall that the form $\phi(z)dz^2$ is well defined independently of the choice of a lift to U (or of a choice of local chart), which implies that the norm above is also well defined.

Recall from Theorem 10.1.15 that any tangent vector at (Σ, f) is represented by (a Beltrami differential) $\mu = \dfrac{\bar{\psi}}{|\psi|}$ for some holomorphic quadratic differential ψ on Σ. Now, since $\mu(\gamma z) = \dfrac{\gamma'(z)}{\bar{\gamma}'(z)}\mu(z)$ and $\phi(\gamma z) = \dfrac{\phi(\gamma)}{\gamma'(z)^2}$, we see that $\mu(z)\phi(z)|dz|^2$ is invariant under the action of γ, when this expression is identified with its lift to U. Therefore $(\phi, \mu) := \displaystyle\int_\Sigma \phi(z)\mu(z)|dz|^2$ is well defined. Under this pairing, $QH(\Sigma)$ is identified with the cotangent space $\mathcal{T}(S)$ at (Σ, f).

Now the following result, due to Royden, is what we call the "infinitesimal rigidity".

Theorem 10.2.4 (Royden's Infinitesimal Rigidity) *Let Σ_1, Σ_2 be two Riemann surfaces both homeomorphic to S. Let $k: QH(\Sigma_1) \to QH(\Sigma_2)$ be a complex linear isometry with respect to the norm $\| \cdot \|$ defined above. Then, there are a bi-holomorphic homeomorphism $h: \Sigma_2 \to \Sigma_1$ and a complex number α with $|\alpha| = 1$ such that $k(\phi) = \alpha h^*(\phi)$.*

The proof of this theorem by Royden relies on his analysis of the differentiability of the unit sphere in $QH(S)$. Let us sketch his argument. For $z \in R_1$, we consider a holomorphic differential ϕ_z on R_1 which has a zero of the highest possible order at z. Then we consider $k(\phi_z)$, which is a holomorphic quadratic differential on Σ_2, and it turns out that there is a unique point z' on Σ_2 on which $k(\phi_z)$ has a zero of the highest possible order. Royden showed that by taking z to z', we obtain a bi-holomorphic map from Σ_1 to Σ_2 which induces k between their spaces of holomorphic quadratic differentials.

Next, we shall see how Theorem 10.2.4 implies Theorem 10.2.3. If $F: \mathcal{T}(S) \to \mathcal{T}(S)$ is an isometry with respect to the Teichmüller metric, some general argument implies that it must be differentiable, and hence induces a real linear isometry F^* from the cotangent space $QH(\Sigma_1)$ to $QH(\Sigma_2)$. From the linearity and the isometricity we have $F^*(i\phi) = cF^*(\phi)$ for some $c \in \mathbb{C}$ with $|c| = 1$, and by continuity c is a constant. The isometricity implies $|1 + c| = |1 + i|$, and hence $c = \pm i$. If $c = i$, then F^* is a complex linear isometry, and by Theorem 10.2.4, F^* is induced from a mapping class. When $c = -i$, by considering another complex structure on Σ_2 obtained by complex conjugation, we get a Riemann surface $\bar{\Sigma}_2$, and a real linear isometry $QH(\Sigma_2) \to QH(\bar{\Sigma}_2)$ corresponding to complex conjugation. Then by Theorem 10.2.4, the complex linear isometry $r \circ F^*$ is induced from a mapping class. Since r can be realised by an orientation reversing homeomorphism, this shows that F^* is induced from an extended mapping class.

10.3   Curve Complexes and Their Automorphisms

The notion of curve complex (or complex of curves), which is a simplicial complex whose vertices are isotopy classes of simple closed curves on a surface, was first introduced by Harvey [12]. It has turned out that this tool is very helpful to understand both mapping class groups and Kleinian groups as the work of Masur-Minsky [21, 22] typically demonstrates.

We now define curve complexes formally.

Definition 10.3.1 Let S be a closed orientable surface of genus greater than 1. The *curve complex* of S, denoted by $\mathcal{CC}(S)$, is a simplicial complex whose vertices are the isotopy classes of non-contractible simple closed curves on S, and where $q + 1$ distinct vertices $v_0, \ldots, v_q$ span a q-simplex if and only if they can be represented by disjoint simple closed curves.

The extended mapping class group $\mathrm{MCG}^e(S)$ acts naturally on the vertex set, by letting $[f][c] = [f(c)]$ for a self-homeomorphism $f : S \to S$ and a simple closed curve c. It is obvious that f preserves the condition that simple closed curves are disjoint, and hence it acts on $\mathcal{CC}(S)$ as a simplicial automorphism. Ivanov [15], and then Korkmaz [18] and Luo [20] for some special cases which Ivanov could not deal with, proved the following theorem, which says that this action of $\mathrm{MCG}^e(S)$ constitutes the entire group of simplicial automorphisms.

Theorem 10.3.2 (Ivanov)   *Every simplicial automorphism is induced by an extended mapping class. Furthermore if $g \geq 3$, then the action of the extended mapping class group is effective, and if $g = 2$, then the kernel of the action is represented by a hyper-elliptic involution. Therefore* $\mathrm{MCG}^e(S) = \mathrm{Aut}(\mathcal{CC}(S))$ *if* $g \geq 3$, *and* $\mathrm{Aut}(\mathcal{CC}(S)) = \mathrm{MCG}^e(S)/\mathbb{Z}_2$ *if* $g = 2$.

We now present a sketch of the proof of this important theorem given by Ivanov in [15]. Since his argument is fairly easy to follow, and very illuminating at the same time, we are going to give it in some detail. We first show that every simplicial automorphism of $\mathcal{CC}(S)$ is induced by an extended mapping class. Let ϕ be a simplicial automorphism of $\mathcal{CC}(S)$. By the definition of $\mathcal{CC}(S)$, the disjoint simple closed curves are sent to disjoint ones by ϕ. We can show as follows that the condition that the intersection number is 1 is also preserved by ϕ.

Lemma 10.3.3 *Let c_1 and c_2 be (isotopy classes) of simple closed curves with $i(c_1, c_2) = 1$, then $i(\phi(c_1), \phi(c_2)) = 1$.*

Proof Two simple closed curves c_1, c_2 satisfy $i(c_1, c_2) = 1$ if and only if there are simple closed curves c_3, c_4 and c_5 with $i(c_1, c_3) = i(c_1, c_4) = i(c_2, c_4) = i(c_2, c_5) = i(c_3, c_5) = 0$ and other intersection numbers being positive as depicted in Fig. 10.1. (As illustrated there, the simple closed curve c_4 bounds a one-holed torus containing c_1 and c_2.) Since the nullity of the intersection number is preserved by both ϕ and ϕ^{-1}, so is the existence of c_3, c_4 and c_5. Therefore, ϕ preserves the condition that the intersection number is 1.   □

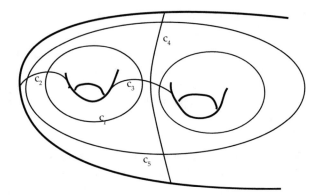

Fig. 10.1 Choice of c_3, c_4 and c_5

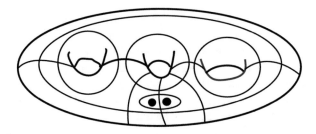

Fig. 10.2 Decomposition into simply-connected domains and two annuli

This in particular shows that ϕ takes (the isotopy class of) a non-separating simple closed curve to a non-separating simple closed curve. It is easy to see that the mapping class group acts on the set of isotopy classes of non-separating simple closed curves transitively. Therefore, by composing the action of a mapping class on $\mathcal{CC}(S)$, we have only to consider the case when ϕ fixes the isotopy class of some non-separating simple closed curve C. Let S' be a surface with two boundary components obtained by cutting S along C. Then ϕ is also regarded as a simplicial automorphism of $\mathcal{CC}(S')$.

Now, it is easy to observe, as in Fig. 10.2, that we can decompose S' into discs and two annuli around the boundary components, by drawing a chain of non-separating (isotopically distinct) simple closed curves and a separating one surrounding the two boundary components depicted in Fig. 10.2. By Lemma 10.3.3, we see that ϕ must preserve the conditions that two simple closed curves have intersection number 1 and that a simple closed curve bounds a one-holed torus. Combining this with the fact that ϕ preserves disjointness, we also see that the configuration given in Fig. 10.2 must be preserved by ϕ. It then follows that there is a homeomorphism $f : S \to S$ fixing C such that f takes each simple closed curve K in the figuration to a curve representing $\phi([K])$.

By considering the automorphism $f^{-1}\phi$ on $\mathcal{CC}(S)$, we see that we have only to consider the case when ϕ fixes every isotopy class of the curves in the configuration of Fig. 10.2. Now, we turn to considering the action of ϕ on the arc complex of S'. The arc complex $\mathcal{AC}(S')$ is defined to be a simplicial complex whose vertices are proper isotopy classes of properly embedded essential arcs. In the same way as the curve complex, $q+1$ vertices span a q-simplex if and only if they can be represented by pairwise disjoint arcs.

In contrast to the curve complex, the arc complex has the property that every codimension-1 simplex is shared by exactly two top-dimensional simplices. For any properly embedded essential arc a, let B be either the union of the two boundary components when a connects the two boundary components, or a boundary component on which the endpoints of a lie when both endpoints are on the same component. Then, by considering the boundary component of a tubular neighbourhood of $a \cup B$, we can associate either a vertex in $\mathcal{CC}(S')$ or a pair of vertices in $\mathcal{CC}(S')$ with $[a] \in \mathcal{AC}(S')$. In the former case, the corresponding simple closed curve separates a one-holed torus containing C from S, and conversely such a simple closed curve always corresponds to an arc connecting the two boundary components of S'. In the latter case, such a pair of simple closed curves, say A and A', is characterised by the property that every simple closed curve on S' which has non-zero intersection number with A also has non-zero intersection number with A'. It is easy to observe that both properties are preserved by ϕ. Therefore, the automorphism ϕ of $\mathcal{CC}(S)$, which is regarded as an automorphism of $\mathcal{CC}(S')$ at the moment, also acts on $\mathcal{AC}(S')$ as a simplicial automorphism.

Choosing one of the boundary components, say b, we can draw on S' disjoint arcs $a_1, \ldots, a_k$ with the following conditions:

(i) Except for a_1, which connects the two boundary components of S', every arc has both endpoints on b.
(ii) Cutting S' along $a_1 \cup \cdots \cup a_k$, we get discs.
(iii) By the correspondence between an arc and (one or two) simple closed curves defined above, a_1 gives a simple closed curve encircling the two boundary components drawn in Fig. 10.2, and each of the others gives two simple closed curves (at least) one of which is among those drawn in Fig. 10.2.

If a_i ($i \neq 1$) gives two simple closed curves c_i, c_i', then by the condition (iii), one of them, say c_i, is among the curves in Fig. 10.2, and hence is fixed by ϕ. The other simple closed curve c_i' is characterised by the property that it cannot be homotoped so as to be disjoint from the one encircling the two boundary components, and that for every curve c drawn in Fig. 10.2 other than this encircling one, c_i' is disjoint from c if and only if c_i is, and $i(c_i', c) = 1$ if and only if $i(c_i, c) = 1$, and is not disjoint from the one encircling the two boundary components. Since ϕ preserves these conditions, we see that ϕ also fixes the isotopy class of c_i'. Therefore ϕ fixes the isotopy classes of the arcs $a_1, \ldots, a_k$.

By adding arcs corresponding to disjoint diagonals of complement of $a_1 \cup \cdots \cup a_k$, we can construct a top-dimensional simplex σ of $\mathcal{AC}(S')$. By considering the intersection with the simple closed curves drawn in Fig. 10.2 as before, we can verify that ϕ also fixes the isotopy classes of these diagonals. It follows then that ϕ fixes the top-dimensional simplex σ. Now, it is known that $\mathcal{AC}(S')$ is path-connected, and hence for every top-dimensional simplex σ', there is a path consisting of "diagonal changes" on quadrilaterals (corresponding to passing through codimension-one faces), from σ to σ'. Since two top-dimensional simplices can share only one codimension-one face, we see inductively that ϕ fixes all top-dimensional simplices, hence acts on $\mathcal{AC}(S')$ by the identity. Now going back to the correspondence between $\mathcal{AC}(S')$ and $\mathcal{CC}(S')$, we see that ϕ acts on $\mathcal{CC}(S')$, hence on $\mathcal{CC}(S)$, by the identity. This completes the proof of Theorem 10.3.2.

10.4 An Alternative Approach to Royden's Rigidity

Ivanov proves in [16] that his theorem which we explained in the previous section gives rise to an alternative proof of Royden's theorem. In this section, we shall sketch his argument given there.

Before starting to explain his argument, we start with defining such notions as measured foliation and horizontal foliation which he used in an essential way, and introduce results of Thurston and Hubbard-Masur.

As we explained in Sect. 10.1.3, for any point $(\Sigma, f) \in \mathcal{T}(S)$ and any holomorphic quadratic differential ϕ on Σ, there is a Teichmüller geodesic ray derived from Teichmüller maps associated with $\mu_t = \dfrac{e^{2t}-1}{e^{2t}+1}\dfrac{\bar{\phi}}{|\phi|}$. To understand the asymptotic behaviour of Teichmüller geodesic rays, the notion of horizontal foliations of holomorphic quadratic differentials is essential.

Recall from Sect. 10.1.3 that $\phi(z)dz^2$ as a $(0,2)$-tensor is well defined on Σ_1. We consider a line field on Σ_1 determined by tangent vectors with the condition $\phi(z)dz^2(v_z, v_z) > 0$ for every $z \in \Sigma_1$ except at the zeroes of ϕ. (Note that this condition determines v_z only up to non-zero scalar multiples.) This gives rise to a singular codimension-one foliation $\mathcal{F}_\phi$ whose singularities are the zeroes of ϕ. Furthermore, ϕ gives a transverse measure for the foliation $\mathcal{F}_\phi$ as follows. Let a be an arc transverse to the leaves of $\mathcal{F}_\phi$. Then we define the measure on a by $\Im(\sqrt{\phi}dz)$, which is well defined independently of the choice of a coordinate. This gives a measure μ_ϕ invariant under homotopies along the leaves, which is called the transverse measure of $\mathcal{F}_\phi$. The pair $(\mathcal{F}_\phi, \mu_\phi)$ is called the horizontal foliation of ϕ. This is an example of a measured foliation, a notion introduced by Thurston. We now turn to explaining Thurston's theory of measured foliations.

In the process of defining a compactification of Teichmüller space, which is now called the Thurston compactification, Thurston introduced and studied measured foliations on surfaces. (See Thurston [31] and Fathi-Laudenbach-Poénaru [8].) A measured foliation on S is a codimension-one foliation allowed to have singularities of negative indices, equipped with a transverse measure invariant under homotopies along its leaves. We denote a measured foliation by a pair $(\mathcal{F}, \mu)$, where $\mathcal{F}$ is a foliation and μ its transverse measure. Two measured foliations on S are said to be (Whitehead)-equivalent if one is obtained from the other by a finite number of operations consisting of collapsing two singularities into one and the inverse operation, and of isotopy. The set of equivalence classes of measured foliations on S with the empty set included as the point of origin, is denoted by $\mathcal{MF}(S)$. For an isotopy class of simple closed curves s on S, we define $i((\mathcal{F}, \mu), s)$ to be $\displaystyle\inf_{c \in s} \int_c d\mu$, which is called the intersection number between c and $(\mathcal{F}, \mu)$. It is easy to check that this number depends only on the equivalence class of $(\mathcal{F}, \mu)$. The intersection number can also be defined between two measured foliations $(\mathcal{F}_1, \mu_1)$ and $(\mathcal{F}_2, \mu_2)$, by $i((\mathcal{F}_1, \mu_1), (\mathcal{F}_2, \mu_2)) := \displaystyle\inf_{(\mathcal{F}_1', \mu_1') \simeq (\mathcal{F}_1, \mu_1), (\mathcal{F}_2', \mu_2') \simeq (\mathcal{F}_2, \mu_2)} \int_S d\mu_1' d\mu_2'$, where the infimum is taken among all measured foliations $(\mathcal{F}_1', \mu_1')$ equivalent to $(\mathcal{F}_1, \mu_1)$ and $(\mathcal{F}_2', \mu_2')$ equivalent to $(\mathcal{F}_2, \mu_2)$. This is an extension of the intersection number between a simple closed curve and a measured foliation if we identify a simple closed curve c with a measured foliation having the following special properties. We consider a measured foliation all of whose non-singular leaves are homotopic to c. When c is non-separating there is only one singular leaf, which contains all singularities. (See Fig. 10.3.) When c is separating, there are exactly two singular leaves, one on each component of $S \setminus c$. (See Fig. 10.4.) In both cases, the total transverse measure of the parallel non-singular leaves is equal to 1.

Thurston proved that the map $\iota: \mathcal{MF}(S) \rightarrow \mathbb{R}^S$, defined by $\iota((\mathcal{F}, \mu)) = (i((\mathcal{F}, \mu), s))_{s \in S}$, where S denotes the set of isotopy classes of non-contractible simple closed curves, is injective. We endow $\mathcal{MF}(S)$ with the topology induced from

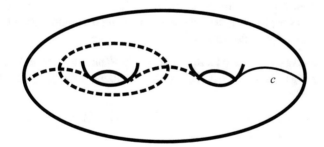

Fig. 10.3 The case when c is non-separating. The dotted graph is the only singular leaf

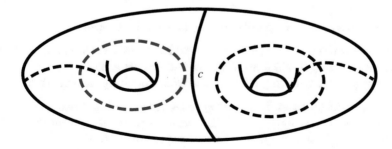

Fig. 10.4 The case when c is separating. The two dotted graphs are the singular leaves

the weak topology on $\mathbb{R}^\mathcal{S}$. A fundamental property of $\mathcal{MF}(S)$ proved by Thurston is the following.

Theorem 10.4.1 (Thurston [31], Fathi–Laudenbach–Poénaru [8]) *For a closed orientable surface S with genus g greater than 1, the space of measured foliations $\mathcal{MF}(S)$ is homeomorphic to $\mathbb{R}^{6g-6}$.*

A disjoint union of non-contractible simple closed curves on S is called a multi-curve. When each component of a multi-curve is given a positive number, it is called a weighted multi-curve. Let $\gamma = w_1 c_1 \sqcup \cdots \sqcup w_k c_k$ be a weighted multi-curve, where $c_1, \ldots, c_k$ denote simple closed curves and $w_1, \ldots, w_k$ weights given to them. Then, there is a unique (up to equivalence) measured foliation $(\mathcal{F}_\gamma, \mu_\gamma)$ such that if we cut S along the leaves containing singular points, we get a disjoint union of open annuli $A_1, \ldots, A_k$ foliated by compact leaves isotopic to $c_1, \ldots, c_k$ respectively, and μ_γ induces a transverse measure giving the width $w_1, \ldots, w_k$ for these annuli. By identifying γ with $(\mathcal{F}_\gamma, \mu_\gamma)$, we can regard the set of weighted multi-curves on S as a subset of $\mathcal{MF}(S)$.

In particular, when $k = 1$, a weighted multi-curve is called a weighted simple closed curve. By the identification above, we regard the set of weighted simple closed curves $\mathbb{R}_+ \mathcal{S}$ as a subset of $\mathcal{MF}(S)$. Thurston proved the following density result.

Theorem 10.4.2 *The set of weighted simple closed curves $\mathbb{R}_+ \mathcal{S}$ is dense in $\mathcal{MF}(S)$.*

Now we return to considering horizontal foliations. The following result due to Hubbard–Masur [14] is quite fundamental for understanding the behaviour of geodesic rays in Teichmüller space.

Theorem 10.4.3 (Hubbard–Masur [14]) *For a Riemann surface Σ homeomorphic to S, let $H_\Sigma \colon QH(\Sigma) \to \mathcal{MF}(S)$ be a map taking ϕ to $(\mathcal{F}_\phi, \mu_\phi)$. Then H_Σ is a homeomorphism.*

Now we turn to Ivanov's argument. For two Teichmüller geodesic rays, there are two typical asymptotic behaviours as follows.

Definition 10.4.4 Let $r(t)$ and $r'(t)$ $(t \in [0, \infty))$ be two geodesic rays in $(\mathcal{T}(S), d_T)$, issued from possibly different points. We say that r and r' are *divergent* if $d_T(r(t), r'(t))$ goes to ∞ as $t \to \infty$.

We say that r and r' are *parallel* if $d_T(r(t), r'(t))$ is bounded.

We note that these two conditions are exclusive but logically there is another possibility that $d_T(r(t).r'(t))$ is unbounded but does not go to ∞.

Ivanov gave the following sufficient conditions for divergence and parallelism.

Theorem 10.4.5 (Ivanov) *Let r_1 and r_2 be geodesic rays issued from (Σ_1, f_1) and (Σ_2, f_2), and directed by $\phi_1 \in QH(\Sigma_1)$ and $\phi_2 \in QH(\Sigma_2)$ respectively. Suppose that $i((\mathcal{F}_{\phi_1}, \mu_{\phi_1}), (\mathcal{F}_{\phi_1}, \mu_{\phi_2})) > 0$. Then r_1 and r_2 are divergent.*

Theorem 10.4.6 (Ivanov) *Let γ_1, γ_2 be the same multi-curve with possibly distinct (positive) weights. Let r_1 and r_2 be geodesic rays issued from (Σ_1, f_1) and (Σ_2, f_2), and directed by ϕ_1 and ϕ_2 respectively, such that $H_{\Sigma_1}(\phi_1) = \gamma_1$ and $H_{\Sigma_2}(\phi_2) = \gamma_2$. Then r_1 and r_2 are parallel.*

Moreover, for any measured foliation $m \in \mathcal{MF}(S)$ and any two points (Σ_1, f_1) and (Σ_2, f_2), the geodesic rays issued from (Σ_1, f_1) and (Σ_2, f_2), and directed by $H_{\Sigma_1}^{-1}(m)$ and by $H_{\Sigma_2}^{-1}(m)$ respectively, are parallel.

Based on these two theorems, Ivanov introduced the following relation between two measured foliations.

Definition 10.4.7 For two measured foliations $m, n \in \mathcal{MF}(S)$, we use the notation $m \sim n$ if there are sequences $\{m_i\}$, $\{n_i\} \subset \mathcal{MF}(S)$ converging to m, n respectively such that the geodesic rays r_{m_i}, r_{n_i} issued from $x = (\Sigma, f) \in \mathcal{T}(S)$ and directed by $H_x^{-1}(m_i)$, $H_x^{-1}(n_i)$ are *not divergent* for each i. By the second statement of Theorem 10.4.6, the definition does not depend on the choice of the basepoint x. We denote the set $\{n \in \mathcal{MF}(S) \mid m \sim n\}$ by $\mathcal{R}(m)$.

Now we can see that there is a characterisation of simple closed curves as follows. Let γ be a simple closed curve regarded as a point in $\mathcal{MF}(S)$. Recall that γ is a measured foliation consisting of compact leaves homotopic to γ which constitutes a foliated open annulus of transverse measure 1 and one or two singular leaves. Then it is easy to see that a measured foliation m satisfies $i(m, \gamma) = 0$ if and only if m is the union of a (possibly empty) foliated annulus with core curve γ and a measured foliation whose leaves are disjoint from γ. Therefore the subset $\mathcal{N}(\gamma) = \{n \in \mathcal{MF}(S) \mid i(n, \gamma) = 0\}$ has codimension 1 in $\mathcal{MF}(S)$.

By Theorem 10.4.5 and the continuity of intersection number, we see that $\mathcal{R}(m) \subset \mathcal{N}(m)$ for any $m \in \mathcal{MF}(S)$. For a simple closed curve $\gamma \in \mathcal{S}$ and any measured foliation m satisfying $i(m, \gamma) = 0$, by the observation above, we can construct a sequence of weighted multi-curves $c_i = w_i \gamma \sqcup \delta_i$ which converges to m for any weighted multi-curve δ_i with $i(\gamma, \delta_i) = 0$. On the other hand $d_i = (1 + 1/i)\gamma + (1/i)\delta_i$ converges to γ. By Theorem 10.4.6 $H_x^{-1}(c_i)$ and $H_x^{-1}(d_i)$ are parallel, and hence are not divergent. This implies that m is contained in $\mathcal{R}(\gamma)$, and hence $\mathcal{R}(\gamma) = \mathcal{N}(\gamma)$ has codimension 1. We can easily verify that the property that $\mathcal{R}(\gamma)$ has codimension 1 characterises γ as a simple closed curve. Furthermore, for two simple closed curves γ, δ, the condition $i(\gamma, \delta) = 0$ is equivalent to the condition that $\gamma \sim \delta$.

Let $I: (\mathcal{T}(S), d_T) \to (\mathcal{T}(S), d_T)$ be an isometry, and for $x \in \mathcal{T}(S)$, let $I^*: QH(y) \to QH(x)$ be the induced linear isometry between cotangent spaces, where $y = I(x)$. Since the definition of the relation $\sim$ is independent of the basepoint, we see that I^* takes the set of simple closed curves to itself and preserves disjointness. Therefore, I^* induces a simplicial automorphism of $\mathcal{CC}(S)$, and by Theorem 10.3.2, an extended mapping class ϕ. By Theorem 10.4.2, I^* and the action of ϕ_* must coincide on the entire $QH(y)$.

In the argument above, the choice of x, and hence y, was arbitrary. We are now going to choose a special y. Let γ and δ be simple closed curves on S which fill up S, that is, such that each component of $S \setminus (\gamma \cup \delta)$ is simply connected. We can construct mutually transverse measured foliations $\mathcal{F}_\gamma$ and $\mathcal{F}_\delta$ on S whose non-singular leaves are all homotopic to γ and δ respectively. These measured foliations determine a conformal structure on S, and hence a point $y = (\Sigma, f)$ in $\mathcal{T}(S)$. There is a holomorphic quadratic differential ϕ on Σ such that $H_y(\phi) = \mathcal{F}_\gamma$. Then since $\mathcal{F}_\delta$ is the vertical foliation of ϕ, we have $H_y(-\phi) = \mathcal{F}_\delta$. Now, by the argument of the preceding paragraph, we have an extended mapping class f such that $\psi := I^* \circ f^{-1}$ acts on $\mathcal{CC}(S)$ by the identity. It remains to show that ψ is the identity.

Since ψ fixes both $\mathcal{F}_\gamma$ and $\mathcal{F}_\delta$, it fixes the geodesic line, which we denote by ℓ_γ, directed by $H_y^{-1}(\mathcal{F}_\gamma)$, and it coincides with the one directed by $H_y^{-1}(\mathcal{F}_\delta)$ with the orientation reversed. Since ψ is an isometry and preserves the positive direction, it acts on ℓ_γ by a translation.

Now we choose another simple closed curve β with $i(\beta, \delta) > 0$ such that $\beta \cup \gamma$ fills up S. Then, by the same argument as above, we can construct a measured foliation $\mathcal{F}_\beta$ corresponding to β which is transverse to $\mathcal{F}_\gamma$, and they determine a point $z = (\Sigma', f') \in \mathcal{T}(S)$. In the same way as ℓ_γ, there is a geodesic line ℓ'_γ through z directed by $H_z^{-1}(\mathcal{F}_\gamma)$ whose negative direction is directed by $H_z^{-1}(\mathcal{F}_\beta)$. Since $i(\beta, \delta) > 0$, by Theorem 10.4.5, the geodesic lines $\ell_\gamma, \ell'_\gamma$ are divergent in the negative direction. Also, since the lines are both directed by holomorphic quadratic differentials whose horizontal foliations are $\mathcal{F}_\gamma$ in the positive directions, they are parallel in the positive directions. This implies that the negative direction of ℓ_γ and the positive direction of ℓ'_γ are also divergent.

Now, if the translation defined by ψ on ℓ_γ has non-zero distance, then by considering the iteration of ψ^{-1} or ψ (depending the direction of the translation), we can move y to the negative direction as far as we want, which makes the image of y have as big a distance as we want from ℓ'_γ. This contradicts the fact that y is within bounded distance from ℓ'_γ and ψ is an isometry. Therefore the translation distance is 0 and ψ fixes y.

Since $\mathbb{R}_+ S$ is dense in $\mathcal{MF}(S)$, the isometry ψ fixes every point of $QH(\Sigma)$. This means that ψ^* is the identify on the cotangent space of $\mathcal{T}(S)$ at y, and hence ψ is the identity.

10.5   Thurston's Asymmetric Metric

Thurston introduced in [30] a new metric on Teichmüller space, which is now called Thurston's asymmetric metric. The Teichmüller metric on $\mathcal{T}(S)$ measures the distance of two complex structures on S using quasi-conformal maps. Thurston's metric measures the distance of two hyperbolic metrics using Lipschitz maps, regarding $\mathcal{T}(S)$ as the space of isotopy classes of hyperbolic metrics on S.

By the uniformisation theorem, every complex structure has unique compatible hyperbolic metric. (Recall that we are only considering closed surfaces.) Therefore, the definition of Teichmüller space in Sect. 10.1 can be reinterpreted as follows.

Definition 10.5.1 The Teichmüller space of S is defined to be

$$\mathcal{T}(S) = \{(\Sigma, f) \mid \Sigma \text{ is an oriented hyperbolic surface,}$$

$$\text{and } f \colon S \to \Sigma \text{ is an orientation-preserving homeomorphism}\}/\sim,$$

where $(\Sigma_1, f_1) \sim (\Sigma_2, f_2)$ if and only if there is an isometry from Σ_1 to Σ_2 homotopic to $f_2 \circ f_1^{-1}$.

Throughout this section, we use this definition of Teichmuüller space, and hence we regard a point in $\mathcal{T}(S)$ as a hyperbolic surface homeomorphic to S with a marking.

It is Lipschitz maps that play the role of quasi-conformal maps in the Teichmüller theory.

Definition 10.5.2 Let (X, d_X) and (Y, d_Y) be metric spaces, and $h \colon X \to Y$ a continuous map. Then we define

$$\mathrm{Lip}(h) := \sup_{x_1, x_2 \in X \ (x_1 \neq x_2)} \frac{d_Y(h(x_1), h(x_2))}{d_X(x_1, x_2)},$$

for which the value ∞ is also allowed.

It is easy to check the following:

Lemma 10.5.3 *Let $h \colon X \to Y$ and $k \colon Y \to Z$ be continuous maps between metric spaces. Then we have $\mathrm{Lip}(k \circ h) \leq \mathrm{Lip}(k) \cdot \mathrm{Lip}(h)$.*

Now we define Thurston's asymmetric metric. Note that we call a metric asymmetric when symmetry fails to hold.

Definition 10.5.4 (Thurston [30]) Let $x_1 = (\Sigma_1, f_1)$, $x_2 = (\Sigma_2, f_2)$ be two points in $\mathcal{T}(S)$. We define $d_L(x_1, x_2) := \inf_{g \simeq f_2 \circ f_1^{-1}} \log \mathrm{Lip}(g)$.

Thurston proved this function d_L satisfies the identity of indiscernibles and the triangle inequality without satisfying the symmetry.

Lemma 10.5.5 *The function $d_L \colon \mathcal{T}(S) \times \mathcal{T}(S) \to \mathbb{R}$ defined above is an asymmetric metric on $\mathcal{T}(S)$.*

Proof The non-negativity can be obtained by comparing area: if there were $h \colon \Sigma_1 \to \Sigma_2$ with $\mathrm{Lip}(h) < 0$, it would give $\mathrm{Area}(\Sigma_2) < \mathrm{Area}(\Sigma_1)$, a contradiction. (See Proposition 2.1 of [30] for details.)

If $d(x_1, x_2) = 0$, then there is a sequence of continuous maps $h_i \colon \Sigma_1 \to \Sigma_2$ homotopic to $f_2 \circ f_1^{-1}$ with $\mathrm{Lip}(h_i) \to 1$. By the theorem of Ascoli–Arzelà, this should converge uniformly to an isometry. Therefore, we have $x_1 = x_2$ then.

The triangle inequality is derived from Lemma 10.5.3.   □

In general $d_L(x_1, x_2)$ and $d_L(x_2, x_1)$ may differ.

Thurston's asymmetric metric d_L has another expression as follows. We note that the Teichmüller metric has a similar expression in terms of "extremal length", which is due to Kerckhoff [17]. In fact the proofs of Theorems 10.4.5 and 10.4.6 which we did not explain use this Kerckhoff formula.

Proposition 10.5.6 *For* $x_1 = (\Sigma_1, f_1), x_2 = (\Sigma_2, f_2)$ *in* $\mathcal{T}(S)$, *we have*

$$d_L(x_1, x_2) = \log \sup_{s \in \mathcal{S}} \frac{\text{length}_{\Sigma_2}(f_2(s))}{\text{length}_{\Sigma_1}(f_1(s))},$$

where length *denotes the length of the closed geodesic in the isotopy class.*

An analogue of a measured foliation in the hyperbolic setting is a measured (geodesic) lamination, which we now define.

Definition 10.5.7 On a hyperbolic surface Σ, a closed subset consisting of disjoint simple geodesics is called a *geodesic lamination*, and each constituting geodesic is called a *leaf*. A geodesic lamination is called a *measured lamination* when it has a transverse measure (with full support) which is invariant under homotopy along leaves. A measured lamination is denoted as a pair (λ, μ), where λ is the supporting geodesic lamination and μ is its transverse measure.

In the same way as for measured foliations, for two measured laminations $(\lambda_1, \mu_1), (\lambda_2, \mu_2)$, we can define their intersection number by $i((\lambda_1, \mu_1), (\lambda_2, \mu_2)) = \int_\Sigma d\mu_1 d\mu_2$. A weighted simple closed geodesic $w\gamma$ on Σ can be regarded as a measured lamination by giving the transverse Dirac measure of weight w to γ. Now, we can define the space of measured laminations and its projectivisation as topological spaces in the following way.

Definition 10.5.8 Let $\mathcal{ML}(\Sigma)$ be the set of measured laminations on Σ. We consider the map $\iota \colon \mathcal{ML}(\Sigma) \to \mathbb{R}^{\mathcal{S}}$ defined by $\iota((\lambda, \mu)) = (i(s, (\lambda, \mu)))_{s \in \mathcal{S}}$, which can be proved to be injective. We endow the *measured lamination space* $\mathcal{ML}(S)$ with the topology induced from $\mathbb{R}^{\mathcal{S}}$ by identifying $\mathcal{ML}(S)$ with its image under ι.

By identifying two (non-empty) measured laminations with the same support and transverse measures which are multiples of each other, and putting the quotient topology, we get the *projective measured lamination space*, which is denoted by $\mathcal{PML}(\Sigma)$. We call each element of $\mathcal{PML}(\Sigma)$ a *projective lamination*.

Although this definition depends on the hyperbolic metric on Σ, if there is a homeomorphism h between two hyperbolic surfaces Σ_1 and Σ_2, then there is natural way to identify $\mathcal{ML}(\Sigma_1)$ with $\mathcal{ML}(\Sigma_2)$ by taking (λ, μ) to a geodesic lamination homotopic to $h(\lambda)$ with the pushed-forward transverse measure. In this way, we can talk about $\mathcal{ML}(S)$ and regard each $\mathcal{ML}(\Sigma)$ for $(\Sigma, f) \in \mathcal{T}(S)$ as a copy of $\mathcal{ML}(S)$ via f.

There is a homeomorphism between $\mathcal{ML}(S)$ and $\mathcal{MF}(S)$ (where S is the topological surface homeomorphic to Σ) preserving the intersection numbers. This fact was first proved by Thurston, but his proof has not been published. Levitt's paper [19] contains a part of Thurston's argument, but does not deal with transverse measures. A proof for the correspondence of transverse measures can be found in Papadopoulos [24, II.1.1].

In particular $\mathcal{ML}(S)$ is homeomorphic to $\mathbb{R}^{6g-6}$ and $\mathcal{PML}(S)$ is homeomorphic to S^{6g-7}.

The following holds in the same way as the measured foliation space.

Theorem 10.5.9 *The set of weighted simple closed curves* $\mathbb{R}_+\mathcal{S}$ *is dense in* $\mathcal{ML}(S)$.

The length function defined on the set of weighted simple closed geodesics can be extended to $\mathcal{ML}(\Sigma)$ continuously in the following way.

Definition 10.5.10 For a measured lamination (λ, μ) on Σ, we define its length by

$$\mathrm{length}_\Sigma(\lambda, \mu) = \int_\Sigma d\mathrm{length}_\Sigma(\lambda)d\mu,$$

where $d\mathrm{length}_\Sigma(\lambda)$ denotes the length element along the leaves of λ.

It follows that for a weighted simple closed geodesic $w\gamma$, its length defined as above is equal to $w\mathrm{length}_\Sigma(\gamma)$.

For each point (Σ, f) in $\mathcal{T}(S)$ and $\lambda \in \mathcal{ML}(S)$ (with its transverse measure included), we can consider $\mathrm{length}_\Sigma(f(\lambda))$. Fixing λ, we can define the function $\mathrm{Length}_\lambda : \mathcal{T}(S) \to \mathbb{R}$. In [30], Thurston showed the following infinitesimal form of his asymmetric metric, which gives $\mathcal{T}(S)$ a Finsler structure.

Proposition 10.5.11 *Let* $x = (\Sigma, f)$ *be a point in* $\mathcal{T}(S)$, *and* v *a tangent vector of* $\mathcal{T}(S)$ *at* x. *Then the (asymmetric) norm of* v *giving a Finsler structure compatible with* d_L *is expressed in the form*

$$\|v\|_L = \sup_{\lambda \in \mathcal{ML}(S)\setminus\{\emptyset\}} \frac{(d\mathrm{Length}_\lambda)_x(v)}{\mathrm{length}_\Sigma(\lambda)}.$$

Now, we turn to cotangent spaces. We define the norm $\|\cdot\|_L^*$ on the cotangent space of $\mathcal{T}(S)$ at x to be the dual of the norm $\|\cdot\|_L$ defined above. For each $\lambda \in \mathcal{ML}(S) \setminus \{\emptyset\}$, the differential $(d\mathrm{Length}_\lambda)_x$ is a linear function defined on the tangent space $T_x\mathcal{T}(S)$, hence can be regarded as a cotangent vector at x. Therefore, $(d\mathrm{Length})_x$ takes $\mathcal{ML}(S) \setminus \{\emptyset\}$ to the cotangent space. The projective lamination space of Σ can be identified with the unit sphere in the cotangent space of $\mathcal{T}(S)$ at (Σ, f) as follows.

Theorem 10.5.12 *For each* $x = (\Sigma, f)$ *in* $\mathcal{T}(S)$, *the map* $\iota_x := d(\log \mathrm{Length})_x$ *embeds* $\mathcal{PML}(\Sigma)$ *onto the unit sphere (with respect to* $\|\cdot\|_L^*$) *of the cotangent space* $T_x^*\mathcal{T}(S)$, *and its image bounds a ball which is convex with respect to* $\|\cdot\|_L^*$.

This theorem suggests that we can characterise cotangent vectors using properties of projective laminations. Indeed Huang, Papadopoulos and the present author [13] introduced the following notions to understand the structure of $\iota_x(\mathcal{PML}(S))$.

We first define an operation of taking the support closure for measured laminations.

Definition 10.5.13 Let λ be a measured or a projective lamination on a hyperbolic surface Σ, and let $|\lambda|$ denote the geodesic lamination which is the support of λ. For each component λ_i of λ, its *minimal supporting surface* is an incompressible subsurface of Σ containing λ_i and minimal among such surfaces up to isotopy. We add to $|\lambda|$ the closed geodesics homotopic to the boundary components of the minimal supporting surfaces of the components of λ that are not already contained in $|\lambda|$, and get the *support closure* $\widehat{|\lambda|}$ of λ.

Next we define some notions on a convex sphere.

Definition 10.5.14 Let S be a convex sphere, i.e., a topological sphere bounding a convex ball B in the Euclidean space $\mathbb{E}^n$.

(a) For $P \in S$, we let A_P be the affine subspace of $\mathbb{E}^n$ of the highest dimension such that P is an interior point of $A_P \cap S$ as a subset of A_P. It can be proved that such an affine surface is unique for each P. The intersection $A_P \cap S$ is called the *face* for P. The dimension of the face $A_P \cap S$ is defined to be the *dimension* of A_P. We have a decomposition of S into the interiors (again as subsets of A_P as above) of faces.

(b) A subface F of a face F' is said to be *adherent* to F' if for any face F'' containing F as a subface, there is a face $\hat{F}$ which contains both F' and F'' as subfaces.

(c) Every face F has a unique maximal face to which it is adherent. Such a face is called the *adherence closure* of F. The dimension of the adherence closure of the face of P is defined to be the *face dimension* of P.

(d) A face whose adherence closure is itself is said to be adherence closed. For an adherence closed face F, the union of the interiors of all faces whose adherence closure is F is called the *adherence core* of F.

Having prepared these notions and terms, we can now describe the image of the projective lamination space as a convex sphere.

Theorem 10.5.15 *For $x \in \mathfrak{T}(S)$, let $\iota_x \colon \mathcal{PML}(S) \to T_x^*\mathfrak{T}(S)$ be the embedding given in Theorem 10.5.12. Then for any $[\lambda] \in \mathcal{PML}(S)$ the following hold. Let $\hat{F}$ be the adherence closure of the face of $\iota_x([\lambda])$.*

(1) *The face $\hat{F}$ is the image under ι_x of the set of all projective laminations whose supports are contained in $\widehat{|\lambda|}$.*

(2) *The interior of $\hat{F}$ is the image under ι_x of a set consisting of projective laminations with the same support as $\widehat{|\lambda|}$.*

(3) *The image under ι_x of the set of all projective laminations whose support closures are $\widehat{|\lambda|}$ coincides with the adherence core of F.*

(4) *The face dimension of $\iota_x([\lambda])$ is the dimension of the space of transverse measures on $\widehat{|\lambda|}$ minus 1.*

Making use of this theorem, and observing that adherence and adherence closure are preserved by a linear isometry, we can prove the following analogue of the first half of Ivanov's theorem, which we explained in Sect. 10.4, in the setting of Thurston's asymmetric metric.

Theorem 10.5.16 *Let $f: T_x^* \mathcal{T}(S) \to T_y^* \mathcal{T}(S)$ be a linear isometry with respect to $\| \cdot \|_L^*$. Then there is an extended mapping class g such that $\iota_y^{-1} \circ f \circ \iota_x: \mathcal{PML}(S) \to \mathcal{PML}(S)$ coincides with the natural action of g on $\mathcal{PML}(S)$.*

To obtain the infinitesimal rigidity theorem corresponding to Royden's theorem from this theorem, the essential step is the following theorem.

Theorem 10.5.17 *Let $f: T_x \mathcal{T}(S) \to T_y \mathcal{T}(S)$ be a linear isometry between tangent spaces with respect to $\| \cdot \|_L$, and g the extended mapping class given in Theorem 10.5.16 for the dual of f. Then for any essential simple closed curve s on S, we have $\mathrm{length}_x(g(s)) = \mathrm{length}_y(s)$.*

Now, let $\iota: \mathcal{T}(S) \to \mathbb{R}^{\mathcal{S}}$ be the map taking x to $(\mathrm{length}_x(s))_{s \in \mathcal{S}}$. It is known (via the work of Fenchel–Nielsen) that ι is injective. Therefore, Theorem 10.5.17 implies the following infinitesimal rigidity theorem.

Theorem 10.5.18 *Let $f: T_x \mathcal{T}(S) \to T_y \mathcal{T}(S)$ be a linear isometry between tangent spaces. Then there is an extended mapping class g such that $y = g_*(x)$ and $f = d(g_*)_x$.*

The proof of Theorem 10.5.17 is quite involved, and we refer the reader to [13].

The infinitesimal rigidity theorem for Thurston's asymmetric metric was also obtained independently by Pan [23] using a method quite different from ours.

References

1. A. A'Campo-Neuen, N. A'Campo, V. Alberge, A. Papadopoulos,   A commentary on Teich-müller's paper Bestimmung der extremalen quasikonformen Abbildungen bei geschlossenen orientierten Riemannschen Flächen, in *Handbook of Teichmüller theory. Vol. V*, vol. 26 of IRMA Lect. Math. Theor. Phys. (European Mathematical Society, Zürich, 2016), pp. 569–580
2. V. Alberge, A. Papadopoulos, On five papers by Herbert Grötzsch, in *Handbook of Teichmüller Theory, Volume VII*, ed. by A. Papadopoulos, vol. 30 of IRMA Lectures in Mathematics and Theoretical Physics (European Mathematical Society, Zürich, 2020)
3. V. Alberge, A. Papadopoulos, W. Su,   A commentary on Teichmüller's paper Extremale quasikonforme Abbildungen und quadratische Differentiale, in *Handbook of Teichmüller Theory. Vol. V*, vol. 26 of IRMA Lect. Math. Theor. Phys. (European Mathematical Society, Zürich, 2016), pp. 485–531
4. L. Bers, An extremal problem for quasiconformal mappings and a theorem by Thurston. Acta Math. **141**(1–2), 73–98 (1978)
5. L. Bers, F.P. Gardiner, Fricke spaces. Adv. Math. **62**(3), 249–284 (1986)
6. C.J. Earle, I. Kra,   On isometries between Teichmüller spaces. Duke Math. J. **41**, 583–591 (1974)

7. C.J. Earle, V. Markovic,   Isometries between the spaces of L^1 holomorphic quadratic differentials on Riemann surfaces of finite type. Duke Math. J. **120**(2), 433–440 (2003)
8. A. Fathi, V. Poénaru, F. Laudenbach,   *Travaux de Thurston sur les surfaces*, vol. 66 of Astérisque (Société Mathématique de France, Paris, 1979). Séminaire Orsay
9. R. Fricke, F. Klein,   *Vorlesungen über die Theorie der elliptischen Modulfunctionen* (B.G. Teubner, Leipzig, 1890–1892)
10. H. Grötzsch, Über möglichst konforme Abbildungen von schlichten Bereichen. Ber. Verh. Sächs. Akad. Leipzig **84**, 114–120 (1932)
11. H. Grötzsch, On closest-to-conformal mappings (translated by Melkana Brakalova-Trevithick), in *Handbook of Teichmüller Theory, Volume VII*, ed. by A. Papadopoulos, vol. 30 of IRMA Lectures in Mathematics and Theoretical Physics (IRMA) (European Mathematical Society Publishing House, Zurich, 2020)
12. W.J. Harvey, Boundary structure of the modular group, in *Riemann Surfaces and Related Topics: Proceedings of the 1978 Stony Brook Conference (State Univ. New York, Stony Brook, N.Y., 1978)*, vol. 97 of Ann. of Math. Stud. (Princeton University Press, Princeton, NJ, 1981), pp. 245–251
13. Y. Huang, K. Ohshika, A. Papadopoulos, The infinitesimal and global Thurston geometry of Teichmüller space, arXiv 2111.13381
14. J. Hubbard, H. Masur, Quadratic differentials and foliations. Acta Math. **142**(3–4), 221–274 (1979)
15. N.V. Ivanov, Automorphism of complexes of curves and of Teichmüller spaces. Int. Math. Res. Notices **14**, 651–666 (1997)
16. N.V. Ivanov, Isometries of Teichmüller spaces from the point of view of Mostow rigidity, in *Topology, Ergodic Theory, Real Algebraic Geometry*, vol. 202 of Amer. Math. Soc. Transl. Ser. 2 (American Mathematical Society, Providence, RI, 2001), pp. 131–149
17. S.P. Kerckhoff, The asymptotic geometry of Teichmüller space. Topology **19**(1), 23–41 (1980)
18. M. Korkmaz, Automorphisms of complexes of curves on punctured spheres and on punctured tori. Topology Appl. **95**(2), 85–111 (1999)
19. G. Levitt, Foliations and laminations on hyperbolic surfaces. Topology **22**(2), 119–135 (1983)
20. F. Luo, Automorphisms of the complex of curves. Topology **39**(2), 283–298 (2000)
21. H.A. Masur, Y.N. Minsky, Geometry of the complex of curves. I. Hyperbolicity. Invent. Math. **138**(1), 103–149 (1999)
22. H.A. Masur, Y.N. Minsky,   Geometry of the complex of curves. II. Hierarchical structure. Geom. Funct. Anal. **10**(4), 902–974 (2000)
23. H. Pan, Local rigidity of teichmüller space with Thurston metric. Preprint. arxiv 2005.11762 (2020)
24. A. Papadopoulos, Réseaux ferroviaires, difféomorphismes pseudo-Anosov et automorphismes symplectiques de l'homologie d'une surface. Publications Mathématiques d'Orsay, 1983
25. H.L. Royden, Automorphisms and isometries of Teichmüller space, in *Advances in the Theory of Riemann Surfaces (Proc. Conf., Stony Brook, N.Y., 1969)*. Ann. of Math. Studies, No. 66 (Princeton University Press, Princeton, NJ, 1971), pp. 369–383
26. O. Teichmüller, Extremale quasikonforme Abbildungen und quadratische Differentiale. Abh. Preuß. Akad. Wiss., Math.-Naturw. Kl. 1940 **22**, 197 (1940)
27. O. Teichmüller, Bestimmung der extremalen quasikonformen Abbildungen bei geschlossenen orientierten Riemannschen Flächen. Abh. Preuß. Akad. Wiss., Math.-Naturw. Kl. 1943 **4**, 1–42 (1943)
28. O. Teichmüller, Determination of extremal quasiconformal mappings of closed oriented Riemann surfaces (translated by Annette A'Campo-Neuen), in *Handbook of Teichmüller Theory. Volume V* (European Mathematical Society (EMS), Zürich, 2016), pp. 533–567

29. O. Teichmüller, Extremal quasiconformal mappings and quadratic differentials (translated by Guillaume Théret), in *Handbook of Teichmüller theory. Volume V* (European Mathematical Society (EMS), Zürich, 2016), pp. 321–483
30. W.P. Thurston, Minimal stretch maps between hyperbolic surfaces. https://arxiv.org/pdf/math/9801039.pdf
31. W.P. Thurston, On the geometry and dynamics of diffeomorphisms of surfaces. Bull. Amer. Math. Soc. (N.S.) **19**(2), 417–431 (1988)

Chapter 11
Holomorphic *G*-Structures and Foliated Cartan Geometries on Compact Complex Manifolds

Indranil Biswas and Sorin Dumitrescu

Abstract This is a survey dealing with holomorphic *G*-structures and holomorphic Cartan geometries on compact complex manifolds. Our emphasis is on the foliated case. We investigate holomorphic foliations with a transverse holomorphic Cartan geometry, and also with the more general structure of branched transverse holomorphic Cartan geometry.

The first part of the chapter presents the geometric notion of holomorphic *G*-structure whose origin was motivated by various classical examples. We explain some classification results for compact complex manifolds endowed with holomorphic *G*-structures highlighting the special case of GL(2, $\mathbb{C}$)-structures and SL(2, $\mathbb{C}$)-structures. The second part of the survey deals with holomorphic Cartan geometry in the classical case and in the branched and generalized case. Two definitions of foliated (branched, generalized) Cartan geometry are described and shown to be equivalent. We provide some classification results of compact complex manifolds with foliated (branched or generalized) Cartan geometries. At the end some related open problems are formulated.

Keywords Holomorphic geometric structure · Holomorphic GL(2)-structure · Kähler manifold · Foliated principal bundle · Foliated Cartan geometry

1991 Mathematics Subject Classification 53C07, 53C10, 32Q57

I. Biswas
School of Mathematics, Tata Institute of Fundamental Research, Mumbai, India
e-mail: indranil@math.tifr.res.in

S. Dumitrescu (✉)
Université Côte d'Azur, CNRS, LJAD, Nice, France
e-mail: dumitres@unice.fr

11.1   Introduction

In this survey paper we deal with holomorphic geometric structures on compact complex manifolds. We focus on holomorphic G-structures and on holomorphic Cartan geometries.

This subject of geometric structures has its roots in the study of the symmetry group of various geometric spaces. Important seminal work which shaped the subject was done by F. Klein and S. Lie. In particular, in his famous Erlangen address (1872), F. Klein established the program of a unifying background for all geometries (including the Euclidean, affine and projective geometries) as being that of homogeneous spaces $X = G/H$ under (the transitive) action of a symmetry group G; here G is a finite dimensional connected Lie group, and H is the stabilizer of a given point in the model space X.

The special case of the complex projective line $\mathbb{CP}^1$ seen as a homogeneous space for the Möbius group $\mathrm{PSL}(2, \mathbb{C})$ appears naturally in the study of second order linear differential equations on the complex domain. More precisely, the quotient of two local linearly independent solutions of an equation of the above type provides a local coordinate in $\mathbb{CP}^1$ on which the monodromy of the equation (obtained by analytic continuation of the local solution along a closed curve avoiding the singularities) acts by elements in the Möbius group $\mathrm{PSL}(2, \mathbb{C})$. This procedure naturally leads to a *complex projective structure* on the Riemann surface M bearing the second order linear differential equation, meaning that M admits a holomorphic atlas with local coordinates in $\mathbb{CP}^1$ such that the transition maps are in $\mathrm{PSL}(2, \mathbb{C})$.

Inspired by Fuchs' work on second order linear differential equations, Poincaré was the first to understand that the above procedure provides a promising background which could lead to a general uniformization theorem for any Riemann surface M: one should find a second order linear differential equation on M which is *uniformizing*, in the sense that, when pull-backed to the universal cover of M, the quotient of two local solutions provides a biholomorphic identification between the universal cover of M and an open set U in the complex projective line $\mathbb{CP}^1$. The monodromy of the equation defined on M furnishes a group homomorphism from the fundamental group of M into the subgroup of the Möbius group $\mathrm{PSL}(2, \mathbb{C})$ preserving $U \subset \mathbb{CP}^1$ and acting properly and discontinuously on U. The above identification (between the universal cover of M and $U \subset \mathbb{CP}^1$) is equivariant with respect to the monodromy morphism; this identification map is called the *developing map*. Consequently, the Riemann surface is uniformized as a quotient of $U \subset \mathbb{CP}^1$ by a subgroup of $\mathrm{PSL}(2, \mathbb{C})$ acting properly and discontinuously on U. Of course, in the generic (hyperbolic) case the open subset U is the upper-half plane and the corresponding subgroup in the Möbius group is a discrete subgroup of $\mathrm{PSL}(2, \mathbb{R})$. A complete proof based on the above considerations and also other proofs of the uniformization theorem for surfaces together with the historical background are presented in [23] (see also [34]).

Another extremely important root of the subject is Elie Cartan's broad generalization of Klein's homogeneous model spaces to the corresponding infinitesimal notion, namely that of *Cartan geometries* (or *Cartan connections*) [61]. Those geometric structures are infinitesimally modeled on homogeneous spaces G/H. A Cartan geometry on a manifold M is naturally equipped with a curvature tensor which vanishes exactly when M is locally modeled on G/H in the following sense described by Ehresmann [28]. In such a situation the Cartan geometry is called *flat*.

A manifold M is said to be *locally modeled on a homogeneous space G/H* if M admits an atlas with charts in G/H satisfying the condition that the transition maps are given by elements of G using the left-translation action of G on G/H. In this way M is locally endowed with the G/H-geometry and all G-invariant geometrical features of G/H have intrinsic meaning on M [28].

In the same spirit as for complex projective structures on Riemann surfaces, homogeneous spaces are useful models for geometrization of topological manifolds of higher dimension. This was Thurston's point of view when he formulated the geometrization conjecture for threefolds, using three-dimensional Riemannian homogeneous spaces G/H (or equivalently, H is compact).

Also Inoue, Kobayashi and Ochiai studied holomorphic projective connections on compact complex surfaces. A consequence of their work is that compact complex surfaces bearing holomorphic projective connections also admit a (flat) holomorphic projective structure such that the corresponding developing map is injective. In particular, such complex surfaces are uniformized as quotients of open subsets U in the complex projective plane $\mathbb{C}P^2$ by a discrete subgroup in PSL(3, $\mathbb{C}$) acting properly and discontinuously on U [38, 46, 48]. Of course, most of the complex compact surfaces do not admit any holomorphic projective connection. It appears to be useful to allow more flexibility in the form of orbifold (or other mild) singularities.

In this spirit [54, 55] Mandelbaum introduced and studied *branched projective structures* on Riemann surfaces. A branched projective structure on a Riemann surface is given by some holomorphic atlas whose local charts are finite branched coverings of open subsets in $\mathbb{C}P^1$ while the transition maps lie in PSL(2, $\mathbb{C}$).

Inspired by the above mentioned articles of Mandelbaum, we defined in [12] a more general notion of *branched holomorphic Cartan geometry* on a complex manifold M which is valid also for complex manifolds in higher dimension and for non-flat Cartan geometries. We defined and studied a more general notion of *generalized holomorphic Cartan geometry* in [9] (see also, [1]).

These notions of generalized and branched Cartan geometry are much more flexible than the usual one: branched holomorphic Cartan geometry are stable under pull-back through holomorphic ramified maps, while generalized holomorphic Cartan geometries are stable by pull-back through holomorphic maps. Also all compact complex projective manifolds admit branched flat holomorphic projective structure [12].

A foliated version of a Cartan geometry (meaning Cartan geometry transverse to a foliation) was worked out in [19, 57]. We defined and studied the notion of a foliated holomorphic branched Cartan geometry (meaning a branched holomorphic Cartan geometry transverse to a holomorphic foliation) in [10]. We define here also the foliated version of generalized Cartan geometry.

We show that these notions are rigid enough to enable one to obtain classification results.

The organization of this paper is as follows. After this introduction, Sect. 11.2 presents the geometric notion of holomorphic G-structure and a number of examples. Section 11.3 focuses on the special case of GL$(2, \mathbb{C})$-structures and SL$(2, \mathbb{C})$-structures and surveys some classification results. Section 11.4 is about holomorphic Cartan geometry in the classical case and in the branched and generalized case. Section 11.5 deals with two equivalent definitions of foliated (branched, generalized) Cartan geometry and provides some classification results. At the end in Sect. 11.6 some related open problems are formulated.

11.2 Holomorphic G-Structure

Let M be a complex manifold of complex dimension n. We denote by TM (respectively, T^*M) its holomorphic tangent (respectively, cotangent) bundle. Let $R(M)$ be the holomorphic *frame bundle* associated with TM. We recall that for any point $m \in M$, the fiber $R(M)_m$ is identified with the set of $\mathbb{C}$-linear maps from $\mathbb{C}^n$ to $T_m M$. The pre-composition with elements in GL$(n, \mathbb{C})$ defines a (right) GL$(n, \mathbb{C})$-action on $R(M)$. This action is holomorphic, free and transitive in the fibers; therefore the quotient of $R(M)$ by GL$(n, \mathbb{C})$ is the complex manifold M. The frame bundle is the typical example of a holomorphic principal GL$(n, \mathbb{C})$-bundle over M.

Let G be a complex Lie subgroup of GL$(n, \mathbb{C})$. Then we have the following definition.

Definition 11.2.1 A holomorphic G-structure on M is a holomorphic principal G-subbundle $\mathcal{R} \longrightarrow M$ in $R(M)$. Equivalently, $\mathcal{R}$ is a holomorphic reduction of the structure group of $R(M)$ to the subgroup $G \subset$ GL$(n, \mathbb{C})$.

It may be mentioned that the above definition is equivalent to the existence of a GL$(n, \mathbb{C})$-equivariant holomorphic map $\Psi : R(M) \longrightarrow$ GL$(n, \mathbb{C})/G$. The pre-image, under Ψ, of the class of the identity element in GL$(n, \mathbb{C})/G$ is the holomorphic G-subbundle in the above definition. Notice that the map Ψ is a holomorphic section of the bundle $R(M)/G$ (with fiber type GL$(n, \mathbb{C})/G$), where the G action on $R(M)$ is the restriction of the principal GL$(n, \mathbb{C})$-action.

The G-structure $\mathcal{R} \subset R(M)$ is said to be *flat* (or, *integrable*) if for every $m \in M$, there exists an open neighborhood U of $0 \in \mathbb{C}^n$ and a holomorphic local biholomorphism

$$f : (U, 0) \longrightarrow (M, m)$$

such that the image of the differential df of f, lifted as a map $df : R(U) \longrightarrow R(M)$, satisfies $df(s(U)) \subset \mathcal{R}$, where $s : U \longrightarrow R(U)$ is the section of the projection $R(U) \longrightarrow U$ given by the standard frame of $\mathbb{C}^n$.

For more details about this concept the reader is referred to [45]. Let us now describe some important examples which are at the origin of this concept.

$SL(n, C)$-**Structure** An $SL(n, \mathbb{C})$-structure on M is equivalent with giving a non-vanishing holomorphic section of the canonical bundle $K_M = \bigwedge^n T^*M$. This holomorphic section $\omega \in H^0(M, K_M)$ defines the holomorphic principal $SL(n, \mathbb{C})$-subbundle in $R(M)$ with fiber above $m \in M$ given by the subset $\mathcal{R}_m \subset R(M)_m$ consisting of all frames $l : \mathbb{C}^n \longrightarrow T_m M$ such that $l^*\omega_m$ is the n-form on $\mathbb{C}^n$ given by the determinant. Therefore, $\mathcal{R} \subset R(M)$ is defined by those frames which have volume 1 with respect to ω.

Recall that with respect to holomorphic local coordinates $(z_1, \ldots, z_n)$ on M, a holomorphic section $\omega \in H^0(M, K_M)$ is given by an expression $f(z_1, \ldots, z_n)dz_1 \wedge dz_2 \wedge \ldots \wedge dz_n$, with f a non-vanishing local holomorphic function. We can always operate a biholomorphic change of local coordinates in order to have f identically equal to 1. This implies that an $SL(n, \mathbb{C})$-structure is always flat.

$Sp(2n, C)$-**Structure** This structure is equivalent with giving a holomorphic pointwise non-degenerate holomorphic 2-form on M. Such a section defines an isomorphism between TM and T^*M and implies that M has even dimension $2n$. Equivalently, an $Sp(2n, \mathbb{C})$-structure is a section $\omega_0 \in H^0(M, \bigwedge^2 T^*M)$ such that ω_0^n is a nowhere vanishing section of $\bigwedge^{2n} T^*M$. The corresponding $Sp(2n, \mathbb{C})$-structure is flat if there exists holomorphic local coordinates with respect to which the local expression of ω_0 is $dz_1 \wedge dz_2 + \ldots + dz_{2n-1} \wedge dz_{2n}$. By Darboux theorem this is equivalent to the condition that $d\omega_0 = 0$, i.e., ω_0 is a holomorphic symplectic form (or equivalently, a holomorphic non-degenerate closed 2-form).

If M is a compact Kähler manifold, then any holomorphic form on M is automatically closed. In this case any holomorphic $Sp(2n, \mathbb{C})$-structure on M is flat: it endows M with a holomorphic symplectic form.

Holomorphic symplectic structures on compact complex non Kähler manifolds were constructed in [32, 33].

$O(n, C)$-**Structure** Giving an $O(n, \mathbb{C})$-structure on M is equivalent to giving a pointwise nondegenerate (i.e., of rank n) holomorphic section g of the bundle $S^2(T^*M)$ of symmetric complex quadratic forms in the holomorphic tangent bundle of M. This structure is also known as a *holomorphic Riemannian metric* [10, 24, 26, 30, 52]. Notice that the complexification of any real-analytic Riemannian or pseudo-Riemannian metric defines a local holomorphic Riemannian metric which coincides with the initial (pseudo-)Riemannian metric on the real locus.

The fiber above $m \in M$ of the corresponding subbundle $\mathcal{R}_m \subset R(M)_m$ is formed by all frames $l : \mathbb{C}^n \longrightarrow T_m M$ such that $l^*(g_m) = dz_1^2 + \ldots + dz_n^2$. In local holomorphic coordinates $(z_1, \ldots, z_n)$ on M the local expression of the section $g \in H^0(M, S^2(T^*M))$ is $\sum_{ij} g_{ij} dz_i dz_j$, with g_{ij} a nondegenerate $n \times n$ matrix with holomorphic coefficients. There exists local coordinates with respect to which the matrix g_{ij} is locally constant if and only if the holomorphic Riemannian metric is flat (e.g., has vanishing curvature tensor). Hence the $O(n, \mathbb{C})$-structure is flat if and only if the corresponding holomorphic Riemannian metric is flat (i.e, there exists local holomorphic coordinates with respect to which the local expression of g is $dz_1^2 + \ldots + dz_n^2$).

Homogeneous Tensors The three previous examples are special cases of the following construction. Consider a holomorphic linear $GL(n, \mathbb{C})$-action on a finite dimensional complex vector space V given by $i : GL(n, \mathbb{C}) \longrightarrow GL(V)$. This action together with $R(M)$ define a holomorphic vector bundle with fiber type V by the usual quotient construction: two elements $(l, v), (l', v') \in R(M) \times V$ are equivalent if there exists $g \in GL(n, \mathbb{C})$ such that

$$(l', v') = (l \cdot g^{-1}, i(g) \cdot v).$$

The holomorphic vector bundle constructed this way will be denoted $R(M) \ltimes_i V$.

Note that for the linear representation

$$V^{p,q} = (\mathbb{C}^n)^{\otimes p} \otimes ((\mathbb{C}^n)^*)^{\otimes q},$$

where p, q are nonnegative integers, the corresponding vector bundle $R(M) \ltimes V$ coincides with the bundle of holomorphic tensor $TM^{\otimes p} \otimes (TM^*)^{\otimes q}$. Moreover any irreducible algebraic representation $i : GL(n, \mathbb{C}) \longrightarrow GL(V)$ is known to be a factor of the representation $(\mathbb{C}^n)^{\otimes p} \otimes ((\mathbb{C}^n)^*)^{\otimes q}$ for some integers (p, q),

Now consider a holomorphic section of $TM^{\otimes p} \otimes (TM^*)^{\otimes q}$ (namely, a holomorphic tensor of type (p, q)) given by a $GL(n, \mathbb{C})$-equivariant map

$$\Theta : R(M) \longrightarrow V^{p,q}.$$

Assume that the image of Θ lies in a $GL(n, \mathbb{C})$-orbit of $V^{p,q}$, with stabilizer $G_0 \subset GL(n, \mathbb{C})$. Such a tensor is called homogeneous (of order 0). Under this assumption, the map $\Theta : R(M) \longrightarrow GL(n, \mathbb{C})/G_0$ defines a holomorphic G_0-structure on M. Conversely, any G_0-structure comes in the above way from a unique (up to a scalar constant) holomorphic tensor Θ.

$CO(n, C)$-Structure Here $CO(n, \mathbb{C}) \subset GL(n, \mathbb{C})$ is the subgroup generated by the homotheties and $O(n, \mathbb{C})$; it is known as the *orthogonal similarity group*. It is the stabilizer of the line generated by the standard holomorphic Riemannian metric $dz_1^2 + \ldots + dz_n^2$ in the linear representation of $GL(n, \mathbb{C})$ on $S^2((\mathbb{C}^n)^*)$. Therefore a $CO(n, \mathbb{C})$—structure is *a holomorphic conformal structure* on M: it is given by a holomorphic line $\mathcal{L} \subset S^2(T^*M)$ such that any nonvanishing local holomorphic section of $\mathcal{L}$ is pointwise a nondegenerate complex quadratic form on the holomorphic tangent bundle. A holomorphic conformal structure is flat if and only if a nontrivial holomorphic local section of $\mathcal{L}$ can be expressed in local holomorphic coordinates as $f(z_1, \ldots, z_n)(dz_1^2 + \ldots + dz_n^2)$, with f a holomorphic function. By a famous result of Gauss, this is always the case when the complex dimension n equals 2. In higher dimensions the flatness of the holomorphic conformal structures is equivalent to the vanishing of a certain curvature tensor. The curvature tensor in question is known as the Weyl conformal tensor.

CSp(2*n*, *C*)-Structure The subgroup $CSp(2n, \mathbb{C}) \subset GL(2n, \mathbb{C})$ is generated by $Sp(2n, \mathbb{C})$ and the homotheties; it is known as the *symplectic similarity group*. This $CSp(2n, \mathbb{C})$ is the stabilizer of the line generated by the standard symplectic form $dz_1 \wedge dz_2 + \ldots + dz_{2n-1} \wedge dz_{2n}$ in the linear representation of $GL(n, \mathbb{C})$ on $\bigwedge^2((\mathbb{C}^n)^*)$. On a complex manifold M, a $CSp(2n, \mathbb{C})$-structure is a holomorphic line subbundle $\mathcal{L} \subset \bigwedge^2(TM^*)$ such that any nonvanishing local holomorphic section is pointwise nondegenerate. This structure is called a *nondegenerate holomorphic twisted 2-form*. It is flat exactly when a nontrivial holomorphic local section is locally expressed as $f(z_1, \ldots, z_n)(dz_1 \wedge dz_2 + \ldots + dz_{2n-1} \wedge dz_{2n})$, with f a local holomorphic function. Moreover, f can be chosen to be constant if and only if the local section is closed. In this case we say that M is endowed with a *twisted holomorphic symplectic structure*.

GL(2, *C*)-Structure Here the corresponding group homomorphism

$$GL(2, \mathbb{C}) \longrightarrow GL(n, \mathbb{C})$$

is defined by the $(n - 1)$-th symmetric product $S^{(n-1)}(\mathbb{C}^2)$ of the standard representation of $GL(2, \mathbb{C})$. Recall that $S^{(n-1)}(\mathbb{C}^2)$ gives the unique, up to tensoring with a one-dimensional representation, irreducible $GL(2, \mathbb{C})$-linear representation in complex dimension n. This representation is the induced $GL(2, \mathbb{C})$-action on the homogeneous polynomials of degree $(n - 1)$ in two variables.

Therefore giving a holomorphic $GL(2, \mathbb{C})$-structure on a complex manifold M of complex dimension $n \geq 2$ is equivalent to giving a holomorphic rank two vector bundle E over M together with a holomorphic isomorphism $TM \overset{\simeq}{\longrightarrow} S^{n-1}(E)$, where $S^{n-1}(E)$ is the $(n - 1)$-th symmetric power of E.

SL(2, *C*)-Structure In this case the group homomorphism $SL(2, \mathbb{C}) \longrightarrow GL(n, \mathbb{C})$ is defined by the $(n - 1)$-th symmetric product $S^{(n-1)}(\mathbb{C}^2)$ of the standard representation of $SL(2, \mathbb{C})$. This is the restriction to $SL(2, \mathbb{C}) \subset GL(2, \mathbb{C})$ of the homomorphism in the previous example.

An $SL(2, \mathbb{C})$-structure on a complex manifold M of complex dimension $n \geq 2$ is the same data as a holomorphic rank two vector bundle E with trivial determinant over M together with a holomorphic isomorphism $TM \overset{\simeq}{\longrightarrow} S^{n-1}(E)$.

11.3   GL(2)-Geometry and SL(2)-Geometry

In this section we study the holomorphic $GL(2, \mathbb{C})$-structures (also called GL(2)-geometry) and the holomorphic $SL(2, \mathbb{C})$-structures (also called SL(2)-geometry) on complex manifolds.

An shown before, holomorphic $GL(2, \mathbb{C})$-structures and $SL(2, \mathbb{C})$-structures are in fact particular cases of holomorphic irreducible reductive G-structures [37, 45]. Recall that they correspond to the holomorphic reduction of the structure group of the frame bundle $R(M)$ of the manifold M from $GL(n, \mathbb{C})$ to $GL(2, \mathbb{C})$ and $SL(2, \mathbb{C})$ respectively. Also, recall that for a $GL(2, \mathbb{C})$-structure, the corresponding group homomorphism $GL(2, \mathbb{C}) \longrightarrow GL(n, \mathbb{C})$ is given by the $(n-1)$-th symmetric product of the standard representation of $GL(2, \mathbb{C})$. This n-dimensional irreducible linear representation of $GL(2, \mathbb{C})$ is also given by the induced action on the homogeneous polynomials of degree $(n-1)$ in two variables. For an $SL(2, \mathbb{C})$-geometry, the corresponding homomorphism $SL(2, \mathbb{C}) \longrightarrow GL(n, \mathbb{C})$ is the restriction of the above homomorphism to $SL(2, \mathbb{C}) \subset GL(2, \mathbb{C})$.

Proposition 11.3.1 *Let M be a complex manifold endowed with a holomorphic $SL(2, \mathbb{C})$-structure (respectively, a holomorphic $GL(2, \mathbb{C})$-structure). Then the following hold:*

(i) *If the complex dimension of M is odd, then M is endowed with a holomorphic Riemannian metric (respectively, a holomorphic conformal structure).*

(ii) *If the complex dimension of M is even, then M is endowed with a nondegenerate holomorphic 2-form (respectively, a nondegenerate holomorphic twisted 2-form). Moreover, if M is Kähler then M is endowed with a holomorphic symplectic form (respectively, a holomorphic twisted symplectic form).*

Proof A holomorphic $SL(2, \mathbb{C})$-structure on a complex surface M is a holomorphic trivialization of the canonical bundle $K_M = \bigwedge^2 T^*M$. Indeed, the standard action of $SL(2, \mathbb{C})$ on $\mathbb{C}^2$ preserves the holomorphic volume form $dz_1 \wedge dz_2$.

Notice that in the case of complex dimension two the holomorphic volume form $dz_1 \wedge dz_2$ coincides with the standard holomorphic symplectic structure. Hence the (flat) $SL(2, \mathbb{C})$-structure coincide with the (flat) $Sp(2, \mathbb{C})$-structure.

This produces a nondegenerate complex quadratic form on $S^{n-1}(\mathbb{C}^2)$, for any $n - 1$ even. Also, for any odd integer $n-1$, this endows $S^{n-1}(\mathbb{C}^2)$ with a nondegenerate complex alternating 2-form on the symmetric product $S^{n-1}(\mathbb{C}^2)$. Consequently, the corresponding irreducible linear representation $SL(2, \mathbb{C}) \longrightarrow GL(n, \mathbb{C})$ preserves a nondegenerate complex quadratic form on $\mathbb{C}^n$, if n is odd and a nondegenerate 2-form on $\mathbb{C}^n$, if n is even (see, for example, Proposition 3.2 and Sections 2 and 3 in [27]; see also [51]).

The above linear representation $GL(2, \mathbb{C}) \longrightarrow GL(n, \mathbb{C})$ preserves the line in $(\mathbb{C}^n)^* \otimes (\mathbb{C}^n)^*$ spanned by the above tensor. The action of $GL(2, \mathbb{C})$ on this line is nontrivial.

(i) Assume that the complex dimension of *M* is odd. The above observations imply that the SL(2, $\mathbb{C}$)-geometry (respectively, GL(2, $\mathbb{C}$)-geometry) induces a *holomorphic Riemannian metric* (respectively, a holomorphic conformal structure) on M.

(ii) Assume that the complex dimension of *M* is even. The SL(2, $\mathbb{C}$)-geometry (respectively, GL(2, $\mathbb{C}$)-geometry) induces a holomorphic nondegenerate 2-form (respectively, a holomorphic twisted nondegenerate 2-form).

Moreover, if *M* is Kähler, any holomorphic differential form on *M* is closed. This implies that the holomorphic nondegenerate 2-form is a holomorphic symplectic form.   □

The simplest nontrivial examples of GL(2, $\mathbb{C}$) and SL(2, $\mathbb{C}$) structures are provided by the complex threefolds. In this case we have the following:

Proposition 11.3.2 *Let M be a complex threefold.*

(i) *A holomorphic* SL(2, $\mathbb{C}$)*-structure on M is a holomorphic Riemannian metric.*
(ii) *A holomorphic* GL(2, $\mathbb{C}$)*-structure on M is a holomorphic conformal structure.*

Proof

(i) Recall that $S^2(\mathbb{C}^2)$ is the unique irreducible SL(2, $\mathbb{C}$)-representation on the three-dimensional complex vector space. It coincides with the natural action of SL(2, $\mathbb{C}$) on the vector space of homogeneous quadratic polynomials in two variables

$$\{aX^2 + bXY + cY^2 \mid a, b, c \in \mathbb{C}\}. \tag{11.3.1}$$

The above (coordinate changing) action preserves the discriminant

$$\Delta := b^2 - 4ac \tag{11.3.2}$$

which is a nondegenerate complex quadratic form. Notice also that the action of $-\mathrm{Id} \in \mathrm{SL}(2, \mathbb{C})$ is trivial. Consequently, we obtain a holomorphic isomorphism between PSL(2, $\mathbb{C}$) and the complex orthogonal group SO(3, $\mathbb{C}$), the connected component of the identity in the complex orthogonal group O(3, $\mathbb{C}$).

The discriminant Δ in (11.3.2) induces a holomorphic Riemannian metric on the complex threefold *M*. This holomorphic Riemannian metric is given by a reduction of the structural group of the frame bundle $R(M)$ to the orthogonal group O(3, $\mathbb{C}$). On a double unramified cover of *M* there is a reduction of the structural group of $R(M)$ to the connected component of the identity, namely the subgroup SO(3, $\mathbb{C}$).

A SL(2, $\mathbb{C}$)-geometry on a complex threefold *M* is exactly the same data as a holomorphic Riemannian metric on *M*. Moreover, a PSL(2, $\mathbb{C}$)-geometry is the same data as a holomorphic Riemannian metric with a compatible holomorphic orientation (i.e., a nowhere vanishing holomorphic section *vol* of $\bigwedge^3 T^*M$

such that any pointwise basis of TM which is orthogonal with respect to the holomorphic Riemannian metric has volume one).

(ii) Notice that the $GL(2, \mathbb{C})$-representation (by coordinate changing) on the vector space defined in (11.3.1) globally preserves the line generated by the discriminant Δ in (11.3.2)(and acts nontrivially on it). Recall that the action of -Id is trivial.

This gives an isomorphism between $GL(2, \mathbb{C})/(\mathbb{Z}/2\mathbb{Z})$ and the conformal group

$$CO(3, \mathbb{C}) = (O(3, \mathbb{C}) \times \mathbb{C}^*)/(\mathbb{Z}/2\mathbb{Z}) = SO(3, \mathbb{C}) \times \mathbb{C}^*.$$

Consequently, the $GL(2, \mathbb{C})$-structure coincides with a holomorphic reduction of the structure group of the frame bundle $R(M)$ to $CO(3, \mathbb{C})$. This holomorphic reduction of the structure group defines a holomorphic conformal structure on M. Notice that $CO(3, \mathbb{C})$ being connected, the two different holomorphic orientations of a three dimensional holomorphic Riemannian manifold are conformally equivalent. □

11.3.1 Models of the Flat Holomorphic Conformal Geometry

Recall that flat conformal structures in complex dimension $n \geq 3$ are locally modeled on the quadric

$$Q_n := \{[Z_0 : Z_1 : \cdots : Z_{n+1}] \mid Z_0^2 + Z_1^2 + \ldots + Z_{n+1}^2 = 0\} \subset \mathbb{CP}^{n+1}.$$

The holomorphic automorphism group of Q_n is $PSO(n + 2, \mathbb{C})$.

Let us mention that Q_n is identified with the real Grassmannian of oriented 2-planes in $\mathbb{R}^{n+2}$ (see, for instance, Section 1 in [42]). As a real homogeneous space we have the following identification:

$$Q_n = SO(n + 2, \mathbb{R})/(SO(2, \mathbb{R}) \times SO(n, \mathbb{R})).$$

Moreover, the standard action of $SO(n + 2, \mathbb{R})$ on Q_n is via holomorphic automorphisms. To see this let us describe the complex structure on Q_n from the (real) Lie group theoretic point of view.

Note that the real tangent bundle of $Q_n = SO(n+2, \mathbb{R})/(SO(2, \mathbb{R}) \times SO(n, \mathbb{R}))$ is identified with the vector bundle associated with the principal $SO(2, \mathbb{R}) \times SO(n, \mathbb{R})$-bundle

$$SO(n + 2, \mathbb{R}) \longrightarrow SO(n + 2, \mathbb{R})/(SO(2, \mathbb{R}) \times SO(n, \mathbb{R}))$$

for the adjoint action of $SO(2, \mathbb{R}) \times SO(n, \mathbb{R})$ on the quotient of real Lie algebras $so(n + 2, \mathbb{R})/(so(2, \mathbb{R}) \oplus so(n, \mathbb{R}))$. The complex structure of the tangent space $so(n+2, \mathbb{R})/(so(2, \mathbb{R}) \oplus so(n, \mathbb{R}))$ at the point $(SO(2, \mathbb{R}) \times SO(n, \mathbb{R}))/(SO(2, \mathbb{R}) \times$

SO$(n, \mathbb{R})$) of Q_n is given by the almost-complex operator J such that its exponential $\{\exp(tJ)\}_{t \in \mathbb{R}}$ is the adjoint action of the factor SO$(2, \mathbb{R})$. Since SO$(2, \mathbb{R})$ lies in the center of SO$(2, \mathbb{R}) \times$ SO$(n, \mathbb{R})$, this almost complex structure J is preserved by the action of the adjoint action of SO$(2, \mathbb{R}) \times$ SO$(n, \mathbb{R})$ on so$(n + 2, \mathbb{R})/($so$(2, \mathbb{R}) \oplus$ so$(n, \mathbb{R}))$. Consequently, translating this almost complex structure J by the action of SO$(n + 2, \mathbb{R})$ we get an almost complex structure on Q_n. This almost complex structure is integrable.

The quadric Q_n is an irreducible Hermitian symmetric space of type **BD I** [6, p. 312]. For more about its geometry and that of its noncompact dual D_n (as a Hermitian symmetric space) the reader is referred to [42, Section 1]. A detailed description of D_3 and Q_3 will be given in Sect. 11.3.2.

Let M be a compact complex manifold of complex dimension n endowed with a flat holomorphic conformal structure. Then the pull-back of the flat holomorphic conformal structure to the universal cover $\widetilde{M}$ of M gives rise to a local biholomorphism dev : $\widetilde{M} \longrightarrow Q_n$ which is called the *developing map* and also gives rise to a group homomorphism from the fundamental group of M to PSO$(n + 2, \mathbb{C})$ which is called the *monodromy homomorphism*. The developing map is uniquely defined up to the post-composition by an element of PSO$(n + 2, \mathbb{C})$, and the monodromy morphism is well-defined up to an inner automorphism of PSO$(n+2, \mathbb{C})$. Moreover, the developing map is equivariant with respect to the action of the fundamental group of M on $\widetilde{M}$ (by deck transformations) and on Q_n (through the monodromy homomorphism).

For more details about this classical description of flat geometric structures, due to C. Ehresmann, the reader is referred to [28, 61] (see also Sect. 11.4). This also leads to the following characterizations of the quadric Q_n.

Proposition 11.3.3 ([37, 59, 65]) *Let M be a compact complex simply connected manifold of complex dimension n, endowed with a flat holomorphic conformal structure. Then M is biholomorphic with Q_n and the conformal structure on it is the standard one.*

Proof Since M is compact and simply connected, the developing map dev : $\widetilde{M} = M \longrightarrow Q_n$ is a biholomorphism. Moreover, this developing map intertwines the flat holomorphic conformal structure on M with the standard conformal structure of Q_n. □

All quadrics Q_n are *Fano manifolds*. Recall that a Fano manifold is a compact complex projective manifold M such that its anticanonical line bundle K_M^{-1} is ample. Fano manifolds are intensively studied. In particular, it is known that they are rationally connected [21, 50], and simply connected [20].

An important invariant for Fano manifolds is their *index*. By definition, the index of a Fano manifold M is the maximal positive integer l such that the canonical line bundle K_M is divisible by l in the Picard group of M (i.e., the group of holomorphic line bundles on M). In other words, the index l is the maximal positive integer such that there exists a holomorphic line bundle L over M such that $L^{\otimes l} = K_M$.

Theorem 11.3.4 below, proved in [14], shows, in particular, that among the quadrics Q_n, $n \geq 3$, only Q_3 admits a holomorphic $GL(2, \mathbb{C})$-structure.

Theorem 11.3.4 ([14]) *Let M be a Fano manifold, of complex dimension $n \geq 3$, that admits a holomorphic $GL(2, \mathbb{C})$-structure. Then $n = 3$, and M is biholomorphic to the quadric Q_3 (the $GL(2, \mathbb{C})$-structure being the standard one).*

Proof Let

$$TM \overset{\sim}{\longrightarrow} S^{n-1}(E)$$

be a holomorphic $GL(2, \mathbb{C})$-structure on M, where E is a holomorphic vector bundle of rank two on X. A direct computation shows that

$$K_M = (\bigwedge^2 E^*)^{\frac{n(n-1)}{2}}.$$

Hence the index of M is at least $\frac{n(n-1)}{2}$. It is a known fact that the index of a Fano manifold N of complex dimension n is at most $n+1$. Moreover, the index is maximal ($= n + 1$) if and only if N is biholomorphic to the projective space $\mathbb{CP}^n$, and the index equals n if and only if N is biholomorphic to the quadric [46]. These imply that $n = 3$ and M is biholomorphic to the quadric Q_3.

The $GL(2, \mathbb{C})$-structure on the quadric Q_3 must be flat [37, 44, 47, 65]. Since the quadric is simply connected this flat $GL(2, \mathbb{C})$-structure coincides with the standard one [37, 59, 65] (see also Proposition 11.3.3). □

Recall that a general result of Borel on Hermitian irreducible symmetric spaces shows that the noncompact dual is always realized as an open subset of its compact dual.

We will give below a geometric description of the noncompact dual D_3 of Q_3 as an open subset in Q_3 which seems to be less known (it was explained to us by Charles Boubel whom we warmly acknowledge).

11.3.2 Geometry of the Quadric Q_3

Consider the complex quadric form $q_{3,2} := Z_0^2 + Z_1^2 + Z_2^2 - Z_3^2 - Z_4^2$ of five variables, and let

$$Q \subset \mathbb{CP}^4$$

be the quadric defined by the equation $q_{3,2} = 0$. Then Q is biholomorphic to Q_3. Let $O(3, 2) \subset GL(5, \mathbb{R})$ be the real orthogonal group for $q_{3,2}$, and denote by $SO_0(3, 2)$ the connected component of $O(3, 2)$ containing the identity element. The quadric Q admits a natural holomorphic action of the real Lie group $SO_0(3, 2)$, which is not transitive, in contrast to the action of $SO(5, \mathbb{R})$ on Q_3. The orbits of the $SO_0(3, 2)$-action on Q_3 coincide with the connected components of the complement $Q_3 \setminus S$, where S is the real hypersurface of Q_3 defined by the equation

$$\mid Z_0 \mid^2 + \mid Z_1 \mid^2 + \mid Z_2 \mid^2 - \mid Z_3 \mid^3 - \mid Z_4 \mid^2 = 0.$$

Notice that the above real hypersurface S contains all real points of Q. In fact, it can be shown that $S \cap Q$ coincides with the set of point $m \in Q$ such that the complex line $(m, \overline{m})$ is isotropic (i.e., it actually lies in Q). Indeed, since $q_{3,2}(m) = 0$, the line generated by $(m, \overline{m})$ lies in Q if and only if m and $\overline{m}$ are perpendicular with respect to the bilinear symmetric form associated to $q_{3,2}$, or equivalently $m \in S$.

For any point $m \in Q \setminus S$, the form $q_{3,2}$ is nondegenerate on the line $(m, \overline{m})$. To prove this first observe that the complex line generated by $(m, \overline{m})$, being real, may be considered as a plane in the real projective space $\mathbb{RP}^4$. The restriction of the (real) quadratic form $q_{3,2}$ to this real plane $(m, \overline{m})$ vanishes at the points m and $\overline{m}$ which are distinct (because all real points of Q lie in S). It follows that the quadratic form cannot have signature $(0, 1)$ or $(1, 0)$ when restricted to the real plane $(m, \overline{m})$. Consequently, the signature of the restriction of $q_{3,2}$ to this plane is either $(2, 0)$ or $(1, 1)$ or $(0, 2)$. Each of these three signature types corresponds to an $SO_0(3, 2)$ orbit in Q.

Take the point $m_0 = [0 : 0 : 0 : 1 : \sqrt{-1}] \in Q$. The noncompact dual D_3 of Q_3 is the $SO_0(3, 2)$-orbit of m_0 in Q. It is an open subset of Q biholomorphic to a bounded domain in $\mathbb{C}^3$; it is the three dimension Lie ball (the bounded domain IV_3 in Cartan's classification).

The signature of $q_{3,2}$ on the above line $(m_0, \overline{m}_0)$ is $(0, 2)$. The signature of $q_{3,2}$ on the orthogonal part of $(m_0, \overline{m}_0)$, which is canonically isomorphic to $T_{m_0}Q$, is $(3, 0)$. Then the $SO_0(3, 2)$-orbit of m_0 in Q inherits an $SO_0(3, 2)$-invariant Riemannian metric. The stabilizer of m_0 is $SO(2, \mathbb{R}) \times SO(3, \mathbb{R})$. Here $SO(2, \mathbb{R})$ acts on $\mathbb{C}^3$ through the one parameter group $\exp(tJ)$, with J being the complex structure, while the $SO(3, \mathbb{R})$ action on $\mathbb{C}^3$ is given by the complexification of the canonical action of $SO(3, \mathbb{R})$ on $\mathbb{R}^3$. Consequently, the action of $U(2) = SO(2, \mathbb{R}) \times SO(3, \mathbb{R})$ is the natural irreducible action on the symmetric product $S^2(\mathbb{C}^2) = \mathbb{C}^3$ constructed using the standard representation of $U(2)$. This action coincides with the holonomy representation of this Hermitian symmetric space D_3; as mentioned before, D_3 is the noncompact dual of Q_3. The holonomy representation for Q_3 is the same.

Recall that the automorphism group of the noncompact dual D_3 is $PSO_0(3, 2)$; it is the subgroup of the automorphism group of Q that preserves D_3 (which lies in Q as the $SO_0(3, 2)$-orbit of $m_0 \in Q$). Consequently, any quotient of D_3 by a lattice in $PSO_0(3, 2)$ admits a flat holomorphic conformal structure induced by that of the quadric Q.

Note that the compact projective threefolds admitting a holomorphic conformal structure (a $GL(2, \mathbb{C})$-structure) were classified in [41]. There are in fact only the standard examples: finite quotients of three dimensional abelian varieties, the smooth quadric Q_3 and quotients of its noncompact dual D_3. In [42], the same authors classified also the higher dimensional compact projective manifolds admitting a flat holomorphic conformal structure; they showed that the only examples are the standard ones.

11.3.3 Classification Results

Theorem 11.3.5 shows that the only compact non-flat Kähler–Einstein manifolds bearing a holomorphic $GL(2, \mathbb{C})$-structure are Q_3 and those covered by its non-compact dual D_3.

Theorem 11.3.5 ([14]) *Let M be a compact Kähler–Einstein manifold, of complex dimension at least three, endowed with a holomorphic $GL(2, \mathbb{C})$-structure. Then we are in one of the following (standard) situations:*

(1) *M admits a finite unramified covering by a compact complex torus and the pull-back of the $GL(2, \mathbb{C})$-structure on the compact complex torus is the (translation invariant) standard one.*

(2) *M is biholomorphic to the three dimensional quadric Q_3 equipped with its standard $GL(2, \mathbb{C})$-structure.*

(3) *M admits an unramified cover by the three-dimensional Lie ball D_3 (the noncompact dual of the Hermitian symmetric space Q_3) and the pull-back of the $GL(2, \mathbb{C})$-structure on D_3 is the standard one.*

In order to explain the framework of Theorem 11.3.4 and Theorem 11.3.5 let us recall that Kobayashi and Ochiai proved in [49] a similar result on holomorphic conformal structures. More precisely, they showed that compact Kähler–Einstein manifolds bearing a holomorphic conformal structure are the standard ones: quotients of tori, the smooth n-dimensional quadric Q_n and the quotients of the noncompact dual D_n of Q_n.

Also Kobayashi and Ochiai proved in [48] that all holomorphic G-structures, modeled on an irreducible Hermitian symmetric space of rank ≥ 2 (in particular, a holomorphic conformal structure), on compact Kähler–Einstein manifolds are *flat*. The authors of [37] proved that all holomorphic irreducible reductive G-structures on uniruled projective manifolds are flat. Since uniruled projective manifolds are simply connected, this implies that a uniruled projective manifold bearing a holomorphic conformal structure is biholomorphic to the quadric Q_n with its standard structure (see also [65] and Proposition 11.3.3). In [18], the following generalization was proved:

All holomorphic Cartan geometries (see [61] or Sect. 11.4) on manifolds admitting a rational curve are flat.

The methods used in [14] to prove Theorem 11.3.4 and Theorem 11.3.5 does not use the results in [18, 37, 48, 49, 65]: they are specific to the case of $GL(2, \mathbb{C})$-geometry and unify the twisted holomorphic symplectic case (even dimensional case) and the holomorphic conformal case (odd dimensional case).

While every compact complex surface of course admits a holomorphic $GL(2, \mathbb{C})$-structure, the situation is much more stringent in higher dimensions. The following result proved in [13] shows that a compact Kähler manifold of even dimension $n \geq 4$ bearing a holomorphic $GL(2, \mathbb{C})$-structure has trivial holomorphic tangent bundle (up to finite unramified cover).

Theorem 11.3.6 *Let M be a compact Kähler manifold of even complex dimension $n \geq 4$ admitting a holomorphic $GL(2, \mathbb{C})$-structure. Then M admits a finite unramified covering by a compact complex torus.*

We will give below two different proofs of Theorem 11.3.6.

Proof of Theorem 11.3.6 Let E be a holomorphic vector bundle on M such that

$$TM \simeq S^{n-1}(E)$$

is an isomorphism with the symmetric product, defining the $GL(2, \mathbb{C})$-structure on M. Then

$$TM = S^{n-1}(E) = S^{n-1}(E)^* \otimes (\bigwedge\nolimits^2 E)^{\otimes(n-1)} = (T^*M) \otimes L ,$$

where $L = (\bigwedge^2 E)^{\otimes(n-1)}$.

The above isomorphism between TM and $(T^*M) \otimes L$ produces, when n is even, a holomorphic section

$$\omega \in H^0(M, \Omega_M^2 \otimes L)$$

which is a fiberwise nondegenerate 2-form with values in L. Writing $n = 2m$, the exterior product

$$\omega^m \in H^0(M, K_M \otimes L^m)$$

is a nowhere vanishing section, where $K_M = \Omega_M^n$ is the canonical line bundle of M.

Consequently, we have $K_M \simeq (L^*)^m$, in particular, for the first real Chern class of M we have $c_1(M) = mc_1(L)$. Any Hermitian metric on TM induces an associated Hermitian metric on L^m, and hence produces an Hermitian metric on L.

We now use a result of Istrati, [39, 40, p. 747, Theorem 2.5], which says that $c_1(M) = c_1(L) = 0$.

Hence the manifold M has vanishing first Chern class, in other words, it is a Calabi–Yau manifold. Recall that, Yau's proof of Calabi's conjecture, [64], endows M with a Ricci flat Kähler metric g.

Using de Rham decomposition theorem and Berger's classification of the irreducible holonomy groups of nonsymmetric Riemannian manifolds (see [43], Section 3.2 and Theorem 3.4.1 in Section 3.4) we deduce that the universal

Riemannian cover $(\widetilde{M}, \widetilde{g})$ of (M, g) splits as a Riemannian product

$$(\widetilde{M}, \widetilde{g}) = (\mathbb{C}^l, g_0) \times (M_1, g_1) \times \cdots \times (M_p, g_p), \qquad (11.3.3)$$

where $(\mathbb{C}^l, g_0)$ is the standard flat complete Kähler manifold and (M_i, g_i) is an irreducible Ricci flat Kähler manifold of complex dimension $r_i \geq 2$, for every $1 \leq i \leq p$. The holonomy of each (M_i, g_i) is either $\mathrm{SU}(r_i)$ or the symplectic group $\mathrm{Sp}(\frac{r_i}{2})$, where $r_i = \dim_{\mathbb{C}} M_i$ (in the second case r_i is even). Notice, in particular, that symmetric irreducible Riemannian manifolds of (real) dimension at least two are never Ricci flat. For more details, the reader is referred to [5, Theorem 1] and [43, p. 124, Proposition 6.2.3].

As a consequence of Cheeger–Gromoll theorem one can deduce the Beauville–Bogomolov decomposition theorem (see [5, Theorem 1] or [8]) which says that there is a finite unramified covering

$$\varphi : \widehat{M} \longrightarrow M,$$

such that

$$(\widehat{M}, \varphi^* g) = (T_l, g_0) \times (M_1, g_1) \times \cdots \times (M_p, g_p), \qquad (11.3.4)$$

where (M_i, g_i) are as in (11.3.3) and (T_l, g_0) is a flat compact complex torus of dimension l. Of course the pull-back Kähler metric $\varphi^* g$ is Ricci-flat because g is so.

We obtain that the initial holomorphic $\mathrm{GL}(2, \mathbb{C})$-structure on M induces a holomorphic $\mathrm{SL}(2, \mathbb{C})$-structure on a finite unramified covering of $\widehat{M}$ in (11.3.4). Indeed, $T\widehat{M} = S^{n-1}(\varphi^* E)$ and since the canonical bundle $K_{\widehat{M}}$ is holomorphically trivial, the holomorphic line bundle $\bigwedge^2 E$ admits a finite multiple which is trivial (in this situation we also say that $\bigwedge^2 E$ is a *torsion* line bundle). Hence on a finite unramified cover of $\widehat{M}$ (still denoted by $\widehat{M}$ for simplicity) we can work with the assumption that $\bigwedge^2 E$ is holomorphically trivial.

This $\mathrm{SL}(2, \mathbb{C})$-structure on $\widehat{M}$ provides a holomorphic reduction $R'(\widehat{M}) \subset R(\widehat{M})$ of the structure group of the frame bundle $R(\widehat{M})$ (from $\mathrm{GL}(n, \mathbb{C})$) to $\mathrm{SL}(2, \mathbb{C})$.

There is a finite set of holomorphic tensors $\theta_1, \cdots, \theta_s$ on $\widehat{M}$ satisfying the condition that the $\mathrm{SL}(2, \mathbb{C})$-subbundle $R'(\widehat{M}) \subset R(\widehat{M})$ consists of those frames that pointwise preserve all tensors θ_i. This is deduced from Chevalley's theorem which asserts that there exists a finite dimensional linear representation W of $\mathrm{GL}(n, \mathbb{C})$, and an element

$$\theta_0 \in W,$$

such that the stabilizer of the line $\mathbb{C}\theta_0$ is the image of the homomorphism $\mathrm{SL}(2, \mathbb{C}) \longrightarrow \mathrm{GL}(n, \mathbb{C})$ defining the $\mathrm{SL}(2, \mathbb{C})$-structure (see [36, p. 80, Theorem

11.2], [22, p. 40, Proposition 3.1(b)]; since $SL(2, \mathbb{C})$ does not have a nontrivial character, the line $\mathbb{C}\theta_0$ is fixed pointwise.

The group $GL(n, \mathbb{C})$ being reductive, we decompose W as a direct sum $\bigoplus_{i=1}^{s} W_i$ of irreducible representations. Now, since any irreducible representation W_i of the reductive group $GL(n, \mathbb{C})$ is a factor of a representation $(\mathbb{C}^n)^{\otimes p_i} \otimes ((\mathbb{C}^n)^*)^{\otimes q_i}$, for some integers $p_i, q_i \geq 0$ [22, p. 40, Proposition 3.1(a)], the above element θ_0 gives rise to a finite set $\theta_1, \ldots, \theta_s$ of holomorphic tensors

$$\theta_i \in H^0(\widehat{M}, (T\widehat{M})^{\otimes p_i} \otimes (T^*\widehat{M})^{\otimes q_i}) \tag{11.3.5}$$

with $p_i, q_i \geq 0$. By construction, $\theta_1, \ldots, \theta_s$ are simultaneously stabilized exactly by the frames lying in $R'(\widehat{M})$.

It is known that the parallel transport on $\widehat{M}$ for the Levi–Civita connection associated with the Ricci-flat Kähler metric φ^*g in (11.3.4) preserves any holomorphic tensor on $\widehat{M}$ [53, p. 50, Theorem 2.2.1]. In particular, θ_i in (11.3.5) are all parallel with respect to the Levi–Civita connection of φ^*g. We conclude that the subbundle $R'(\widehat{M}) \subset R(\widehat{M})$ defining the holomorphic $SL(2, \mathbb{C})$-structure is invariant under the parallel transport of the Levi–Civita connection for φ^*g. This implies that the holonomy group of φ^*g lies in the maximal compact subgroup of $SL(2, \mathbb{C})$. Consequently, the holonomy group of φ^*g lies in $SU(2)$.

From (11.3.4) it follows that the holonomy of φ^*g is

$$\text{Hol}(\varphi^*g) = \prod_{i=1}^{p} \text{Hol}(g_i), \tag{11.3.6}$$

where $\text{Hol}(g_i)$ is the holonomy of g_i. As noted earlier,

- either $\text{Hol}(g_i) = SU(r_i)$, with $\dim_{\mathbb{C}} M_i = r_i \geq 2$, or
- $\text{Hol}(g_i) = Sp(\frac{r_i}{2})$, where $r_i = \dim_{\mathbb{C}} M_i$ is even.

Therefore, the above observation, that $\text{Hol}(\varphi^*g)$ is contained in $SU(2)$, and (11.3.6) together imply that

(1) either $(\widehat{M}, \varphi^*g) = (T_l, g_0)$, or
(2) $(\widehat{M}, \varphi^*g) = (T_l, g_0) \times (M_1, g_1)$, where M_1 is a K3 surface equipped with a Ricci-flat Kähler metric g_1.

If $(\widehat{M}, \varphi^*g) = (T_l, g_0)$, then proof of the theorem evidently is complete. Therefore, we assume that

$$(\widehat{M}, \varphi^*g) = (T_l, g_0) \times (M_1, g_1), \tag{11.3.7}$$

where M_1 is a K3 surface equipped with a Ricci-flat Kähler metric g_1. Note that $l \geq 2$ (because $l + 2 = n \geq 4$) and l is even (because n is so).

At this stage there are two different ways to terminate the proof. We will give below the two of them.

(I) *First Proof.*

Since $\mathrm{Hol}(g_1) = \mathrm{SU}(2)$, we get from (11.3.6) that $\mathrm{Hol}(\varphi^*g) = \mathrm{SU}(2)$. The holonomy of φ^*g is the image of the homomorphism

$$h_0 : \mathrm{SU}(2) \longrightarrow \mathrm{SU}(n) \tag{11.3.8}$$

given by the $(n - 1)$-th symmetric power of the standard representation. The action of $h_0(\mathrm{SU}(2))$ on $\mathbb{C}^n$, obtained by restricting the standard action of $\mathrm{SU}(n)$, is irreducible. In particular, there are no nonzero $\mathrm{SU}(2)$-invariants in $\mathbb{C}^n$.

On the other hand we have:

- the direct summand of $T\widehat{M}$ given by the tangent bundle TT_l is preserved by the Levi–Civita connection on $T\widehat{M}$ corresponding to φ^*g, and
- this direct summand of $T\widehat{M}$ given by TT_l is generated by flat sections of TT^l.

Since $T\widehat{M}$ does not have any flat section, we conclude that $l = 0$: a contradiction. This terminates the first proof.

(II) *Second proof.*

Fix a point $t \in T_l$. The holomorphic vector bundle over M_1 obtained by restricting φ^*E to $\{t\} \times M_1 \subset T_l \times M_1$ will be denoted by F. Restricting the isomorphism

$$T\widehat{M} \xrightarrow{\;\sim\;} S^{l+1}(\varphi^*E)$$

to $\{t\} \times M_1 \subset T_l \times M_1$, from (11.3.7) we conclude that

$$S^{l+1}(F) = TM_1 \oplus \mathcal{O}_{M_1}^{\oplus l} ; \tag{11.3.9}$$

note that $(T\widehat{M})|_{\{t\} \times M_1} = TM_1 \oplus \mathcal{O}_{M_1}^{\oplus l}$. The vector bundle TM_1 is polystable of degree zero (we may use the Kähler structure g_1 on M_1 to define the degree), because M_1 admits a Kähler–Einstein metric. Hence $TM_1 \oplus \mathcal{O}_{M_1}^{\oplus (l+1)}$ is polystable of degree zero. This implies that $\mathrm{End}(TM_1 \oplus \mathcal{O}_{M_1}^{\oplus (l+1)})$ is polystable of degree zero, because for any Hermitian–Einstein structure on $TM_1 \oplus \mathcal{O}_{M_1}^{\oplus (l+1)}$ (which exists by [62]), the Hermitian structure on $\mathrm{End}(TM_1 \oplus \mathcal{O}_{M_1}^{\oplus (l+1)})$ induced by it is also Hermitian–Einstein. On the other hand, the holomorphic vector bundle $\mathrm{End}(F)$ is a direct summand of $\mathrm{End}(S^{l+1}(F))$. Since $\mathrm{End}(S^{l+1}(F)) = \mathrm{End}(TM_1 \oplus \mathcal{O}_{M_1}^{\oplus (l+1)})$ is polystable of degree zero, this implies that the holomorphic vector bundle $\mathrm{End}(F)$ is also polystable of degree zero.

Since $\mathrm{End}(F)$ is polystable, it follows that F is polystable [2, p. 224, Corollary 3.8].

Let H be a Hermitian–Einstein metric on F. The holonomy of the Chern connection on F associated with H is contained in SU(2), because $\bigwedge^2 F = \mathcal{O}_{M_1}$ (since $\bigwedge^2 E$ is trivial). Let Hol(H) denoted the holonomy of the Chern connection on F associated with H. We note that H is connected because M_1 is simply connected. Therefore, either Hol(H) $=$ SU(2) or Hol(H) $=$ U(1) (since TM_1 is not trivial, from (11.3.9) it follows that F is not trivial and hence the holonomy of H cannot be trivial).

First assume that Hol(H) $=$ SU(2). Let $\mathcal{H}$ denote the Hermitian–Einstein metric on $S^{n-1}(F)$ induced by H. Since Hol(H) $=$ SU(2), the holonomy Hol($\mathcal{H}$) of the Chern connection on $S^{n-1}(F)$ associated with $\mathcal{H}$ is the image of the homomorphism

$$h_0 : \mathrm{SU}(2) \longrightarrow \mathrm{SU}(n) \tag{11.3.10}$$

given by the $(n-1)$-th symmetric power of the standard representation. The action of $h_0(\mathrm{SU}(2))$ on $\mathbb{C}^n$, obtained by restricting the standard action of SU(n), is irreducible. From this it follows that

$$H^0(M_1, S^{n-1}(F)) = 0$$

because any holomorphic section of $S^{n-1}(F)$ is flat with respect to the Chern connection on $S^{n-1}(F)$ associated with $\mathcal{H}$ [53, p. 50, Theorem 2.2.1]. On the other hand, from (11.3.9) we have

$$H^0(M_1, S^{n-1}(F)) = \mathbb{C}^l . \tag{11.3.11}$$

In view of this contradiction we conclude that Hol(H) $\neq$ SU(2).

So assume that Hol(H) $=$ U(1). Then

$$F = \mathcal{L} \oplus \mathcal{L}^* , \tag{11.3.12}$$

where $\mathcal{L}$ is a holomorphic line bundle on M_1. Note that degree($\mathcal{L}$) $= 0$, because F is polystable. From (11.3.12) it follows that

$$S^{n-1}(F) = \bigoplus_{j=0}^{n-1} \mathcal{L}^{1-n+2j} .$$

Now using (11.3.11) we conclude that $H^0(M_1, \mathcal{L}^k) \neq 0$ for some nonzero integer k. Since degree($\mathcal{L}$) $= 0$, this implies that the holomorphic line bundle $\mathcal{L}$ is trivial; note that the Picard group Pic(M_1) is torsionfree. Hence $S^{n-1}(F)$ is trivial. Since TM_1 is not trivial, this contradicts (11.3.9) (see [3, p. 315, Theorem 2]).

Therefore, (11.3.7) can't occur. This completes the second proof of the theorem. □

Recall that a compact Kähler manifold of odd complex dimension bearing a holomorphic $\mathrm{SL}(2, \mathbb{C})$-structure also admits a holomorphic Riemannian metric and inherits of the associated holomorphic (Levi–Civita) connection on the holomorphic tangent bundle. Those manifolds are known to have vanishing Chern classes [4]. Indeed, following Chern–Weil theory one classically computes real Chern classes of the holomorphic tangent bundle of a manifold M using a Hermitian metric on it. This provides representatives of the Chern class $c_k(TM) \in \mathrm{H}^{2k}(M, \mathbb{R})$ which are smooth forms on M of type (k, k). Starting with a holomorphic connection on TM and doing the same formal computations, one gets another representative of $c_k(TM)$ which is a holomorphic form: in particular, it is of type $(2k, 0)$. Classical Hodge theory says that on Kähler manifolds nontrivial real cohomology classes do not have representatives of different types. This implies the vanishing of the real Chern class: $c_k(TM) = 0$ for all k.

But a compact Kähler manifold M with vanishing first two Chern classes $(c_1(TM) = c_2(TM) = 0)$ is known to be covered by a compact complex torus [38]. Indeed, using the vanishing of the first Chern class of M, Yau's proof of Calabi conjecture (which is the key ingredient to obtain the result) [64] endows M with a Ricci flat Kähler metric. The vanishing of the second Chern class implies that the Ricci flat metric is flat (i.e., it has vanishing sectional curvature). Hence M admits a flat Kähler metric. Since M is compact, Hopf–Rinow theorem implies that the flat metric is complete, meaning the universal cover of M is isometric with $\mathbb{C}^n$ endowed with its standard translation invariant Kähler metric. By Bieberbach's theorem M admits a finite cover which is a quotient of $\mathbb{C}^n$ by a lattice of translations.

Therefore, Theorem 11.3.6 has the following corollary.

Corollary 11.3.7 *Let M be a compact Kähler manifold of complex dimension $n \geq 3$ bearing a holomorphic $\mathrm{SL}(2, \mathbb{C})$-structure. Then M admits a finite unramified covering by a compact complex torus.*

11.4 Holomorphic Cartan Geometry

In his celebrated Erlangen address (delivered in 1872), F. Klein defined a *geometry* as a manifold X equipped with a transitive action of a Lie group G. The Lie group G is the symmetry group of the geometry. Notice that any choice of a base point $p_0 \in X$ identifies the manifold X with the homogeneous space G/H, where H is the closed subgroup of G stabilizing p_0.

The Euclidean geometry is a first important example of geometry in the sense of Klein. The Euclidean symmetry group is the group of all motions. The stabilizer of the origin is the orthogonal group $H = \mathrm{O}(n, \mathbb{R})$ and the group of Euclidean motions $G = \mathrm{O}(n, \mathbb{R}) \ltimes \mathbb{R}^n$ is a semi-direct product of the orthogonal group with the translation group.

The holomorphic version of the above geometry is the *complex Euclidean space* $(X = \mathbb{C}^n, dz_1^2 + \ldots + dz_n^2)$. Here $dz_1^2 + \ldots + dz_n^2$ is the nondegenerate complex quadratic form (of maximal rank n) and $H = O(n, \mathbb{C})$ is the corresponding complex orthogonal group. The symmetry group $G = O(n, \mathbb{C}) \ltimes \mathbb{C}^n$ is the group of complex Euclidean motions. With the point of view of G-structures, the complex quadratic form $dz_1^2 + \ldots + dz_n^2$ defines a flat holomorphic $O(n, \mathbb{C})$-structure in the sense of Sect. 11.2 and hence a flat holomorphic Riemannian metric (see [13, 24, 26, 30, 52]). Any flat holomorphic Riemannian metric is locally isomorphic to the complex Euclidean space.

Another classical geometry encompassed in Klein's definition is the *complex affine space* $X = \mathbb{C}^n$. The symmetry group is the complex affine group $G = GL(n, \mathbb{C}) \ltimes \mathbb{C}^n$. This symmetry group preserves lines parametrized at constant speed.

Klein's definition contains also the *complex projective space*. The symmetry group of the complex projective space $\mathbb{C}P^n$ is the complex projective group $PGL(n + 1, \mathbb{C})$. This symmetry group preserves lines.

As a unifying generalization of the above concept of Klein geometry and of the Riemannian geometry, E. Cartan elaborated the framework of what we now call *Cartan geometry*. A Cartan geometry is an infinitesimal version of a Klein geometry: this generalizes Riemann's construction of a Riemannian metric infinitesimally modeled on the Euclidean space. We will see that any Cartan geometry has an associated curvature tensor measuring the infinitesimal variation of the Cartan geometry with respect to the corresponding Klein model. The curvature tensor vanishes exactly when the Cartan geometry is locally isomorphic to a Klein geometry. This generalizes to the Cartan geometry background the Riemannian notion of curvature: recall that the Riemannian curvature vanishes exactly when the metric is flat and hence locally isomorphic to the Euclidean space.

The Cartan geometries for which the curvature tensor vanishes identically are called *flat*.

11.4.1   Classical Case

Let us now introduce the classical definition of a Cartan geometry in the complex analytic category (the reader is referred to [61] for more details).

The model of the Cartan geometry is a Klein geometry G/H, where G is a connected complex Lie group and $H \subset G$ is a connected complex Lie subgroup. The complex Lie algebras of G and H are denoted by $\mathfrak{g}$ and $\mathfrak{h}$ respectively.

Then we have the following definition due to Cartan and formalized by Ehresmann.

Definition 11.4.1 A holomorphic Cartan geometry with model (G, H) on a complex manifold M is a holomorphic principal H-bundle $\pi : E_H \longrightarrow M$ equipped with a $\mathfrak{g}$-valued holomorphic 1-form $\omega \in H^0(E_H, \Omega^1_{E_H} \otimes \mathfrak{g})$ satisfying:

(1) $\omega : TE_H \longrightarrow E_H \times \mathfrak{g}$ defines a vector bundle isomorphism;
(2) ω is H-equivariant with H acting on $\mathfrak{g}$ via conjugation;

(3) the restriction of ω to each fiber of π coincides with the Maurer–Cartan form associated with the action of H on E_H.

A ω-constant vector field on E_H is a section of TE_H which is the preimage through ω of a fixed element in $\mathfrak{g}$. Condition (1) in Definition 11.4.1 means that the holomorphic tangent bundle TE_H is trivialized by ω-constant vector fields. This condition implies, in particular, that the complex dimension of the model space G/H is the same as the complex dimension of M.

Condition (3) in Definition 11.4.1 implies that the ω-constant vector fields corresponding to preimages of elements in $\mathfrak{h} \subset \mathfrak{g}$ form a family of holomorphic vector fields on E_H trivializing the holomorphic vertical tangent space (the kernel of $d\pi$ in TE_H). The restriction of ω to the vertical tangent bundle identifies the vertical tangent bundle with the trivial bundle $E_H \times \mathfrak{h} \longrightarrow E_H$.

Moreover, condition (2) in Definition 11.4.1 implies that if $Z_1, Z_2 \in H^0(E_H, TE_H)$ are two ω-constant vector fields such that one of them, say Z_1, is vertical, then $\omega([Z_1, Z_2]) = [\omega(Z_1), \omega(Z_2)]_{\mathfrak{g}}$.

In fact, ω is a Lie algebra isomorphism from the family of ω-constant vector fields to $\mathfrak{g}$ if and only if the holomorphic 2-form $d\omega + \frac{1}{2}[\omega, \omega]_{\mathfrak{g}}$ vanishes identically. In this case the Cartan geometry defined by ω is called *flat*. The holomorphic 2-form

$$K(\omega) := d\omega + \frac{1}{2}[\omega, \omega]_{\mathfrak{g}} \in H^0(E_H, \Omega^2_{E_H} \otimes \mathfrak{g})$$

is called the curvature tensor of the Cartan geometry. Note that the above observation implies that $K(\omega)$ vanishes on all pairs (Z_1, Z_2), where Z_1 is a ω-constant vertical holomorphic vector field and Z_2 is any ω-constant holomorphic vector field. Consequently, $K(\omega)$ is the pull-back, through π, of an element of $H^0(M, \Omega^2(M) \otimes \mathrm{ad}(E_G))$; here $\mathrm{ad}(E_G)$ is the vector bundle associated with E_H through the action of $H \subset G$ on $\mathfrak{g}$ via the adjoint representation. In particular, any holomorphic Cartan geometry on a complex manifold of dimension one (complex curve) is necessarily flat.

Definition 11.4.1 should be seen as an infinitesimal version of the principal H-bundle given by the quotient map $G \longrightarrow G/H$. The form ω generalizes the left-invariant Maurer–Cartan form ω_G of G. It should be recalled here that the Maurer–Cartan form ω_G is the tautological one-form on G which identifies left-invariant vector fields on G with elements in the Lie algebra $\mathfrak{g}$. It satisfies the so-called Maurer–Cartan equation $d\omega + \frac{1}{2}[\omega, \omega]_{\mathfrak{g}} = 0$ (this is a straightforward consequence of the Lie–Cartan derivative formula, see, for instance, [61]).

We recall a classical result of Cartan (see [61]):

Theorem 11.4.2 (Cartan) *Let (E_H, ω) be a Cartan geometry with model (G, H). The curvature $K(\omega) = d\omega + \frac{1}{2}[\omega, \omega]_{\mathfrak{g}}$ vanishes identically if and only if (E_H, ω) is locally isomorphic to $(G \longrightarrow G/H, \omega_G)$, where ω_G is the left-invariant Maurer–Cartan form on the (right) principal H-bundle $G \longrightarrow G/H$.*

In [28] Ehresmann studied manifolds M equipped with a flat Cartan geometry. More precisely, a flat Cartan geometry on M, gives a $(G, X = G/H)$-structure on M in the following sense described by Ehresmann (in [28] the terminology is that of a *locally homogeneous space*).

Definition 11.4.3 A holomorphic (G, X)-structure on a complex manifold M is given by an open cover $(U_i)_{i \in I}$ of M with holomorphic charts $\phi_i : U_i \longrightarrow X$ such that the transition maps $\phi_i \circ \phi_j^{-1} : \phi_j(U_i \cap U_j) \longrightarrow \phi_i(U_i \cap U_j)$ are given (on each connected component) by the restriction of an element $g_{ij} \in G$.

Recall that the G-action on X is holomorphic. Also note that any geometric feature of X which is invariant by the symmetry group G has an intrinsic meaning on the manifold M equipped with a (G, X)-structure.

Consider a (G, X)-structure on a manifold M (as in Definition 11.4.3). Then the pull-back of the (G, X)-structure to the universal cover $\widetilde{M}$ of M gives rise to a local biholomorphism dev $: \widetilde{M} \longrightarrow X$ which is called the *developing map* (of the (G, X)-structure) and to a group homomorphism from the fundamental group of M to G called the *monodromy morphism*. The developing map is uniquely defined up to a post-composition by an element in G (acting on X) and the monodromy morphism is well-defined up to an inner automorphism of G. Moreover, the developing map is equivariant with respect to the action of the fundamental group of M on $\widetilde{M}$ (by deck transformations) and on X (through the monodromy morphism). A more detailed construction of the developing map of a flat Cartan geometry will be given in Sect. 11.5.1 focusing on the foliated case.

By Cartan's equivalence principle, many important geometric structures can be seen as G-structures and also as Cartan geometries. For example, a holomorphic Riemannian metric on a complex manifold of complex dimension n is a holomorphic $O(n, \mathbb{C})$-structure on M and also a holomorphic Cartan geometry with model $(G = O(n, \mathbb{C}) \ltimes \mathbb{C}^n, H = O(n, \mathbb{C}))$. Those two points of view are equivalent; in particular the notion of flatness is the same in these two descriptions of the holomorphic Riemannian metric.

Also a holomorphic Cartan geometry whose model is the complex affine space (meaning $G = GL(n, \mathbb{C}) \ltimes \mathbb{C}^n$ and $H = GL(n, \mathbb{C})$) is actually a holomorphic affine connection on M. This Cartan geometry is flat if and only if the holomorphic affine connection is flat and torsion-free. In this case M is endowed with a holomorphic affine structure (i.e., a holomorphic (G, X)-structure, with $X = G/H = \mathbb{C}^n$ being the complex affine space and $G = GL(n, \mathbb{C}) \ltimes \mathbb{C}^n$ its symmetry affine group).

A holomorphic Cartan geometry whose model is the complex projective space (meaning $G = PGL(n + 1, \mathbb{C})$ and H is its parabolic subgroup stabilizing a point in $\mathbb{C}P^n$) defines a holomorphic projective connection on M. When the connection is flat, this Cartan geometry defines a holomorphic projective structure on M (i.e, a holomorphic (G, X)-structure, with $X = G/H = \mathbb{C}P^n$ being the complex projective space while $G = PGL(n + 1, \mathbb{C})$ is its automorphism group).

11.4.2 Branched Cartan Geometry

In [54, 55] Mandelbaum defined and studied complex *branched affine and projective structures* on Riemann surfaces. A holomorphic branched affine (respectively, projective) structure on a Riemann surface is given by some open cover (as in Definition 11.4.3) and local charts $\phi_i : U_i \longrightarrow \mathbb{C}$ which are finite branched coverings such that the transition maps $\phi_i \circ \phi_j^{-1} : \phi_j(U_i \cap U_j) \longrightarrow \phi_i(U_i \cap U_j)$ are given by restrictions of an element g_{ij} lying in the complex affine group of $\mathbb{C}$ (respectively, in PSL$(2, \mathbb{C})$).

Generalizing Mandelbaum's definition, we introduced in [12] the notion of *branched holomorphic Cartan geometry* on a complex manifold M. This notion is well-defined for manifolds M of any complex dimension and not only in the flat case.

The precise definition is the following [12]:

Definition 11.4.4 A branched holomorphic Cartan geometry with model (G, H) on a complex manifold M is a holomorphic principal H-bundle $\pi : E_H \longrightarrow M$ equipped with a $\mathfrak{g}$-valued holomorphic 1-form $\omega \in H^0(E_H, \Omega^1_{E_H} \otimes \mathfrak{g})$ satisfying:

(1) $\omega : TE_H \longrightarrow E_H \times \mathfrak{g}$ defines a vector bundle morphism which is an isomorphism on an open dense set in E_H;
(2) ω is H-equivariant with H acting on $\mathfrak{g}$ via conjugation.
(3) the restriction of ω to each fiber of π coincides with the Maurer–Cartan form associated with the action of H on E_H.

Condition (1) implies that the complex dimension of the model G/H is the same as the complex dimension of M.

Condition (2) implies that the open dense set U in E_H over which ω is an isomorphism is H-invariant. More precisely, there exists a divisor $D \subset M$ such that $U = \pi^{-1}(M \setminus D)$.

The divisor D is called *the branching divisor* of the branched Cartan geometry (E_H, ω).

The branched Cartan geometry (E_H, ω) is called *flat* if its curvature tensor, defined again as $K(\omega) = d\omega + \frac{1}{2}[\omega, \omega]_\mathfrak{g}$, vanishes identically. In this case it was proved in [12] that there exists a holomorphic developing map dev : $\widetilde{M} \longrightarrow X$ and a group homomorphism form the fundamental group of M to G (the monodromy morphism of the branched flat Cartan geometry). The developing map is uniquely defined up to a post-composition by an element in G (acting on X) and the monodromy morphism is well-defined up to inner conjugacy in G. Moreover, the developing map is equivariant with respect to the action of the fundamental group of M on $\widetilde{M}$. In the branched case the developing map is a holomorphic dominant map; its differential is invertible exactly over the open dense set $M \setminus D$ (away from the branching divisor D).

In particular, M is endowed with a branched holomorphic (G, X)-structure in the sense of the following definition:

Definition 11.4.5 A branched holomorphic (G, X)-structure on a complex manifold M is given by an open cover $(U_i)_{i \in I}$ of M with branched holomorphic maps $\phi_i : U_i \longrightarrow X$ such that the transition maps $\phi_i \circ \phi_j^{-1} : \phi_j(U_i \cap U_j) \longrightarrow \phi_i(U_i \cap U_j)$ are given (on each connected component) by the restriction of an element $g_{ij} \in G$.

Notice that the element g_{ij} in the above definition is unique (since any element of G is uniquely determined by the restriction of its action on a nontrivial open set in X).

In the most basic example, where G is the complex Lie group $\mathbb{C}$ and $H = \{0\}$, the associated Cartan geometry is simply a nontrivial holomorphic 1-form ω on a Riemann surface. The branching divisor D is the divisor of zeros of ω. The developing map is a primitive of ω and the image of the monodromy morphism is the group of periods of ω.

A branched holomorphic Cartan geometry whose model is the complex projective space (recall that $G = \mathrm{PGL}(n + 1, \mathbb{C})$ and H is the parabolic subgroup of $\mathrm{PGL}(n + 1, \mathbb{C})$ stabilizing a point in $\mathbb{CP}^n$) defines a branched holomorphic projective connection on M. When the projective connection is flat this Cartan geometry defines a branched holomorphic projective structure on M (i.e., a branched holomorphic (G, X)-structure, with $X = G/H = \mathbb{CP}^n$ being the complex projective space and $G = \mathrm{PGL}(n + 1, \mathbb{C})$ its symmetry projective group).

This notion of branched Cartan geometry is much more flexible than the standard one. It is stable by pull-back through holomorphic ramified maps. This was used in [12] to prove that all compact complex projective manifolds admit a branched flat holomorphic projective structure.

A more general version of Cartan geometry was defined in [1, 15] where this concept is referred as a *generalized Cartan geometry*. A holomorphic generalized Cartan geometry (E_H, ω) with model (G, H) on a complex manifold M satisfies conditions (2) and (3) in Definitions 11.4.1 and 11.4.4, but conditions (1) are dropped. In particular, there is no relation between the complex dimension of M and the complex dimension of the model space G/H. In the flat case (defined by the vanishing of the curvature tensor $K(\omega) = d\omega + \frac{1}{2}[\omega, \omega]_{\mathfrak{g}}$), the developing map (of a holomorphic generalized Cartan geometry) is still a holomorphic map $\widetilde{M} \longrightarrow X = G/H$ which is equivariant with respect to the monodromy morphism; but here there is no condition on the rank of the developing map (see [15]).

Consider again the basic example, where G is the complex Lie group $\mathbb{C}$ and $H = \{0\}$. The associated Cartan geometry on a complex manifold M is a holomorphic 1-form $\omega \in H^0(\Omega^1(M, \mathbb{C}))$. The branching divisor D is the divisor of zeros of ω. The Cartan geometry is flat if and only if $d\omega = 0$. In particular, all those geometries are flat on compact Kähler manifolds and on compact complex surfaces. In this flat case, the developing map is a primitive of ω and the image of the monodromy morphism is the group of periods of ω.

11.5 Transverse Cartan Geometry

The definition of a transverse Cartan geometry was first given by Blumenthal in [19] in the context of foliated differential bundles (see, for instance, [58]). In [10] we worked out a definition of a transverse branched holomorphic Cartan geometry using the formalism of Atiyah's bundle. We present here this notion of transverse branched Cartan geometry in the complex analytic category.

Let M be a complex manifold and $\mathcal{F}$ a holomorphic foliation on M of complex codimension n. For simplicity we assume first that $\mathcal{F}$ does not admit singularities. We will explain later how holomorphic foliations with singularities can be endowed with a transverse Cartan geometry.

Recall that the associated tangent space $T\mathcal{F} \subset TM$ is a holomorphic distribution of codimension n which is stable by Lie bracket. Conversely, any holomorphic distribution stable by Lie bracket is *integrable*, meaning it coincides with the tangent space of a foliation $\mathcal{F}$.

The normal bundle of the foliation, defined as the quotient $\mathcal{N}_{\mathcal{F}} = TM/T\mathcal{F}$, admits a canonical holomorphic flat connection $\nabla^{\mathcal{F}}$ along the leaves of the foliation $\mathcal{F}$. To define analytically this connection one chooses two local sections X and N of $T\mathcal{F}$ and $\mathcal{N}_{\mathcal{F}}$ respectively and defines the derivative of N in the direction of X as being

$$\nabla^{\mathcal{F}}_X N = q([X, \widetilde{N}]),$$

where $q : TM \longrightarrow \mathcal{N}_{\mathcal{F}} = TM/T\mathcal{F}$ is the quotient map and $\widetilde{N}$ is any local section of TM such that $q(\widetilde{N}) = N$.

The above connection is well-defined along $\mathcal{F}$. Indeed, if $\widehat{N}$ is another choice of lift for N to TM, then $q([X, \widehat{N}]) = q([X, \widetilde{N}])$, since $\widehat{N} - \widetilde{N} \in T\mathcal{F}$ the Lie bracket stability of $T\mathcal{F}$ implies that $[\widehat{N} - \widetilde{N}, X] \in T\mathcal{F}$. Consequently, $q([\widehat{N} - \widetilde{N}, X]) = 0$ which implies $q([X, \widehat{N}]) = q([X, \widetilde{N}])$.

Notice that the local section $N = q(\widetilde{N})$ of $\mathcal{N}_{\mathcal{F}}$ is parallel with respect to $\nabla^{\mathcal{F}}$ if and only if $[\widetilde{N}, X] \in \mathcal{F}$, for all $X \in \mathcal{F}$.

Moreover, Jacobi's identity for the Lie bracket implies the vanishing of the curvature tensor of $\nabla^{\mathcal{F}}$ and hence its flatness.

A transverse holomorphic Cartan geometry on $(M, \mathcal{F})$ with model (G, H) (with G a complex connected Lie group and $H \subset G$ a connected complex Lie subgroup) is defined by the following data (summarized below in the conditions I and II).

I. A holomorphic principal H-bundle $\pi : E_H \longrightarrow M$ over M which admits a flat partial holomorphic connection in the direction of $\mathcal{F}$: this means there exists a H-invariant holomorphic foliation $\widetilde{\mathcal{F}}$ of E_H such that $d\pi(T\widetilde{\mathcal{F}}) = T\mathcal{F}$ and the restriction of $d\pi$ to $\widetilde{\mathcal{F}}$ is a submersion over $\mathcal{F}$. Hence $\widetilde{\mathcal{F}}$ is a H-invariant lift of $\mathcal{F}$ to E_H.

II. A holomorphic $\mathfrak{g}$-valued one–form $\omega \in H^0(E_H, \Omega^1_{E_H} \otimes \mathfrak{g})$ satisfying the following conditions:

(1) ω is H-equivariant for the adjoint action of H on $\mathfrak{g}$;
(2) the restriction of ω to any fiber of π coincides with the Maurer–Cartan form ω_H;
(3) ω vanishes on the foliation $T\widetilde{\mathcal{F}} \subset TE_H$;
(4) the induced morphism $\omega : (TE_H)/T\widetilde{\mathcal{F}} \longrightarrow E_H \times \mathfrak{g}$ is constant on parallel sections of $(TE_H)/T\widetilde{\mathcal{F}}$ (with respect to the canonical connection $\nabla^{\widetilde{F}}$ along the foliation $\widetilde{\mathcal{F}}$).
(5) the homomorphism $\omega : (TE_H)/T\widetilde{\mathcal{F}} \longrightarrow E_H \times \mathfrak{g}$ is an isomorphism.

Condition (4) can be formulated, as in [19], as a vanishing of a Lie derivative. More precisely, it is equivalent to $L_X\omega = 0$, for any X tangent to $\widetilde{\mathcal{F}}$ (this comes from our previous description of flat sections of the normal bundle of the foliation). Notice that by the Lie–Cartan formula, namely $L_X\omega = i_X d\omega + d(i_X\omega)$, we get $L_X\omega = i_X d\omega = 0$.

A *branched* transverse holomorphic Cartan geometry is defined by the same data, except that condition (5) is satisfied *over a nonempty open subset of* E_H (see [10]). Condition (1) implies that this nonempty open subset of E_H is the pull-back through π of a nonempty open subset in M (the complement of a divisor in M [10]).

Condition (5) implies in both cases (classical and branched) that the complex dimension of the model G/H is the same as the complex codimension of $\mathcal{F}$.

The transverse curvature $K(\omega) = d\omega + \frac{1}{2}[\omega, \omega]_\mathfrak{g}$ is a $\mathfrak{g}$-valued two-form on E_H. Conditions (1), (2), (3) and (4) imply that the curvature $K(\omega)$ vanishes on any pair $(Z_1, Z_2) \in TE_H$ such that one of them is vertical or tangent to $\widetilde{\mathcal{F}}$. Hence the transverse curvature is the pull-back to E_H of a holomorphic section of $\Lambda^2(\mathcal{N}_{\mathcal{F}})^* \otimes \mathrm{ad}(E_G)$ (recall that $\mathrm{ad}(E_G)$ is the vector bundle associated with E_H through the action of $H \subset G$ on $\mathfrak{g}$ by adjoint representation).

When the transverse curvature vanishes identically, we have a (branched) transverse holomorphic (G, X)-structure. In this case the pull-back of the (branched) flat transverse Cartan geometry on the universal cover $\widetilde{M}$ of M is given by a holomorphic developing map $\mathrm{dev} : \widetilde{M} \longrightarrow X = G/H$ which is constant on the leaves of $\mathcal{F}$. Moreover, the developing is a submersion in the classical case and a submersion away from the branching divisor in the branched case. The developing map is equivariant with respect to the monodromy homomorphism from the fundamental group of M into G (see [10, 58]).

In the simplest case where $G = \mathbb{C}$ and $H = \{0\}$, the bundle E_H coincides with M and a branched transverse structure for this model (G, H) for a complex codimension one holomorphic foliation is given by a holomorphic one-form ω on M whose kernel coincides with $\mathcal{F}$. In particular, the form ω satisfies the integrability condition $\omega \wedge d\omega = 0$. This is equivalent to the condition that $d\omega$ vanishes in restriction to $\mathcal{F}$.

Moreover, ω satisfies condition (4) above which is $L_X\omega = i_X d\omega = 0$, for any local holomorphic tangent vector field to $\mathcal{F}$. Since $\mathcal{F}$ is of complex codimension one, this implies $d\omega = 0$.

Hence codimension one foliations admit branched holomorphic transverse structure exactly when they are defined as the kernel of a global holomorphic closed one-form ω. The branching divisor is the divisor of zeros of ω. The developing map of the transverse translation structure is a primitive of ω. The monodromy group of the transverse translation structure is the additive subgroup of $\mathbb{C}$ generated by the periods of ω. In general this group is not a lattice.

A branched transverse holomorphic Riemannian metric is a branched transverse Cartan geometry with model $(G = O(n, \mathbb{C}) \ltimes \mathbb{C}^n,\ H = O(n, \mathbb{C}))$. In the flat case we get a branched transverse (G, X)-structure, with $G = O(n, \mathbb{C}) \ltimes \mathbb{C}^n$ being the group of complex Euclidean motions and X the complex Euclidean space.

When the model is that of the complex affine space (respectively, that of the complex projective space) we obtain the notion of branched transverse holomorphic affine connection (respectively, that of branched transverse holomorphic projective connection). In the flat case, we get a branched transverse affine structure (respectively, that of a branched transverse holomorphic projective structure).

In the classical (unbranched) case, transversely affine (respectively transversely projective) foliations of complex dimension one were studied by several authors (see, for instance [60] and references therein).

One could define a more general notion of a *generalized transverse holomorphic Cartan geometry* with model (G, H) by dropping condition (5) (see Sect. 11.5.1). In this case there is no relation anymore between the complex codimension of $\mathcal{F}$ and the complex codimension of the model G/H. In the flat case the developing map of such a generalized transverse holomorphic Cartan geometry is a holomorphic map from the universal cover $\widetilde{M}$ to the model space $X = G/H$ which is constant on the leaves of $\mathcal{F}$. In general, this map is not a submersion at the generic point.

Notice that for the trivial foliation $\mathcal{F}$ (given by points in M) the definition of a transverse Cartan geometry is the same as the definition of a Cartan geometry over M. The same holds in the branched case (respectively, in the generalized case).

11.5.1 Foliated Atiyah Bundle Description

Consider again $\pi : E_H \longrightarrow M$ a holomorphic (right) principal H-bundle over M. The kernel of the differential $d\pi$ defines a holomorphic (vertical) subbundle in the holomorphic tangent bundle TE_H which is holomorphically isomorphic to $E_H \times \mathfrak{h}$ (this isomorphism is realized by the identification of the fundamental vector fields of the H-action with the Lie algebra of the left-invariant vector fields on H; or equivalently, using the Maurer–Cartan form ω_H of H). Notice that this isomorphism is not invariant for the lifted (right) H action on TE_H, but equivariant with respect to the adjoint action of H on its Lie algebra $\mathfrak{h}$.

Following Atiyah [4], let us define the holomorphic quotient bundle over M as

$$\mathrm{ad}(E_H) := \mathrm{kernel}(d\pi)/H \longrightarrow M .$$

As explained above, $\mathrm{ad}(E_H)$ is holomorphically isomorphic to the twisted vector bundle $E_H \times^H \mathfrak{h}$ associated with the principal H-bundle E_H via the adjoint action of H on its Lie algebra $\mathfrak{h}$. Recall that $\mathrm{ad}(E_H)$ is known as the adjoint vector bundle of E_H. Notice that, since the adjoint action of H on its Lie algebra preserves the Lie algebra structure of $\mathfrak{h}$, the fiber of $\mathrm{ad}(E_H)$ has the Lie algebra structure of $\mathfrak{h}$: it is identified with $\mathfrak{h}$ up to a conjugation.

One can check that the quotient

$$(TE_H)/H \longrightarrow M$$

has also the structure of a holomorphic vector bundle over M (see [4]), classically denoted by $\mathrm{At}(E_H)$ and known as the *Atiyah bundle* of E_H [4].

The quotient by H of the short exact sequence of holomorphic vector bundles over E_H

$$0 \longrightarrow \mathrm{kernel}(\mathrm{d}\pi) \longrightarrow TE_H \xrightarrow{\mathrm{d}\pi} \pi^*TM \longrightarrow 0\,.$$

leads to the following short exact sequence of

$$0 \longrightarrow \mathrm{ad}(E_H) \xrightarrow{\iota''} \mathrm{At}(E_H) \xrightarrow{\widehat{\mathrm{d}\pi}} TM \longrightarrow 0\,, \tag{11.5.1}$$

where $\widehat{\mathrm{d}\pi}$ is constructed from $\mathrm{d}\pi$; this is known as the Atiyah exact sequence for E_H.

A holomorphic connection in the principal H-bundle E_H is a splitting of the Atiyah exact sequence [4].

Let us now consider the foliated case: the basis M of the principal H-bundle E_H is a complex manifold endowed with a holomorphic foliation $\mathcal{F}$.

Define the foliated Atiyah bundle as the subbundle

$$\mathrm{At}_{\mathcal{F}}(E_H) := (\widehat{\mathrm{d}\pi})^{-1}(T\mathcal{F}) \subset \mathrm{At}(E_H)\,. \tag{11.5.2}$$

So from (11.5.1) we get the short exact sequence

$$0 \longrightarrow \mathrm{ad}(E_H) \longrightarrow \mathrm{At}_{\mathcal{F}}(E_H) \xrightarrow{\mathrm{d}'\pi} T\mathcal{F} \longrightarrow 0\,, \tag{11.5.3}$$

where $\mathrm{d}'\pi$ is the restriction of $\widehat{\mathrm{d}\pi}$ in (11.5.1) to the subbundle $\mathrm{At}_{\mathcal{F}}(E_H)$.

The principal H-bundle E_H admits a partial holomorphic connection in the direction of $\mathcal{F}$ if the exact sequence (11.5.3) splits, meaning there exists a holomorphic homomorphism

$$\lambda : T\mathcal{F} \longrightarrow \mathrm{At}_{\mathcal{F}}(E_H)$$

such that $\mathrm{d}'\pi \circ \lambda = \mathrm{Id}_{T\mathcal{F}}$, where $\mathrm{d}'\pi$ is the projection homomorphism in (11.5.3).

Notice that the splitting of (11.5.3) given by the homomorphism λ can be also defined using a projection homomorphism $p : \mathrm{At}_{\mathcal{F}}(E_H) \longrightarrow \mathrm{ad}(E_H)$ such that p is the identity map in restriction $\mathrm{ad}(E_H)$ (the canonical inclusion of $\mathrm{ad}(E_H)$ in

$\mathrm{At}_{\mathcal{F}}(E_H)$ is the injective homomorphism in (11.5.3)). The homomorphism p is uniquely determined by λ (and conversely) by the condition that the image of λ in $\mathrm{At}_{\mathcal{F}}(E_H)$ is the kernel of p. In general the image of λ is not a foliation in TE_H and, as first proved by Ehresmann, the curvature of the partial connection $\lambda : T\mathcal{F} \longrightarrow \mathrm{At}_{\mathcal{F}}(E_H)$ is the obstruction to the integrability of the image of λ.

To make this statement precise, consider two local holomorphic sections X_1 and X_2 of $T\mathcal{F}$ and compute the locally defined holomorphic section $p([\lambda(X_1), \lambda(X_2)])$ of $\mathrm{ad}(E_H)$ (notice that the Lie bracket is well defined: $\lambda(X_1)$ and $\lambda(X_2)$ representing H-invariant sections of TE_H, their Lie bracket is also a H-invariant section of TE_H). One can easily check that this defines an $\mathcal{O}_M$-linear homomorphism

$$\mathcal{K}(\lambda) \in H^0(M, \mathrm{Hom}(\bigwedge^2 T\mathcal{F}, \mathrm{ad}(E_H))) = H^0(M, \mathrm{ad}(E_H) \otimes \bigwedge^2 T\mathcal{F}^*),$$

which is the *curvature* of the connection λ. The connection λ is called *flat* if $\mathcal{K}(\lambda)$ vanishes identically.

A (partial) connection on the principal H-bundle E_H induces a canonical (partial) connection on any bundle associated with E_H via a representation of H. In particular, a (partial) connection on E_H induces a (partial) connection on the adjoint bundle $\mathrm{ad}(E_H)$.

Since $\mathrm{At}_{\mathcal{F}}(E_H)$ is a subbundle of $\mathrm{At}(E_H)$, any partial connection $\lambda : T\mathcal{F} \longrightarrow \mathrm{At}_{\mathcal{F}}(E_H)$ induces a unique associated homomorphism $\lambda' : T\mathcal{F} \longrightarrow \mathrm{At}(E_H)$ and from (11.5.1) we get the following exact sequence

$$0 \longrightarrow \mathrm{ad}(E_H) \overset{\iota'}{\longrightarrow} \mathrm{At}(E_H)/\lambda'(T\mathcal{F}) \overset{\widehat{d\pi}}{\longrightarrow} TM/T\mathcal{F} = \mathcal{N}_{\mathcal{F}} \longrightarrow 0,$$
$$(11.5.4)$$

where ι' is given by ι'' in (11.5.1).

Let λ be a flat partial connection on E_H. In this case the image of λ in the foliated Atiyah bundle $\mathrm{At}_{\mathcal{F}}(E_H)$ is a foliation. It uniquely defines an H-invariant foliation $\widetilde{\mathcal{F}}$ in TE_H such that $d\pi(T\widetilde{\mathcal{F}}) = T\mathcal{F}$ and the restriction of $d\pi$ to $\widetilde{\mathcal{F}}$ is a submersion over $\mathcal{F}$. Consequently, the definition of a flat partial connection on E_H agrees with the one given in the context of foliated differential bundles in Sect. 11.5.

Recall that the normal bundle $TM/T\mathcal{F}$ is endowed with a canonical partial flat connection along $\mathcal{F}$, namely $\nabla^{\mathcal{F}}$, defined in Sect. 11.5.

Lemma 11.5.1 ([10]) *The flat partial connection λ on E_H induces a unique flat partial connection on $\mathrm{At}(E_H)/\lambda'(T\mathcal{F})$ (along $\mathcal{F}$) such that the homomorphisms in the exact sequence (11.5.4) are connection preserving (where $\mathrm{ad}(E_H)$ is endowed with the canonical connection induced from (E_H, λ) and $TM/T\mathcal{F}$ is endowed with $\nabla^{\mathcal{F}}$).*

Proof We have seen that the image of λ defines an H-invariant holomorphic foliation $\widetilde{\mathcal{F}}$ on E_H, such that the differential $d\pi(T\widetilde{\mathcal{F}}) = T\mathcal{F}$ and the restriction of $d\pi$ to $\widetilde{\mathcal{F}}$ is a submersion over $\mathcal{F}$.

Consider the canonical partial connection $\nabla^{\widetilde{\mathcal{F}}}$ on the normal bundle $TE_H/T\widetilde{\mathcal{F}}$. Notice that $\nabla^{\widetilde{\mathcal{F}}}$ is H-invariant. Since $\mathrm{At}(E_H) = (TE_H)/H$ we have

$(TE_H/T\widetilde{\mathcal{F}})/H = \mathrm{At}(E_H)/\lambda'(\mathcal{F})$. By H-invariance, the natural connection $\nabla^{\widetilde{\mathcal{F}}}$ on $TE_H/T\widetilde{F}$ in the direction of $\widetilde{\mathcal{F}}$ descends to a flat partial connection on $\mathrm{At}(E_H)/\lambda(T\mathcal{F})$ in the direction on $\mathcal{F}$.

Let us show that the morphism ι' in (11.5.4) is connection preserving. Consider a local holomorphic vector field X on M tangent to $\mathcal{F}$, defined on an open set $U \subset M$. Let X' be the unique holomorphic vector field in $\pi^{-1}(U) \subset E_H$, tangent to $\widetilde{\mathcal{F}}$, which lifts X, meaning that $d\pi(X') = X$. Let N be a holomorphic section of $\mathrm{kernel}(d\pi) \subset TE_H$ over $\pi^{-1}(U)$. Then the Lie bracket $[X', N]$ is such that $d\pi([X', N]) = 0$, meaning that $[X', N]$ is a vertical vector field: a section of $\mathrm{kernel}(d\pi)$. But $\mathrm{ad}(E_H) = \mathrm{kernel}(d\pi)/H$, and, consequently, the inclusion ι' of $\mathrm{ad}(E_H)$ in $\mathrm{At}(E_H)/\lambda'(T\mathcal{F})$ in (11.5.4) intertwines the partial connections on $\mathrm{ad}(E_H)$ and $\mathrm{At}(E_H)/\lambda'(T\mathcal{F})$ in the direction of $\mathcal{F}$. It also follows that the projection homomorphism $\widehat{d\pi}$ in (11.5.4) is partial connection preserving as well. □

Now we give the foliated Atiyah bundle theoretical definition of a transverse branched Cartan geometry as introduced in [10].

Let G be a connected complex Lie group and $H \subset G$ a complex Lie subgroup with Lie algebras $\mathfrak{g}$ and $\mathfrak{h}$ respectively.

Again $\pi : E_H \longrightarrow M$ is a holomorphic principal H-bundle over M. Define

$$E_G = E_H \times^H G \longrightarrow M \tag{11.5.5}$$

to be the principal G-bundle over M obtained by extending the structure group of E_H using the inclusion of H in G. Denote by $\mathrm{ad}(E_G) = E_G \times^G \mathfrak{g}$ the adjoint bundle for E_G.

The Lie algebra inclusion of $\mathfrak{h}$ in $\mathfrak{g}$ induces an injective homomorphism of holomorphic vector bundles

$$\iota : \mathrm{ad}(E_H) \longrightarrow \mathrm{ad}(E_G). \tag{11.5.6}$$

Consider a flat partial connection λ on E_H in the direction of $\mathcal{F}$. We have seen that λ induces flat partial connection on all the associated bundles, in particular, λ induces partial connections on E_G and $\mathrm{ad}(E_H)$ and $\mathrm{ad}(E_G)$.

A transverse *branched* holomorphic Cartan geometry with model (G, H) on the foliated manifold $(M, \mathcal{F})$ is given by the following data

(I) A holomorphic principal H-bundle E_H on M equipped with a flat partial connection λ, and

(II) A holomorphic homomorphism

$$\beta : \mathrm{At}(E_H)/\lambda'(T\mathcal{F}) \longrightarrow \mathrm{ad}(E_G), \tag{11.5.7}$$

such that the following three conditions hold:

(1) β is partial connection preserving,
(2) β is an isomorphism *over a nonempty open subset of M*, and

(3) the following diagram is commutative:

$$
\begin{array}{ccccccc}
0 & \longrightarrow & \mathrm{ad}(E_H) & \xrightarrow{\iota'} & \mathrm{At}(E_H)/\lambda'(T\mathcal{F}) \longrightarrow & \mathcal{N}_{\mathcal{F}} & \longrightarrow 0 \\
& & \| & & \downarrow{\beta} & \downarrow{\overline{\beta}} & \\
0 & \longrightarrow & \mathrm{ad}(E_H) & \xrightarrow{\iota} & \mathrm{ad}(E_G) & \longrightarrow \mathrm{ad}(E_G)/\mathrm{ad}(E_H) \longrightarrow 0
\end{array}
$$

$$(11.5.8)$$

where the top exact sequence is the one in (11.5.4), and ι is the homomorphism in (11.5.6).

From the commutativity of (11.5.8) it follows immediately that the homomorphism $\overline{\beta} : \mathcal{N}_{\mathcal{F}} \longrightarrow \mathrm{ad}(E_G)/\mathrm{ad}(E_H)$ in (11.5.8) is an isomorphism over a point $m \in M$ if and only if $\beta(m)$ is an isomorphism. Notice that the classical case of an unbranched transverse Cartan geometry is that where the open subset in condition II) (2) is the entire manifold M. This corresponds to condition II) (5) in the equivalent definition given in Sect. 11.5. Notice also that the condition II) (1) here is equivalent to the condition II (5) in Sect. 11.5.

Let n be the complex dimension of $\mathfrak{g}$. Consider the homomorphism of n-th exterior products

$$\bigwedge^n \beta : \bigwedge^n (\mathrm{At}(E_H)/\lambda'(T\mathcal{F})) \longrightarrow \bigwedge^n \mathrm{ad}(E_G)$$

induced by β. The homomorphism β fails to be an isomorphism precisely over the divisor of the section $\bigwedge^n \beta$ of the line bundle $\mathrm{Hom}(\bigwedge^n (\mathrm{At}(E_H)/\lambda'(T\mathcal{F})), \bigwedge^n \mathrm{ad}(E_G))$.

The *branching* set $D \subset M$ defined earlier in Sect. 11.5 coincides with the vanishing set of the holomorphic section $\bigwedge^n \beta$. This divisor $\mathrm{div}(\bigwedge^n \beta)$ is called the *branching divisor* for $((E_H, \lambda), \beta)$.

The triple $((E_H, \lambda), \beta)$ characterizes a classical (unbranched) holomorphic Cartan geometry if and only if β is an isomorphism over the entire M. In this case the branching divisor is trivial.

In order to define the more general notion of a *transverse generalized Cartan geometry* (which is a foliated version of a generalized Cartan geometry, as defined in [1, 15]) one drops condition II) (2) in the above definition of a transverse branched Cartan geometry.

In order to define the developing map of a transverse (branched or generalized) holomorphic Cartan geometry the following result is useful.

Lemma 11.5.2 *A transverse generalized holomorphic Cartan geometry $((E_H, \lambda), \beta)$ over the foliated manifold $(M, \mathcal{F})$ defines a holomorphic connection on the principal G-bundle E_G which is flat in the direction of $\mathcal{F}$.*

The proof below is that given in [10]. It does not use condition II) (2) in the definition of the transverse Cartan geometry and works as well for transverse branched and generalized Cartan geometries.

Proof Consider the homomorphism

$$\mathrm{ad}(E_H) \longrightarrow \mathrm{ad}(E_G) \oplus \mathrm{At}(E_H), \quad v \longmapsto (\iota(v), -\iota''(v)) \tag{11.5.9}$$

(see (11.5.6) and (11.5.1) for ι and ι'' respectively).

The corresponding quotient $(\mathrm{ad}(E_G) \oplus \mathrm{At}(E_H))/\mathrm{ad}(E_H)$ is identified with the Atiyah bundle $\mathrm{At}(E_G)$. The inclusion of $\mathrm{ad}(E_G)$ in $\mathrm{At}(E_G)$ as in (11.5.1) is given by the inclusion $\mathrm{ad}(E_G) \hookrightarrow \mathrm{ad}(E_G) \oplus \mathrm{At}(E_H)$, $w \longmapsto (w, 0)$, while the projection $\mathrm{At}(E_G) \longrightarrow TM$ is given by the composition

$$\mathrm{At}(E_G) \hookrightarrow \mathrm{ad}(E_G) \oplus \mathrm{At}(E_H) \overset{(0,\widehat{\mathrm{d}\pi})}{\longrightarrow} TM,$$

with $\widehat{\mathrm{d}\pi}$ the projection in (11.5.1).

Consider the subbundle $\lambda'(T\mathcal{F}) \subset \mathrm{At}(E_H)$ in (11.5.4). The composition

$$\mathrm{At}(E_H) \longrightarrow \mathrm{At}(E_H)/\lambda'(T\mathcal{F}) \overset{\beta}{\longrightarrow} \mathrm{ad}(E_G),$$

where the first homomorphism is the quotient map, will be denoted by β'. The homomorphism

$$\mathrm{ad}(E_G) \oplus \mathrm{At}(E_H) \longrightarrow \mathrm{ad}(E_G), \quad (v, w) \longmapsto v + \beta'(w) \tag{11.5.10}$$

vanishes on the image of $\mathrm{ad}(E_H)$ by the map in (11.5.9). Therefore, the homomorphism in (11.5.10) produces a homomorphism

$$\varphi : \mathrm{At}(E_G) = (\mathrm{ad}(E_G) \oplus \mathrm{At}(E_H))/\mathrm{ad}(E_H) \longrightarrow \mathrm{ad}(E_G). \tag{11.5.11}$$

The composition

$$\mathrm{ad}(E_G) \hookrightarrow \mathrm{At}(E_G) \overset{\varphi}{\longrightarrow} \mathrm{ad}(E_G)$$

is the identity map of $\mathrm{ad}(E_G)$. Consequently, φ defines a holomorphic connection on the principal G-bundle E_G (see above and also [4]).

It remains to prove that φ is flat along $\mathcal{F}$. Since the homomorphism β in (11.5.7) is partial connection preserving, it follows that the restriction of the connection φ in the direction of $\mathcal{F}$ coincides with the partial connection along $\mathcal{F}$ induced on E_G by λ. Since λ is flat along $\mathcal{F}$, the corresponding induced connection on E_G is flat as well. $\qquad\square$

Denote by

$$\mathrm{Curv}(\varphi) \in H^0(M, \mathrm{ad}(E_G) \otimes \Omega_M^2)$$

the curvature of the connection φ.

Since β is connection preserving, the contraction of $\mathrm{Curv}(\varphi)$ with any tangent vector of $T\mathcal{F}$ vanishes. This implies that $\mathrm{Curv}(\varphi)$ is actually a section of $\mathrm{ad}(E_G) \otimes \bigwedge^2 \mathcal{N}_{\mathcal{F}}^*$.

The transverse (branched or generalized) Cartan geometry $((E_H, \theta), \beta)$ is called *flat* if the above curvature tensor $\mathrm{Curv}(\varphi)$ vanishes identically.

Assume that the transverse Cartan geometry is flat and M is simply connected. Then the flat bundle E_G is trivial over M, isomorphic to $M \times E_G$. The subbundle $E_H \subset E_G$ is described as a holomorphic reduction of the structure group of E_G to H and hence by a holomorphic map $M \longrightarrow G/H$. This is the developing map dev of the transverse (branched or generalized) flat Cartan geometry.

The differential of this developing map dev is given by the homomorphism $\overline{\beta}$ in(11.5.8). Consequently, in the branched case, the developing map is *a submersion away from the branching divisor*. There is no condition on the differential of the developing map for a transverse generalized Cartan geometry.

Since the connection φ on E_G along $\mathcal{F}$ is induced from the flat connection λ on E_H, the developing map dev is constant on each (connected) leaf of $\mathcal{F}$.

If M is not simply connected the monodromy morphism of the flat connection φ on E_G is the monodromy morphism of the transverse branched Cartan geometry. When pulled-back on the universal cover $\widetilde{M}$, the flat bundle E_G becomes trivial and the associated developing map obtained as above is equivariant with respect to the monodromy morphism. For more details about the construction of the developing map the reader is referred to [10, 15, 19, 57].

Singular Foliations. Let us consider now a complex manifold $\widehat{M}$ endowed with a holomorphic singular foliation $\mathcal{F}$. It is classically known that there exists a maximal open dense set $M \subset \widehat{M}$ such that in restriction to M the foliation $\mathcal{F}$ is a nonsingular foliation. Moreover, this maximal open set M is the complement of an analytic subset in $\widehat{M}$ which is of complex codimension at least two in $\widehat{M}$.

We will say that $(\widehat{M}, \mathcal{F})$ admits a transverse (branched or generalized) holomorphic Cartan geometry with model (G, H) if $(M, \mathcal{F})$ admits a transverse (branched or generalized) holomorphic Cartan geometry with model (G, H).

An easy example of this situation is given by fibrations over homogeneous spaces. More precisely, consider a complex manifold $\widehat{M}$ which admits a holomorphic map ρ to a complex homogeneous space G/H which is a submersion on a nontrivial open dense set in $\widehat{M}$. Then the fibers of ρ define a singular holomorphic foliation $\mathcal{F}$ on $\widehat{M}$ bearing a transverse branched holomorphic flat Cartan geometry with model (G, H) in the sense of the above definition.

More generally, it was proved in [10, Proposition 3.2] that if $\rho : \widehat{M} \longrightarrow N$ is a holomorphic map which is a submersion of a nontrivial open dense set in $\widehat{M}$ and N admits a holomorphic Cartan geometry with model (G, H), then the holomorphic (singular) foliation $\mathcal{F}$ defined by the fibers of ρ bears a transverse branched holomorphic Cartan geometry with model (G, H). This transverse geometry is flat if and only if the Cartan geometry on N is flat.

11.5.2   Flatness Results

11.5.2.1   Rationally Connected Manifolds

Let us recall that a complex projective manifold $\widehat{M}$ is called *rationally connected* if for any pair of points $m, n \in \widehat{M}$ there exists a (maybe singular) rational curve $C \subset \widehat{M}$ such that $m, n \in C$.

Examples of rationally connected projective manifolds are given by Fano projective manifolds [21, 50]. Recall that a projective manifold $\widehat{M}$ is Fano if its anti-canonical bundle $-K_{\widehat{M}}$ is ample.

In this context the following result was proved in [10].

Theorem 11.5.3 *Let $\widehat{M}$ be a rationally connected projective manifold endowed with a (possibly singular) holomorphic foliation $\mathcal{F}$. Then the following statements hold:*

(1) Any transverse generalized holomorphic Cartan geometry with model (G, H) on $(\widehat{M}, \mathcal{F})$ is flat and, if G/H is an analytic affine variety, defined by a holomorphic map $\widehat{M} \longrightarrow G/H$ (constant on the leaves of $\mathcal{F}$);

(2) There is no transverse branched holomorphic Cartan geometry on $(\widehat{M}, \mathcal{F})$ with model a nontrivial analytic affine variety G/H. In particular, there is no transverse branched holomorphic affine connection on $(\widehat{M}, \mathcal{F})$.

Notice that if G is a complex linear algebraic group and H a closed reductive algebraic subgroup, then G/H is an affine analytic variety (see Lemma 3.32 in [56]).

Proof Let $\widehat{M}$ be a complex projective rationally connected manifold. Consider $M \subset \widehat{M}$ to be the maximal open subset such that the foliation $\mathcal{F}$ is nonsingular on M. Then the complex codimension of the complement of M in $\widehat{M}$ is at least two.

(1) Let $((E_H, \lambda), \beta)$ be a transverse generalized holomorphic Cartan geometry with model (G, H) on the foliated manifold $(M, \mathcal{F})$.

By Lemma 11.5.2 the bundle E_G (obtained from E_H by extension of the structure group) inherits a holomorphic connection φ over M (flat in the direction of $\mathcal{F}$). On a smooth curve any holomorphic connection is flat. In particular, the restriction of E_G to any smooth rational curve lying in M is flat. Moreover, since a rational curve is simply connected, the flat holomorphic bundle E_G over the rational curve is isomorphic to the trivial bundle endowed with the trivial connection.

There is a nonempty open subset of M which is covered by smooth rational curves $C \subset M$ such that the restriction $(TM)|_C$ of the holomorphic tangent bundle to C is ample. Consequently, $H^0(C, (\Omega_M^2)|_C) = 0$. In particular, the curvature of the holomorphic connection φ vanishes identically on M. By definition of its curvature, the transverse generalized holomorphic Cartan geometry $((E_H, \lambda), \beta)$ is flat.

Rationally connected manifolds are known to be simply connected [20]. Hence $\widehat{M}$ is simply connected. Since M is a dense open subset of complex codimension at least 2, its fundamental group is isomorphic to that of $\widehat{M}$. It follows that M is simply connected. Therefore, the developing map dev : $M \longrightarrow G/H$ of the transverse generalized flat Cartan geometry $((E_H, \lambda), \beta)$ is defined on M. Recall that the developing is constant on the leaves of $\mathcal{F}$.

By Hartog's extension theorem dev extends as a holomorphic map defined on $\widehat{M}$.

(2) In the branched case, the above developing map dev : $\widehat{M} \longrightarrow G/H$ is a holomorphic submersion on an open dense set and the generic leaves of $\mathcal{F}$ coincide with the connected components of the fibers of the developing map. But if G/H is an analytic affine variety any holomorphic map from the compact manifold $\widehat{M}$ to G/H must be constant: a contradiction. In particular, this holds if G/H is the complex affine space (the model of an affine connection transverse to $\mathcal{F}$). $\square$

11.5.2.2 Simply Connected Calabi–Yau Manifolds

Let $\widehat{M}$ be a simply connected compact Kähler manifold with vanishing real first Chern class $c_1(T\widehat{M}) = 0$.

Those manifolds are known as simply connected *Calabi–Yau manifolds*. Recall that by Yau's proof of Calabi conjecture they are endowed with Ricci flat Kähler metrics.

In this context, using the main result in [12] we proved in [10] the following result:

Theorem 11.5.4 *Let $\widehat{M}$ be a simply connected Calabi–Yau manifold endowed with a (possible singular) holomorphic foliation $\mathcal{F}$. Let G be a complex Lie group which is simply connected or complex semi-simple and let $H \subset G$ be a closed complex Lie subgroup. Then:*

(1) Any transverse generalized holomorphic Cartan geometry with model (G, H) on $(\widehat{M}, \mathcal{F})$ is flat and, if G/H is an analytic affine variety, defined by a holomorphic map $\widehat{M} \longrightarrow G/H$ (constant on the leaves of $\mathcal{F}$).

(2) If the model G/H is a nontrivial analytic affine variety G/H, then there is no transverse branched Cartan geometry on $(\widehat{M}, \mathcal{F})$ with model (G, H). In particular, there is no transverse branched holomorphic affine connection on $(\widehat{M}, \mathcal{F})$.

Proof Let $\widehat{M}$ be a complex projective rationally connected manifolds. Consider $M \subset \widehat{M}$ be the maximal open subset such that the foliation $\mathcal{F}$ is nonsingular on M. Then the complex codimension of the complement of M in $\widehat{M}$ is at least two.

(1) Let $((E_H, \lambda), \beta)$ be a transverse generalized holomorphic Cartan geometry with model (G, H) on the foliated manifold $(M, \mathcal{F})$.

By Lemma 11.5.2 the bundle E_G (obtained from E_H by extension of the structure group) inherits a holomorphic connection φ over M (flat in the direction of $\mathcal{F}$).

We prove now that φ is flat on the entire M.

The principal G-bundle E_G extends to a holomorphic principal G-bundle $\widehat{E}_G$ over $\widehat{M}$, and the connection φ extends to a holomorphic connection $\widehat{\varphi}$ on $\widehat{E}_G$ [7, Theorem 1.1].

Consider $\alpha : G \longrightarrow \mathrm{GL}(N, \mathbb{C})$, a linear representation of G with discrete kernel. The corresponding Lie algebra representation α' is an injective Lie algebra homomorphism from $\mathfrak{g}$ to $\mathfrak{gl}(N, \mathbb{C})$. For G simply connected those representations exist by Ado's theorem. For G complex semi-simple those representations do also exist (see Theorem 3.2, chapter XVII in [35]).

Consider the holomorphic vector bundle E_α over $\widehat{M}$ with fiber type $\mathbb{C}^N$ associated with $\widehat{E}_G$ via the representation α. We have seen that E_α inherits from $\widehat{\varphi}$ a holomorphic connection $\widehat{\varphi}_\alpha$. By Theorem [12, Theorem 6.2], the holomorphic connection $\widehat{\varphi}_\alpha$ is flat over $\widehat{M}$. Since the curvature of $\widehat{\varphi}_\alpha$ is the image of the curvature of $\widehat{\varphi}$ through the Lie algebra homomorphism α' and α' is injective, it follows that $\widehat{\varphi}$ is also flat. Therefore φ is flat and the transverse generalized holomorphic Cartan geometry $((E_H, \lambda), \beta)$ is flat.

Since M is simply connected, the developing map dev $: M \longrightarrow G/H$ of the transverse flat generalized Cartan geometry $((E_H, \lambda), \beta)$ is defined on M. Recall that the developing is constant on the leaves of $\mathcal{F}$.

By Hartog's extension theorem dev extends as a holomorphic map defined on $\widehat{M}$.

(2) In the branched case, the above developing map dev $: \widehat{M} \longrightarrow G/H$ is a holomorphic submersion on an open dense set and the generic leaves of $\mathcal{F}$ coincide with the connected components of the fibers of the developing map. But if G/H is an analytic affine variety any holomorphic map from the compact manifold $\widehat{M}$ to G/H must be constant: a contradiction. In particular, this holds if G/H is the complex affine space (the model of an affine connection transverse to $\mathcal{F}$). $\qquad\square$

11.5.3 A Topological Criterion

Let M be a compact connected Kähler manifold of complex dimension n equipped with a Kähler form ω.

For any holomorphic vector bundle V over M we define

$$\mathrm{degree}(V) := (c_1(V) \cup \omega^{n-1}) \cap [M] \in \mathbb{R}, \qquad (11.5.12)$$

with $c_1(V)$ the real first Chern class of V. The degree of a divisor D on M is defined as being $\mathrm{degree}(\mathcal{O}_M(D))$.

The degree is a topological invariant.

Fix an effective divisor D on X. Fix a holomorphic principal H-bundle E_H on X.

Theorem 11.5.5 *Let M be a Kähler manifold endowed with a nonsingular holomorphic foliation $\mathcal{F}$. Assume that $(M, \mathcal{F})$ admits a transverse branched holomorphic Cartan geometry $((E_H, \lambda), \beta)$ with model (G, H) and branching divisor $D \subset M$.*

Then $\mathrm{degree}(\mathcal{N}_{\mathcal{F}}^*) - \mathrm{degree}(D) = \mathrm{degree}(\mathrm{ad}(E_H))$.

In particular, if $D \neq 0$, then $\mathrm{degree}(\mathcal{N}_{\mathcal{F}}^*) > \mathrm{degree}(\mathrm{ad}(E_H))$.

Proof Let k be the complex dimension of the transverse model geometry G/H.

Recall that the homomorphism $\overline{\beta} : \mathcal{N}_{\mathcal{F}} \longrightarrow \mathrm{ad}(E_G)/\mathrm{ad}(E_H)$ in (11.5.8) is an isomorphism over a point $m \in M$ if and only if $\beta(m)$ is an isomorphism.

The branching divisor D coincides with the vanishing divisor of the holomorphic section $\bigwedge^k \overline{\beta}$ of the holomorphic line bundle $\bigwedge^k (\mathcal{N}_{\mathcal{F}}^*) \otimes \bigwedge^k (\mathrm{ad}(E_G)/\mathrm{ad}(E_H))$. We have

$$\mathrm{degree}(D) = \mathrm{degree}(\bigwedge^k (\mathrm{ad}(E_G)/\mathrm{ad}(E_H)) \otimes \bigwedge^k (\mathcal{N}_{\mathcal{F}}^*))$$

$$= \mathrm{degree}(\mathrm{ad}(E_G)) - \mathrm{degree}(\mathrm{ad}(E_H)) + \mathrm{degree}(\mathcal{N}_{\mathcal{F}}^*). \qquad (11.5.13)$$

Recall that E_G has a holomorphic connection ϕ (see (11.5.11)) which induces a holomorphic connection on $\mathrm{ad}(E_G)$. Hence we have $c_1(\mathrm{ad}(E_G)) = 0$ [4, Theorem 4], which implies that $\mathrm{degree}(\mathrm{ad}(E_G)) = 0$. Therefore, from (11.5.13) it follows that

$$\mathrm{degree}(\mathcal{N}_{\mathcal{F}}^*) - \mathrm{degree}(D) = \mathrm{degree}(\mathrm{ad}(E_H)). \qquad (11.5.14)$$

If $D \neq 0$, then $\mathrm{degree}(D) > 0$. Hence in that case (11.5.14) yields $\mathrm{degree}(\mathcal{N}_{\mathcal{F}}^*) > \mathrm{degree}(\mathrm{ad}(E_H))$. $\qquad\square$

Corollary 11.5.6 *The hypothesis and notation of Theorem 11.5.5 is used.*

(i) *If* $\mathrm{degree}(\mathcal{N}_{\mathcal{F}}^*) < 0$, *then there is no transverse branched holomorphic affine connection on $(M, \mathcal{F})$.*

(ii) *If* $\mathrm{degree}(\mathcal{N}_{\mathcal{F}}^*) = 0$, *then every transverse branched holomorphic affine connection on $(M, \mathcal{F})$ has a trivial branching divisor on M.*

Proof Recall that the model of a transverse holomorphic affine connection on $(M, \mathcal{F})$ is (G, H), with $H = \mathrm{GL}(k, \mathbb{C})$ and $G = \mathbb{C}^k \rtimes \mathrm{GL}(k, \mathbb{C})$. The homomorphism

$$\mathrm{M}(k, \mathbb{C}) \otimes \mathrm{M}(k, \mathbb{C}) \longrightarrow \mathbb{C}, \quad A \otimes B \longmapsto \mathrm{trace}(AB)$$

is nondegenerate and GL(k, $\mathbb{C}$)-invariant. In other words, the Lie algebra $\mathfrak{h}$ of $H = $ GL(k, $\mathbb{C}$) is self-dual as an H-module. Hence we have ad(E_H) $=$ ad(E_H)*, in particular, the equality

$$\text{degree(ad}(E_H)) = 0$$

holds. Hence from Theorem 11.5.5,

$$\text{degree}(\mathcal{N}_{\mathcal{F}}^*) = \text{degree}(D) . \tag{11.5.15}$$

For the effective divisor D we have degree(D) $\geq$ 0. Moreover, for a *nonzero* effective divisor D we have degree(D) $>$ 0. Therefore, the two points of the corollary follow from (11.5.15) and Theorem 11.5.5. $\qquad\qquad\square$

11.6 Some Related Open Problems

We present here some open questions dealing with holomorphic *G*-structures and holomorphic (foliated) Cartan geometries of compact complex manifolds.

SL(2, $\mathbb{C}$)-Structures in Odd Dimension and Holomorphic Riemannian Metrics
We have seen that any holomorphic SL(2, $\mathbb{C}$)-structure on a complex manifold of odd dimension M produces an associated *holomorphic Riemannian metric* on M. Moreover, for the complex dimension three, these two structures are equivalent.

Exotic compact complex threefolds endowed with holomorphic Riemannian metrics (or equivalently, SL(2, $\mathbb{C}$)-structures) were constructed by Ghys in [30] using deformations of quotients of SL(2, $\mathbb{C}$) by normal lattices. Moreover, Ghys proved in [30] that all holomorphic Riemannian metrics on those exotic deformations are *locally homogeneous* (meaning that local holomorphic vector fields preserving the holomorphic Riemannian metric on M span the holomorphic tangent bundle TM).

It was proved in [24] that all holomorphic Riemannian metrics on compact complex threefolds are locally homogeneous. This means that any compact complex threefold M bearing a holomorphic Riemannian metric admits a flat holomorphic Cartan geometry, or equivalently, a holomorphic (G, X)-structure such that the holomorphic Riemannian metric on M comes from a global G-invariant holomorphic Riemannian metric on the model space X. The classification of all possible models (G, X) was done in [26] where it was deduced that all compact complex threefolds endowed with a (locally homogeneous) holomorphic Riemannian metric also admit a finite unramified cover equipped with a holomorphic Riemannian metric of constant sectional curvature.

We conjecture that SL(2, $\mathbb{C}$)-*structures on compact complex manifolds of odd dimension are always locally homogeneous.* More generally we conjecture that *holomorphic Riemannian metrics on compact complex manifolds are locally homogeneous.*

Of course, these conjectures generalize to the non Kähler framework, known results in the Kähler context (see Theorem 11.3.7 and also the discussion below about the Fujiki case).

Some evidence toward these conjectures is given by the recent result in [13] proving that simply connected compact complex manifolds do not admit holomorphic Riemannian metrics. Notice that in the locally homogeneous case, this would be a direct consequence of the fact that the developing map dev : $M \longrightarrow X$ should be a submersion from a compact manifold to a complex affine model: a contradiction.

It should be clarified that the above mentioned conjectures are an important *partial* step toward the classification of compact complex manifolds with holomorphic Riemannian metrics. Indeed, even the classification of compact complex manifolds endowed with *flat* holomorphic Riemannian metrics is still an open problem. To understand the flat case one should first prove that M is a quotient of the complex Euclidean space $(\mathbb{C}^n, \, dz_1^2 + \ldots + dz_n^2)$ by a discrete subgroup of the group of complex Euclidean motions $O(n, \mathbb{C}) \ltimes \mathbb{C}^n$ (this is a special holomorphic version of Markus conjecture which asserts that compact manifolds with unimodular affine structures, meaning here locally modeled on (G, X), with $X = \mathbb{R}^{2n}$ and $G = SL(2n, \mathbb{R}) \ltimes \mathbb{R}^{2n}$, are complete: they are quotients of the model space X by a discrete subgroup of G.

The second step would be to classify discrete subgroups of $G = O(n, \mathbb{C}) \ltimes \mathbb{C}^n$ acting properly discontinuously and with a compact quotient on the model $(\mathbb{C}^n, \, dz_1^2 + \ldots + dz_n^2)$. Notice that a special case of Auslander's conjecture predicts that those subgroups are virtually solvable (i.e., they admit a finite index subgroup which is solvable). More precisely, the general Auslander's conjecture states that *a flat complete unimodular affine compact manifold has virtually solvable fundamental group*. Notice that compactness is essential in the statement of Auslander's conjecture; non compact flat complete unimodular affine manifolds with non abelian free fundamental group were constructed by Margulis.

Holomorphic Affine Connections on Fujiki Class $\mathcal{C}$ manifolds

Recall that a compact Kähler manifold bearing a holomorphic connection in its holomorphic tangent bundle TM has vanishing real Chern classes [4]. We have seen that this result can be obtain following Chern–Weil theory and computing real Chern classes of the holomorphic tangent bundle TM first using a Hermitian metric on it and then the holomorphic connection. This provides representatives of the Chern class $c_k(TM) \in H^{2k}(M, \mathbb{R})$ which are smooth forms on M of type (k, k) (given by the first computation) and other of type $(2k, 0)$ (when computed via the holomorphic connection). Classical Hodge theory says that on Kähler manifolds nontrivial real cohomology classes do not have representatives of different types. This implies the vanishing of the real Chern class: $c_k(TM) = 0$, for all k. This part of the argument directly adapts to compact complex manifolds manifolds which are bimeromorphic to Kähler manifolds; those manifolds were studied in [29] and are now said to be *in the Fujiki class $\mathcal{C}$*. All images of compact Kähler manifolds through holomorphic maps are known to belong to the Fujiki class $\mathcal{C}$ [63].

We have also seen that a compact Kähler manifold M with vanishing first two Chern classes $(c_1(TM) = c_2(TM) = 0)$ admits a finite unramified covering by a compact complex torus [38, 64].

We conjecture that the same is true for Fujiki class $\mathcal{C}$ manifolds, namely *a compact complex manifold M in the Fujiki class $\mathcal{C}$ satisfying the condition $c_1(TM) = c_2(TM) = 0$ admits a finite unramified covering by a compact complex torus.*

In particular, we conjecture that *a complex compact manifold M in the Fujiki class $\mathcal{C}$ whose holomorphic tangent bundle TM has a holomorphic connection admits a finite unramified covering by a compact complex torus.* This was proved in [9, Proposition 4.2] for *Moishezon manifolds* (i.e., compact complex manifolds bimeromorphic to projective manifolds) [57]. Also this was proved in [16, Theorem C] for holomorphic Riemannian metrics on Fujiki class $\mathcal{C}$ manifolds under some technical assumption (which could probably be removed).

GL$(2, \mathbb{C})$-Structures on Compact Kähler Manifolds of Odd Dimension
Recall that complex manifolds of odd dimension bearing a GL$(2, \mathbb{C})$-structure also admit a holomorphic conformal structure. Flat conformal structures on compact projective manifolds were classified in [42]: beside some projective surfaces, there are only the standard examples.

As for the conclusion in Theorem 11.3.4 and in Theorem 11.3.5, we conjecture that *Kähler manifolds of odd complex dimension ≥ 5 and bearing a holomorphic GL$(2, \mathbb{C})$-structure are covered by compact tori.*

Holomorphic Cartan Geometries on Compact Complex Surfaces
In a series of papers, Inoue, Kobayashi and Ochiai studied holomorphic affine and projective connections on compact complex surfaces. A consequence of their work is that compact complex surfaces bearing holomorphic affine connections (respectively, holomorphic projective connections) also admit *flat* holomorphic affine connections (respectively, flat holomorphic projective connections) with corresponding injective developing map. In particular, those complex surfaces are uniformized as quotients of open subsets in the complex affine plane (respectively, complex projective plane) by a discrete subgroup of affine transformations (respectively, projective transformations) acting properly and discontinuously [38, 46, 48].

We conjecture that those results can be generalized to all holomorphic Cartan geometries; namely, that compact complex surfaces bearing holomorphic Cartan geometries with model (G, H) also admit *flat* holomorphic Cartan geometries with model (G, H) and with corresponding injective developing maps into G/H. This would provide uniformization result for compact complex surfaces bearing holomorphic Cartan geometries with model (G, H) as compact quotients of open subsets U in G/H by discrete subgroups in G preserving $U \subset G/H$ and acting properly and discontinuously on U.

In order to address this problem locally (in the neighborhood of a Cartan geometry in the deformation space), the deformation theory for holomorphic (non necessarily flat) Cartan geometries was recently worked out in [17].

One could also naturally ask the analogous question in the framework of *branched* holomorphic Cartan geometries on compact complex surfaces. Recall that the branched framework is much broader since all projective surfaces admit branched flat holomorphic projective connections [12]. Moreover, branched torsion free holomorphic affine connections on compact complex surfaces which are *non projectively flat* were constructed in [12]: these branched affine connections are essential (meaning they are not obtained as pull-back of unbranched ones by a ramified holomorphic map) since it is known that (unbranched) torsion free holomorphic affine connections on compact complex surfaces are *projectively flat* [25].

Foliated Cartan Geometries on Compact Complex Tori

This question is about holomorphic foliations on complex tori with transverse holomorphic Cartan geometry.

Our motivation for this question came from Ghys' classification of codimension one holomorphic nonsingular foliations on complex tori [31] which be briefly describe here.

(I) The simplest examples of codimension one holomorphic foliations are those given by the kernel of some holomorphic 1-form ω. Since holomorphic 1-forms on complex tori are necessarily translation invariant, the foliation given by the kernel of ω is also translation invariant.

(II) Now assume that $T = \mathbb{C}^n/\Lambda$, with Λ a lattice in $\mathbb{C}^n$ such that there exists a linear form $L : \mathbb{C}^n \longrightarrow \mathbb{C}$ sending Λ to a lattice Λ' in $\mathbb{C}$. Then L descends to a map $\widehat{L} : T \longrightarrow \mathbb{C}/\Lambda'$. Pick a nonconstant meromorphic function u on the elliptic curve $\mathbb{C}/\Lambda'$ and consider the meromorphic closed 1-form $\Omega = \widehat{L}^*(udz) + \omega$ on T, with ω as above and dz a uniformizing holomorphic 1-form on $\mathbb{C}/\Lambda'$. It is easy to check that the foliation given by the kernel of Ω extends to all of T as a nonsingular holomorphic codimension one foliation. This foliation is not invariant by all translations in T, but only by those which lie the kernel of L. They act on T as a subtorus of symmetries of codimension one.

Ghys' theorem asserts that all codimension one (nonsingular) holomorphic foliations on complex tori are constructed as in (I) or (II) above. In particular, they are invariant by a subtorus of complex codimension one. Moreover, for generic complex tori, there are no nonconstant meromorphic functions and, consequently the construction (II) does not apply. All holomorphic codimension one foliations on generic tori are translation invariant.

We recently proved in [11] that all holomorphic Cartan geometries on complex tori are translation invariant.

We conjecture that the foliated analogous also holds, namely that *all holomorphic foliations bearing transverse holomorphic Cartan geometries on complex tori are translation invariant.*

Acknowledgments We thank Athanase Papadopoulos for his kind invitation and for his careful reading of the manuscript. We thank Charles Boubel who kindly explained to us the geometric construction of the noncompact dual D_3 of the three dimensional quadric Q_3 presented in Sect. 11.2. This work has been supported by the French government through the UCAJEDI Investments in the Future project managed by the National Research Agency (ANR) with the reference number ANR2152IDEX201. The first-named author is partially supported by a J. C. Bose Fellowship, and school of mathematics, TIFR, is supported by 12-R&D-TFR-5.01-0500.

References

1. D.V. Alekseevsky, P.W. Michor, Differential geometry of Cartan connections. Publ. Math. Debrecen **47**, 349–375 (1995)
2. B. Anchouche, I. Biswas, Einstein-Hermitian connections on polystable principal bundles over a compact Kähler manifold. Am. J. Math. **123**, 207–228 (2001)
3. M.F. Atiyah, On the Krull-Schmidt theorem with application to sheaves. Bull. Soc. Math. France **84**, 307–317 (1956)
4. M.F. Atiyah, Complex analytic connections in fibre bundles. Trans. Amer. Math. Soc. **85**, 181–207 (1957)
5. A. Beauville, Variétés kähleriennes dont la première classe de Chern est nulle. J. Diff. Geom. **18**, 755–782 (1983)
6. A.L. Besse, *Einstein Manifolds* (Springer, Berlin, 2008)
7. I. Biswas, A criterion for holomorphic extension of principal bundles. Kodai Math. J. **34**, 71–78 (2011)
8. F.A. Bogomolov, Kähler manifolds with trivial canonical class. Izv. Akad. Nauk. SSSR **38**, 11–21 (1974). English translation in Math. USSR Izv. **8**, 9–20 (1974)
9. I. Biswas, S. Dumitrescu, Holomorphic affine connections on non-Kähler manifolds. Int. J. Math. **27** (2016)
10. I. Biswas, S. Dumitrescu, Transversely holomorphic branched Cartan geometry. J. Geom. Phys. **134**, 38–47 (2018)
11. I. Biswas, S. Dumitrescu, Holomorphic Cartan geometries on complex tori. C. R. Acad. Sci. **356**, 316–321 (2018)
12. I. Biswas, S. Dumitrescu, Branched Holomorphic Cartan geometries and Calabi–Yau manifolds. Int. Math. Res. Not. **23**, 7428–7458 (2019)
13. I. Biswas, S. Dumitrescu, Holomorphic Riemannian metric and fundamental group. Bull. Soc. Math. France **147**, 455–468 (2019)
14. I. Biswas, S. Dumitrescu, Holomorphic $GL_2(\mathbb{C})$-geometry on compact complex manifolds. Manuscripta Math. **166**(1), 251–269 (2021)
15. I. Biswas, S. Dumitrescu, Generalized holomorphic Cartan geometries. Eur. J. Math. **6**(3), 661–680 (2020)
16. I. Biswas, S. Dumitrescu, H. Guenancia, A Bochner principle and its applications to Fujiki class $\mathcal{C}$ manifolds with vanishing first Chern class. Comm. Contemp. Math. **22**(6), 1950051 (2020)
17. I. Biswas, S. Dumitrescu, G. Schumacher, Deformation theory of holomorphic Cartan geometries. Indag. Math. **31**, 512–524 (2020)
18. I. Biswas, B. McKay, Holomorphic Cartan geometries and rational curves. Complex Manifolds **3**, 145–168 (2016)
19. R.A. Blumenthal, Cartan connections in foliated bundles. Mich. Math. J. **31**, 55–63 (1984)
20. F. Campana, On twistor spaces of the class $\mathcal{C}$. J. Diff. Geom. **33**, 541–549 (1991)
21. F. Campana, Connexité rationnelle des variétés de Fano. Ann. Sci. École Norm. Sup. **25**, 539–545 (1992)

22. P. Deligne, J.S. Milne, A. Ogus, K-y. Shih, *Hodge Cycles, Motives, and Shimura Varieties.* Lecture Notes in Mathematics, vol. 900 (Springer, Berlin, 1982)
23. H.P. de Saint Gervais, *Uniformization of Riemann Surfaces. Revisiting a Hundred Year Old Theorem* (European Mathematical Society, Zurich, 2016)
24. S. Dumitrescu, Homogénéité locale pour les métriques riemanniennes holomorphes en dimension 3. Ann. Inst. Fourier **57**, 739–773 (2007)
25. S. Dumitrescu, Connexions affines et projectives sur les surfaces complexes compactes. Math. Zeitschrift **264**, 301–316 (2010)
26. S. Dumitrescu, A. Zeghib, Global rigidity of holomorphic Riemannian metrics on compact complex 3-manifolds. Math. Ann. **345**, 53–81 (2009)
27. M. Dunajski, M. Godlinski, $GL_2(\mathbb{R})$-structures, G_2 geometry and twistor theory. Q. J. Math. **63**, 101–132 (2012)
28. C. Ehresmann, Sur les espaces localement homogènes. L'Enseign. Math. **35**, 317–333 (1936)
29. A. Fujiki, On the structure of compact manifolds in $\mathcal{C}$. Adv. Stud. Pure Math. **1**, 231–302 (1983)
30. E. Ghys, Déformations des structures complexes sur les espaces homogènes de $SL(2, \mathbb{C})$. J. Reine Angew. Math. **468**, 113–138 (1995)
31. E. Ghys, Feuilletages holomorphes de codimension un sur les espaces homogènes complexes. Ann. Fac. Sci. Toulouse **5**, 493–519 (1996)
32. D. Guan, Examples of compact holomorphic symplectic manifolds which are not Kählerian II. Invent. Math. **121**, 135–145 (1995)
33. D. Guan, Examples of compact holomorphic symplectic manifolds which are not Kählerian III. Int. J. Math. **6**, 709–718 (1995)
34. R.C. Gunning, *On Uniformization of Complex Manifolds: The Role of Connections* (Princeton University Press, Princeton, 1978)
35. G.P. Hochschild, *Basic Theory of Algebraic Groups and Lie Algebras* (Springer, Berlin, 1981)
36. J.E. Humphreys, *Linear Algebraic Groups.* Graduate Texts in Mathematics, vol. 21 (Springer, New York, 1975)
37. J.-M. Hwang, N. Mok, Uniruled projective manifolds with irreducible reductive G-structures. J. Reine Angew. Math. **490**, 55–64 (1997)
38. M. Inoue, S. Kobayashi, T. Ochiai, Holomorphic affine connections on compact complex surfaces. J. Fac. Sci. Univ. Tokyo **27**, 247–264 (1980)
39. N. Istrati, Twisted holomorphic symplectic forms. Bull. London Math. Soc. **48**, 745–756 (2016)
40. N. Istrati, Conformal structures on compact complex manifolds, Ph. D. Thesis, Université Sorbonne Paris Cité (2018)
41. P. Jahnke, I. Radloff, Projective threefolds with holomorphic conformal structures. Int. J. Math. **16**, 595–607 (2005)
42. P. Jahnke, I. Radloff, Projective manifolds modeled after hyperquadrics. Int. J. Math. **29** (2018)
43. D.D. Joyce, *Compact Manifolds with Special Holonomy.* Oxford Mathematical Monographs (Oxford University Press, Oxford, 2000)
44. B. Klingler, Un théorème de rigidité non métrique pour les variétés localement symétriques hermitiennes. Comment. Math. Helv. **76**, 200–217 (2001)
45. S. Kobayashi, *Transformation Groups in Differential Geometry* (Springer, Berlin, 1995)
46. S. Kobayashi, T. Ochiai, Characterizations of complex projective spaces and hyperquadrics. J. Math. Kyoto Univ. **13**, 31–47 (1973)
47. S. Kobayashi, T. Ochiai, Holomorphic projective structures on compact complex surfaces. Math. Ann. **249**, 75–94 (1980)
48. S. Kobayashi, T. Ochiai, Holomorphic structures modeled after compact hermitian symmetric spaces, in *Manifolds and Lie Groups*, ed. by Hano et al., Papers in honor of Yozo Matsushima. Progress in Mathematics, vol. 14 (1981), pp. 207–221
49. S. Kobayashi, T. Ochiai, Holomorphic structures modeled after hyperquadrics. Tohoku Math. J. **34**, 587–629 (1982)
50. J. Kollár, Y. Miyaoka, S. Mori, Rational connectedness and boundedness of Fano manifolds. J. Diff. Geom. **36**, 765–779 (1992)
51. W. Krynski, GL(2)-geometry and complex structures, arXiv:1910.12669 (2019)

52. C.R. LeBrun, $\mathcal{H}$-space with a cosmological constant. Proc. R. Soc. Lond. A **380**(1778), 171–185 (1982)
53. M. Lübke, A. Teleman, *The Kobayashi-Hitchin Correspondence* (World Scientific Publishing Co., Inc., River Edge, 1995)
54. R. Mandelbaum, Branched structures on Riemann surfaces. Trans. Amer. Math. Soc. **163**, 261–275 (1972)
55. R. Mandelbaum, Branched structures and affine and projective bundles on Riemann surfaces. Trans. Amer. Math. Soc. **183**, 37–58 (1973)
56. B. McKay, Extension phenomena for holomorphic geometric structures. SIGMA **5**, 58 (2009)
57. B. Moishezon, On n dimensional compact varieties with n independent meromorphic functions. Amer. Math. Soc. Transl. **63**, 51–77 (1967)
58. P. Molino, *Riemannian Foliations*, Translated from the French by Grant Cairns. With appendices by Cairns, Y. Carrière, É. Ghys, E. Salem, V. Sergiescu, Progress in Mathematics, vol. 73 (Birkhäuser Boston, Inc., Boston, 1988)
59. T. Ochiai, Geometry associated with semisimple flat homogeneous spaces. Trans. Amer. Math. Soc. **152**, 159–193 (1970)
60. B.A. Scárdua, Transversely affine and transversely projective holomorphic foliations. Ann. Sci. Ecole Norm. Sup. **30**, 169–204 (1997)
61. R. Sharpe, *Differential Geometry: Cartan's Generalization of Klein's Erlangen Program*. Graduate Texts in Mathematics, vol. 166 (Springer, New York, 1997)
62. K. Uhlenbeck, S.-T. Yau, On the existence of Hermitian-Yang-Mills connections in stable vector bundles. Comm. Pure Appl. Math. **39**, 257–293 (1986)
63. J. Varouchas, Kähler spaces and proper open morphisms. Math. Ann. **283**, 13–52 (1989)
64. S.-T. Yau, On the Ricci curvature of a compact Kähler manifold and the complex Monge-Ampère equation. I. Comm. Pure Appl. Math. **31**, 339–411 (1978)
65. Y. Ye, Extremal rays and null geodesics on a complex conformal manifold. Int. J. Math. **5**, 141–168 (1994)

Index

Symbols
$(d − 1)$-dimensional ball, 9
$(d − 1)$-dimensional hemisphere, 9
A-orthogonality, 101
C-orthocenter, 64
D-orthogonality, 100
δ-hyperbolic, 210
δ-thin, 210
CAT(0)-space, 252, 255, 256, 258

A
Abelian differential, 353
Absolute period, 359
Additive orthogonality, 101
Adherence core, 412
Adherent, 412
    closure, 412
Affine orthogonality, 164
Alexander horned sphere, 272
Alexandrov–Fenchel inequality, 188, 201
Algebraically primitive, 385
Algebraic Entropy, 259–261
Anti-norm, 56
Antipodal point, 9
Arc, 9
Arc-length curvature, 73
Arc-length orthogonality, 65
Arc-length parametrization, 69
Arc of symmetry, 11
Area orthogonality, 101, 138, 163
Asymmetric distance, 51
Average width, 175

B
Ball, 9
Banach space theory, 51
Beltrami differential, 393
Betti number, 221, 222
Birkhoff orthogonality, 56, 98, 101, 103–110,
        112, 114, 128, 130, 133, 134, 136,
        139, 141–144, 146, 157–163
    approximate, 157, 159–161
    approximate symmetry, 160
    in Hilbert C^*-modules, 158
    of operators, 157
Bisector of segment, 58
Bishop–Gromov, 212, 213, 226, 230
Blaschke characterization, 107
Blaschke selection theorem, 11, 181
Body of constant width, 86
Boussouis orthogonality, 100, 149, 150, 161
Branched holomorphic Cartan geometry, 440
Branched holomorphic (G, X)-structure, 441
Branched holomorphic projective connection,
        441
Branched transverse holomorphic Cartan
        geometry, 443
Brunn–Minkowski inequality, 190
Busemann space, 226, 227, 229, 230, 239,
        242, 249, 250, 252, 253, 255
Butterfly, 21

C
Carlsson orthogonality, 99, 100, 145–150, 161,
        163

© The Author(s), under exclusive license to Springer Nature Switzerland AG 2022   463
A. Papadopoulos (ed.), *Surveys in Geometry I*,
https://doi.org/10.1007/978-3-030-86695-2

Cauchy formula, 177
Caustic, 84
Cayley graph, 211
Center of minimal enclosing circle, 61
Centroid curve, 74
Chebyshev set, 61
Chordal orthogonality, 164
Circular curvature, 73
Collar, 239, 240
Complete convex body, 38
Completely periodic, 356
Complete set, 38, 86
Complex dilatation, 393
Complex geodesic, 369
Complex structures
    moduli space, 360
Concept of area, 55
Concept of curvature of curves in normed
        planes, 71
Constant diameter, 31
Constant width, 33, 75
Continuous orthogonality, 101
Convex body, 10, 172
Convex geometry, 50
Convex hull, 10
Convex polyhedron, 172
Convex polytope, 172
Convex set, 9, 172
Coordinates
    period, 359
Corner, 12
Covering, 227, 228
Covering conjecture of Hadwiger, 87
Crofton formula, 177
Curve complex, 401
Curve of constant width, 70
Curve theory in normed planes, 69

D
Degree of a divisor, 453
Degree of a holomorphic vector bundle, 453
Diameter, 9
Differential
    Abelian, 353
    holomorphic, 353
    quadratic, 354
Diminnie orthogonality, 100, 163
Discrete and computational geometry, 50
Discrete spherical Laplacian, 202
Disk, 9
Distance, 9
    Gromov–Hausdorff, 207
Distance geometry problem, 68

Distance to a set, 9
Divergent rays, 406
Dual norm, 55

E
Ellipse, 71
Entropy, 212
    algebraic, 259–261
Equiframed curve, 85
Equilateral set, 67
Euler line, 62
Extension of geodesics, 227, 252, 256
Extreme point, 11
Extreme supporting hemisphere, 21

F
Face, 412
    dimension, 412
Fano manifold, 427
Feuerbach hypersphere, 62
Finitely presented group, 246
Finiteness theorem, 249, 251, 255, 258
Finsler geometry, 50
Finsler space, 76
Flow
    vertical, 355
Foliated Atiyah bundle, 444
Four-vertex theorem, 74
Frenet formulas, 73
Functional analysis, 50

G
Gauges, 51
Geodesic lamination, 410
Geometrically primitive, 385
Geometric dilation, 69
Geometric intersection matrix, 312, 313, 324
Geometric simple connectivity, 273, 275, 310,
        313
    brutal violation of, 286, 310, 319, 325
Geometry of the Minkowskian simplex, 62
Girth of a normed space, 79
GL(2)-geometry, 424
Great circle, 9
Great subsphere, 9
Gromov–Hausdorff topology, 207
Group
    finitely presented, 246
    marked, 242, 243, 246–249
    veech, 365
Growth, 231–233, 255, 259, 261

H
Half-translation surface, 347
Halving pair, 69
Handlebody decomposition, 311–315
Hausdorff distance, 11, 181
Height orthogonality, 101
Hemisphere, 9
Hilbert modular surface, 384
Hilbert space, 123, 132, 158–161
Holomorphic G-structure, 420
Holomorphic Cartan geometry, 437
Holomorphic conformal structure, 422
Holomorphic differential, 353
Holomorphic quadratic differential, 396
Holomorphic Riemannian metric, 421
Holomorphic symplectic form, 421
Homogeneous orthogonality, 101
Homology group, 220, 221

I
Inequality
  Bishop–Gromov, 212, 213, 226, 230
  quadrangle, 214, 265
Infinitesimal rigidity theorem
  Teichmüller metric, 400
  Thurston's metric, 413
Inner product space, 65, 97, 98, 100, 101, 103,
    104, 107, 109, 112–114, 119–123,
    127–136, 139–148, 150–152,
    154–157, 160–163
Isoperimetric inequality, 197
Isoperimetric problem in Minkowski spaces,
    70
Isosceles orthogonal, 58
Isosceles orthogonality, 99, 100, 102, 103,
    114, 116, 118–124, 127, 132, 133,
    145, 146, 151, 157, 158, 160–163
Ivanov's theorem, 401

J
Jacobian torus, 382
James constant, 161
James orthogonality, 64
Joly's construction, 136

K
Kakutani's theorem, 107

L
Löwner ellipse, 134

Left supporting hemisphere, 21
Legendre curve, 75
Lemma
  collar, 239, 240
  Margulis, 231, 233
  Ping-Pong, 263–265
Lune, 12

M
Mahler conjecture, 83
Mapping class group, 398
  extended, 398
Marchaud's characterization, 113
Margulis Lemma, 231, 233
Marked group, 242, 243, 246–249
Maximal dilatation, 393
Measured lamination, 410
Measured lamination space, 410
  projective, 410
Metric capacity, 68
Minimum chord, 65
Minkowskian geometry of circles and systems
    of circles, 63
Minkowski billiard, 80
Minkowski curvature, 72
Minkowski functional, 52
Minkowski geometry, 49
Minkowski inequality, 188
  first, 188
  second, 188
Minkowski minimal surface, 79
Minkowski sum, 180
Minsum/median hyperplane problem, 68
Miquel's theorem, 64
Mixed volume, 186
Möbius geometry, 65
Module, 376
Moduli space
  complex structures, 360
  translation surfaces, 360
Moduli space of complex structures, 360
Moduli space of translation surfaces, 360
Monge point, 62

N
Non-commutative space, 310, 325
Non-degenerate orthogonality, 101
Norm, 52
Normal curvature, 72
Normal parametrization, 194

O

Open ball, 9
Open hemisphere, 9
Opposite hemispheres, 9
Opposite side, 24
Optimization, 50
Orbit
    singular, 355
Order of hemispheres, 20
Origami, 347
Orthocenter, 60
Orthogonal, 9

P

Packing, 223, 228, 229, 255
Pair of arms, 21
Parallel rays, 406
Part of a ball, 18
Partial connection on a principal bundle, 445
Perfect norm, 86
Period
    absolute, 359
    coordinates, 359
    relative, 359
Periodic, 399
Ping-Pong Lemma, 263–265
Poincaré conjecture, 270
Point of touching, 10
Positively homogeneous orthogonality, 102
Primitive
    algebraically, 385
    geometrically, 385
Projective lamination, 410
Pseudo-Anosov, 399
Pythagorean orthogonality, 99, 100, 102,
        124–128, 134, 138, 143, 146, 148,
        150, 161–163
    Hermite–Hadamard, 162

Q

Quadrangle Inequality, 214, 265
Quadratic differential, 354
Quarter of disk, 18
Quasi-conformal map, 391
Quermassintegral, 177

R

Radon curve, 158, 159
Radon norm, 108, 159
Radon plane, 56, 159, 160
Real multiplication, 383

Rectangular constant, 140
Reduced body, 18, 62
Reducible, 399
Regular polygon, 63
Regular spherical k-gon, 10
Relative interior, 9
Relative period, 359
Reuleaux polygon, 11
Right supporting hemisphere, 21
Roberts orthogonal, 58
Roberts orthogonality, 98, 99, 102–104, 116,
        150, 157, 158, 160
    approximate, 157
    in C^*-algebra, 158
    in Hilbert C^*-modules, 158
    of complex double row matrices, 158
Rotation around an arc, 11
Rotational body, 11
Royden's rigidity theorem, 399

S

Saddle surface, 80
Schoenflies ball (smooth n-dimensional), 272
Schoenflies ball (smooth 4-dimensional), 269,
        324
Schoenflies problem (4-dimensional
        differentiable), 270, 273
Schoenflies problem (higher-dimensional
        topological), 272
Schwartz rounding, 191
Self-dual shape, 37
Semi-circle, 9
Singer orthogonality, 100, 123, 150, 151, 161,
        163
Singular orbit, 355
Smooth norm, 52
Smooth Poincaré conjecture, 273
Smooth point, 10
Sooth body, 10
Space
    Busemann, 226, 227, 229, 230, 239, 242,
            249, 250, 252, 253, 255
    CAT(0), 252, 255, 256, 258
Spherical ball, 9
Spherical distance, 9
Spherical polygon, 10
Square-tiled surface, 347
Steiner formula, 173
Steiner symmetrization, 192
Stratum, 357
Strictly convex norm, 53
Strongly isomorphic polytopes, 200
Support closure, 412

Support function, 178
Support hyperplane, 178
Supporting between, 21
Supporting great sphere, 10
Supporting hemisphere, 10
Supporting lune, 13
Surface
   Veech, 376
Symmetric, 11
Symmetric orthogonality, 101
Systole, 231, 236, 238, 239, 256, 258

T
Tautological subspace, 380
Teichmüller curve, 369
Teichmüller disk, 369
Teichmüller distance, 394
Teichmüller map, 395
Teichmüller space, 391
Thickness, 17
Thickness of lune, 12
Thin-thick decomposition, 236, 239
Thurston's asymmetric metric, 409
Topological Finiteness, 258
Total mean curvature, 175
Touch from inside, 10
Touch from outside, 10
Trace field, 380

Translation surfaces, 346
   moduli space, 360
Transverse holomorphic Cartan geometry, 442
Twisted holomorphic symplectic structure, 423

U
Unique orthogonality, 101

V
Veech dichotomy, 356
Veech group, 365
Veech surface, 376
Vertex, 10
Vertical flow, 355
Viterbo conjecture, 83
Voronoi diagram, 58

W
Whitehead manifold, 275, 314
Width, 14
Wirtinger inequality, 196
Wulff shape, 37

Z
Zindler curve, 70

Printed in the United States
by Baker & Taylor Publisher Services